CELL MEMBRANES

Methods and Reviews

Volume 2

CELL MEMBRANES

Methods and Reviews

Volume 2

Edited by

Elliot Elson
William Frazier
and
Luis Glaser

Washington University School of Medicine
St. Louis, Missouri

PLENUM PRESS • NEW YORK AND LONDON

ISBN 978-1-4684-4825-2   ISBN 978-1-4684-4823-8 (eBook)
DOI 10.1007/978-1-4684-4823-8

© 1984 Plenum Press, New York
A Division of Plenum Publishing Corporation
233 Spring Street, New York, N.Y. 10013

CONTRIBUTORS

Steven K. Akiyama   Membrane Biochemistry Section, Laboratory of Molecular Biology, National Cancer Institute, National Institutes of Health, Bethesda, Maryland 20205

Vann Bennett   Department of Cell Biology and Anatomy, Johns Hopkins University School of Medicine, Baltimore, Maryland 21205

Ingrid Blikstad   Division of Biology, California Institute of Technology, Pasadena, California 91125

Susan E. Coombs   Section of Biochemistry, Molecular and Cell Biology, Division of Biological Sciences, Cornell University, Ithaca, New York 14853

William A. Frazier   Departments of Biological Chemistry and Neurobiology, Division of Biology and Biomedical Sciences, Washington University School of Medicine, St. Louis, Missouri 63110

Nancy J. Galvin   Department of Biological Chemistry and Neurobiology, Division of Biology and Biomedical Sciences, Washington University School of Medicine, St. Louis, Missouri 63110

Alfred G. Gilman   Department of Pharmacology, University of Texas Health Science Center at Dallas, Dallas, Texas 75235

George P. Hess   Section of Biochemistry, Molecular and Cell Biology, Division of Biological Sciences, Cornell University, Ithaca, New York 14853

Gordon A. Jamieson, Jr.   Department of Biological Chemistry and Neurobiology, Division of Biology and Biomedical Sciences, Washington University School of Medicine, St. Louis, Missouri 63110

Algirdas J. Jesaitis   Department of Immunology, Scripps Clinic and Research Foundation, La Jolla, California 92037

Elias Lazarides   Division of Biology, California Institute of Technology, Pasadena, California 91125

Beth L. Meyers—Hutchins   Department of Biological Chemistry, Division of Biology and Biomedical Sciences, Washington University School of Medicine, St. Louis, Missouri 63110. *Present address*: Syva, Department of Immunochemistry, Palo Alto, California 94303

Randall T. Moon   Division of Biology, California Institute of Technology, Pasadena, California 91125

W. James Nelson   Division of Biology, California Institute of Technology, Pasadena, California 91125

Richard G. Painter   Department of Immunology, Scripps Clinic and Research Foundation, La Jolla, California 92037

Elliott M. Ross   Department of Pharmacology, University of Texas Health Science Center at Dallas, Dallas, Texas 75235

Larry A. Sklar   Department of Immunology, Scripps Clinic and Research Foundation, La Jolla, California 92037

Murray D. Smigel   Department of Pharmacology, University of Texas Health Science Center at Dallas, Dallas, Texas 75235

Kenneth M. Yamada   Membrane Biochemistry Section, Laboratory of Molecular Biology, National Cancer Institute, National Institutes of Health, Bethesda, Maryland 20205

PREFACE

We are extremely pleased that all of the chapters in this volume provide up-to-date information on a variety of topics of interest to scientists working on membrane biology. As in the past, we have attempted to expedite the transition from submission of the manuscripts to publication in order to make the reviews as timely as possible. Cell biology and molecular biology are increasingly becoming concerned with the study of structural elements in cells and their assembly. The rules which govern membrane synthesis, assembly and interaction of membrane components with other cellular elements, notably the cytoskeleton, are at the center of research in these fields. We will continue in subsequent volumes of this series to focus on these areas. We would welcome suggestions of topics which would benefit from a review at the present time.

We thank all of the contributors for providing these very excellent reviews and for doing so in a timely fashion.

Elliot L. Elson
William A. Frazier
Luis Glaser

St. Louis, Missouri

CONTENTS

Chapter 1

Chemotactic Transduction in the Cellular Slime Molds

William A. Frazier, Beth L. Meyers–Hutchins, Gordon A. Jamieson, Jr., and Nancy J. Galvin

Chapter 2

Leukocyte Chemotaxis: Mobilization of the Motile Apparatus by *N*-Formyl Chemotactic Peptides

Richard G. Painter, Algirdas J. Jesaitis, and Larry A. Sklar

Chapter 3

The Interactions of Cells with Extracellular Matrix Components

Kenneth M. Yamada and Steven K. Akiyama

Chapter 4

The Human Erythrocyte as a Model System for Understanding
Membrane Cytoskeleton Interactions

Vann Bennett

Chapter 5

Regulation of Assembly of the Spectrin-Based Membrane Skeleton
in Chicken Embryo Erythroid Cells

Randall T. Moon, Ingrid Blikstad, and Elias Lazarides

Chapter 6

Assembly and Establishment of Membrane–Cytoskeleton Domains
during Differentiation: Spectrin as a Model System

W. James Nelson and Elias Lazarides

Chapter 7

Role of the β-Adrenergic Receptor in the Regulation of Adenylate Cyclase

Murray D. Smigel, Elliott M. Ross, and Alfred G. Gilman

Chapter 8
Acetylcholine Receptor: Some Methods Developed to Study a
Membrane-Bound Regulatory Protein
Susan E. Coombs and George P. Hess

CELL
MEMBRANES
Methods and Reviews

CHEMOTACTIC TRANSDUCTION IN THE CELLULAR SLIME MOLDS

William A. Frazier, Beth L. Meyers–Hutchins, Gordon A. Jamieson, Jr., and Nancy J. Galvin

1. INTRODUCTION—CHEMOTAXIS IN THE CELLULAR SLIME MOLDS

Two eukaryotic cell types have been extensively studied as paradigms for the chemotaxis of freely motile cells: the leukocytes such as PMNs and macrophages (Neidel and Cuatrecasas, 1980; Schiffman, 1982; Zigmond, 1978) and the cellular slime molds, particularly the species *Dictyostelium discoideum* (Devreotes, 1982; Frazier *et al.*, 1982; Gerisch, 1982). Both these systems have inherent complications for studying the mechanism(s) of chemotaxis *per se,* that is, the process or processes by which the occupancy of chemotactic receptors is sensed by the inner workings of the cell, specifically those components involved in cellular motility, and translated into directional information. The white cells respond to chemoattractants such as complement peptides and *N*-formylated peptides (e.g., F-Met-Leu-Phe, or FMLP) with a barrage of killer mechanisms such as superoxide and peroxide production, lysosomal enzyme secretion, increased cell–cell adhesiveness, and enhanced phagocytosis (Neidel and Curatrecasas, 1980; Schiffmann, 1982; Zigmond, 1978). Many or all of these responses may have nothing to do with the chemotactic response or effect on cell

William A. Frazier, Beth L. Meyers–Hutchins, Gordon A. Jamieson, and Nancy J. Galvin • Departments of Biological Chemistry and Neurobiology, Division of Biology and Biomedical Sciences, Washington University School of Medicine, St. Louis, Missouri 63110. Beth L. Meyers–Hutchins' present address is Syva, Department of Immunochemistry, Palo Alto, California 94303. Work supported by grants from the National Institutes of Health (NS 13269) and the National Science Foundation (PCM 83 02640). William A. Frazier is an Established Investigator of the American Heart Association. Gordon A. Jamieson, Jr., is an NIH postdoctoral fellowship recipient (GM 07864).

motility. In general, most chemoattractants also cause the response of chemo-kinesis, or an increase in rate of motility with no imposed directional component.

The cellular slime molds would, at first glance, seem to be a much simpler "model system" in which to study chemotaxis free of these complications. How-ever, the slime molds present complications of their own. First, there are two kinds of chemotaxis which differ in some respects that may or may not have fundamental mechanistic implications. Growing or vegetative cells of many species respond chemotactically to folic acid and some of its derivatives (Pan *et al.*, 1972, 1975; Nandini–Kishore and Frazier, 1981). This type of chemotaxis is thought to be used in finding bacterial foodstuffs and, as far as is currently known, does not involve other complex cellular responses such as cell aggre-gation or signal relay as adjuncts to the effect on motility. Upon starvation, all slime mold species develop the ability to secrete their own endogenously pro-duced chemoattractant or acrasin (Bonner *et al.*, 1969) and respond to it chem-otactically. Furthermore, to promote aggregation of cells over wide areas (Tom-chik and Devreotes, 1981), e.g., hundreds of cell diameters in width, all but one species (van Haastert *et al.*, 1982a) concomitantly develop the means to produce and secrete the acrasin in response to its reception (Schaffer, 1962, 1975; Roos *et al.*, 1975) thus creating an endless feedback loop of re-sponse–secretion–response (Gerisch and Wick, 1975). In the case of *D. dis-coideum* the acrasin is cyclic AMP (cAMP) (Bonner *et al.*, 1969), and the relay enzyme is adenylate cyclase (Bonner *et al.*, 1972; Devreotes *et al.*, 1979; De-vreotes and Steck, 1979). The coupling of the cAMP receptor, apparently the same one that is involved in chemotaxis, to this adenylate cyclase is of crucial importance for the cells to aggregate properly. However, it seems that this coupling is irrelevant for the cellular response of chemotaxis *per se* (Theibert and Devreotes, 1983). The receptor–adenylate cyclase coupling has recently been addressed in an excellent review by Devreotes (1982) and will not be further considered here. Suffice it to say that it is important to sort out events which follow receptor stimulation that are related to adenylate cyclase activation from those which are directly on the pathway to chemotaxis. This may be facilitated by the recent finding by Theibert and Devreotes (1983) that caffeine blocks the adenylate cyclase response without affecting chemotaxis to cAMP, the cAMP receptor, or the guanylate cyclase response, which is a leading candidate for a transduction effector (Brenner and Thoms, 1984).

Here the advantage of vegetative chemotaxis becomes apparent, since any intracellular response that is solely involved in chemotaxis should be observed in both vegetative and differentiating cells stimulated with the appropriate che-moattractant. In addition, species comparisons are possible now that the acrasins of two other species have recently been identified, neither of which uses cAMP as the acrasin. *Polysphondylium pallidum* employs a completely blocked glutamic acid–ornithine dipeptide called "glorin" as its acrasin (Shimomura *et al.*, 1982), and *D. minutium* uses a pterin derivative similar to 6-carboxy-pterin (van Haastert *et al.*, 1982a). Hence genuine chemotactic postreceptor events should be found in each of these species, whereas those specific to formation of the particular acrasin and its relay should not be common. Such a "consensus" postreceptor response which occurs in many different species in vegetative and developmental stages is a brief rise in the intracellular cGMP concentration which returns to basal levels by 20 sec after receptor stimulation with the appropriate attractant (Mato *et al.*, 1977a; Mato and Malchow, 1978).

Much of the biochemical work thus far has concentrated on the identification and characterization of receptors for these chemoattractants. A property that all of them seem to share is the very rapid rate of the binding and dissociation reaction. The latter is a particularly difficult problem when it comes to assaying the receptors in any other state than on the intact cell. This is because assays that are commonly used for the separation of bound and free ligand when the receptor is on membranes or in the detergent-solubilized state are simply too slow. We have developed two assays that cope with this problem in the case of the detergent-solubilized receptor (Meyers and Frazier, 1981; Meyers–Hutchins, 1983) and, with these assays to follow the cAMP-binding activitity of the receptor, have purified and studied some of the properties of the cAMP receptor from aggregating *D. discoideum* cells. These results and a summary of what is known about receptors for other slime mold attractants are presented in Section 2.

One way of looking at chemotactic transduction is as a bias or asymmetry imposed on the cell's motility apparatus. Thus it is important to consider our current knowledge of slime mold motility and how similar and different the process may be in other, more extensively studied cells. The analysis by Futrelle *et al.* (1982) of cell motility during chemotaxis *in situ* may be particularly revealing in terms of the underlying cellular mechanisms. Great strides have been made in identifying the component molecules of the motility system of slime molds (see review by Spudich and Spudich, 1982; Taylor and Fechheimer, 1982). Only the salient components as they relate to our conception of chemotaxis will be dealt with in Section 4. In addition, it seems that the structure operationally defined as a cytoskeleton is intimately involved in cellular motility and its directionality. Virtually all the components of the motility or force-generating apparatus are found associated with the detergent-resistant cytoskeleton, with the probable exception of negative modulators or disassembly proteins such as the protein severin (Yamamoto *et al.*, 1982), which depolymerizes actin filaments. In other cell types evidence has implicated microtubules and their organizing centers in the formation of cell polarity or directionality (Malech *et al.*, 1977; Gottlieb *et al.*, 1981; Kupfer *et al.*, 1982; Anderson *et al.* 1982).

In the last part of this chapter (Section 5), we will present some data and hypotheses that may serve to establish a testable link between the receptors for chemoattractants, the guanylate cyclase, and the cytoskeleton.

2. RECEPTORS FOR CHEMOATTRACTANTS OF THE CELLULAR SLIME MOLDS

2.1. The Folate Receptor of Vegetative D. discoideum

In the vegetative state, D. discoideum and other species of slime mold are able to chemotax toward higher concentrations of folate (Pan et al., 1972). Early studies from Bonner's laboratory (Pan et al., 1975) used a variety of compounds that are structural analogs for the three regions of the folate molecule (the pterin ring, the p-aminobenzoate ring, and the glutamate moiety) in an attempt to determine the region of the folate structure that might be the site recognized by the receptor. Based on the activity of pterin derivatives and the inactivity of p-aminobenzoate and glutamate, these authors concluded that the pterin moiety was the "pharmacophore" for the chemotaxis receptor. Since the use of radio-labeled probes such as [3H]-folate (Wurster and Butz, 1980; van Driel, 1981) or the more stable [3H]-methotrexate (MTX) (Nandini–Kishore and Frazier, 1981), it has become clear that the folate receptor is quite specific for folate, MTX, and aminopterin, all compounds that contain all three pterin, PABA, and glutamate regions. The activity of pterin appears to be due to presence of a distinct receptor for pterin compounds including pterin-6-carboxylate, which may be the acrasin in D. minutium (van Haastert et al., 1982a). Thus pterins do not compete for the binding of either radiolabeled MTX or folate (Wurster and Butz, 1980; Nandini–Kishore and Frazier, 1981). The fact that we now know of the existence of chemotactic receptors for three different compounds on D. discoideum cells leads to the speculation that receptors for many more compounds may exist.

Studies using [3H]-folate to characterize the receptor were limited by the presence of a very active folate deaminase on vegetative cells and secreted by them (Pan and Wurster, 1978; Kakebeeke et al., 1980). Several groups have been unable to find a useful inhibitor of the enzyme. Nonetheless, Wurster and Butz (1980) were able to find specific binding sites for folic acid on vegetative Ax-2 cells. The rates of association and dissociation were very fast, and the affinity was in the range of Kd = 150 nM folate. Surprisingly, these authors found that deaminofolate, the product of the folate deaminase, was able to compete for the binding of folate, whereas pterin and lumazine (deaminopterin) were not. Van Driel (1981), in a similar study, found a Kd of 300 nM, but the particular assay method used severely limited the precision of the data. In the same study, an attempt was made to identify folate-binding proteins using affinity chromatography on columns of folate–Sepharose. Seven proteins were found that eluted with folate, but it is likely that none of these represents the chemotactic receptor since the very rapid off-rate of the receptor would prevent its being retained on the column. We have found that under similar conditions, the cAMP receptor is not retained on cAMP–Sepharose (Meyers–Hutchins, 1983).

In our search for a suitable inhibitor of the folate deaminase, we found that methotrexate and aminopterin, both diaminopterin analogs of folate, were competitive inhibitors of the enzyme. Spectral and HPLC analyses indicated that neither compound was altered in any way upon incubation with *D. discoideum* cells (Nandini–Kishore and Frazier, 1981). Furthermore, chemotaxis assays using a gradient of compound established in an agar dish showed that both MTX and aminopterin were as potent chemoattractants of vegetative cells as folate itself, indicating that degradation of the attractant is not essential for chemotaxis. In addition, both MTX and aminopterin were able to block the response of cells to a gradient of folate when they were present at a constant high concentration in the agar. On the basis of these data we tested [3H]-MTX as a ligand for the folate receptor. The stability of the compound to enzymatic degradation was confirmed by binding studies in which a steady-state level was maintained for 30 min. In addition, MTX is much more stable to light and oxygen than folate; hence nonspecific binding is quite low. We find that the wild type NC-4 cells have 50,000–100,000 sites per cell with an affinity (Kd) of 20–100 nM MTX. During differentiation of the cells in suspension, the number of sites drops at 9 hr to about 30% of the number on vegetative cells, but the affinity remains constant (Nandini–Kishore and Frazier, 1981). It is during this period that the cAMP receptor appears in the membrane and seems to supplant the folate receptor in its ability to stimulate a directional response. At the same time the response to folate is becoming less directional, until when cells can respond in a highly directional way to cAMP, they respond to folate with an increase in random motion or chemokinesis (Nandini–Kishore and Frazier, 1981). When cells in this state are refed to "erase" the developmental program (Soll and Waddell, 1975), they rapidly switch the coupling of these two receptors such that the folate receptor is once again coupled to chemotaxis and the cAMP receptor becomes chemokinetic (Varnum and Soll, 1981). This behavior would be explained if both receptors used a common chemotaxis transducer apparatus that had a limited number of coupling sites such that the two receptors had to compete for them.

2.2. The cAMP Receptor of Aggregating *D. discoideum*

The cAMP receptor that participates in the chemotactic response of *D. discoideum* (Green and Newell, 1975; Henderson, 1975; King and Frazier, 1979) has appeared to many investigators a simple model system amenable to the study of the details of the structure–function relationships of a receptor for a small molecule. Its function as a chemotactic receptor also attracted much attention at about the time when the mechanisms of bacterial chemotaxis were being revealed by the power of genetics (Springer *et al.*, 1979). The promise of the genetic approach to understanding eukaryotic chemotaxis using *Dictyostelium* as a model has remained largely unfulfilled, for two reasons. First, it has turned out to be very difficult to do genetics in a technical sense, because the parasexual cycle, which relys on low-frequency events, must be used for mapping. This is because the sexual cycle involves the formation and germination of the macrocyst, an impenetrable giant cell in which sexual events take place (Newell, 1982). Only recently have some hints emerged as to how macrocyst formation and breakdown might be controlled (Saga and Yanagisawa, 1982). Second, even the most cunning selection strategies produce unexpected results owing to the nature of the highly interdependent sequence of events that must occur for any development to take place at all. One must be sure that unrelated functions are appearing at their normal times while the function of interest is truly mutated and not just masked. The difficulties here are compounded by the lack of understanding as to what subsequent events might actually depend on a mutated protein. For example, what later processes require the presence of a functional cAMP receptor? In this vein, there has been great confusion as to whether the so-called developmental effects of cAMP, such as precocious onset of cell adhesion or other markers which occurs upon pulsing with cAMP, are mediated via the chemotaxis receptor or some other pathway such as the intracellular cAMP-binding proteins. This aspect of the problem should be simplified somewhat by the realization that there are now cAMP derivatives available such as N6-(aminohexyl)-cAMP which are able to produce the developmental stimulation and yet are totally inert at the chemotactic receptor on the cell surface (Juliani *et al.*, 1981). This suggests that many of the proposed pleiotypic effects of cAMP are due to the stimulation of intracellular cAMP-binding proteins such as the regulatory subunit of cAMP-dependent protein kinase (Leichtling *et al.*, 1981b, 1982, 1983; Majerfeld *et al.*, 1983). The recent reports of transformation in *Dictyostelium* (Hirth *et al.*, 1982; Barclay and Meller, 1983) indicate that techniques such as site-directed mutagenesis (Barclay and Meller) and complementation may be extremely powerful tools in approaching the problems of chemotaxis and the developmental sequence of gene expression in slime mold (Chung *et al.*, 1981; Mehdy *et al.*, 1983). Furthermore, the discovery of a possible transposable element (Rosen *et al.*, 1983) and an endogenous plasmid (Metz *et al.*, 1983) in *Dictyostelium* offers additional possibilities for developmental analysis.

The direct biochemical approach to the study of the cAMP receptor has been difficult as well, owing primarily to a property that may be common to all slime mold chemotaxis receptors, namely the very rapid rate of dissociation of bound attractant. This intriguing property may well relate to the mechanism of signal reception or spatial decoding of the concentration signal. When the receptor is assayed on intact cells, this problem is easily overcome by using an assay in which the cells with bound cAMP are centrifuged through an oil layer. This occurs very rapidly (in less than a second) in a machine such as a Microfuge. When the receptor on plasma membranes is being characterized, however, this method is not as useful. Real difficulties are encountered when the receptor is solubilized with detergent. Until a few years ago, no methods were available to separate cAMP from a soluble macromolecule rapidly enough to measure binding to the receptor. We have developed two methods for the assay of detergent-solubilized cAMP receptors which are based on different structural properties common to many receptors. The first of these takes advantage of the hydrophobic nature of the lipid-binding region of the receptor protein. The detergent solution containing receptor is diluted into a slurry of decyl–agarose containing labeled cAMP and DTT as a phosphodiesterase inhibitor (Meyers and Frazier, 1981). Many proteins including the receptor bind hydrophobically to the decyl matrix and become immobilized on the large agarose beads, which can then be filtered very rapidly. The limitation of this assay is the porosity of the matrix, which entraps large amounts of the [^{3}H]-cAMP thus making the nonspecific binding very high unless a wash step is used. Since variable washing rates will lead to variable amounts of cAMP dissociation from the receptor, the precision of the assay suffers, and it is useful primarily as means of rapidly determining a lower-limit estimate of the amount of cAMP-binding activity present in a given sample. More recently we have found that the receptor is a Con-A-binding glycoprotein (see below), and we have used this information to design a more precise assay for the solubilized receptor. Vegetative *D. discoideum* cells, which bind large amounts of Con A, are fixed with glutaraldehyde and coated with Con A, thus producing an inert carrier for the receptor. The receptor is bound to these coated cells, a state in which it can still bind cAMP. The cells are then centrifuged through an oil layer just as in the receptor assay for intact cells. The very rapid separation of bound and free cAMP yields data that are quite precise. We have used this assay extensively in characterizing the cAMP receptor in detergent extracts of plasma membranes and during its purification. Recently, van Haastert and Kien (1983) reported the use of an assay for cAMP binding to intact cells in which cells with bound cAMP are mixed with saturated ammonium sulfate.

In this 80% saturated ammonium sulfate solution, the cAMP is apparently "locked on" the receptor and dissociates very slowly. It remains to be seen whether this assay is also useful in the assay of cAMP receptors on plasma membranes and in the detergent-solubilized state. No explanation for the mechanism of the effect of ammonium sulfate was offered. It is interesting that Neufeld et al. (1983) have reported that deoxycholate can "lock" agonists onto the β-adrenergic receptor.

Some peculiar but quite interesting properties of the cAMP receptor have come to light during our attempts to purify it (Meyers-Hutchins and Frazier, 1984). One of the more problematic aspects of these studies was the consistent finding that when plasma membranes were made from intact cells, these membranes contained at most only a few percent of the initial cAMP-binding activity. Furthermore, when the membranes were solubilized with detergent, the yields of soluble cAMP receptor activity found with the hydrophobic immobilization assay (Meyers and Frazier, 1981) were extremely variable. Being aware of the tremendous amount of proteolytic activity in these cells, we initially ascribed this low activity to proteolysis. Subsequently, we realized that upon separation of the crude detergent extract on DEAE Sephadex, substantial amounts of cAMP-binding activity elute at high (0.6 M) NaCl concentration. When protein eluting from the column at lower salt (0.2 M) is added back to these active fractions, the binding activity is immediately inhibited. This putative inhibitor of cAMP binding may also account for the low activity in membrane preparations. Further purification and characterization of this inhibitory substance is in progress. The active cAMP-binding protein eluting from DEAE Sephadex with high salt has been further purified using hydrophobic chromatography on decyl–agarose, the same matrix used in the cAMP-binding assay. Photoaffinity labeling with 8-azido-[^{32}P]cAMP was used to identify proteins with a cAMP-binding site. On two-dimensional PAGE analysis, only the most acidic M_r 70,000 protein is labeled, and this labeling is prevented by cAMP but not by cGMP at 200 μM. Thus the most acidic species was purified by nondenaturing preparative gel electrophoresis. This cAMP-binding protein is present only in detergent extracts of membranes from 6-hr differentiated cells. When vegetative cell membranes are used in this preparation, no cAMP-binding activity is recovered, even after the DEAE and hydrophobic column steps. Thus no cryptic cAMP-binding activity is present in vegetative membranes. Interestingly, [^{3}H]-MTX-binding activity, which may represent the folate receptor (Nandini–Kishore and Frazier, 1981), is recovered.

Earlier photoaffinity labeling studies using the azido–cAMP reagent (Wallace and Frazier, 1979) identified an M_r-40,000 protein as a potential receptor candidate, and Juliani and Klein (1981) reported the labeling of a receptor candidate of M_r 45,000 and another of M_r 47,000. Both groups determined that the affinity of the 8-azido derivative of cAMP for the chemotaxis receptor is very low, about 100-fold less than that of cAMP itself. This fact and the realization that other proteins have been identified in the M_r-40,000- to M_r-50,000-molecular-weight range which have a reasonable affinity for the photolabel suggest that it is possible to account for all these species as due to slight contamination of the intact cells by these soluble proteins. For example, the regulatory subunit of the *D. discoideum* cAMP-dependent protein kinase has a high affinity for the azido reagent and has an M_r of 40,000 on SDS gels (Leichtling *et al.*, (1981b, 1982). There is also an intracellular protein of M_r 185,000 (Leichtling *et al.*, 1981c) with a subunit M_r of 47,000 which binds cAMP and adenosine compounds and is labeled with 8-azido-cAMP. This protein may also carry S-adenosylhomocysteine hydrolase activity (de Gunzburg *et al.*, 1983). Furthermore, the membrane phosphodiesterase is readily labeled by the reagent, and it has an M_r of 48,000 (Wallace and Frazier, 1979). It should be pointed out that we have not been able to label the M_r-70,000 protein on intact cells with the azido–cAMP reagent, and in labeling experiments with plasma membranes, the nonspecific labeling is overwhelming, thus precluding detection of the protein even if it were labeled. It is only after solubilization and partial purification that we have been able to detect the specifically labeled M_r-70,000 species. These results could be due to the very low affinity of the azido reagent for the chemotactic receptor (Wallace and Frazier, 1979).

The purified M_r-70,000 cAMP-binding protein (Meyers-Hutchins and Frazier, 1984) has the same nucleotide specificity, kinetics, and affinity as the receptor on intact cells (Green and Newell, 1975; Henderson, 1975; King and Frazier, 1979; Mullens and Newell, 1978; Coukell, 1981). It appears to have one binding site per 70,000 Daltons, and the molecular size indicates that it is predominantly a monomer in detergent solution. In the solubilized state, PMSF irreversibly inhibits the cAMP-binding activity of the receptor, whereas in plasma membranes, the activity is reversibly inhibited, and in intact cells, PMSF has no effect. It thus appears as if some site on the receptor that reacts with PMSF is available only from the inner face of the membrane, implying that the receptor is a transmembrane protein. The difference between the reversible inhibition in membranes and the irreversible inhibition in the solubilized state is at present not understood. It may be that some intracellular domain of the receptor has an enzymatic activity that is sensitive to PMSF such as a protease or a kinase. The view that the receptor may have associated enzymatic activity, or may in fact be an enzyme itself, is an idea that dates to the discovery of the membrane PDE. It was somewhat later proven that the substrate specificity of the mPDE and receptor specificity for chemotaxis were distinguishable (Malchow *et al.*, 1973). Thus the idea that the receptor is an enzyme fell into disfavor. Recently, however, van Haastert and Kien (1983) have proposed, based on cAMP analog studies and quantum–chemical calculations, that a nucleophile on the receptor forms a covalent bond with the phosphorus atom of cAMP thus stabilizing a trigonal bipyramidal configuration of a pentacovalent phosporus. The cyclic 3'-5' ring is not opened but is proposed to adopt a favorable diequatorial position. If this scheme is correct, either the receptor–phosphorus bond must be extremely labile or the receptor must enzymatically reverse the bond to account for the rapid dissociation of cAMP under normal conditions. The ability of high ammonium sulfate concentrations to lock cAMP onto the receptor may be related to the stabilization of the putative covalent bond under these conditions, perhaps owing to a dehydration of the receptor binding-site region. It may be that the PMSF effect on the receptor that we have found is related to this proposed ability of the receptor to act as an enzyme to the extent that it can provide a nucleophile which reacts with the phosphorus atom of cAMP.

A property that has been useful in designing a new assay for the receptor and which serves to identify this cAMP-binding protein as a cell surface molecule is that it binds to Con A in a sugar-specific manner. Thus the glycoprotein nature of the receptor and its ability to bind to the hydrophobic decyl agarose indicate that it behaves like a membrane glycoprotein and hence reinforce the idea that this Mr-70,000 cAMP-binding protein, which has not been previously identified, is the cAMP receptor that is exposed at the cell surface (Meyers-Hutchins and Frazier, 1984). In this context it is interesting that Con A enhances cGMP production (Mato et al., 1978a) and inhibits the cAMP-elicited secretion of cAMP (relay response) (Devreotes, 1982). The PMSF data and other experiments presented below indicate that the receptor is also a transmembrane glycoprotein. This feature allows one to consider the process of transduction in terms of a direct physical linkage between the receptor protein and the next protein component of the system, which may be either an inner-face membrane protein or an intracellular, soluble protein. In the remainder of this review, we will consider the current state of studies on the putative chemotaxis transduction systems of Dictyostelium and describe our working hypothesis for signal transduction, which is based on recent data (Galvin et al., 1984) regarding the association of cAMP receptors with the cytoskeleton.

3. PUTATIVE TRANSDUCTION EVENTS IN *D. DISCOIDEUM*

The suggestion that both the folate and the cAMP receptors can interact with the same transduction apparatus focuses attention on an important criterion for the identification of a transducer function. That is, it should be stimulated by any appropriate chemoattractant, not just the acrasin of that particular species. Implicit in this statement is, of course, a faith in the conservatism of speciation so that one expects that a fundamental process such as the linkage between chemotaxis receptors and the force-generating components of the cell has been retained and not newly devised at each necessity. In only a few cases can this criterion of generality be applied. Mato, Konijn and their co-workers have done a thorough job of testing the guanylate cyclase response to chemoattractants in a variety of species (Mato and Konijn, 1977), and it is largely due to their efforts that cGMP is now recognized as a likely component in the transduction system. A more serious problem for our current analysis of available data is the lack of a causal relationship in most cases between some change that ensues after stimulation of the cell with a chemoattractant and the process of chemotaxis itself. Until some causal link can be forged, any change in the cell after stimulation is fair game for consideration as a transduction mechanism, and no response can be proven to be necessary for transduction. Processes that have thus far been seen to change in stimulated *Dictyostelium* cells include (1) cAMP levels and adenylate cyclase activity, (2) cGMP levels and guanylate cyclase activity, (3) extracellular and presumably intracellular pH, (4) Ca^{2+} influx or intracellular redistribution, (5) protein methylation, (6) lipid methylation, and (7) protein phosphorylation (Devreotes, 1982). It is now generally agreed that the adenylate cyclase response is a species-specific function whose sole purpose is signal relay (Devreotes, 1982; Theibert and Devreotes, 1983). Presumably, other species in which acrasin relay occurs also have specific synthetic and/or secretion mechanisms to propagate the signal (Wurster *et al.*, 1976).

3.1. cGMP and Guanylate Cyclase

Only the transient increase in intracellular cGMP due to a brief stimulation of guanylate cyclase has been seen in a number of species with different attractants. Recently, Ross and Newell (1981) isolated an interesting streamer (StmF) mutant (StmF-406) whose phenotype can be explained by a prolonged elevation of intracellular cGMP. The mutant continues to respond to cAMP for much longer periods of time than the wild type and forms long streams during chemotaxis *in situ*. The streamer mutants have either altered or missing cGMP-specific PDE, thus slowing the degradation of the cGMP bolus produced upon chemotactic stimulation (van Haastert *et al.*, 1982b). The generality of the cGMP response and the streamer mutants with altered cGMP metabolism both support the involvement of cGMP in the transduction response. A pulse of cAMP applied to *Dictyostelium* cells in suspension causes a measurable rise in intracellular cGMP within as short a time as 5 sec (Mato and Konijn, 1977; Mato *et al.* 1977b). Wurster and Butz (1983) recently showed that the level of intracellular cGMP adapts during the continuous application of cAMP to responsive cells. The cGMP level reaches a maximum value at about 20 sec and then decreases to about 50% of this level as long as a 5-nM cAMP stimulus is present. Upon removal of the cAMP, its level decays rapidly owing to mPDE, and the cGMP level inside the cells decays with a lag of about 1 min probably owing to the cGMP PDE (Dicou and Brachet, 1980). Recently it has been reported that the intracellular cGMP response displays adaptation in response to cAMP stimulation of cells (van Haastert, 1983a,b,c; van Haastert and van der Heijden, 1983).

Very little is known about the guanylate cyclase since, like the adenylate cyclase (Devreotes, 1982), it is very unstable in cell homogenates (Ward and Brenner, 1977). Some characterization has been accomplished by Ward and Brenner (1977) and by Mato and Malchow (1978). The enzyme has the expected high Km for GTP and is moderately activated by ATP (Mato, 1979). It is inactive with Ca^{2+} as the only divalent metal and appears to absolutely require Mg^{2+} or Mn^{2+}. Since the Km for Mn^{2+} activation is 700 uM and the total cellular Mn^{2+} concentration is about 10 uM (Brenner, personal communication), Mg^{2+} which is present in the cells at 3 mM is probably the physiological activator although in vitro it is less effective than Mn^{2+}. Ca^{2+} is inhibitory at concentrations above 100 μM, and calmodulin has no effect on the enzyme when added to lysates. The cyclase has a pH dependence similar to that of the cGMP PDE, going from less than 10% of maximal activity at pH 6 to more than 90% at pH 7 (Ward and Brenner, 1977). The cellular distribution of the cyclase is unclear since it is found in both particulate and soluble form in lysates (Ward and Brenner, 1977; Mato and Malchow, 1978; Brenner, personal communication).

If an increase in the steady-state level of cGMP is a primary transduction event, then there must be a cGMP "receptor" in the sense that the R subunit of cAMP-dependent protein kinase is the receptor for cAMP produced in cells in response to activators of adenylate cyclase. So far the only cGMP-binding protein found in Dictyostelium is the cGMP-specific phosphodiesterase (Dicou and Brachet, 1980; Mato et al., 1978b), which appears to have an allosteric site for cGMP-mediated activation of the enzyme and a separate catalytic site of distinguishable specificity for cGMP analogs (van Haastert et al., 1982c; Bulgakov and van Haastert, 1983). If cGMP has a second messenger role in chemotactic transduction, it is difficult to imagine how this hydrolytic enzyme might directly participate in the effector pathway. In all studies in the literature, cGMP-binding assays were performed using a procedure in which the cyclic-nucleotide-binding protein is bound to a Millipore filter in order to separate bound and free ligand (Mato et al., 1978b; Rahmsdorf and Gerisch, 1978; van Haastert et al., 1982c). If a cGMP-binding protein exists that does not bind to the filter, it would have thus far gone undetected. We have attempted to identify an intracellular cGMP-binding protein using 8-azido-cIMP generated from radiolabeled 8-azido-cAMP and found only the regulatory subunit of the cAMP-dependent protein kinase to be specifically labeled by this reagent. The reason for this may now be explained by the finding that cIMP does not act as a cGMP analog for either cGMP binding or cGMP PDE "activation" (van Haastert *et al.*, 1982c), but it is a good substrate for the cGMP PDE catalytic site.

This cGMP-specific PDE, first described by Dicou and Brachet (1980) in mutant cells lacking the "nonspecific" cyclic nucleotide PDE, has the following properties: (1) it is activated by cGMP in the range of 10 nM–1 μM, i.e., displays positive cooperativity (van Haastert *et al.*, 1982c), (2) the activation occurs as well at pH 6 as at pH 8, (3) the pH optimum for the catalytic activity is 7.5–8 (Dicou and Brachet, 1980; Bulgakov and van Haastert, 1983), (4) the enzyme is inhibited at mM Cu^{2+} (Dicou and Brachet, 1980), and (5) it is not a Con-A-binding glycoprotein (Bulgakov and van Haastert, 1983) although the nonspecific cell surface and extracellular PDE is bound to Con A agarose. Furthermore, it is this enzyme which is altered or missing in the streamer mutants isolated by Ross and Newell (van Haastert *et al.*, 1982b). When these mutants are pulsed with cAMP, intracellular cGMP reaches a higher level and remains elevated for longer than in the wild type. The fact that in these mutants the initial level of cGMP is much higher indicates that the guanylate cyclase and the cGMP PDE are both active at the same time, and that the PDE is probably not activated in a feedback step. If it is, the feedback is so rapid that there is no discernible time delay.

An alternative to the classical second-messenger role envisioned for cGMP in the transduction process is suggested by recent work from Goldberg's group (Walseth *et al.*, 1983; Goldberg *et al.*, 1983) on cGMP metabolism in the rod outer segment of retina. When cells are labeled with [^{18}O] water, the stable isotope is selectively incorporated into the alpha phosphoryl of AMP and GMP upon hydrolysis of the cyclic nucleotide by phosphodiesterase. After isolation of the nucleotides, the abundance of the label is determined in each phosphoryl position by conversion to glycerol phosphate followed by gas chromatography–mass spectroscopy. Labeling rates are determined and from these the flux of nucleotide through the pathway is calculated. Upon stimulation of the retina with light, the steady-state level of cGMP (and other guanyl nucleotides) is unchanged, but the net flux of cGMP produced by the guanylate cyclase and hydrolyzed by the cGMP PDE is increased as much as 125-fold. This means that the entire pool of cGMP is being hydrolyzed once every 20 msec. The metabolic sequelae of this tremendous activation of the cGMP pathway are the production of two protons and one molecule of pyrophosphate for each cGMP formed and hydrolyzed in addition to the liberation of about 11 Kcal/mol free energy from the hydrolysis of cGMP. Goldberg *et al.* (1983) suggest that it is some aspect of this increased flux which serves as the message to mobilize calcium which then acts on the sodium channel of the rod outer-segment membranes. Since no cGMP-binding protein other than the PDE has as yet been found in *Dictyostelium*, it is an intriguing possibility that cGMP hydrolysis is also the "message" in chemotactic transduction. We are currently testing this idea in collaboration with Dr. Goldberg's group.

3.2. pH and Calcium Ion Regulation during the Chemotactic Response

In 1978 Malchow *et al.* (1978a,b) reported that recordings of extracellular pH from unbuffered suspensions of *Dictyostelium* cells showed transient acidification during a chemotactic response to cAMP. The data were complicated by the fact that hydrolysis of the cAMP by the nonspecific PDE contributed to the acidification. However, at cAMP doses below 0.1 μM this effect was minimal, and a concentration of 0.3 nM cAMP produced a half-maximal response. The response was specific for cAMP analogs that are active as chemoattractants, and about 3000 protons were extruded per molecule of bound cAMP. An important question is where do these protons come from? At saturating levels of chemoattractant with 100,000 receptors/cell, enough protons would be removed from the cell's interior to increase the pH inside the cell to 10. Even though substantial buffering of this change would occur owing to intracellular contents, this large a proton depletion could cause a significant alkalinization of the cytoplasm.

In collaboration with Dr. Paul Schlesinger, we have used the pH dependence of the fluorescence excitation of fluorescein (Thomas *et al.*, 1979) to measure the intracellular pH (pHi) of *D. discoideum* cells in the vegetative state and during the first 3 hr of differentiation (Jamieson *et al.*, 1983). Surprisingly, the "resting" pHi of the cells just after harvesting from the HL-5 growth medium is 6.2 (+/−0.1, n = 4) and remains at this level for approximately 2 hr. At this time there is a very reproducible, transient alkalinization of the cells during which the pHi increases to 7.1 +/− 0.3 (n = 4) and within 10 min falls to the previous level of about 6.2. The basis for this rise in pHi appears to be the activation of a sodium/H+ exchange process since it is blocked by 200 μM amiloride or the removal of extracellular sodium. This mechanism is common in many higher cell types as a means of pHi control (Rothenberg *et al.*, 1983). It is perhaps significant that in other systems undergoing changes in differentiative state, transient pHi changes have been detected (Webb and Nuccitelli, 1981; Steinhardt and Morisawa, 1982).

The importance of the pHi increase which we have found for the initiation of or "commitment" to a developmental program is suggested by the finding that the same concentrations of amiloride that block the pHi change at 2 hr of starvation also slow morphogenesis in populations of cells on agar. If cells are treated with amiloride before 2 hr, preventing the alkalinization, their differentiation is delayed. However, if treatment is started after 2 hr, the cells have already alkalinized and they also proceed with development normally. Thus the presence of amiloride *per se* does not block development; the drug must be present before the pHi change to have an effect. The role of pHi in regulating development has been suggested by the work of Gross *et al.* (1983), who presented evidence that the choice of stalk or spore cell differentiation pathways is controlled by DIF (a low-molecular-weight factor) (Kay and Jermyn, 1983) and ammonia, which work against each other. The predominant effect is apparently determined by pHi. Weak acids, such as propionate, diethylstilbesterol (which blocks the ATP-dependent proton pump), and low extracellular pH itself can all shift the cell population toward an overabundance of stalk cell production. Being a weak base, ammonia acts by tending to raise pHi which produces more spores. Gross *et al.* (1983) suggest that DIF, like diethylstilbesterol, promotes stalk cell formation by inhibiting the ATP-dependent proton pump, thus causing a lower pHi. We have found that diethylstilbesterol at 10 μM has the same effect on the progression of development as 200 μM amiloride when added prior to the alkalinization at 2 hr of starvation (Jamieson *et al.*, manuscript in preparation). This suggests that *Dictyostelium* has at least two mechanisms for regulating pHi, the amiloride-sensitive sodium/proton exchanger and the diethylstilbesterol-sensitive ATP-dependent proton pump.

An immediate consequence of the low pHi of *Dictyostelium* for our discussion of chemotactic transduction events is that both the guanylate cyclase (Ward and Brenner, 1977) and the cGMP PDE (Dicou and Brachet, 1980; Bulgakov and van Haastert, 1983) are essentially *inactive* at the resting pHi of the cell. Although it is possible that both are sequestered in a more alkaline intracellular compartment, their immediate release into the soluble fraction upon cell lysis by a number of methods makes a physical compartmentation unlikely. It seems more likely that the extrusion of protons seen by Malchow *et al.* (1978a,b) is the result of a transient alkalinization of the cells' interior after stimulation by cAMP. This increase in pHi could then be the primary switch that turns on the production of cGMP by the guanylate cyclase and also attenuates it by activating the cGMP PDE. It seems that localized changes in pHi in response to receptor stimulation could establish intracellular pH gradients which would serve to polarize the cell, effectively producing a gradient inside the cell which follows the gradient of attractant present outside the cell. To test for variations in pHi in different regions of the cell it is necessary to make fluorescence video records of cells containing the fluorescein probe after excitation at the two wavelengths used in the pH measurements. The fluorescence images are then digitized by computer, and the ratio of fluorescence at the two wavelengths is calculated. The resulting map of ratios is directly related to the pHi at each point in the cell. In collaboration with Dr. Elliot Elson's laboratory, we have begun to collect single-cell data and to test this methodology. Although no data on the response to attractants are available at this writing, the method seems to be feasible and yields pHi values within the range found with populations of cells (discussed previously).

In the neutrophil system, changes in pHi have been detected in response to chemotactic peptides, and amiloride blocks the alkalinization (Sha'afi *et al.* 1982). It is not yet clear whether substantial parallels exist between the mechanisms at work in bacterial chemotaxis and the transduction mechanism which is functional in eucaryotic chemotaxis. However, it is intriguing that pHi has recently been found to be a primary mediator of bacterial chemotaxis which affects the methylation pathway (Kihara and MacNab, 1981; Repaske and Adler, 1981). It may be that the rapid alkalinization we observe at about 2 hr of development is the first of the autonomous cAMP responses which begin to occur at about this time, and which continue with increasing frequency until the cells aggregate (Devreotes, 1982; Gerisch, 1982). If a rapid alkalinization is found to be an immediate response to chemoattractants, this would have fundamental implications for the transduction mechanism. The obvious effects on the activity of the guanylate cyclase and the cGMP PDE could easily explain the observed intracellular cGMP accumulation and its time course in cells pulsed with a degradable attractant and with a "square-wave" stimulus (Wurster and Butz, 1983).

The change in pHi would also have a large effect on the binding of Ca^{2+} by all Ca^{2+}-binding proteins. Since Ca^{2+}-binding sites tend to be acidic, carboxylate-containing regions of proteins, the affinity of these sites for Ca^{2+} is a very sensitive function of pH (Taylor and Fechheimer, 1981). Protonation of the binding site lowers the affinity for Ca^{2+}; thus the large increase in pHi upon stimulation would promote the binding of Ca^{2+}. This effect in itself may be the driving force for the Ca^{2+} influx upon stimulation reported by Wick *et al.* (1978) and more recently by Bumann *et al.* (1984). However, it is clear that extracellular Ca^{2+} is not the primary source of Ca^{2+} during chemotaxis, since the cells perform quite well in EGTA. In this case, an increase in the affinity for Ca^{2+} of cytoplasmic Ca^{2+}-binding proteins would tend to pull Ca^{2+} from other cellular compartments such as mitochondria or nuclei. If the free cytosplasmic Ca^{2+} concentration in *Dictyostelium* is as low (1 μM or less) as in other cells (Taylor and Fechheimer, 1981), the change in pHi from 6 to 7 or higher could drastically alter the locally available Ca^{2+}. In addition, effects of pH on the binding of Ca^{2+} to actomyosin and its regulatory proteins may have substantial consequences for the contractile apparatus. It is reasonable to expect that a large change in pHi will have a pleiotypic effect on a variety of cellular processes. This may provide an explanation for many of the changes observed such as those in protein and lipid methylation after stimulation with a chemoattractant. As will be seen in Section 4, calcium is an important regulatory ion for the processes of actin and actomyosin gelation and contraction, and several calcium-sensitive proteins have been identified which appear to regulate contractile processes in *Dictyostelium* (Condeelis and Vahey, 1982; Taylor and Fechheimer, 1981, 1982).

The best known protein mediator of calcium regulation is calmodulin (CaM), a small (Mr-17,000), acidic, stable protein which is nearly ubiquitous in the plant and animal kingdoms (Klee *et al.*, 1980). Recently CaM was purified using conventional, nonaffinity methods from *Dictyostelium* cells by Bazari and Clarke (1981). The protein was found to have some of the same functional properties in terms of mediating calcium-dependent processes as CaMs from higher organisms. The molecular weight reported by Bazari and Clarke (1981) was similar to that of bovine CaM, but the extinction coefficient and amino acid composition were quite different from other CaMs. We have isolated *Dictyostelium* CaM using an improved extraction protocol and an immobilized phenothiazine affinity method (Jamieson and Vanaman, 1979). This affinity-purified CaM has an extinction coefficient which reflects the lack of tryptophan, a property of all other CaMs (Klee *et al.*, 1980), and an amino acid composition similar to that of bovine CaM (Jamieson and Frazier, 1983). As first determined by Bazari and Clarke (1981), we also find that *Dictyostelium* CaM does not contain a trimethyllysine residue, a property of many but not all CaMs. However, Rowe *et al.* (1983) have found that *Dictyostelium* CaM can be *N*-methylated presumably on one or more lysine residues by enzymes present in a variety of rat tissues. It appears that *N*-methyltransferases exist in *Dictyostelium* (Mato and Marin–Cao, 1979; van Waarde, 1983; our unpublished data); thus it remains to be determined why CaM is not methylated. Since the role of the trimethyllysine residue in CaM is obscure, the relevance of its absence in *Dictyostelium* is at present unclear.

It now remains to identify the intracellular target proteins for calcium and CaM regulation. Some of these are undoubtedly part of the contractile apparatus such as myosin light-chain and heavy-chain kinases (Maruta *et al.*, 1983). In addition, CaM-regulated kinases have been found in other systems (Woodgett *et al.*, 1983) which may link the roles of calcium and phosphorylation. Of particular interest are the high-molecular-weight spectrinlike proteins such as fodrin (Glenney and Glenney, 1983), which is found directly beneath membranes (Levine and Willard, 1981) and in association with actin filaments (Glenney *et al.*, 1982b) and which binds CaM (Carlin *et al.*, 1983; Glenney *et al.*, 1982a). We have identified a large number of CaM-binding proteins using affinity chromatography of *Dictyostelium* extracts on columns of CaM–Sepharose (Jamieson and Frazier, 1982). Several of these proteins display dramatic changes in their levels of CaM-binding activity during the first 8 hr of differentiation. We have also used a radioiodinated CaM overlay method to detect CaM-binding proteins of the membrane and cytoskeletal fractions. Of particular interest for further investigation are two high-molecular-weight CaM-binding proteins that associate with the cytoskeleton.

3.3. Methylation Reactions of Lipids and Proteins

Protein methylation, and particularly carboxylmethylation, has been established as an important reaction in the control of bacterial chemotactic responses (Springer *et al.*, 1979). Recently, with the use of inhibitors of cellular methylation reactions, evidence has been obtained implicating these reactions in chemotaxis of leukocytes (Schiffmann, 1982) and macrophages (Pike and Snyderman, 1982). After stimulation of responsive *Dictyostelium* cells, Mato and Marin–Cao (1979) found an increase in the methylation of an M_r-120,000 protein that was maximal at 15–30 sec. Even this must be considered slow on the time scale of the chemotactic response. The protein of M_r 120,000 has not been identified, but it is interesting that a protein has been isolated from *Dictyostelium* which has potent actin gelation activity and which has an M_r of 120,000 (Condeelis *et al.*, 1982). This protein, in contrast to an M_r-95,000 protein that cross-links actin filaments (Condeelis and Vahey, 1982; Brier *et al.*, 1983), is not regulated by Ca^{2+}. A decrease in phospholipid methylation was also seen, but this occurred at 2 min after stimulation. In whole cells stimulated with 8-bromo-cGMP or cell lysates stimulated with cGMP, phospholipid methylation was found to increase with a two- to threefold stimulation being reached by 1 min (Alemany *et al.*, 1980). These data were taken to suggest that the transient rise in intracellular cGMP after chemotactic stimulation might mediate changes in lipid methylation. The relatively long lag for this response tends to suggest that it is probably not involved in the motility response. Cells have been seen to extend pseudopodia within 5 sec after chemotactic stimulation (Swanson and Taylor, 1982); hence any mediator of this process must act very rapidly.

More recently, van Waarde (1982) has reported that a protein of M_r 46,000 becomes carboxylmethylated within 3 sec after stimulation of differentiated cells with cAMP. Moreover, a protein of this Mr is methylated in vegetative *D. discoideum* cells in response to folate and in *D. lacteum* in response to monapterin (van Waarde, 1983), a pterin derivative that is equipotent with the as-yet-unidentified pterinlike acrasin of this species (van Haastert *et al.*, 1982a). Thus the methylation of this protein is rapid enough to make it a candidate for a transduction component, and the response occurs in more than one species with multiple chemoattractants. These studies were all performed with wild-type cells which grow on bacteria containing very active methyltransferases, and the results were seen to differ with cells grown on two different bacterial strains. Thus it would be desirable to confirm these results in axenic strains of *Dictyostelium*.

A further complication is that these methylation experiments were performed by loading live cells with radiolabeled methionine and blocking protein synthesis with large amounts (250 μg/ml) of cycloheximide. Mutzel *et al.* (1983) have found that an M_r-17,000 protein is aminomethylated (alkali-stable methyl groups) in cell lysates using S-adenosylmethionine as the methyl donor, and that the methylation of this protein increases sixfold during the first 6 hr of development. This is thought to be due to an increase in methyl transferase rather than substrate activity. Although the low-molecular-weight methyl acceptor has not been identified, its size and heat stability suggest that it may be calmodulin. Both Bazari and Clarke (1981) and Jamieson and Frazier (1984) have found that *Dictyostelium* calmodulin is not methylated, raising the possibility that Mutzel *et al.* (1983) may have created conditions in their *in vitro* methylation experiments which allowed the methylation of calmodulin. We have not been able to detect any *N*-methylation of purified *Dictyostelium* CaM using S-adenosylmethionine as a methyl donor *in vitro* (Skogen, Jamieson, and Frazier, unpublished observations). The effect of chemoattractants on the methylation of the M_r-17,000 protein has not yet been reported.

3.4. Phosphorylation Reactions

The involvement of intracellular cAMP in the regulation of phosphorylation reactions has spurred the efforts of many investigators who have tried to detect a change in phosphorylation reactions inside *Dictyostelium* cells upon chemotactic stimulation. However, it has only recently been established that *Dictyostelium* contains a cAMP-dependent protein kinase (Leichtling *et al.*, 1982, 1984; Majerfeld *et al.*, 1983), and that its regulatory subunit can regulate the mammalian kinase (Leichtling *et al.*, 1981a) even though the R subunit molecular weight is only 40,000 in slime mold. This enzyme is presumably responsive to intracellular cAMP levels, and it is unlikely that it would be a means of signal transduction for chemotaxis. It is possible, however, that the chemotaxis receptor is able to stimulate phosphorylation by a cAMP-independent kinase such as the tyrosine kinase, which is part of the EGF and other receptors (Ushiro and Cohen, 1980). As of this writing, tyrosine phosphorylation has not been reported to occur in *Dictyostelium*. Lubs–Haukeness and Klein (1982) reported that cAMP stimulation of aggregation competent cells causes the rapid (5 sec) stimulation of phosphorylation of an M_r-47,000 protein. The relationship of this protein to the chemotactic response is unclear. An increase of the *in vitro* phosphorylation of myosin heavy chains has been observed after chemotactic stimulation (Rahmnsdorf *et al.*, 1978; Maruta *et al.*, 1983), but this is thought to be due to stimulation *in vivo* of heavy-chain dephosphorylation during the response thus creating more sites for subsequent phosphorylation in the *in vitro* assay (Gerisch, 1982). One of the two myosin heavy-chain kinases from *Dictyostelium* is inactivated by calcium and calmodulin (Maruta *et al.*, 1983), which could presumably lead to a lower level of myosin heavy-chain phosphorylation. Dephosphorylated myosin heavy chains form filaments more readily and have a higher actin-activated Mg^{2+}–ATPase activity than the phosphorylated form of myosin heavy chains (Kuczmarski and Spudich, 1980).

Coffman *et al.* (1982) examined the *in vitro* phosphorylation of *Dictyostelium* proteins during development, finding substantial changes in their patterns on two-dimensional gels. Rapid changes of the phosphorylation of proteins in response to chemoattractants were not reported. We have searched for *in vitro* phosphorylation of *Dictyostelium* cell lysates using a variety of different buffers and ionic conditions in the reaction. The level of total phosphate incorporation varies widely with different buffer conditions, but in no case was a difference found in the protein species phosphorylated in the presence and absence of cAMP (unpublished observations). King and Frazier (1977) reported the transfer of the gamma phosphate of extracellular ATP to an acceptor compound of low molecular weight associated with cell membranes (King, 1979). This reaction does not appear to involve protein phosphorylation, however, and the acceptor has not been identified. Thus, at present, no persuasive evidence exists for the involvement of protein phosphorylation in chemotactic transduction. It may, however, have a regulatory role in subsequent intracellular responses such as regulation of myosin–actin interactions during movement. This effect may be more related to strength or extent of motility than to its direction.

4. THE MOTILITY APPARATUS OF *DICTYOSTELIUM*

4.1. Traction Models of Motility

Cell motility in *Dictyostelium* has recently been reviewed by Spudich and Spudich (1982). Chemotaxis represents a special case of motility, which perhaps can be thought of as a bias imposed on what normally appears to be a random or at least nondirected process. This view is supported by the morphological observations of Schliwa *et al.* (1982) and Zigmond *et al.* (1981) on PMN motility and chemotaxis. The way in which one thinks about the process of motility itself determines in large degree the model that one constructs to explain chemotaxis. In the two limits, divergent models are plausible to explain motility. In one case, referred to here as a traction model, actomyosin in some form exerts a pull or a push on one or more fixed points in the cell or on its surface. The very active nature of the plasma membrane along with the well-recognized submembranous cortical actin meshwork (Spudich and Spudich, 1982; Giffard *et al.*, 1983) implies that a large amount of the force generated should be on the plasma membrane of the cell, and that the membrane in turn should have the property of at least focal adherence to a substratum, the ultimate fixed point against which the push or pull is exerted. The properties of the cytoskeleton of eukaryotic cells, its remarkably plastic nature on the one hand and its great stability on the other, provide a means of visualizing how force generated near the membrane or the nucleus of a cell might be propagated to other regions such as might occur during the shape change that occurs in PMNs preparatory to chemotaxis (Zigmond, 1978; Zigmond *et al.*, 1981). Thus the traction model stresses the similarity of the motile cell's contractile machinery to the contractile apparatus of muscle in which the principle of movement against a fixed point or plane is so readily appreciated. To be sure, most of the same proteins are found in nonmuscle cells, and they have the same properties in terms of their general mechanisms of action. What is lacking is the clear demonstration that the supramolecular organization exists in a totally analogous way. For example, Spudich and Spudich (1982) point out that the actual structural state of nonmuscle myosin is unknown. Thus uncertainties of this type regarding the structural and mechanical relationship of the contractile proteins to one another prevent the whole-hearted acceptance of a model for motility that is based solely on a global organization of the motility apparatus. This view is clearly an extreme, and it is certain that the force generated by actin and/or myosin is in some way responsible for the locomotory activity of cells. The debate appears to center on the extent or level of organization required. That a high degree of organization can exist in cells is demonstrated by the preparation of detergent-insoluble cytoskeletons from a variety of cells (Brown *et al.*, 1976; Heuser and Kirschner, 1980; Schliwa and van Blerkom, 1981; Galvin *et al.*, 1984) which contain the contractile as well as other associated proteins. The cytoskeletal model is one which will be quite useful in studying the regulation of processes related to motility (see below). It is also true that the structures that make up the cytoskeleton are all labile to varying degrees and each requires special conditions for stabilization. Furthermore, there are many proteins that serve the function of altering (and thus presumably regulating *in vivo*) the filamentous systems of the cytoskeleton in response to changes in calcium, pH, nucleotide levels, and receptor stimulation.

4.2. The Solation—Contraction Coupling Model of Motility

A rather different view of cell motility derives from the work of Taylor and co-workers as stated in the "solation—contraction coupling hypothesis" (Condeelis and Taylor, 1977; Hellewell and Taylor, 1979; Taylor and Fechheimer, 1982). This model focuses more on the interactions that occur among contractile proteins and their regulators as they influence the viscoelastic properties of the cytoplasm. The gelation, solation, and contraction of cell extracts has provided the basis for this model, and more recent work has focused on the identification and purification of the protein factors involved in these reactions, their distribution in the cell, and the effects of Ca^{2+} ion and pH on these processes (Taylor and Fechheimer, 1981). In this model, local changes in the cytoplasm lead to regions of low (sol) and high (gel) viscosity. Contractions of actin and myosin that are thought to occur in solated regions can then exert force on nearby gelled regions, or, by collapsing in upon itself, the solated pocket can squirt its liquid contents, thus deforming other areas of cytoplasm or of membrane. A key feature of this model is that a highly organized or gelled state is inhibitory to contraction, and partial solation is required to allow contraction to occur. By studying the effects of pH and Ca^{2+} on the viscosity, light scattering, and strain birefringence of soluble extracts of *Dictyostelium,* Hellewell and Taylor (1979) found that at a pH less than 6.6 the system was in a low viscosity or sol state but that contraction did not occur. At pH 6.8 a gel state was favored and no contraction was found, whereas at a pH greater than 7.0 solation occurred as well as contraction. These results were obtained at low Ca^{2+} concentrations (30 nM). At pH 6.8, raising the free Ca^{2+} to near 1 μM would solate a gel formed at low Ca^{2+}, thus in addition to the pH regulation of the system, it is very sensitive to the free Ca^{2+} concentration as well. One can imagine the exquisitely fine control that could be exercised by subtle variations in the pH and Ca^{2+} levels in different regions of the cell.

In view of the low pH in *Dictyostelium* cells (see above), it seems likely that the ground state of the cytoplasm is that described as solated and noncontractile below pH 6.6 (Hellewell and Taylor, 1979). Conversely, the cytoskeleton of *Dictyostelium* is most stable at pH 6.1 (below) and would serve to maintain the shape of the cell. It is not clear whether solated cytoplasmic extracts and a highly stabilized cytoskeleton are mutually exclusive or whether they should indeed have a reciprocal relationship. That is, recruitment of actin and myosin into the cytoskeleton may leave adjacent areas of cytoplasm relatively poor in actin, thus preventing it from assuming a tightly gelled state. At a constant low Ca^{2+} level, as the pH is increased the cytoskeleton becomes less stable and the region would gel; then as it increases further, the region of cytoplasm should solate and contract. Of course, it is possible that concomitant increases in Ca^{2+} could serve to offset the effects of rapidly changing pH, but as noted previously, the increasing pH would also serve to buffer the free Ca^{2+} concentration since Ca^{2+}-binding sites will, in general, bind Ca^{2+} more tightly at a higher pH. In specific cases, however, this may also mean that the Ca^{2+}-binding proteins that confer the Ca^{2+} sensitivity on the cytoplasm bind Ca^{2+} with higher affinity, thus magnifying the effect of Ca^{2+} on the gelation and contraction process. The rapid advances being made in the study of specific proteins that regulate the assembly and cross-linking of cytoskeletal proteins should soon allow the formulation of testable models for modulation of cytoskeletal and cytoplasmic structure. It seems quite probable that the distinction between cytoskeleton and cytoplasm will become less sharp as more complex control mechanisms are worked out.

4.3. Regulatory Proteins

The cytoplasmic components responsible for the gelation and contraction observed in these extracts undoubtedly include actin and myosin (Taylor and Fechheimer, 1982). The control of the polymerization of actin and its interaction with myosin and the cross-linking of actin filaments appear to be due to a group of actin-binding proteins of molecular weights 120,000 and 95,000 as well as low-molecular-weight proteins of M_r 30,000 and 18,000. These low-molecular-weight species also appear to be major components of the cytoskeleton of *Dictyostelium* prepared by dissolving cells in nonionic detergent under appropriate conditions (see below). The 120,000- and the 95,000-M_r proteins are both actin filament cross-linkers (Condeelis *et al.*, 1982; Condeelis and Vahey, 1982; Brier *et al.*, 1983), the M_r-95,000 protein being Ca^{2+} sensitive and the M_r-120,000 protein not. The M_r-30,000 protein is monomeric in solution and also cross-links actin filaments. This activity is inhibited by very low concentrations (100 nM) of calcium. In contrast to the actin-cross-linking activity of the M_r-95,000 protein, the activity of this M_r-30,000 protein is relatively insensitive to changes in pH (Fechheimer and Taylor, 1983). Spudich and co-workers have purified an M_r-40,000 protein named severin, which depolymerizes actin filaments in the presence of Ca^{2+} (Yamamoto *et al.*, 1982). Interestingly, Spudich and co-workers (Giffard *et al.*, 1983) have found that the cortical actin matrix is stabilized by low Ca^{2+} conditions, and we have found that the detergent-resistant cytoskeleton is stabilized by low pH (6.1) and low Ca^{2+} (Galvin *et al.*, 1984).

4.4. Interaction of Contractile Proteins with the Surface Membrane

A key feature of the traction model of motility is that the filamentous-force-generating components must be anchored or physically attached in some way to the membrane. Support for this idea is found in the many experiments which demonstrate transmembrane associations between cell surface proteins and submembranous actomyosin which result in capping of the cell surface protein in an energy-dependent reaction (Bourguignon and Singer, 1977; Painter and Gins-

berg, 1982; Schreiner *et al.,* 1977). In *Dictyostelium,* Condeelis (1979) has
found that Con A caps can be isolated after Triton X-100 (TX-100) treatment
of capped cells and that these caps are markedly enriched in actin and myosin.
As in other cells, capping is inhibited by azide and would thus appear to be
energy dependent. Thus transmembrane associations would appear to exist between the Con A receptors on the surface and the actin and myosin in the
submembranous caps. Giffard *et al.* (1983) report that TX-100 treatment of cells
results in a cortical actin meshwork which in many spots appears to be attached
to patches that appear in thin sections as short segments of membrane. Luna *et
al.* (1981) observed that actin-and-myosin-free (stripped) membranes cross-linked
muscle F actin filaments assayed as an increase in the viscosity of the solution
with falling ball viscometry. The cross-linking activity appeared to be due to
protein components of the membranes which were resistant to TX-100 solubilization. This suggests that one or more membrane proteins may serve the function
of directly binding membrane proteins to the cortical actin–myosin meshwork.
Jacobson (1980) has used surface membranes of *Dictyostelium* attached to po-
lylysine-derivatized beads to examine the interaction of F-actin with the cytoplasmic surface of the membrane. Thus evidence exists for the association of
components of the contractile system with the inner face of the surface membrane.
Fodrin- or spectrinlike molecules, not yet identified in slime molds, would appear
to be attractive candidates for mediating actin–membrane interactions in a calcium-dependent way (Levine and Willard, 1981; Glenney *et al.,* 1982a,b; Carlin
et al., 1983).

   Some observations support and corroborate the solation–contraction hypothesis as well. For example, the gelation of myosin-free *D. discoideum* extracts, which reflects primarily actin, is reversed by free calcium levels approaching 1 μM (Condeelis and Taylor, 1977). Giffard *et al.* (1983) found that
low calcium was also necessary for recovery of a cortical actin matrix from
detergent-treated cells, and Galvin *et al.* (1984) found that the optimum stability
of the cytoskeleton required EGTA. Thus these observations suggest that the
properties of the gel determined by Taylor's group correlate well with the actin
meshwork of cells exposed *in situ.* Solation of the gel or perhaps contraction of
the gel when it contains myosin may equate with breakdown of an isolatable
cytoskeleton at higher pH and calcium levels (Giffard *et al.,* 1983; Galvin *et
al.,* 1984). Swanson and Taylor (1982) stimulated motile *Dictystelium* amebae
with cAMP released from micropipettes and observed that membrane pseudopodia were extended toward the stimulus very rapidly. A short time later, the
organelles appeared to flow toward the "front" of the cell, which could be
explained by the solation of the cytoplasm and pulling or pushing the organelles
forward. In fact, when viewing cells *in vivo,* one cannot tell whether the blebs
and extensions of membrane are caused by forces exerted on the membrane by

anchored filaments pushing or pulling the membrane (traction model) or by solated cytoplasm being squirted against the inner face of the membrane. It would appear that the amplitude of the movements observed in cytoplasmic extracts by Taylor and Fechheimer (1982) is large enough to account for the rate and magnitude of the movements of cell boundaries seen during chemotactic migration. The feature lacking in the cytoplasmic extract model system is, of course, the attachment of the actin filament mass to the plasma membrane, which, as noted previously, is amply documented by morphological and biochemical criteria. It seems obvious at this point that the traction model and the sola-tion–contraction coupling hypothesis are not exclusive and, in fact, describe different aspects of the same system. As will be seen below, a different aspect of cell motility is offered by the perspective obtained in studying a cytoskeletal preparation. It seems that a synthesis of all three models will be required to adequately describe the function and regulation of the motility apparatus during chemotaxis.

5. CYTOSKELETAL INTERACTIONS OF RECEPTORS AND EFFECTORS IN D. DISCOIDEUM

5.1. What Is the Cytoskeleton?

The cytoskeleton is an operationally defined structure that represents a collection of many of the proteins involved in cell motility plus more or less of the purposefully and adventitiously associated cell components. Depending on the detergent extraction conditions employed, the major filamentous systems of the cell can be retained, along with the proteins that have some functional association with the filaments (Heuser and Kirschner, 1980; Schliwa and van Blerkom, 1981). In nucleated cells the nuclear remnant is retained (Osborn and Weber, 1977) which includes a diverse array of proteins. Ribosomes and mRNA have been reported in cytoskeletal preparations (Lenk et al., 1977), as well as enzymes whose presence is difficult to rationalize given our naïve view of cellular or-ganization (Dang et al., 1983). In many studies, the components obtained under the particular conditions used have not been well characterized, thus leading to confusion about the significance of the reported associations of various macro-molecules with these cytoskeletons. Recently, concern has been raised by several studies that nonfilamentous residues of membranes can be left behind under some conditions, and the generality of these observations is not yet clear (Ben-Ze'ev et al., 1979; Dang et al., 1983). These problems notwithstanding, the detergent-extracted cell or cytoskeletal preparation has proven to be extremely useful in studying processes that prove intractable in the whole cell. However, care must be exercised in assuming that because a cellular component is detergent insoluble, it must be associated in some meaningful way with part of the filamentous cytoskeleton (Ben-Ze'ev *et al.*, 1979; Dang *et al.*, 1983).

5.2. The Role of the Cytoskeleton in Chemotaxis

The discussion of cell motility (Section 4) leaves one with the feeling that chemotaxis must in some way involve the organization and correct marshaling of the cell's contractile apparatus. Thus beyond the obvious consideration that much of the cell's actin and myosin reside in the cytoskeleton at any given time, it is reasonable to inquire whether other cytoskeletal components give direction to this machinery. In PMNs, changes are seen in the cellular location of the microtubule organizing centers (MTOC) very rapidly after either a chemotactic or chemokinetic stimulus. This suggests that although the actomysin system may provide the force for cell motility, the microtubule system has a critical role in directing that force to produce directed locomotion. Some controversy exists concerning just what position the MTOC occupies, but all studies have found a decidedly nonrandom location. For example, in neutrophils chemotactically oriented in Millipore filters in a Boyden chamber, Malech *et al.* (1977) reported that a significant percentage of the cells have their MTOC positioned between the nucleus and the front or advancing edge of the cell. However, using freely motile neutrophils in Zigmond (1977) chambers, Anderson *et al.* (1982) found that in 65% of the oriented cells the MTOC was between nuclear lobes, in 35% it was "behind" the nucleus, and in less than 1% was the MTOC "anterior" to the nucleus. Schliwa *et al.* (1982) have observed that when PMNs experience a jump in chemoattractant concentration with no imposed gradient (conditions for chemokinesis), the MTOCs tend to split as though the cell were orienting in two directions at the same time.

5.3. cAMP Receptor Association with the Cytoskeleton

It seemed reasonable to us that since the function of chemotaxis receptors is to modulate some aspect of cellular motility, and since the cytoskeleton contains many of the relevant proteins, it might prove interesting to prepare cytoskeletons from *Dictyostelium* cells and to determine whether components of the chemotactic response were included in this structure. A recent study by McRobbie and Newell (1983) indicated that the cAMP receptor could affect the cytoskeleton directly, since they found that the amount of actin (presumably F-actin) associated with the cytoskeleton is rapidly (3 sec) increased by cAMP treatment of the intact cells.

When aggregation-competent (6-hr-differentiated) cells are dissolved with 1% TX-100 in 17 mM Pi buffer, pH 6.1 with 1 mM EGTA, cAMP-binding sites remain associated with the insoluble residue. We have found (Galvin *et al.*, 1984) that these sites have the same nucleotide specificity, rapid kinetics, approximate affinity, and developmental regulation as the cAMP receptor on the surface of the cells. All the cell surface sites are recovered along with some internal sites as well. These may be involved in cAMP internalization (Klein, 1979) or in receptor recycling. The cytoskeletons prepared with our methods were characterized in terms of the distribution of a battery of marker enzymes and found to contain no cell surface mPDE and less than 10% of the alkaline phosphatase, both surface membrane enzymes. In addition, intralysosomal and intramitochondrial enzymes were released into the detergent supernatant. Of the many Con-A-binding membrane glycoproteins of these cells, only a few discrete proteins accounting for about 15% of the total were retained. These cytoskeletons contained only 10% of the cell protein and 50% of the cellular actin. Figure 1 shows a cytoskeleton prepared under our conditions that has been critical point dried to prevent collapse of the filamentous structures. Little membrane material appears to remain, and the cortical actin meshwork around the periphery is abundant. Thin sections (Fig. 2) reveal a nearly complete lack of membranous material at the perimeter or inside the cell. Thus the cAMP receptor is probably associated with some type of filamentous structure rather than with a membrane residue. cAMP receptors appear to become associated with the cytoskeletal residue as soon as they are synthesized at about 2 hr of development. They are attached whether or not the receptor is occupied with cAMP, since cells can be dissolved with TX-100 first and the [^{3}H]-cAMP added later.

The conditions of cell lysis were varied in attempts to determine what the receptor might be linked to such that it remained insoluble in detergent. When cell membranes are prepared by a two-phase polymer method (Siu *et al.*, 1977) at pH 8, the receptor is readily extracted with nonionic detergents (Meyers–Hutchins and Frazier, 1984). Thus it is not inherently insoluble. When cells are extracted at pH 8 in either Tris or phosphate buffer, the receptor is also largely soluble. The optimum pH for retention of the receptor on the insoluble material is pH 6.1. This is also the optimum for recovery of cytoskeletal components such as actin, suggesting that the pH dependence is reflecting the stability of the cytoskeleton itself. The recovery of receptor drops off steeply below and above pH 6.1. It is interesting that the intracellular pH of *Dictyostelium* cells is about pH 6.2 (see Section 3.2), very close to the optimum for cytoskeletal stability or receptor attachment. It was pointed out previously that changes in intracellular pH could easily modulate the activity of the actomyosin system as

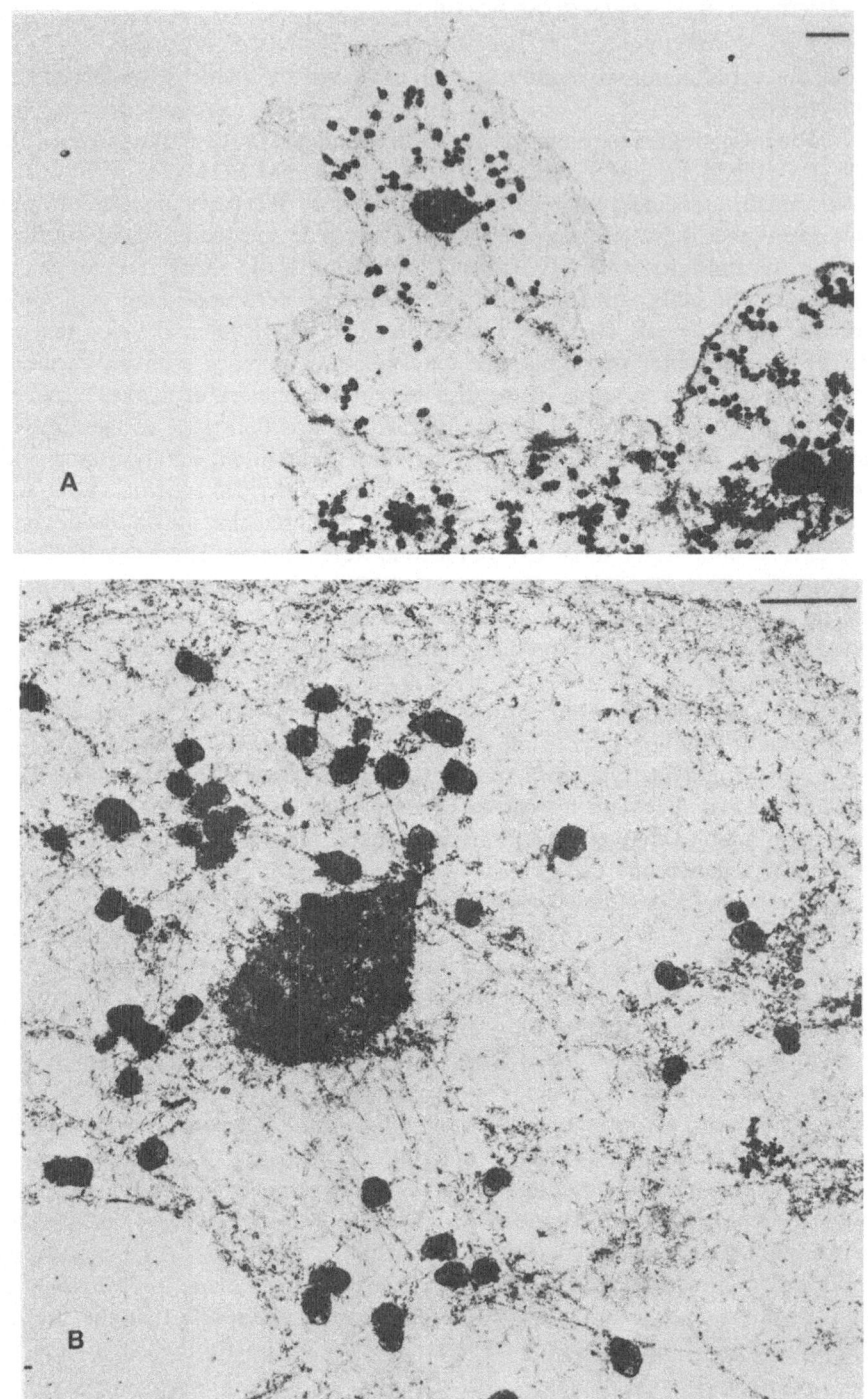

well as the guanylate cyclase and phosphodiesterase. It also appears that pH could determine the degree of association of the receptor with potential effector molecules that could also be associated with the cytoskeleton. Experiments with cytochalasin b, DNAse I treatment (Mannherz et al., 1975; Painter et al., 1982), and colchicine were equivocal and did not allow us to determine if receptors are attached primarily to microfilaments or microtubules. Since nothing is presently known about the properties or indeed the existence of intermediate filaments (Lazarides, 1982) in slime molds, they must be considered by default to be a possibility for attachment of receptors. White et al. (1983) have identified slime mold tubulin and have characterized some of its properties. Since we recover much more cAMP-binding activity with the cytoskeleton of cells incubated at low temperature than at room temperature (Galvin et al., 1984), and White et al. (1983) find that Dictyostelium microtubules are cold sensitive, it is unlikely that the receptor is directly attached to microtubules.

5.4. Folate Receptors of Vegetative Cells

To explore the generality of the cytoskeletal association of chemotaxis receptors, we have also begun to examine the folate receptor of vegetative D. discoideum cells using [^{3}H]-methotrexate (MTX) as the ligand of choice (Nandini–Kishore and Frazier, 1981). The folate receptor also remains associated with the detergent-insoluble residue upon cell lysis under conditions found to be optimal for retention of the cAMP receptor on the cytoskeleton. We have not yet characterized these vegetative cytoskeletons as thoroughly as the 6-hr residues described previously. It seems likely, however, based on an identical distribution of marker enzymes, that these vegetative cytoskeletons represent a very similar complement of cellular components. They also appear similar in electron micrographs. Preliminary evidence suggests that as the cells differentiate, the folate receptor becomes dissociated from the cytoskeleton and the cAMP receptor becomes attached as it is made. This could explain the shift in chemoresponsiveness from folate to cAMP that occurs during early development (Nandini–Kishore and Frazier, 1981; Varnum and Soll, 1981). The folate receptors do not disappear from the membrane but instead remain in a state which can no longer stimulate directional motility but which can stimulate rate of random motion. It is an intriguing possibility that when coupled to the cytoskeleton, a

←

FIGURE 1. Critical-point-dried cytoskeleton prepared in 1% TX-100 in 17 mM Pi, pH 6.1 with 1 mM EGTA and 0.2 mM magnesium chloride. The bar is 1 μm. (A) The whole mount cytoskeleton; (B) higher magnification including the nuclear remnant.

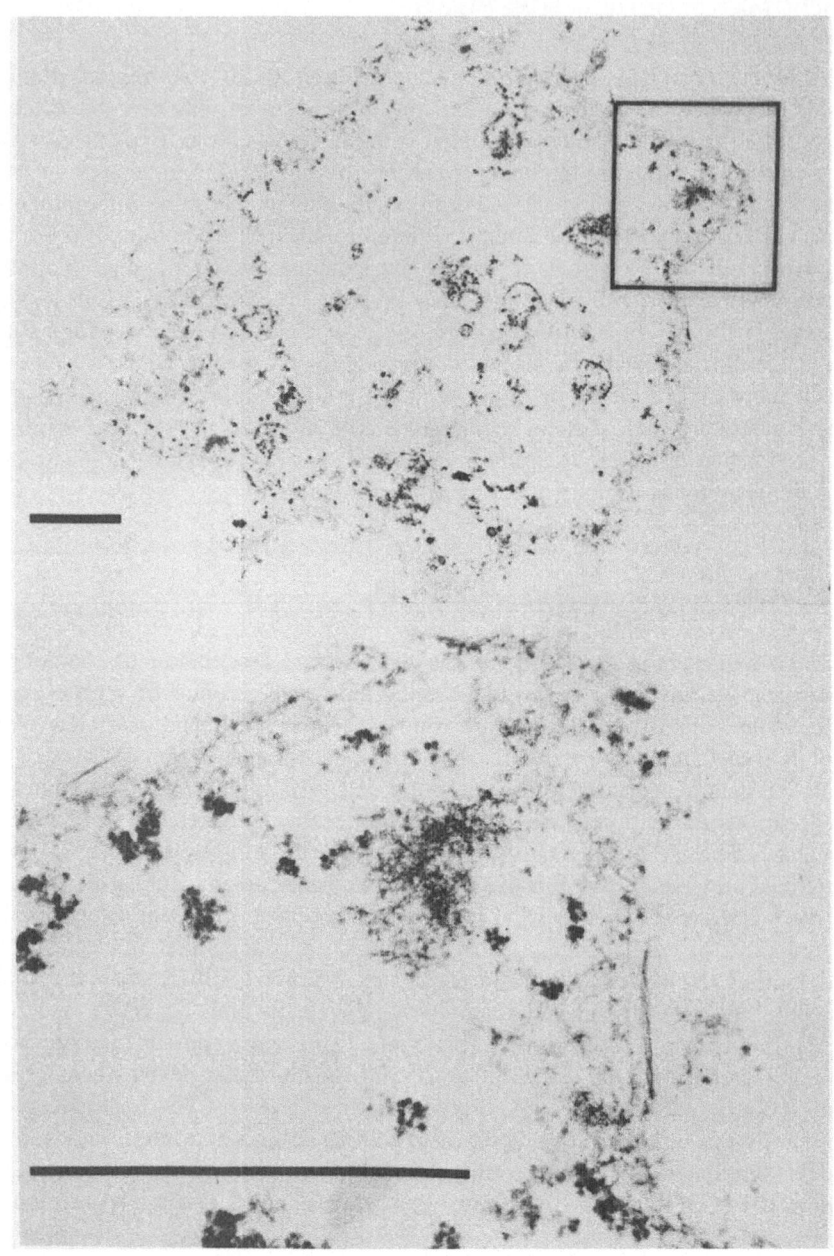

FIGURE 2. Thin section of a cytoskeleton prepared as in Fig. 1 at two magnifications. The bar is 1 μm.

receptor is capable of causing a chemotactic response, but when uncoupled (as the folate receptor after cAMP chemotaxis or the cAMP receptor after refeeding aggregation competent cells), that same receptor can stimulate rate of motion but can impart no directional information. Since spatial polarity at some level of the detection/response system seems imperative for eucaryotic chemotaxis, it is tempting to suggest that the nonrandom association of chemotaxis receptors with the cytoskeleton may provide the basis for this asymmetry.

5.5. Effector Association with the Cytoskeleton

As noted previously, it is only for the transient accumulation of cGMP that a clear correlation exists between stimulation of all slime mold chemotaxis receptors and a postreceptor (effector) response (Mato and Konijn, 1977). All other cellular responses to chemoattractants have not been tested for their universality. Unfortunately, the *D. discoideum* guanylate cyclase is a difficult enzyme to study, since it is very labile and inhibitors of the enzyme apparently are present in cell lysates. Thus the activity cannot be assayed until lysates are diluted or partly purified. We tried to determine whether the guanylate cyclase activity might also be associated, at least in part, with the TX-100-insoluble cytoskeleton which contains the folate and cAMP receptors of vegetative and aggregating cells, respectively (see above).

In an attempt to circumvent some of the problems inherent in the direct assay of guanylate cyclase and at the same time identify the cyclase protein for antibody selection, we have affinity labeled cellular GTP-binding proteins with the photolabile analog 8-azido-GTP containing a [^{32}-P] label in the alpha position to prevent its transfer in phosphorylation reactions. When cell lysates are labeled with 1-μM azido-GTP using short preincubation (15 sec) and photolysis (30 sec) times, no labeled proteins are detected upon autoradiography of SDS gels. If the lysates are diluted five- or tenfold before labeling, a weakly labeled band is seen at an M_r of about 75,000. When the cytoskeletons are prepared by our methods (described previously), collected by centrifugation through the oil layer, and then labeled with azido-GTP, a heavily labeled band is seen at M_r 75,000, and under some conditions, a weakly labeled band is seen at M_r near 45,000. Both bands are specifically labeled since their labeling is competed totally by unlabeled GTP. We have characterized some properties of the labeling of the M_r-75,000 protein for comparison with the known properties of the guanylate cyclase (Ward and Brenner, 1977; Mato and Malchow, 1978; Mato, 1979). The pH optimum for labeling and its buffer ion dependence are the same, that is, pH 8 with the "Good" buffers being superior. The labeling reaction is inhibited by Ca^{2+} and requires Mg^{2+} or Mn^{2+}, with Mn^{2+} being the better divalent ion. The labeling is inhibited by GTP and to a slightly lesser degree by GppNp, the nonhydrolyzable GTP analog. Labeling is not inhibited by cGMP or cAMP and is stimulated by ATP at 0.1 mM and inhibited at 1 mM ATP. All these properties are consistent with the known properties of the guanylate cyclase. Furthermore, the partial association of the enzyme activity with the particulate fraction reported by both groups is readily explained if the enzyme were to have a cytoskeletal attachment. In fact, the distribution between the particulate and soluble fractions correlates well with the known effects of buffers and pH on the stability of the cytoskeleton. The pH dependence of the recovery of the M_r-75,000 band in the cytoskeletal pellet is identical to that found for the recovery of the cAMP receptor in this fraction. Thus it seems likely that the M_r-75,000 species which labels with azido-GTP represents a polypeptide chain of the guanylate cyclase which is attached to the detergent insoluble cytoskeleton. Thus it is probably significant that the molecular weight of this labeled protein is the same as that reported for the purified guanylate cyclase from rat lung (Garbers, 1979). A further comparison of the properties of the *Dictyostelium* guanylate cyclase activity with the photoaffinity-labeled protein and with the enzyme from higher organisms is in progress.

The presence of both the chemotactic receptor and the effector candidate, the guanylate cyclase, on the cytoskeleton would open the way for more direct experiments regarding their coupling than was heretofore possible. For example, affinity-labeled cyclase could be chemically cross-linked to other cytoskeletal proteins in an attempt to identify its molecular associations. It will be interesting to determine whether the functional coupling between receptor and cyclase in the intact cell is preserved in the cytoskeletal preparation. If so, the cytoskeleton would offer a readily manipulated system for the study of receptor–effector coupling. The possibility remains that other potential effectors of the chemotactic response are also associated in some way with the cytoskeleton or work by influencing the association of already identified components with this structure. It has already been noted that the intracellular free calcium level and pH both represent mechanisms of modulating the integrity of the cytoskeleton and perhaps the association of receptors and cyclase with it. Methylation and phosphorylation reactions have been proposed as potential effector mechanisms in chemotaxis. We are currently determining the phosphorylated and methylated components of the cytoskeleton before and after chemotactic stimulation and whether the enzymes that carry out these reactions might themselves be associated with the TX-100-insoluble material. It is our hope that the cytoskeletal system will allow the investigation of transduction reactions under more readily controlled conditions than the intact cell.

6. CONCLUSIONS

Given the rapid progress that has recently characterized the investigation of chemotaxis in the cellular slime molds and in leukocytes, it seems likely that the elucidation of chemotactic transduction will soon be forthcoming. However, it should be realized that very extensively studied receptors such as the insulin receptor are still not completely understood in terms of how their message is processed or transduced by the cell. For chemotaxis receptors, the situation is conceptually even more complex since the process regulated, cellular motility, is incompletely understood. It is hoped that the cytoskeleton will provide both a conceptual and a literal framework on which to build a useful cell-free system for the study of the molecular interactions and events that constitute the reactions responsible for the direction or biasing of cellular motility. Our working hypothesis is that chemotaxis receptors are associated with the cytoskeleton because they have functional interactions with other cytoskeletal components. These may occur directly with transduction proteins such as the guanylate cyclase or methyl-accepting proteins, or indirectly through the mutual association of receptors and effectors with contractile proteins or with large "bridging" proteins such as fodrin or spectrin. A potential purpose for this association may arise from the necessity for rapid communication of receptors or transducers with their counterparts in other regions of the cell surface. For example, if cells read the direction of signal reception *in vivo* by knowing which side of the cell a cAMP wave first encounters, the cell may respond only to the first few percent of activated receptors by shutting off the receptors or transducers in other parts of the cell. This rapid global desensitization could perhaps be mediated by communication through the cytoskeletal network. Such a local perturbation which results in a global effect mediated by the cytoskeleton is the phenomenon of "anchorage modulation." For example, if Con A platelets are applied to lymphocytes in small numbers, such that less than 10% of the cell surface is covered, an immobilization of the cell surface immunoglobulin receptors occurs which is relieved by colchicine and cytochalasin B (Henis and Elson, 1981a,b). Painter and co-workers (Chapter 2, this volume) have found that the FMLP receptor becomes associated with the cytoskeleton after the ligand binds. Further work will attempt to focus on the distribution and localization of the components of the chemotactic response in *Dictyostelium* cells and in their cytoskeletons, and on the relationship of these components to one another and to cytoskeletal filaments.

ACKNOWLEDGMENTS.   We thank Dianne Stockhausen, Mitchell Vance, and Sally Hennessy for excellent technical assistance.

REFERENCES

Alemany, S., Sill, M. G., and Mato, J. M., 1980, Regulation by cyclic GMP of phospholipid methylation during chemotaxis in *Dictyostelium discoideum*, *Proc. Natl. Acad. Sci. USA* **77**:6996–6999.

Anderson, D. C., Wible, L. J., Hughes, B. J., Smith, C. W., and B. R. Brinkley, 1982, Cytoplasmic microtubules in polymorphonuclear leukocytes: Effects of chemotactic stimulation and colchicine, *Cell* **31**:719–729.

Barclay, S. L. and Meller, E., 1983, Efficient transformation of *Dictyostelium discoideum* amoebae, *Mol. Cell. Biol.* **3**:2117–2130.

Bazari, W. L., and Clarke, M., 1981, Characterization of a novel calmodulin from *Dictyostelium discoideum*, *J. Biol. Chem.* **256**:3598–3603.

Ben-Ze'ev, A., Duerr, A. Solomon, S. and Penman, S. 1979, The outer boundary of the cytoskeleton: A lamina derived from plasma membrane proteins, *Cell* **17**:859–865.

Bonner, J. T., Barkley, D. S., Hall, E. M., Konijn, T. M., Mason, J. W., O'Keefe, G., III, and Wolfe, P. B., 1969, Acrasin, acrasinase and sensitivity to acrasin in *Dictyostelium discoideum*, *Devel. Biol.* **20**:72–87.

Bonner, J. T., Hall, E. M., Noller, S., Oleson, F. B. and Roberts, A. B., 1972, Synthesis of cAMP and phosphodiesterase in various species of the cellular slime molds and its bearing on chemotaxis and differentiation, *Dev. Biol.* **29**:402–409.

Bourguignon, L. Y. W., and Singer, S. J., 1977, Transmembrane interactions and mechanism of capping of surface receptors by their specific ligands, *Proc. Natl. Acad. Sci. USA* **74**:5031–5035.

Brenner, M., and Thoms, S. D., 1984, Caffeine blocks activation of cyclic AMP synthesis in *Dictyostelium discoideum*, *Devel. Biol.*, **101**:136–146.

Brier, J., Fechheimer, M., Swanson, J., and Taylor, D. L., 1983, Abundance, relative gelation activity, and distribution of the 95,000 dalton actin-binding protein from *Dictyostelium discoideum*, *J. Cell Biol.* **97**:178–185.

Brown, S., Levinson, W., and Spudich, J. A., 1976, Cytoskeletal elements of duck embryo fibroblasts revealed by detergent extraction, *J. Supramol. Struct.* **5**:119–130.

Bulgakov, R. and van Haastert, J. M., 1983, Isolation and partial characterization of a cyclic GMP-dependent cyclic GMP-specific phosphodiesterase from *Dictyostelium discoideum*, *Biochim. Biophys. Acta* **756**:56–66.

Bumann, J., Wurster, B., and Malchow, D., 1984, Attractant-induced changes and oscillations of the extracellular Ca + + concentration in suspensions of differentiating *Dictyostelium* cells, *J. Cell. Biol.*, **98**:173–178.

Carlin, R. K., Bartlett, D. C., and Siekevitz, P., 1983, Identification of fodrin as a major calmodulin binding protein in post synaptic density preparations, *J. Cell Biol.* **96**:443–448.

Chung, S., Landfear, S. M., Blumberg, D. D., Cohen, N. S., and Lodish, H. F., 1981, Synthesis and stability of developmentally regulated *Dictyostelium* mRNAs are affected by cell-cell contact and cAMP, *Cell* **24**:785–797.

Coffman, D. S., Leichtling, B. H., and Rickenberg, H. V., 1982, The phosphorylation of membranal proteins in *Dictyostelium discoideum* during development, *Devel. Biol.* **93**:422–429.

Condeelis, J. S., 1979, Isolation of Con A caps during various stages of formation and their association with actin and myosin, *J. Cell Biol.* **80**:751–758.

Condeelis, J. S., and Taylor, D. L., 1977, The contractile basis of ameboid movement V. The control of gelatin, solation and contraction in extracts from *Dictyostelium discoideum*, *J. Cell Biol.* **74**:901–927.

Condeelis, J., and Vahey, M., 1982, A calcium and pH-regulated protein from *Dictyostelium discoideum* that cross-links actin filaments, *J. Cell Biol.* **94:**466–471.

Condeelis, J., Geosits, S., and Vahey, M., 1982, Isolation of a new actin-binding protein from *Dictyostelium discoideum*, *Cell Motility* **2:**273–285.

Coukell, M. B., 1981, Apparent positive cooperativity at a surface cAMP receptor in *Dictyostelium*, *Differentiation* **20:**29–35.

Dang, C. V., Yang, D. C. H., and Pollard, T. D., 1983, Association of methionyl tRNA synthetase with detergent-insoluble components of the rough endoplasmic reticulum, *J. Cell Biol.* **96:**1138–1147.

de Gunzburg, J., Hohman, R., Part, D., and Veron, M., 1983, Evidence that a cAMP binding protein from *Dictyostelium discoideum* carries S-adenosyl-L-homocysteine hydrolase activity, *Biochimie* **65:**33–41.

Devreotes, P. N., 1982, Chemotaxis, in: *The Development of Dictyostelium discoideum* (W. F. Loomis, ed.), Academic Press, New York, pp. 117–168.

Devreotes, P. N., and Steck, T. L., 1979, Cyclic AMP relay in *Dictyostelium discoideum*. II. Requirements for the initiation and termination of the response, *J. Cell. Biol.* **80:**300–309.

Devreotes, P. N., Derstine, P. L., and Steck, T. L., 1979, Cyclic AMP relay in *Dictyostelium discoideum*. I. A techique to monitor responses to controlled stimuli, *J. Cell Biol.* **80:**291–299.

Dicou, E., and Brachet, P., 1980, A separate phosphodiesterase for the hydrolysis of cyclic guanosine 3′,5′-monophosphate in growing *Dictyostelium discoideum* amoebae, *Eur. J. Biochem.* **109:**507–514.

Fechheimer, M., and Taylor, D. L., 1983, Isolation and characterization of a calcium-sensitive 30,000 dalton actin binding protein from *Dictyostelium discoideum*, *J. Cell Biol.* **97:**281a.

Frazier, W. A., Nandini–Kishore, S. G., and Meyers, B. L., 1982, Chemotactic receptors of *Dictyostelium discoideum*, *J. Cell. Biochem.* **18:**181–196.

Futrelle, R. P., Traut, J., and McKee, W. G., 1982, Cell behavior in *Dictyostelium discoideum*: Preaggregation response to localized cyclic AMP pulses, *J. Cell Biol.* **92:**807–821.

Galvin, N. J., Stockhausen, D., Meyers–Hutchins, B. L., and Frazier, W. A., 1984, Association of the cAMP chemotaxis receptor with the detergent-insoluble cytoskeleton of *Dictyostelium discoideum*, *J. Cell Biol.*, **98:**584–595.

Garbers, D. L., 1979, Purification of soluble guanylate cyclase from rat lung, *J. Biol. Chem.* **254:**240–243.

Gerisch, G., 1982, Chemotaxis in *Dictyostelium*, *Ann. Rev. Physiol.* **44:**535–552.

Gerisch, G., and Wick, U., 1975, Intracellular oscillations and release of cAMP from *Dictyostelium discoideum* cells, *Biochem. Biophys. Res. Commun.* **65:**364–370.

Giffard, R. G., Spudich, J. A., and Spudich, A., 1983, Ca++-sensitive isolation of a cortical actin matrix from *Dictyostelium* amoebae, *J. Muscle Res. Cell Motility* **4:**115–131.

Glenney, J. R., Jr., and Glenney, P., 1983, Fodrin is the general spectrin-like protein found in most cells whereas spectrin and the TW protein have a restricted distribution, *Cell* **34:**503–512.

Glenney, J. R., Jr., Glenney, P., and Weber, K., 1982a, Erythroid spectrin, brain fodrin and intestinal brush border proteins (TW260–240) are related molecules containing a common calmodulin-binding subunit bound to a variant cell type specific subunit, *Proc. Natl. Acad. Sci. USA* **79:**4002–4005.

Glenney, J. R., Jr., Glenney, P., and Weber, K., 1982b, F-actin binding and cross-linking properties of porcine brain fodrin, a spectrin-related molecule, *J. Biol. Chem.* **257:**9781–9787.

Goldberg, N. D., Ames, A., III, Gander, J. E., and Walseth, T. F., 1983, Magnitude of increase in retinal cGMP metabolic flux determined by O incorporation into nucleotide -phosphoryls corresponds with intensity of photic stimulation, *J. Biol. Chem.* **258:**9213–9219.

Gottlieb, A. I., McBurnie–May, L., Subrahmanyan, L., and Kalnins, V. I., 1981, Distribution of microtubule organizing centers in migrating sheets of endothelial cells, *J. Cell Biol.* **91**:589–594.

Green, A. A., and Newell, P. C., 1975, Evidence for the existence of two types of cAMP binding sites in aggregating cells of *Dictyostelium discoideum, Cell* **6**:129–136.

Gross, J. D., Bradbury, J., Kay, R. R., and Peacey, M. J., 1983, Intracellular pH and the control of cell differentiation in *Dictyostelium discoideum, Nature* **303**:244–245.

Hellewell, S. B., and Taylor, D. L., 1979, The contractile basis of ameboid movement VI. the solation-contraction coupling hypothesis, *J. Cell Biol.* **83**:633–648.

Henderson, E. J., 1975, The cyclic adenosine 3′:5′-monophosphate receptor of *Dictyostelium discoideum, J. Biol. Chem.* **250**:4730–4736.

Henis, Y. I., and Elson, E. L., 1981a, Differences in the response of several cell types of inhibition of surface receptor mobility by local concanavalin A binding, *Exptl. Cell Res.* **136**:189–201.

Henis, Y. I., and Elson, E. L.,, 1981b, Inhibition of the mobility of mouse lymphocyte surface immunoglobulins by locally bond concanavalin A, *Proc. Natl. Acad. Sci. USA* **78**:1072–1076.

Heuser, J. E., and Kirschner, M. W., 1980, Filament organization revealed in platinum replicas of freeze-dried cytoskeletons, *J. Cell Biol.* **86**:212–234.

Hirth, P.–K., Edwards, C. A., and Firtel, R. A., 1982, A DNA-mediated transformation system for *Dictyostelium discoideum, Proc. Natl. Acad. Sci. USA* **79**:7356–7360.

Jacobson, B. S., 1980, Actin binding to the cytoplasmic surface of the plasma membrane isolated from *Dictyostelium discoideum, Biochem. Biophys. Res. Comm.* **97**:1493–1498.

Jamieson, G. A., Jr., and Frazier, W. A., 1982, Calmodulin-binding proteins during *Dictyostelim* differentiation, *J. Cell Biol.* **95**:35a.

Jamieson, G. A., Jr., and Frazier, W. A., 1983, *Dictyostelium* calmodulin: Affinity isolation, and characterization, *Arch. Biochem. Biophys.,* **227**:609–617.

Jamieson, G. A., Jr., and Vanaman, T. C., 1979, Affinity chromatography of calmodulin on an immobilized phenothiazine conjugate, *Biochem. Biophys. Res. Commun.* **90**:1048–1056.

Jamieson, G. A., Jr., Frazier, W. A., and Schlesinger, P. H., 1983, Transient increase in intracellular pH during *Dictyostelium* differentiation, *J. Cell Biol.* **97**:77a.

Juliani, M., Brusca, J., and Klein, C., 1981, cAMP regulation of cell differentiation in *Dictyostelium discoideum* and the role of the cAMP receptor, *Devel. Biol.* **83**:114–121.

Juliani, M. H., and Klein, C., 1981, Photoaffinity labeling of the cell surface cyclic AMP receptor of *Dictyostelium* and its modification in down regulated cells *J. Biol. Chem.* **256**:613–619.

Kakebeeke, P. I., DeWit, R. J., and Konijn, T. M., 1980, Folic acid deaminase activity during development in *Dictyostelium discoideum, J. Bacteriol.* **143**:307–312.

Kay, R. R., and Jermyn, K. A., 1983, A possible morphogen controlling differentiation in *Dictyostelium, Nature* **303**:242–244.

Kihara, M., and MacNab, R. M., 1981, Cytoplasmic pH mediates pH taxis and weak acid repellent taxis of bacteria, *J. Bacteriol.* **145**:1209–1221.

King, A. C., 1979, Regulation of the cyclic adenosine monophosphate receptor of differentiating *Dictyostelium discoideum,* Ph.D. thesis, Washington University, St. Louis, Missouri.

King, A. C., and Frazier, W. A., 1977, Reciprocal periodicity in cAMP binding and phosphorylation of differentiating *Dictyostelium discoideum* cells, *Biochem. Biophys. Res. Commun.* **78**:1093–1099.

King, A. C., and Frazier, W. A., 1979, Properties of the oscillatory cAMP binding component of *Dictyostelium discoideum* cells and isolated plasma membranes, *J. Biol. Chem.* **254**:7168–7176.

Klee, C. B., Crouch, T. H., and Richman, P. G., 1980, Calmodulin, *Ann. Rev. Biochem.* **49**:489–515.

Klein, C., 1979. A slowly dissociating form of the cell surface cyclic adenosine 3′:5′-monophosphate receptor of *Dictyostelium discoideum, J. Biol. Chem.* **254**:12573–12578.

Kuczmarski, E. R., and Spudich, J. A., 1980, Regulation of myosin self-assembly: Phosphorylation of *Dictyostelium* heavy chain inhibits thick filament formation, *Proc. Natl. Acad. Sci. USA* 77:7292–7296.

Kupfer, A., Louvard, D., and Singer, S. J., 1982, Polarization of the golgi apparatus and the microtubule-organizing center in cultured fibroblasts at the edge of an experimental wound, *Proc. Natl. Acad. Sci. USA* 79:2603–2607.

Lazarides, E., 1982, Intermediate filaments: A chemically heterogeneous, developmentally regulated class of proteins, *Ann. Rev. Biochem.* 51:219–250.

Leichtling, B. H., Coffman, D. S., Yaeger, E. S., and Rickenberg, H. V., 1981a, Occurrence of the adenylate cyclase "G protein" in membranes of *Dictyostelium discoideum, Biochem. Biophys. Res. Commun.* 102:1187–1195.

Leichtling, B. H., Spitz, E., and Rickenberg, H. V., 1981b, A cAMP-binding protein from *Dictyostelium discoideum* regulates mammalian protein kinase, *Biochem. Biophys. Res. Commun.* 100:515–522.

Leichtling, B. H., Tihon, C., Spitz, E., and Rickenberg, H. V., 1981c, A cytoplasmic cAMP-binding protein in *Dictyostelim discoideum, Devel. Biol.* 82:150–157.

Leichtling, B. H., Majerfeld, I. H., Coffman, D. S., and Rickenberg, H. V., 1982, Identification of the regulatory subunit of a cAMP dependent protein kinase in *Dictyostelium discoideum, Biochem. Biophys. Res. Commun.* 105:949–955.

Leichtling, B. H., Majerfeld, I. H., Spitz, E, Schaller, K. L., Woffendin, C. Kakinuma, S., and Rickenberg, H. V., 1984, A cytosolic cyclic AMP-dependent protein kinase in *Dictyostelium discoideum*. II. Developmental regulation, *J. Biol. Chem.*, 259:662–668.

Lenk, R., Ransom, L., Kaufmann, Y., and Penman, S., 1977, A cytoskeletal structure with associated polyribosomes obtained from HeLa cells, *Cell* 10:67–78.

Levine, J., and Willard, M., 1981, Fodrin: axonally transported polypeptides associated with the interanal periphery of many cells, *J. Cell Biol.* 90:631–643.

Lubs–Haukeness, J., and Klein, C., 1982, Cyclic nucleotide-dependent phosphorylation in *Dictyostelium discoideum* amoebae, *J. Biol. Chem.* 257:12204–12208.

Luna, E. J., Fowler, V. M., Swanson, J., Branton, D., and Taylor, D. L., 1981, A membrane cytoskeleton from *Dictyostelium discoideum*, I. Identification and partial characterization of an actin-binding activity, *J. Cell Biol.* 88:396–409.

Majerfeld, I. H., Leichtling, B. H., Meligeni, J. A., Spitz, E., and Rickenberg, H. V., 1984, A cytosolic cyclic AMP-dependent protein kinase in *Dictyostelium discoideum:* I. Properties, *J. Biol. Chem.* 259:654–661.

Malchow, D., Fuchila, J., and Jastorff, B., 1973, Correlation of substrate specificity of cAMP-phosphodiesterase in *Dictyostelium discoideum* with chemotactic activity of cAMP analogues, *FEBS Lett.* 34:5–9.

Malchow, D., Nanjundiah, V., and Gerish, G., 1978a, pH oscillations in cell suspensions of *Dictyostelium discoideum*, their relation to cyclic AMP signals, *J. Cell Sci.* 30:319–330.

Malchow, D., Nanjundiah, V., Wurster, B., Eckstein, F., and Gerish, G., 1978b, Cyclic AMP induced pH changes in *Dictyostelium discoideum* and their control by calcium, *Biochim. Biophys. Acta* 538:473–480.

Malech, H. L., Rott, R. K., and Gallin, J. I., 1977, Structural analysis of human neutrophil migration. Centriole, microtubule and microfilament orientation and function during chemotaxis, *J. Cell. Biol.* 75:666–693.

Mannherz, H. G., Barrington–Leigh, J., Leberman, R., and Pfrang, H., 1975, A specific 1:1 G-actin:DNase I complex formed by the action of DNase I on F-actin, *FEBS Lett.* 60:34–38.

Maruta, H., Baites, W., Dieter, P., Marme, D., and Gerisch, G., 1983, Myosin heavy chain kinase inactivated by Ca2+/calmodulin from aggregating cells of *Dictyostelium discoideum, EMBO Journal* **2**:535–542.

Mato, J., 1979. Activation of *Dictyostelium discoideum* guanylate cyclase by ATP, *Biochem. Biophys. Res. Commun.* **88**:569–574.

Mato, J. M., and Konijn, T. M., 1977, Chemotactic signaling and cGMP accumulation in *Dictyostelium*, in: *Development and Differentiation in the Cellular Slime Molds* (P. Cappuccinelli and J. M. Ashworth, eds.), Elsevier/North Holland, Amsterdam, pp. 93–103.

Mato, J. M., and Malchow, D., 1978, Guanylate cyclase activation in response to chemotactic stimulation in *Dictyostelium discoideum, FEBS Lett.* **90**:119–122.

Mato, J. M., and Marin–Cao, D., 1979, Protein and phospholipid methylation during chemotaxis in *Dictyostelium discoideum* and its relationship to calcium movements. *Proc. Natl. Acad. Sci. USA* **76**:6106–6109.

Mato, J., Krens, F., van Haastert, P.J.M., and Konijn, T. M., 1977, cAMP dependent cGMP accumulation in *Dictyostelium discoideum, Proc. Natl. Acad. Sci. USA* **74**:2348–2351.

Mato, J. M., van Haastert, P. J. M., Krens, F. A., and Konijn, T. M., 1978a, Chemotaxis in *Dictyostelium discoideum:* Effect of concanavalin A on chemoattractant mediated cyclic GMP accumulation and light scattering decrease, *Cell Biol. Int. Rep.* **2**:163–170.

Mato, J. M., Woelders, H., van Haastert, P. J. M., and Konijn, T. M., 1978b, Cyclic GMP binding activity in *Dictyostelium discoideum, FEBS Lett.* **90**:261–264.

McRobbie, S. J., and Newell, P. C., 1983, Changes in actin associated with the cytoskeleton following chemotactic stimulation of *Dictyostelium discoideum, Biochem. Biophys. Res. Commun.* **115**:351–359.

Mehdy, M. C., Ratner, D., and Firtel, R. A., 1983, Induction of cell-type-specific gene expression in *Dictyostelium, Cell* **32**:763–771.

Metz, B. A., Ward, T. E., Welker, D. L., and Williams, K. L., 1983, Identification of an endogenous plasmid in *Dictyostelium discoideum. EMBO Journal* **2**:515–519.

Meyers, B. L., and Frazier, W. A., 1981, Solubilization and hydrophobic immobilization assay of a cAMP binding protein from *Dictyostelium discoideum* plasma membranes, *Biochem. Biophys. Res. Commun.* **101**:1011–1017.

Meyers–Hutchins, B. L., and Frazier, W. A., 1984, Purification and characterization of a membrane-associated cAMP-binding protein from developing *Dictyostelium discoideum, J. Biol. Chem.* **259**: 4379–4388.

Mullens, I. A., and Newel, P. C., 1978, cAMP binding to cell surface receptors of *Dictyostelium, Differentiation* **10**:171–176.

Mutzel, R., Wurster, B., and Malchow, D., 1983, Alkali-stable methylation of a 17,000 dalton protein increases during cell differentiation in *Dictyostelium discoideum, FEMS Microbiol. Lett.* **19**:71–75.

Nandini–Kishore, S. G., and Frazier, W. A., 1981, [3H]-Methotrexate as a ligand for the folate receptor of *Dictyostelium discoideum, Proc. Nat. Acad. Sci USA* **78**:7299–7303.

Neidel, J. E., and Cuatrecasas, P., 1980, Formyl peptide chemotactic receptors of leukocytes and macrophages, *Current Topics Cell. Reg.* **17**:137–170.

Neufeld, G., Steiner, S., Korner, M., and Schramm, M., 1983, Trapping of the beta-adrenergic receptor in the hormone-induced state. *Proc. Natl. Acad. Sci. USA* **80**:6441–6445.

Newell, P. C., 1982, Genetics, in: *The Development of* Dictyostelium discoideum, (W. F. Loomis, ed.) Academic Press, New York, pp. 35–70.

Osborn, M., and Weber, K., 1977, The detergent resistant cytoskeleton of tissue culture cells includes the nucleus and the microfilament bundles, *Exp. Cell Res.* **106**:339–349.

Painter, R. G., and Ginsberg, M., with Jaques, B., 1982, Concanavalin A induces interactions between surface glycoproteins and the platelet cytoskeleton, *J. Cell Biol.* **92**:565–573.

Pan, P., and Wurster, B., 1978, Inactivation of the chemoattractant folic acid by cellular slime mold and identification of the reaction product, *J. Bacteriol.* **136:**955–959.

Pan, P., Hall, E. M., and Bonner, J. T., 1972, Folic acid as the second chemotactic substance in the cellular slime moulds, *Nature New Biol.* **237:**181–182.

Pan, P., Hall, E. M., and Bonner, J. T., 1975, Determination of the active portion of the folic acid molecule in cellular slime mold chemotaxis, *J. Bacteriol.* **122:**185–191.

Pike, M. C., and Snyderman, R. 1982, Transmethylation reactions regulate affinity and functional activity of chemotactic factor receptors on macrophages, *Cell* **28:**107–114.

Rahmsdorf, H. J., and Gerisch, G., 1978, Specific binding proteins for cyclic AMP and cyclic GMP in *Dictyostelium discoideum*, *Cell Differentiation* **7:**249–257.

Rahmsdorf, H. J., Malchow, D., and Gerisch, G., 1978, Cyclic AMP induced phosphorylation in *Dictyostelium discoideum* of a polypeptide comigrating with myosin heavy chains, *FEBS Lett.* **88:**322–326.

Repaske, D. R. and Adler, J., 1981, Change in intracellular pH of *Esherichia coli* mediates the chemotactic response to certain attractants and repellents, *J. Bacteriol.* **145:**1196–1208.

Roos, W., Nanjundiah, V., Malchow, D., and Gerisch, G., 1975, Amplification of cAMP signals in aggregating cells of *Dictyostelium discoideum*, *FEBS Lett.* **53:**139–142.

Rosen, E., Sivertsen, A., and Firtel, R. A., 1983, An unusual transposon encoding heat shock inducible and developmentally regulated transcripts in *Dictyostelium*, *Cell* **35:**243–251.

Ross, F. M., and Newell, P. C., 1981, Streamers: Chemotactic mutants of *Dictyostelium discoideum* with altered cyclic GMP metabolism, *J. Gen. Microbiol.* **127:**339–350.

Rothenberg, P., Cassel, D., Reuss, L., and Glaser, L., 1983, Initial events in the interaction of epidermal growth factor with cells. 13th International Cancer Congress, Part C, *Biology of Cancer* (2), Liss, New York, pp. 109–121.

Rowe, P. M., Murtaugh, T. J., Bazari, W. L., Clark, M., and Siegel, F. L., 1983, Radiometric assay of S-adenosylmethionine:calmodulin(lysine)*N*-methyltransferase by calcium-dependent hydrophobic interaction chromatography, *Anal. Biochem.* **133:**394–400.

Saga, Y., and Yanagisawa, K., 1982, Macrocyst development in *Dictyostelium discoideum*. I. Induction of synchronous development by giant cells and biochemical analysis, *J. Cell Sci.* **55:**341–352.

Schaffer, B. M., 1962, The acrasins, *Adv. Morphog.* **2:**109–182.

Schaffer, B. M., 1975, Secretion of cAMP induced by cAMP in the cellular slime mold *Dictyostelium discoideum*, *Nature (London)* **255:**549–552.

Schiffmann, E., 1982. Leukocyte chemotaxis, *Ann. Rev. Physiol.* **44:**553–568.

Schliwa, M., and van Blerkom, J., 1981, Structural interactions of cytoskeletal components, *J. Cell Biol.* **90:**222–235.

Schliwa, M., Pryzwansky, K. B., and van Blerkom, J., 1982, Implications of cytoskeletal interactions for cellular architecture and behaviour, *Phil. Trans. R. Soc. Lond. B* **299:**199–205.

Schreiner, G. F., Fukiwara, K., Pollard, T. D., and Unanue, E. R., 1977, Redistribution of myosin accompanying capping of surface Ig, *J. Exp. Med.* **145:**1393–1398.

Sha'afi, R. I., Nacache, P. H., Molski, T. F. P., and Volpi, M., 1982, Chemotactic stimuli induced changes in the pHi of rabbit neutrophils, in: *Intracellular pH: Its Measurement, Regulation and Utilization in Cellular Functions* (R. Nuccitelli and D. W. Deamer, eds.), Liss, New York, pp. 513–526.

Shimomura, O., Suthers, H. L. B., and Bonner, J. T., 1982, Chemical identity of the acrasin of the cellular slime mold *Polysphondylium violaceum*, *Proc. Nat. Acad. Sci. USA* **79:**7376–7379.

Siu, C. H., Lerner, R. A., and Loomis, W. F., Jr. 1977, Rapid accumulation and disappearance of plasma membrane proteins during development of wild type and mutant strains of *Dictyostelium discoideum*, *J. Mol. Biol.* **116:**469–488.

Soll, D. R., and Waddell, D. R., 1975, Morphogenesis in the slime mold *Dictyostelium discoideum*
    I. The accumulation and erasure of "morphogenic information," *Devel. Biol.* **47**:292–302.
Springer, M. S., Goy, M. F., and Adler, J., 1979, Protein methylation in behavioral control
    mechanisms and signal transduction, *Nature (London)* **280**:279–284.
Spudich, J. A., and Spudich, A. 1982, Cell motility, in: *Development of* Dictyostelium discoideum
    (W. F. Loomis, ed.), Academic Press, New York, pp. 169–194.
Steinhardt, R. A., and Morisawa, M., 1982, Changes in intracellular pH of *Physarum* plasmodium
    during the cell cycle and in response to starvation, in: *Intracellular pH: Its Measurement,
    Regulation and Utilization in Cellular Functions* (R. Nuccitelli and D. W. Deamer, eds.), Liss
    New York, pp. 341–385.
Swanson, J. A., and Taylor, D. L., 1982, Local and spatially coordinated movements in *Dictyos-
    telium discoideum* amoebae during chemotaxis, *Cell* **28**:225–232.
Taylor, D. L., and Fechheimer, M., 1981, Calcium regulation in amoeboid movement, in: *Cal-
    modulin and Intracellular Calcium Receptors* (S. Kakiuchi, H. Hidaka, and A. R. Means,
    eds.), Plenum Press, New York, pp. 349–373.
Taylor, D. L., and Fechheimer, M., 1982, Cytoplasmic structure and contractility: The sola-
    tion–contraction coupling hypothesis, *Phil. Trans. R. Soc. Lond. B* **299**:185–197.
Theibert, A., and Devreotes, P. N., 1983, Cyclic 3',5'-AMP relay in *Dictyostelium discoideum:*
    Adaptation is independent of activation of adenylate cyclase, *J. Cell Biol.* **97**:173–177.
Thomas, J. A., Buchsbaum, R. N., Zimniak, A., and Racker, E., 1979, Intracellular pH measure-
    ments in Ehrlich ascites tumor cells utilizing spectroscopic probes generated *in situ, Biochemistry*
    18:2210–2218.
Tomchik, K. J., and Devreotes, P. N., 1981, Adenosine 3',5'-monophosphate waves in *Dictyostelium
    discoideum:* A demonstration by isotope dilution fluorography, *Science* **212**:443–446.
Ushiro, H., and Cohen, S., 1980, Identification of phosphotyrosine as a product of the EGF-activated
    protein kinase, *J. Biol. Chem.* **255**:8363–8365.
Van Driel, R., 1981, Binding of the chemoattractant folic acid by *Dictyostelium discoideum, Cells
    Eur. J. Biochem.* **115**:391–395.
van Haastert, P. J. M., 1983a, Relationship between adaptation of the folic acid and the cAMP
    mediated cGMP response in *Dictyostelium, Biochem. Biophys. Res. Commun.* **115**:130–136.
van Haastert, P. J. M., 1983b, Binding of cAMP and adenosine derivatives to *Dictyostelium dis-
    coideum* cells: Relationships of binding, chemotactic and antagonistic activities, *J. Biol. Chem.*
    258:9643–9648.
van Haastert, P. J. M., 1983c, Sensory adaptation of *Dictyostelium discoideum* cells to chemotactic
    signals, *J. Cell Biol.* **96**:1559–1565.
van Haastert, P. J. M., and Kien, E., 1983, Binding of cAMP derivatives to *Dictyostelium discoideum*
    cells: Activation mechanism of the cell surface cAMP receptor, *J. Biol. Chem.* **258**:9636–9642.
van Haastert, P. J. M., and van der Heijden, P. R., 1983, Excitation, adaptation and deadaptation
    of the cAMP-mediated cGMP response in *Dictyostelium discoideum, J. Cell Biol.* **96**:347–353.
van Haastert, P. J. M., de Witt, R. J. W., Grijpma, Y., and Konijn, T. M., 1982a, Identification
    of a pterin as the acrasin of the cellular slime mold *Dictyostelium lacteum, Proc. Nat. Acad.
    Sci. USA* **79**:6270–6274.
van Haastert, P. J. M., van Lookeren Campagne, M. M., and Ross, F. M., 1982b, Altered cGMP-
    phosphodiesterase activity in chemotactic mutants of *Dictyostelium discoideum, FEBS Lett.*
    147:149–152.
van Haastert, P. J. M., van Walsum, H., van der Meer, R. C., Bulgakov, R., and Konijn, T. M.,
    1982c, Specificity of the cyclic GMP-binding activity and of a cyclic GMP-dependent cyclic
    GMP phosphodiesterase in *Dictyostelium discoideum, Molec. Cellular Endocrinol.* **25**:171–182.

van Waarde, A., 1982, Rapid, transient methylation of four proteins in aggregative amoebae of *Dictyostelium discoideum* as a response to stimulation with cAMP, *FEBS Lett.* **149**:266–270.

van Waarde, A., 1983, Cyclic AMP, folic acid and pterin-mediated protein carboxymethylation in cellular slime molds, *FEBS Lett.* **161**:45–50.

Varnum, B., and Soll, D. R., 1981, Chemoresponsiveness to cAMP and folic acid during growth, development and dedifferentiation in *Dictyostelium discoideum, Differentiation* **18**:151–160.

Wallace, L. J., and Frazier, W. A., 1979, Photoaffinity labeling of cAMP and AMP-binding proteins of differentiating *Dictyostelium discoideum* cells, *Proc. Natl. Acad. Sci. USA* **76**:4250–4254.

Walseth, T. F., Gander, J. E., Eide, S. J., Krick, T. P., and Goldberg, N. D., 1983, O Labeling of adenine nucleotide-phosphoryls in platelets, *J. Biol. Chem.* **258**:1544–1558.

Ward, A., and Brenner, M., 1977, Guanylate cyclase from *Dictyostelium discoideum, Life Sci.* **21**:997–1008.

Webb, D. J., and Nuccitelli, R., 1981, Direct measurement of intracellular pH changes in *Xenopus* eggs at fertilization and cleavage, *J. Cell Biol.* **91**:562–567.

White, E., Tolbert, E. M., and Katz, E. R., 1983, Identification of tubulin in *Dictyostelium discoideum:* characterization of some unique properties, *J. Cell Biol.* **97**:1011–1019.

Wick, U., Malchow, D., and Gerisch, G. 1978, cyclic-AMP stimulated calcium influx into aggregating cells of *Dictyostelium discoideum, Cell Biol. Int. Rep.* **2**:71–79.

Woodgett, J. R., Davison, M. T., and Cohen, P., 1983, The calmodulin-dependent glycogen synthase kinase from rabbit skeletal muscle: Purification, subunit structure and substrate specificity, *Eur. J. Biochem.* **136**:481–487.

Wurster, B., and Butz, U., 1980, Reversible binding of the chemoattractant folic acid to cells of *Dictyostelium discoideum, Eur. J. Biochem.* **109**:613–618.

Wurster, B., and Butz, U., 1983, A study on sensing and adaptation in *Dictyostelium discoideum:* Guanosine 3′,5′-phosphate accumulation and light scattering responses, *J. Cell Biol.* **96**:1566–1570.

Wurster, B., Pan, P., Tyan, G. G., and Bonner, J. T., 1976, Preliminary characterization of the acrasin of the cellular slime mold *Polysphondylium violaceum, Proc. Natl. Acad. Sci. USA* **73**:795–799.

Yamamoto, K., Pardee, J. D., Reidler, J., Stryer, L., and Spudich, J. A., 1982, Mechanism of interaction of *Dictyostelium* severin with actin filaments, *J. Cell Biol.* **95**:711–719.

Zigmond, S. H., 1977, Ability of polymorphonuclear leukocytes to orient in gradients of chemotactic factors, *J. Cell Biol.* **75**:606–616.

Zigmond, S. H., 1978, Chemotaxis by polymorphonuclear leukocytes, *J. Cell Biol.* **77**:269–287.

Zigmond, S. H., Levitsky, H. I., and Kreel, B. J., 1981, Cell polarity: An examination of its behavioral expression and its consequences for polymorphonuclear leukocyte chemotaxis, *J. Cell Biol.* **89**:585–592.

2

LEUKOCYTE CHEMOTAXIS

Mobilization of the Motile Apparatus by N-Formal Chemotactic Peptides

Richard G. Painter, Algirdas J. Jesaitis, and Larry A. Sklar

1. OVERVIEW

Chemotaxis is a process by which organisms or cells sense and move toward higher concentrations of a chemoattractant or away from noxious agents. Chemotaxis is best understood in simple organisms such as bacteria (see Koshland, 1981; Boyd and Simon, 1982, for reviews) and ameboid cells (see Gerish, 1982, and Chapter 1, by Frazier et al., this volume for a more thorough discussion of chemotaxis in *Dictyostelium discoideum* amebae). In these simple organisms, chemotactic behavior serves a sensory function which allows the cell to find nutrients and to avoid noxious or toxic agents in an ever-changing environment.

In more advanced creatures, the roles of chemotaxis are complex and more diverse. Since the cellular microenvironment in such organisms is more controlled than that of microorganisms, chemotaxis of cells is, as would be expected, not concerned with finding nutrients or avoiding noxious agents. In the developing embryo, for example, chemical gradients apparently exist which aid in the migration of certain cell types to tissue-specific sites and allow neuronal connections to develop. In most higher organisms, the release of agents and growth factors at tissue injury sites causes cells from adjacent unaffected areas

Richard G. Painter, Algirdas J. Jesaitis, and Larry A. Sklar ● Department of Immunology, Scripps Clinic and Research Foundation, La Jolla, California 92037.
Work supported by United States Public Health Service, National Institutes of Health grants GM-31439, AI 17354, HL 16411, and AI 19032. Richard G. Painter is a recipient of NIH RCDA-AM-00437. Larry A. Sklar is an Established Investigator of the American Heart Association. This is publication 3340IMM from the Research Institute of Scripps Clinic.

to migrate into the injury zone thereby accelerating the healing process. The chemotaxis of leukocytes such as neutrophils and monocytes from the bloodstream through the endothelium and the extracellular matrix to sites of infection or tissue inflammation plays a crucial function in host immunity. Because chemotactic behavior covers such a broad spectrum, we will limit this review specifically to leukocyte chemotaxis. In keeping with the theme of this volume, we will particularly emphasize those aspects of the problem which determine how specific receptor–ligand interactions couple (directly or by means of intermediary steps) with the cell motile apparatus or cytoskeleton to produce directed cell migration as we currently understand this process.

A diverse array of agents can induce leukocytes to exhibit directed migration (for a review see Wilkinson, 1974). Some of these act indirectly by causing the production and release of chemical mediators that possess intrinsic chemotactic activity, e.g., complement fragments (C5a) or leukotrienes. For this reason, we will limit our discussion to those agents which act directly upon the leukocyte membrane and which are mediated by interactions with leukocyte receptors. The most thoroughly studied chemoattractants are C5a, leukotrienes, and synthetic N-formylated peptides which are structurally similar to N-formylated peptides found in bacterial cultures (Schiffmann et al., 1975). We will largely focus our attention on this latter system.

2. N-FORMYL PEPTIDES AS A GENERAL MODEL FOR THE STUDY OF CHEMOTACTIC AND SENSORY PROCESS IN LEUKOCYTES

Schiffmann et al. (1975) first reported that filtrates of bacterial cultures were chemoattractants for mammalian leukocytes. Knowing that the active agent was dialysable and protease sensitive, these workers reasoned that mammalian cells might have evolved a means of detecting N-formylated peptides which are, of course, uniquely present in prokaryotic but not eukaryotic proteins.* Subsequently, a series of N-formylated peptides were synthesized and shown to have intrinsic chemoattractant activity. Of these the tripeptide, N-formyl met-leu-phe (FMLP), was the most potent, being biologically active in the nanomolar range.

---

* There is one recent report indicating the presence of N-formylated peptides in lysates of mammalian mitochondria (Carp, 1982). This probably reflects the fact that mitochondria are related evolutionarily to bacteria. In addition, these observations suggest that N-formyl peptides or proteins released at sites of tissue inflammation may further exacerbate the inflammatory response by attraction and activation of leukocytes at these sites.

Several unique features combine to make this system a valuable model system for the study of chemotactic behavior and sensory transduction as well as the general problem of transmembrane signaling mechanisms. These features include the following: (1) The availability of synthetic derivatives of the ligand with agonist- and antagonistlike properties. This allows for the preparation, with relative ease, of well-defined fluorescent or photoaffinity labels (Niedel *et al.*, 1979b, 1980; Niedel and Cuatrecasas, 1980; Sklar *et al.*, 1981a,b, Painter *et al.*, 1982; Schmitt *et al.*, 1983). (2) The high affinity and stereospecificity of the receptor–ligand interaction (subnanomolar in some cases). (3) The rigorous demonstration of a specific surface receptor molecule which appears to be a surface membrane glycoprotein of 50–60 Kdaltons (Niedel *et al.*, 1980; Painter *et al.*, 1982; Schmitt *et al.*, 1983). In addition, some progress toward extraction and isolation of the receptor has recently been reported (Goetzl *et al.*, 1981; Niedel, 1981). (4) Binding of ligand to receptor, producing a variety of biological responses which are rapid (seconds to minutes) and readily quantitated. In addition to shape changes and directed motility, these include the release of preformed proteases, the production of vasoconstrictive leukotrienes, and generation of highly reactive and toxic metabolites of oxygen such as O_2^-, H_2O_2, and $OH^{\cdot}$ among others. The biochemistry of these responses has been extensively reviewed elsewhere (Snyderman and Goetzl, 1981; Schiffmann, 1982; Ward, 1982). The quantitative relationships between the dynamics of ligand–receptor interaction and the biological and biochemical responses of the cell are reviewed in detail by Sklar *et al.*, (1984b). (5) Finally, the response of the neutrophil to *N*-formyl peptides, which behaves as a classical sensory response system showing adaptation to increases in *N*-formyl peptide concentrations (Zigmond and Sullivan, 1979; Seligmann *et al.*, 1982; see also Sklar *et al.*, 1984b, for a more detailed discussion).

3. MOLECULAR PROPERTIES OF THE *N*-FORMYL PEPTIDE RECEPTOR

Elucidation of the relationship between the amino acid sequence and the binding and/or biological responses to the ligand have been carried out by a number of investigators, most notably Becker and Freer and their colleagues (Aswanikumar *et al.*, 1977; Williams *et al.*, 1977; Showell *et al.*, 1976; Becker, 1979; Niedel *et al.*, 1979a; Freer *et al.*, 1980, 1982). Initial studies established that the most active tripeptides were FMLP and *N*-formyl Nleu-leu-phe. Both these *N*-terminal sequences are present in prokaryotic proteins but absent in mammalian proteins consistent with the notion that mammalian organisms evolved this receptor as a means to detect invasion of foreign organisms.* The formyl group is essential for activity, and its replacement with other N-terminal blocking groups (e.g., *tert*-butyloxycarbonyl or carbobenzoxy) actually generates peptides with the properties of pharmacologic antagonists (Freer, 1981).

Synthesis of analogs with varying amino acid substitutions and increasing peptide size have established the following general structural requirements for optimal binding:

1. At least five residues contribute to the binding (and biological) activity of the peptide (Niedel *et al.*, 1979a).
2. An absolute requirement is methionine or norleucine but not hydrophobic residues of longer side-chain lengths at position 1.
3. In general, at position 2, leucine or related hydrophobic amino acids such as valine are required for optimal activity whereas polar side chains reduce binding efficiency.
4. At position 3, phenylalanine is required.
5. At positions 4 (and possibly 5), addition of a variety of hydrophobic amino acids enhances activity. The ability to add a variety of amino acids including tyrosine and lysine in this position without impairing binding has the practical advantage of adding radioiodine, fluorescent, or photoaffinity probes for various experimental purposes.

On the basis of these studies, Freer *et al.* (1982) have proposed a model for the ligand-binding site that features a requirement for hydrogen bonding between the N-formyl group and the receptor in a relatively nonpolar binding pocket about the size of a pentapeptide.

In addition to the wealth of information concerning the structural requirements for ligand binding and bioactivity, recent progress has been made with respect to defining the molecular properties of the receptor itself. Niedel *et al.* (1980) were the first to successfully identify the molecular size of the receptor in sodium dodecylsulfate (SDS) under reduced conditions. Using a hexapeptide (N-formyl-nleu-leu-phe-nle-tyr-lys) labeled with ^{125}I to very high specific activity, these workers were able to cross-link the radioligand to a 50- to 60-kdalton species (M_r) with bifunctional cross-linking reagents. Similar results were obtained in the presence or absence of reducing agents when extracted directly into

---

* There is one recent report indicating the presence of N-formylated peptides in lysates of mammalian mitochondria (Carp, 1982). This probably reflects the fact that mitochondria are related evolutionarily to bacteria. In addition, these observations suggest that N-formyl peptides or proteins released at sites of tissue inflammation may further exacerbate the inflammatory response by attraction and activation of leukocytes at these sites.

SDS indicating that the ligand-binding unit, unlike the insulin receptor (Pilch and Czech, 1979, 1980; Yip *et al.*, 1982), is not disulfide bonded to itself or some other nonligand-binding unit (at least in membranes isolated from unstimulated neutrophils).

These results were confirmed and extended in our own laboratory using a 2-nitro-4-azido phenyl derivative of the same hexapeptide (Painter *et al.*, 1982; Schmitt *et al.*, 1983). As shown in Fig. 1, the photo-cross-linked receptor, when labeled at 0°C under nonstimulatory conditions, was shown to be composed of at least two major species (M_r = 60,000, pI = 6.5; M_r = 50,000, pI = 6.0) separable by 2-D gel electrophoresis by the technique of O'Farrell (1975). In addition the receptor appears to be a wheat germ agglutinin-binding glycoprotein (Painter *et al.*, 1982).

The receptor can be extracted in active form from isolated membranes with nonionic detergents but not urea, high- or low-ionic-strength treatments implying that it is an integral membrane protein (Niedel, 1981). The extracted receptor binds ligand with similar but not identical affinity as the membrane-associated protein and can be affinity labeled with identical results (Niedel, 1981). In addition, the ligand-binding activity of the extracted protein, like that of the membrane-bound form, is inactivated by free thiol-specific reagents suggesting a role for free-SH groups in the maintenance of binding activity.

When extracted with Triton X-100 from membranes isolated from unstimulated neutrophils (Jesaitis *et al.*, 1982a), the photolabeled receptor–ligand

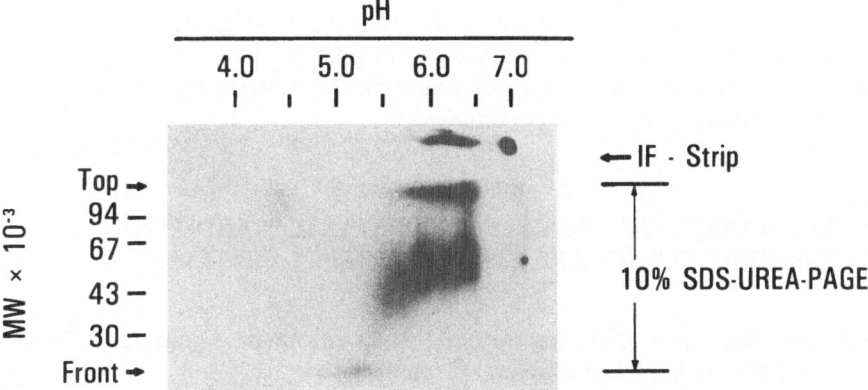

FIGURE 1. Autoradiogram of a two-dimensional polyacrylamide gel of neutrophil membranes labeled with a photoactivatable derivative of *N*-formyl-nle-leu-phe-nle-[^{125}I]-tyr-lys. Electrophoresis was performed according to the method of O'Farrell (1975) with some modification (Schmitt *et al.*, 1983). The specific *N*-formyl peptide labeling is localized in two major spots. Reprinted from Schmitt *et al.* (1983) with permission of the publisher.

complex behaves as a homogeneous species as judged by gel filtration on Sepharose 4B (Painter *et al.*, 1982). More recently, we have shown that this detergent-solubilized form has a Stokes radius of 39 Å, slightly larger than that of bovine serum albumin, indicating that after solubilization with Triton X-100, at least, a fraction of the labeled protein appears to exist as a monomer and is not associated with other accessory proteins large enough to detectably alter the molecular size (Allen *et al.*, in preparation). A larger fraction of specific label is observed that can be reduced with β-mercaptoethanol to monomer, suggesting the presence of intermolecular disulfide bonds. We do not know whether this species is formed after detergent extraction or whether it exists in the cell membrane. When extracted from membranes obtained from neutrophils exposed to *N*-formyl peptide at 37°C, the receptor–ligand complex appears to rapidly associate with a submembranous cytoskeletal matrix (see Section 4.2).

Dolmatch and Niedel (1983) have demonstrated that papain treatment of intact neutrophils converts the photolabeled receptor from 60 kdaltons to 30–35 kdaltons as judged by SDS–PAGE analysis. It is not known whether the missing mass is lost prior to SDS extraction or represents proteolytic nicks in an otherwise intact receptor on the cell surface. Interestingly, the cells were still fully capable of binding *N*-formyl peptides with unaltered affinity or number of binding sites per cell. The cells maintained their biological responsiveness to the peptides as well (Dolmatch and Niedel, 1983).

Some progress has been made toward isolation of the receptor in active form. Goetzl *et al.* (1981) prepared a partly purified preparation from 2–4×10^9 neutrophils by affinity chromatography on FMLP Sepharose, followed by gel filtration. They obtained a preparation that exhibited binding activity for FMLP, a fraction of which exhibited the affinity and ligand specificity of the membrane-associated receptor. SDS–PAGE analysis revealed a major component of 68,000 daltons as well as several other components.

4. RESPONSES OF THE CELL CYTOSKELETON AND THE CONTRACTILE APPARATUS TO *N*-FORMYL PEPTIDES

4.1. Morphological Changes Induced by Chemotactic Agents and Their Relationship to Cytoskeletal Organization

The morphological response of neutrophils to gradients (or uniform solutions) of *N*-formyl peptides has been extensively studied and can be divided into at least three distinct phases. These include, in roughly sequential order, cell membrane ruffling and pseudopod formation, elongation or polarization of the cell toward the source of the stimulus, and finally, directed ameboid movement in a relatively straight line toward the stimulus (Ramsey, 1972; Allan and Wilkinson, 1978). In this section, we will examine the relationship between these morphological responses and the organization of cytoskeletal apparatus.

4.1.1. Membrane Ruffling and Pseudopod Formation

When micropipettes loaded with N-formyl peptides are placed near a neutrophil, the cell responds within a few seconds by extending pseudopods from its surface. With time, pseudopod formation is concentrated at that membrane region nearest the source of attractant (Gerish and Keller, 1981). Allowing for the time it takes for peptide to diffuse from the pipette tip to the cell, this response ranks among the fastest responses of this cell to N-formyl peptides yet observed. The ruffling response can also be observed when adherent cells are exposed to uniform concentrations of N-formyl peptides (Zigmond and Sullivan, 1979; Davis et al., 1982). These studies demonstrated that a twofold *increase* in the concentration of peptide could elicit a ruffling response if the concentration range was within two orders of magnitude of the K_d for the peptide and receptor. Yuli and Snyderman (1983) have also measured changes in the light-scattering properties of neutrophil suspensions that appear to correspond to the morphological changes described.

The pseudopod responses observed in gradients or uniform solutions of peptide appear to be reversible for at least 5–6 min. When the micropipette tip is repositioned, the previously formed pseudopodia will retract and are replaced by new ones which form near the micropipette tip. This can occur even within the uropod region of the cell suggesting that sufficient receptors remain over the entire cell surface to initiate a new response and that the previously apparently nonresponsive regions of the cell are responsive when exposed to sufficient concentrations of ligand.

Zigmond et al. (1981), using a different approach, found that cells preoriented for 15–30 min in a gradient of peptide could indeed reverse their direction of movement if the gradient was reversed. However, in contrast to the results of Gerish and Keller (1981), it was found that the cell could also do so by executing a U-turn maneuver rather than by extending pseudopodia from the uropod region although the latter response was observed occasionally. The reasons for these discrepancies are unclear but could reflect the different methodologies employed by the laboratories or perhaps the precise timing of the first and second stimuli. In view of new evidence showing dramatic changes in surface receptor number, surface receptor topography, and adaptive changes that occur during incubation with N-formyl peptide, the precise timing of the gradient reversal could influence the experimental outcome.

Electron microscopic studies indicate that the membrane blebs and ruffles formed resemble true pseudopods. They are rich in actin filaments and myosinlike filaments (Hartwig et al., 1977; Hoffstein et al., 1977; Berlin and Oliver, 1978; Painter and McIntosh, 1979; Stendahl et al., 1980; Painter et al., 1981) and probably contain actin-binding protein (Hartwig et al., 1977; Stendahl et al., 1980; Valerius et al., 1981, 1982) and gelsolin (Yin et al., 1981). In addition, apparent submembranous Ca^{2+} appears to be depleted from those membrane regions at the advancing pseudopods of chemotaxing cells (Cramer and Gallin, 1979) and at sites of particle phagocytosis (Hoffstein, 1979).

4.1.2. Cell Elongation and Sustained Locomotion

Neutrophils also adopt a polarized or elongated morphology when exposed to a homogeneous concentration of N-formyl peptide in suspension although the cell population as a whole is randomly oriented (Zigmond and Sullivan, 1979; Davis et al., 1982). Although suspension cells obviously cannot locomote through the fluid phase, when exposed to concentrations of N-formyl peptides, near the K_d, cells will change their morphology from a relatively spl ical shape to an elongated cell (Davis et al., 1982). Similarly, exposure of adherent cells to a homogeneous concentration yields a similar result (see Fig. 2 and also Zigmond and Sullivan, 1979). Exposure of such cells to nonchemotactic agents like wheat germ agglutinin or a monoclonal antibody (NMS-1) that can stimulate $O_2^-\bullet$ production, secretion (Painter et al., 1984), or ruffling does not apparently polarize the cell (Fig. 2).

Ultrastructural and immunofluorescent studies indicate that the leading lamellipodial ruffle of such cells is rich in actin microfilaments (Oliver et al., 1978), actin-binding protein, myosin (Stendahl et al., 1980), and gelsolin, the Ca^{+2}-sensitive protein that appears to control, in part, the actin filament length (Yin et al., 1980). In contrast to less motile cells like fibroblasts, F-actin is organized in a random network of filaments with little evidence of stress fiber bundles in neutrophils and macrophages (Oliver et al., 1978; Painter et al., 1981). Overall, the cell orientation is identical to that of any locomoting cell type with microtubules radiating from a perinuclear centriole lying just anterior to the cell nucleus. In addition, the distribution of certain surface membrane components of such cells becomes assymetrically reoriented with Fc receptors,

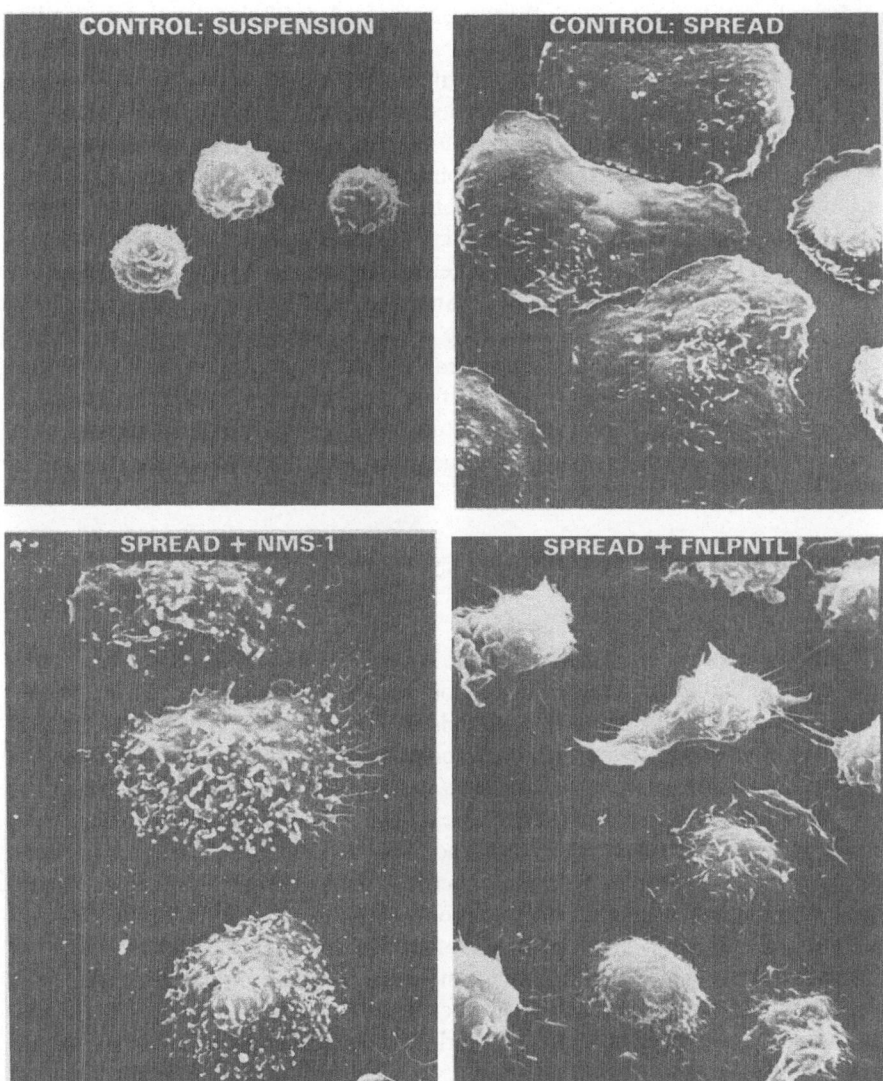

FIGURE 2. Scanning electron micrographs of neutrophils. Neutrophils fixed in suspension (control: suspension) show a relatively spherical shape. If live cells are allowed to spread on a glass substratum (control: spread), they assume a flattened, "pancake"-like morphology. Upon exposure to a uniform solution of 5nM N-formyl-nle-leu-phe-nle-tyr-lys (spread + FNLPNTĽ), the cells assume an elongated morphology typical of motile cells. In contrast, exposure of such cells to a monoclonal antibody (spread + NMS-1), which stimulates O_2- production and secretion (Painter et al., in press) causes the formation of many bleblike projections but does not result in polarization of the cells.

Con A receptors (Walter et al., 1980; Wilkinson et al., 1980), and unoccupied N-formyl peptide receptors (Sullivan and Zigmond, 1982) localized at the leading edge of the locomoting cells.

The various activities associated with the cell surface appear to be segregated into distinct domains in FMLP-oriented cells as well, with there being little distinction between cells oriented in suspension or on a substrate. Davis et al. (1982) have shown that fluid phase pinocytosis occurs largely in the cell tail region. Further studies established that coated and uncoated vesicles are predominantly found in this region of the cell as well (Davis et al., 1982; Hoffstein et al., 1982). Finally, secretory vesicles or granules (Wright and Gallin, 1979) as well as elements of the Golgi Apparatus (Kupfer et al., 1982; Bergmann et al., 1983) appear to preferentially fuse with the leading membrane front.

From these observations, it is clear that a drastic change in the structural and functional organization of cell surface membrane proceeds in parallel with the dramatic changes observed in the underlying cytoskeletal apparatus. This, of course, raises the possibility that one process modulates the other.

4.2. Interactions between the Cell Cytoskeleton and the N-Formyl Peptide Receptor

As mentioned previously, surface membrane N-formyl peptide receptors can be quantitatively extracted with nonionic detergents from cells or isolated membranes of unstimulated neutrophils (Niedel, 1982; Painter et al., 1982; Jesaitis et al., 1984), even if the membranes have first been exposed to bifunctional cross-linking reagents (Niedel, 1981).

After incubation with N-formyl peptide at 37°C the situation changes radically. Within as little as 10 sec after addition of peptide, the receptor converts to a much-higher-affinity form and becomes progressively more resistant with time to extraction with nonionic detergents (Jesaitis et al., 1983, 1984; Sklar et al., 1984a).* Several observations suggest that this conversion to a relatively Triton-insoluble form is due to direct molecular interaction of the receptor with the cell cytoskeleton. First, electron microscopic autoradiography of the isolated cytoskeletons revealed a dense network of microfilaments arranged in a cell-

---

* A small fraction (~5% of the total) of cytoskeletally attached receptors apparently preexist in the resting cell (Jesaitis et al., 1984). This fraction may correspond to the small population of receptors having a 50-fold higher affinity than that of the bulk receptors as reported by Koo et al. (1983) in isolated membranes. This affinity change appears to be regulated by guanyl nucleotide levels as well.

sized and -shaped pattern around a central nucleus. At times prior to internalization, grains arising from bound ^{125}I-hexapeptide (N-formyl-nle-leu-phe-nle-[^{125}I]-tyr-lys) were largely localized at the (former) surface of these structures (Fig. 3) in close proximity to a band of underlying actin filaments. Second, and most significant, dihydrocytochalasin B (1 μg/ml) completely blocked both the conversion of the receptor-ligand complex to the slowly dissociating form and its cosedimentation with the Triton cytoskeleton. In addition, internalization of ligand was blocked, suggesting that these processes may be coupled.

Several observations indicate that the high-affinity, Triton-insoluble complex occurs at the cell surface prior to internalization. First, the receptor–ligand complex is associated with the submembranous actin-rich matrix (Luna et al., 1981; Mescher et al., 1981) of isolated plasma membranes (Jesaitis et al., 1983). Second, the phenomenon can be demonstrated under conditions where internalization of ligand is suppressed (Jesaitis et al., 1984).

This high-affinity conversion and concurrent association of receptors with the cytoskeleton may represent a general phenomenon. Cyclic AMP, a chemoattractant for *Dictyostelium discoideum*, also becomes tightly associated with the Triton-insoluble cytoskeleton of the ameboid form of this organism (Galvin et al., 1984; also see Chapter 1 by Frazier et al. in this volume). Vale and Shooter (1982) have shown that nerve growth factor is converted to a more slowly dissociating form which associated with the Triton cytoskeleton when the cells are exposed to wheat germ agglutinin. There is, of course, a preponderance of evidence showing that many membrane proteins are segregated into a relatively freely mobile fraction and a relatively immobile fraction, as judged by laser photobleaching studies (Schlessinger et al., 1976; Sheetz et al., 1980; Koppel et al., 1981; Webb et al., 1981; Jesaitis and Yguerabide, 1980; Woda et al., 1980). Given this, the results of Vale and Shooter could be explained by WGA-induced cross-binding of mobile NGF receptors being cross-linked by WGA to a small fraction of receptors preanchored to the cytoskeleton. In fact, Koch (1981) has recently demonstrated that Concanavalin A can immobilize erythrocyte membrane proteins by this mechanisms.

In addition to these results, studies in a variety of cell types using numerous receptor–ligand systems suggest that clustering of membrane receptors by ligands also induces cytoskeletal interactions (Nicolson, 1973; Bourguignon and Singer, 1977; Flanagan and Koch, 1978; Schlessinger et al., 1976; Woda et al., 1980; Shelterline and Hopkins, 1981; Painter and Ginsberg, 1982). In reciprocal fashion, clustering of the red-cell cytoskeletal protein spectrin by antispectrin antibodies (Nicolson and Painter, 1973) or by removal of spectrin from the membrane (Elgsater and Branton, 1974) results in the clustering of external membrane glycoproteins.

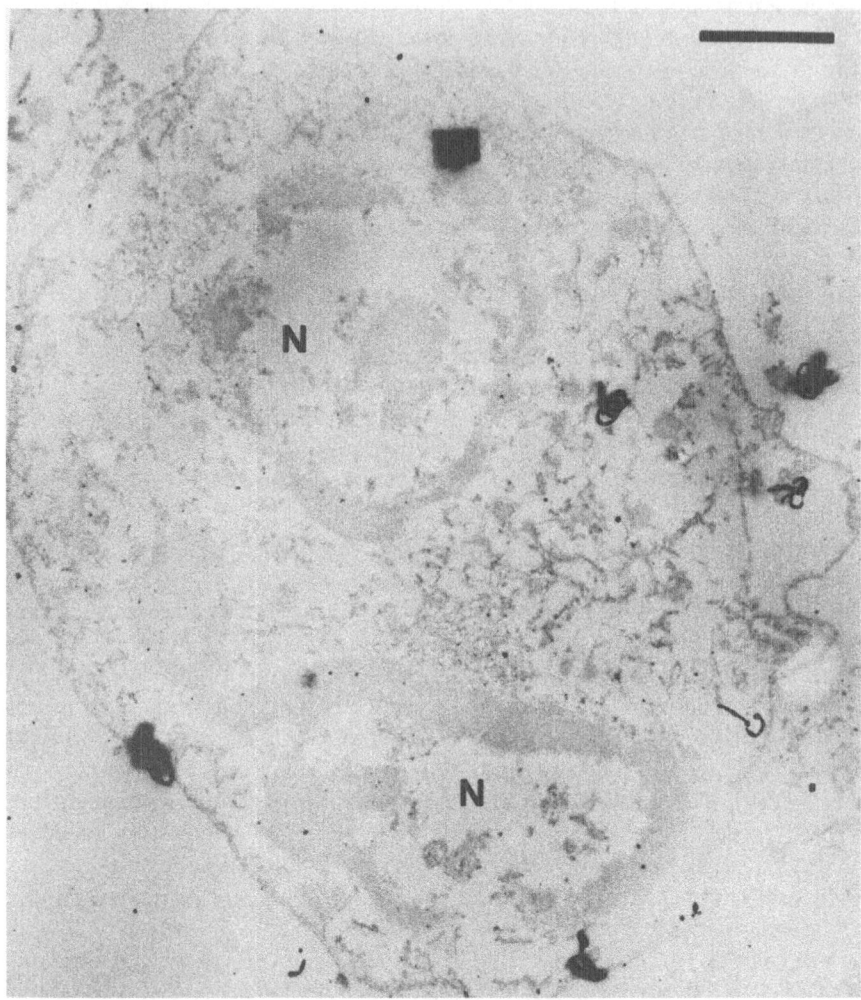

FIGURE 3. EM autoradiograph the Triton X100-insoluble cytoskeleton obtained from neutrophils exposed to 2 nM *N*-formyl-nle-leu-phe-nle-[^{125}I]-tyr-lys for 60 sec at 37°C prior to the preparation and isolation of the cytoskeletons. Note the association of grains arising from the radioactive decay of the bound peptide associated with the surface of these microfilament-rich structures. Bar equivalent to 1 μm. Reprinted from Jesaitis *et al.* (1984) with permission of the publisher.

4.3. Changes in the Organization of Cytoskeletal Proteins Induced by Chemotactic Peptides

In the "resting" neutrophil only 25% of the total cell actin is in the filamentous or F form (Rao and Varani, 1982; Fechheimer and Zigmond, 1983). The remainder is presumably complexed with profilin, a protein that prevents F-actin from polymerizing (Markey *et al.*, 1978). Using the DNAse I inhibition assay for F-actin (Lazarides and Lindberg, 1974), Rao and Varani (1982) and Fechheimer and Zigmond (1983) found that the F-actin content of the cells rapidly and transiently increased about twofold when neutrophils were exposed to FMLP. The observed F-actin increase had plateaued at the earliest time point measured (30 sec). These results are in agreement with those of White *et al.* (1982), who measured the amount of sedimentable actin rather than polymerization state *per se*. This assay could potentially be influenced by functional changes in F-actin cross-linking proteins as well as the G-actin polymerization state. Because the most rapid biochemical changes induced by *N*-formyl peptides (including pseudopod formation and membrane ruffling) occur within 5–10 sec, it would be interesting to know the F-actin polymerization state within the first few seconds. Although such an experiment has not been reported to date for actin polymerization *per se,* the results of White *et al.* (1982) indicated an increase in sedimentable actin within 5 sec. Thus, it appears that alterations in the actin polymerization state and/or filament cross-linkage by actin-binding proteins is a rapid response of neutrophils to *N*-formyl peptide.

Fechheimer and Zigmond (1983) also reported that when FNLP-stimulated cells were shifted back to FNLP-free media, pseudopod formation ceased at a rate that correlated with a partial return of the F-actin content to basal levels.

Chemotaxis, by definition, implies a directional, and therefore localized, response to the cell. Small increases in F-actin content in a bulk cell suspension may therefore actually reflect large localized changes in F-actin concentration. Taylor *et al.* (1980a) have microinjected fluorescently labeled actin into living fibroblasts. The injected actin becomes incorporated into the cytoskeleton in a manner that accurately reflects the actin distribution in noninjected cells. Although a fibroblast is clearly less mobile than a neutrophil, such cells do exhibit directional motility, albeit at a much slower rate. For example, fibroblasts exposed to a probe vibrating with high frequency moved toward the probe (vibrotaxis). When cells microinjected with actin were exposed to such a probe, increases in the actin-associated fluorescence at the cell membrane nearest the probe were observed. Of course, such studies will have to be carried out with chemotaxing leukocytes in order to determine if the observations of Taylor *et al.* (1980a) are generally applicable. Nevertheless, their results do raise the possibility that new actin filaments are recruited and/or polymerized at the cell surface closest to the chemotactic stimulus.

In all fairness, it should be noted that actin polymerization by ligand is not sufficient for directed motility. For example, Concanavalin A, which is not a chemoattractant, caused increases similar to those induced by FMLP in F-actin content in neutrophils (Rao and Varani, 1982). Likewise, agents like thrombin, which is a potent blood platelet stimulus, induced a twofold increase in F-actin as well (Carlsson et al., 1979). It seems likely that increased F-actin content is a general cellular response to any stimulatory ligand regardless of whether it is chemotactic or not.

Like actin, the polymerization rate of tubulin is affected by exposure of neutrophils to chemotactic agents like FMLP. Immunofluorescent localization studies and electron microscopic studies indicate that the numbers of microtubules and their length increase upon exposure to N-formyl peptides and other stimulants like Ca^{2+} ionophores (Hoffstein et al., 1977; Anderson et al., 1982). An accompanying increase in the level of posttranslational tyrosinolation of tubulin has also been reported (Nath et al., 1982). However, tyrosinolation of tubulin per se does not appear to affect the rate or extent of microtubule polymerization (J. Nath, personal communication). A preliminary report indicates that the labeled tubulin may be predominantely localized in plasma membrane-rich fractions (Nath and Flavin, 1983). This interesting result raises the possibility that modification of tubulin mediated by N-formyl peptide stimulation of neutrophils could induce the interaction and assembly of microtubules on the cytoplasmic surface of the plasma membrane.

In spite of changes in microtubule polymerization state, microtubule-disrupting agents like colchicine do not dramatically affect leukocyte chemotaxis (Zigmond et al., 1981). They appear to affect the overall precision and efficiency by which the cells turn and move toward the source of attractant (Zigmond et al., 1981). Colchicine-treated cells move toward the source perfectly well, but their path deviates more from a straight path than the untreated controls. Although some could argue that this "staggering" behavior was entirely due to nonspecific drug "intoxication," it could also mean that microtubules organize the intracellular organelles for more efficient cellular movement (Malawista and Chevance, 1982). This notion is consistent with the fact that microtubules radiate from a microtubule-organizing center near the cell nucleus in a direction parallel to the axis of cellular movement. The organizing role is also consistent with the known association of mitochondria (Ball and Singer, 1982) and intracellular granules (Byers and Porter, 1977) with microtubules. This concept is supported by recent studies that employ viable enucleated neutrophil cell fragments (termed "cytoplasts" or "neutroplasts"). These interesting creatures produced by several differing procedures appear to be deficient in microtubules and intracellular organelles (Keller and Bessis, 1975; Malawista and Chevance, 1982; Roos *et al.*, 1983; Korchak *et al.*, 1983a). Nevertheless, they still appear to be capable of directed motility when exposed to gradients of chemotactic agents, albeit at a considerably reduced rate of migration (Malawista and Chevance, 1982; Roos *et al.*, 1983).

Myosin light chain is phosphorylated within a minute of the addition of *N*-formyl peptides to neutrophils (Fechheimer and Zigmond, 1983). Since light-chain phosphorylation is essential for myosin-mediated contraction in nonskeletal muscle tissues and cells (Adelstein and Eisenberg, 1980), these results suggest that increased actomyosin contractile forces are generated in response to chemotactic factors. As in the case of actin polymerization, increased phosphorylation of myosin light chain is not unique to neutrophils or to chemotactic factors (Fox and Phillips, 1982).

How might these biochemical modifications be controlled? In the case of myosin light-chain phosphorylation, it is well known that a calmodulin-dependent light-chain-specific protein kinase is responsible for this modification (Adelstein and Eisenberg, 1980). These considerations imply that intracellular levels of free Ca^{2+} ion probably regulate this process as well as other cytoskeletal interactions. Furthermore, recent studies have shown that "specific" calmodulin inhibitors like trifluoperazine and N-(6-aminohexyl)-5-chloro-1-naphthalene sulfonamide (W-7) inhibit neutrophil chemotaxis toward *N*-formyl peptides (Elferink *et al.*, 1982). In Section 5 we will consider the implications of free intracellular Ca^{2+} as a prime regulatory ion in chemotactic phenomenon.

5. ROLE OF Ca^{2+} ION IN THE CONTROL OF CELL MOTILITY

5.1. The Effect of Ca^{2+} on the Organization of the Cytoskeleton

In addition to myosin phosphorylation, free Ca^{2+} ion in the micromolar range *in vitro* appears to regulate the interaction of several important actin-binding or cross-linking proteins with actin filaments. Table I summarizes the known effects of Ca^{2+} on the interaction of these proteins with actin.

What is clear from Table I is that Ca^{2+} can either directly or indirectly (through calmodulin-dependent kinases) affect the gel–sol state of the cytoplasm. For some time it has been recognized that the gel–sol state of the cytosol plays an important role in cell motility (Frey–Wyssling, 1957; Taylor and Condeelis, 1979; Yin *et al.*, 1980, 1981). Morphological observations suggest that the

TABLE 1. The Effect of Free Ca^{2+} on Actin Binding or Cross-Linking Proteins

Actin-binding protein	Source	Subunit (mol. wt.)	Effect on actin solutions	Phosphorylation dependence	Ca^{2+} dependence	Reference
1. Myosin	Nonskeletal	2 × 200,000 20,000 16,000	Contraction	Yes	μM levels activate calmodulin, dep phosp. of light chain (20 k)	Sellers et al., 1981; Adelstein and Eisenberg, 1980
2. Actin-binding protein	Chicken gizzard (filamin) Macrophages Platelets	2 × 250,000	Gel formation	No No Yes	None None None	Wang, 1977; Wang et al., 1975; Hartwig and Stossel, 1975 Wallach et al., 1978; Rosenberg et al., 1981; Carroll and Gerrard, 1982
3. Gelsolin (villin)	Macrophages Platelets (intestine)	95,000	Regulates gel–sol state to form 3-D network	Unknown	Ca^{2+} (μM) favors sol state; $< 10^{-6}$ M yields gelation	Yin et al., 1980; Wang and Bryan, 1981; Glenny and Weber, 1981
4. α-Actinin	Platelets, HeLa cells	2 × 100,000	Bundles actin filaments	No	> μM levels block interaction	Lazarides and Burridge, 1975; Rosenberg et al., 1981; Burridge and Feramisco, 1981; Yeltman et al., 1981
5. Profilin	Nonmuscle cells	15,220	Binds to G-actin; inhibits polymerization	No	Yes	Carlsson et al., 1977; Markey et al., 1978
6. Vinculin	Chicken gizzard, HeLa cells	130,000	Bundles actin; may attach stress fiber adhesion plaques	Yes (tyrosine specific)	HeLa cell form requires calcium to interact with actin	Geiger 1979; Geiger et al., 1980; Burridge and Feramisco, 1980, 1981; Sefton et al., 1981

cytoplasm flows freely as the cell streams forward during locomotion indicating that transient, localized alterations in the gel–sol state (and, by inference, actin-cross-linking state) are occurring.

If Ca^{2+} is indeed involved in the regulation of the gel–sol state *in vivo,* three questions must be addressed, namely: (1) What is the effect of chemotactic peptides on *free* intracellular Ca^{2+} concentration? (2) What is the *local* concentration of free cytosolic Ca^{2+} in locomoting cells? (3) How do the observed concentration changes relate to morphological responses of the cell?

5.2. The Effect of *N*-Formyl Peptides on Free Intracellular Ca^{2+} Levels

Several lines of evidence indicate that the *free* Ca^{2+} level increases from basal levels of $< 10^{-7} - 10^{-8}$ M to levels of at least $10^{-6} - 10^{-5}$ M. Although this is accompanied by influx of extracellular calcium and other ions (Naccache *et al.,* 1977a,b), current data indicate that most of the intracellular increase is from intracellular sources. The recent introduction of the fluorescent EGTA analog Quin 2-AM (Tsien *et al.,* 1982) has provided new insight into the extent and kinetics of intracellular Ca^{2+} changes in response to binding of *N*-formyl chemotactic peptides as well as other stimuli. Quin 2 is introduced into resting cells in the form of an ester which allows this nonpolar molecule to pass the cell membrane. Once inside, nonspecific esterases in the cytosol convert the ester to the EGTA-like analog. Since the fluorescence of the molecule increases sixfold upon complexation of calcium, intracellular Ca^{2+} concentration much below the K_d (10^{-7} M) results in low fluorescence. At higher Ca^{2+} levels fluorescence increases. Unstimulated neutrophils preloaded with Quin 2 to cytosolic levels of $0.5 - 1$ mM exhibited 60% of maximal Quin 2 fluorescence, indicative of a cytosolic Ca^{2+} level near 150 nM (Korchak *et al.,* 1983b; Sklar 1984). When *N*-formyl peptide was added to such cells at 37°C in a stirred suspension, Quin 2 fluorescence rapidly increased, with the earliest detectable change occurring within 1–2 sec after peptide addition. Using pulse-binding techniques described by Sklar *et al.* (1981b, 1984b), we estimate that a half-maximal Quin 2 response can be obtained with less than 1% occupancy of the total cell surface *N*-formyl receptors. One can estimate by appropriate calibration of the system that a peak Ca^{2+} concentration of at least 1 μM is attained, although a degree of uncertainty is introduced by the fact that this concentration is close to that which gives a maximum Quin 2 fluorescent change. The changes observed are transient when binding of ligand to receptor is inhibited, returning to near-resting values with a half-time of 20 sec. The interpretation of the Quin 2 studies could be influenced by the presence of mM levels of a Ca^{2+} chelator within the cell.

However, Fechheimer and Zigmond (1983) have shown that enzymes regulated by micromolar levels are in fact activated in the FNLP-stimulated cells. Clearly, these results indicate that intracellular levels of free Ca^{2+} achieve levels that potentially could affect the gel–sol state of the cytosol.

5.3. The Local Concentration of Free Ca^{2+} Ion in Locomoting Cells and Its Relationship to Motility

Taylor *et al.* (1980b) have microinjected the Ca^{2+}-sensitive luminescent protein aqueorin into living amebae. Using quantitative microscopic techniques, they have shown that free Ca^{2+} levels transiently increase from $< 10^{-7}$ to $> 10^{-7}$ at the leading front or lamellipodial edges of locomoting cells and remain elevated in the cell tail. As previously mentioned, this same laboratory has similarly microinjected fluorescent actin into ameboid cells and has shown that actin is found in both sites (Taylor *et al.*, 1980b).

Several reports suggest that the local concentration of Ca^{2+} may change in response to chemotactic peptides. In addition to the Quin 2 studies discussed previously, Naccache *et al.* (1979) have shown that neutrophils labeled with the Ca^{2+}-sensitive fluorescent probe chlortetracycline (CTC) show fluorescence changes induced by *N*-formylated peptide probe that are suggestive of alterations in membrane-bound Ca^{2+}. Recent studies by Korchak *et al.* (1983b) indicate that this probe senses a change in a different pool of intracellular Ca^{2+} than does Quin-2, however. Alternatively, given the difference in the K_Ds of the two probes ($K_{DQuin\ 2} = 10^{-7}$M and $K_{DCTC} = 4 \times 10^{-4}$ M), these results could mean that higher levels of intracellular Ca^{2+} must be achieved to cause a change in CTC fluorescence. Moreover, the exact relationship between CTC fluorescence changes and Ca^{2+} levels is less clear than for Quin-2. It is generally thought that CTC measures changes in membrane-bound Ca^{2+} although this has not been directly shown, whereas Quin-2 very probably does measure changes in free cytosolic Ca^{2+}. Clearly, caution must be exercised when interpreting such data.

More directly, Cramer and Gallin (1979) have histochemically localized plasma-membrane-bound Ca^{2+} using the pyroantimonate method and have demonstrated that Ca^{2+} staining is lost at the leading membrane regions of chemotaxing neutrophils. The lost Ca^{2+} is presumed to transiently raise local concentrations of free and bound Ca^{2+} in the adjacent cytoplasm. Although some doubt remains concerning the specificity and exact interpretation of such methods, the data do suggest the possibility that as in the fibroblast system, free Ca^{2+} ion may transiently increase at the leading edge of chemotaxing neutrophils.

6. THE ROLE OF THE ADAPTATION IN CHEMOTAXIS

The response of neutrophils to chemotactic stimuli, including N-formyl peptides, exhibits many of the characteristics associated with a classical sensory adaptive system. As we have discussed elsewhere (Sklar *et al.*, 1984b), the response of neutrophils to N-formyl peptides appears to exhibit classical adaptation behavior typical of more well-studied sensory functions like animal vision (Autrum, 1981), bacterial chemotaxis (Koshland, 1981), and plant phototropism (Bergman *et al.*, 1969). As in these systems, the neutrophil responses involve relatively short latencies, transiency, relatively short refractory periods, and restoration, within minutes, of sensitivity to *changes* in stimulus level rather than the absolute stimulus concentration. This relationship between changes in stimulus concentration and responsiveness can be expressed mathematically using the Weber–Fechner law of sensory adaptation, which states that within a certain range of stimulus intensity (e.g., concentration of chemoattractant) the response of a system will be proportional to the logarithm of the quotient of the new stimulus intensity and the adaptation level (i.e., the stimulus intensity to which the system was adapted; Delbruck and Reichhardt, 1956; Sklar *et al.*, 1984b). Such analysis shows that response of neutrophils to N-formyl peptide or other chemoattractants, whether they be morphological changes such as pseudopod formation, chemotaxis (Zigmond and Sullivan, 1979; Keller *et al.*, 1977, 1978), or membrane potential changes (Seligmann *et al.*, 1982), obeys the Weber–Fechner law over approximately four orders of magnitude of stimulus concentration (1 nM–10 μM). Thus, neutrophils behave as a classical sensory adaptive system.

One implication of cellular adaptation in neutrophils which may be of great significance for the neutrophil host defensive function derives from the dynamics of adaptation. By definition, the adaptation level of a responsive cell changes depending on the external stimulus intensity (Delbruck and Reichhardt, 1956). This change has a rate characteristic of the cell, and if it is on the order of the rate of change of the stimulus intensity, the cell will *not* respond. Such phenomena have been universally experienced in vision as "sunrise experiments" (Bergman *et al.*, 1969) in which the physiological perception of intensity of light remains constant for approximately one-half hour just after sunset or just before sunset or sunrise, even though the absolute intensity of light decreases or increases exponentially during this time. In a similar fashion, neutrophils will not respond to chemoattractant infused into their external environment at rates of less than 2 pM/sec (Sklar *et al.*, 1981a). Such dynamics would permit directed migration without stimulation of the cells' microbicidal arsenal (e.g., secretory granules, generation of toxic oxygen radicals). In addition, they suggest that the rate of receptor occupancy by the chemotactic peptides must be greater than the intrinsic rate of adaptation in order for the cell to respond.

7. COUPLING THE CHEMOTACTIC SENSORY SYSTEM TO ACHIEVE DIRECTED MOTILITY

How does the cell sense a gradient of N-formyl peptide and convert this information into a vectorial movement toward ever-increasing concentrations of attractant. This is, of course, the central issue and, unfortunately, the one about which we know the least. In spite of this, several attractive hypotheses have been published that are consistent with both the morphological response seen during chemotaxis and the known biochemistry of cytoskeletal proteins. The reader is referred to numerous reviews of these topics (Stossel, 1978; Taylor and Condeelis, 1979; Taylor et al., 1979, 1981; Stossel et al., 1981; Taylor and Fechheimer, 1982) for a more detailed discussion of current concepts concerning cell motility. See Korn (1978, 1982) for reviews of the biochemistry of actin and actin-binding proteins.

Zigmond and Sullivan (1979) have proposed what is in essence an adaptation model to explain how cells continue up a chemotactic gradient. When they exposed adherent neutrophils to a uniform solution of 10^{-9} M FNLP, the cells responded within seconds by putting out blebs and pseudopods in all directions. This ruffling activity ceased within a minute, and the cells gradually assumed a polarized morphology. If the concentration of peptide was then increased, the cells responded as before; i.e., they had adapted to the initial concentration of peptide. However, *lowering* the concentration of peptide *suppressed* blebbing. Thus, they suggested that as long as the cells (or a portion of their membranes) continue to encounter an *increase* in attractant concentration, pseudopod formation is favored and will continue. In contrast, pseudopod formation is suppressed and does not occur when the concentration of peptide is unchanged or lowered. This *adaptive* model suggests that the cells are sensitive to *changes* in concentration, perhaps spatially *and* temporally, rather than absolute concentrations, *per se*.

On a biochemical level, sustained pseudopod extension toward direction of increasing attractant concentration could be explained by assuming that increased occupancy of receptor at the cell front results in a localized, sustained level of free Ca^{2+} ion. This would presumably favor a localized sol state resulting in a continued cytoplasmic flow in the correct direction. This is depicted schematically in Fig. 4, where the vertical arrows represent regions of elevated cytosolic

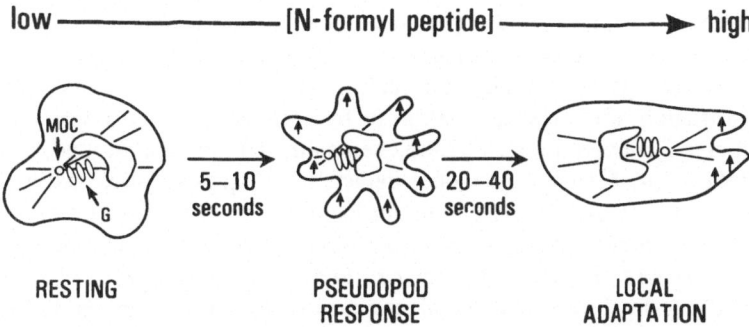

low ———————— [N-formyl peptide] ————————→ high

5–10
seconds

20–40
seconds

RESTING

PSEUDOPOD
RESPONSE

LOCAL
ADAPTATION

FIGURE 4. A schematic description of an adaptive model of chemotactic response which could explain directed cytoplasmic flow in the presence of a gradient of chemotactic peptide. The resting cell senses an increase in receptor occupancy with time at all regions of its membrane surface and responds with local increases in cytosolic-free Ca^{2+} (indicated by the vertical arrows). This local change results in extension of pseudopods by the cell within 5–10 sec. Those pseudopods which extend toward areas of decreasing or unchanging peptide concentration rapidly adapt, and Ca^{2+} falls (no arrows), resulting in suppression of pseudopod extension in these regions. If the pseudopods move into areas sufficiently high in peptide concentration, the Ca^{2+} levels maintained are sufficient to maintain cytosolic flow. Thus, the cell elongates and continues to move in the correct direction. Also shown is the reorientation of the microtubule organizing center (MOC) and its associated tubules and the Golgi (G) apparatus toward the anterior region of the cell. This reorientation occurs prior to overall cell elongation (Malech *et al.*, 1976).

free Ca^{2+}. In contrast, pseudopod extension toward lower or equal concentrations would result in lower receptor occupancy; free calcium levels presumably would rapidly fall in these cellular regions (Fig. 4), leading to a suppression of cytoplasmic flow. Important requirements of such a hypothesis are the following: (1) Free Ca^{2+} levels present at localized sites must be relatively *transient* and *localized*. This must be the case since sustained flow in all directions would not result in directed movement of the cell. Likewise, the signal must be transient to allow suppression of pseudopod formation in the wrong direction. (2) Increasing concentrations of peptide must cause a *sustained* Ca^{2+} response of sufficient rapidity to account for the pseudopod response, which is on the order of seconds. This requirement is necessary in order to allow for continued cytoplasmic flow only in the correct direction. (3) As occupied receptors are downregulated, unoccupied receptors must be continually supplied to the advancing cell membrane by insertion of either latent and/or recycled receptors at the front of the cell membrane. This would ensure that sufficient receptors are maintained in the advancing cell membrane so that stimulation by increasing concentrations can continue over an extended period of time. By no means does this require *exclusive* insertion of new receptors at the cell front. Indeed, maintenance of unoccupied receptors over the entire cell surface would be desirable, either by lateral diffusion after frontal insertion or by a global insertion process, so that the cell could change direction if the chemoattractant gradient abruptly changed its direction. (4) Finally, the model depicted in Fig. 4 requires that localized regions of the cell membrane can *adapt* locally, i.e., that exposure of one part of the cell membrane to given concentration of attractant does not affect the responsiveness of another part of the cell to a lower concentration of attractant.

Evidence lending experimental support to the first requirement has already been discussed. We also have preliminary evidence that indicates that as long as N-formyl peptide concentrations increase with time, the Quin 2 fluorescence remains at levels indicative of high free-Ca^{2+} concentration (Sklar 1984). These data were obtained by slow infusion of subnanomolar levels of peptide into a stirred suspension of Quin-2-loaded cells. In essence, this protocol represents a temporal gradient of chemoattractant. As long as the peptide concentration increased, free Ca^{2+} levels remained elevated. When the peptide concentration no longer increased (after the infusion was ended), Ca^{2+} concentrations fell, decaying with a half-life of about 20 sec, a time similar to that observed when a single pulse of N-formyl peptide is given.

As regards the third point above, several lines of evidence suggest that receptors are unregulated from internal pools within minutes (Zigmond, 1981; Zigmond et al., 1982). The intracellular source(s) of these latent receptors is not entirely resolved, but probably is derived from secretory granule or vesicle fusion with the plasma membrane (Gallin et al., 1978, 1979; Fletcher and Gallin, 1982; Fletcher et al., 1982) and/or surface receptors recycled via the Golgi apparatus (Jesaitis et al., 1982b, 1983; Painter et al., 1984).

The classic studies of Abercrombie et al. (1970) indicated that inert particles bound to the leading cell membrane move backward over the cell surface as location proceeds. They suggested that incorporation of new membrane at the leading plasma membrane edges could account for their observations. In fact, it has been shown that neutrophil granule contents are slowly released during cell locomotion and that granule fusion and release occur at the leading cell front (Wright and Gallin, 1979). The possibility that the receptor may pass through the Golgi or a Golgi-associated compartment, like a variety of other polypeptide hormone receptors, on its way back to the plasma membrane, is of interest in the context of receptor topography in chemotaxing cells. Singer and colleagues have recently demonstrated that viral antigens present initially in the Golgi apparatus of infected fibroblasts moved from the Golgi (presumably in vesicular form) to the membrane surface at the leading edge of the motile cell (Bergmann et al., 1984). The data would predict that N-formyl peptide receptors (and possibly other cell surface components as well) would concentrate at the cell front as the cell chemotaxed through a gradient. In fact, as mentioned earlier, this appears to be the case (Sullivan and Zigmond, 1982). After orientation, fixed cells showed an increased density of FMLP–hemocyanin molecules over the leading lamellipodial regions of the cells and little over the uropod (Sullivan and Zigmond, 1982). On unstimulated fixed cells, the distribution of sites was relatively uniform. Additional explanations of these results are conceivable, however, other than upregulation of new receptors at the cell anterior. It is possible that a "cappinglike" phenomenon has been induced by the peptide stimulation reorientation. Alternatively, receptors could be downregulated preferentially at the uropod region. However, one preliminary report suggests that occupied receptors remaining at the cell surface in living, oriented neutrophils are distributed in a relatively uniform distribution on the surface (Walter and Marasco, 1983). Thus, the results are at least consistent with the notion of latent receptor expression at the cell front.

Thus far no data exist in the neutrophil system which bear directly on the issue of localized adaptation. However, adaptation of the photoreceptor of localized regions of the plasma membrane of *Phycomyces* has been conclusively shown (Delbruck and Varju, 1961; Dennison and Bozof, 1973) suggesting that such a process is possible. In view of a large body of evidence indicating that stimulation of localized regions of leukocyte membranes is possible, it seems reasonable to suppose that local adaptation at such sites might also occur.

8. THE ROLE OF MYOSIN IN CHEMOTAXIS

As indicated earlier, the contractile activity of myosin is activated by a light-chain-specific protein kinase in the presence of micromolar concentrations of calcium ion (Adelstein and Eisenberg, 1980; Fechheimer and Zigmond, 1983). These facts together with the observation that myosin is present in extending pseudopods (Painter and McIntosh, 1979; Stendahl et al., 1980; Painter et al., 1981; Valerius et al., 1981) together with F-actin (Berlin and Oliver, 1978; Oliver et al., 1978; Painter et al., 1981) suggest that contractile forces may be exerted in the local environment of the extending pseudopod. If so, this could allow the pulling of the trailing cell body via the cytoskeleton in the direction of the forward-moving pseudopod extensions (Abercrombie et al., 1970; Goldman et al., 1976; Southwick and Stossel, 1983). In glycerinated fibroblasts, Goldman and colleagues (1976) have clearly shown that tail retraction is mediated by a myosin-dependent process in the presence of Mg^{2+}-ATP. In addition, the polarity of actin filaments attached to the plasma membrane is such that a pulling force could be generated in a direction parallel to the direction of movement of the pseudopod (Mooseker and Tilney, 1975; Burgess and Schroeder, 1977; Edds, 1977; Small *et al.*, 1978).

However, the exact role played by myosin is obscured by a lack of knowledge about the kinetic relationship between the rise in calcium in the cytosol and the subsequent contractile activity. More needs to be known about the relative rates of myosin activation and of gel–sol transformation, *in situ*. In addition, more needs to be known about the rate at which myosin is inactivated by phosphatases after calcium levels have declined as a result of local adaptation. Finally, as Taylor and colleagues have pointed out, the interaction of myosin with actin can be potentially modified by various actin-binding proteins. Clearly, more needs to be determined before the exact role of myosin can be elucidated. Perhaps the development of reagents such as monoclonal antibodies which distinguish activated myosin from its inactive form, used in conjunction with microinjection techniques, could resolve some of these issues.

9. FUTURE PROSPECTS

Clearly, much remains to be determined concerning the precise mechanism of cell chemotaxis and motility in ameboidlike cells. Two new technical developments have recently been described that should greatly improve our understanding of this complex phenomenon.

First is the development of procedures for the microinjection of solutions into cells like fibroblasts (Taylor and Wang, 1978; Taylor *et al.*, 1980a,b) and macrophages (Amato *et al.*, 1983) without apparent impairment of cell viability or physiological responses. As we have tried to point out, activation of the motile apparatus is likely to be localized to specific regions of the cell cytoplasm. The elegant studies already performed by Taylor and his colleagues with aqueorin and fluorescein–actin-injected cells described previously demonstrate the power and utility of these procedures. Microinjection of specific antibodies against selected cytoskeletal proteins like gelsolin, for example, should help to more clearly elucidate the role of these proteins in chemotaxis.

Finally, the recent development of procedures to prepare neutroplasts or enuleated neutrophils should help resolve several long-standing problems. These "bags" of cytosol are largely free of organelles and microtubles (Roos *et al.*, 1983; Korchak *et al.*, 1983a).

As an example, Swanson and Taylor (1982) cut *Dictyostelium discoideum* amebae into a nucleated and an enucleated piece. They found that the nucleated fragment could move toward cAMP-loaded micropipettes in a normal manner.

The enucleated piece responded by extending pseudopods showing that it was viable and responsible to cAMP; however, this cell fragment could not move toward the source of chemoattractant in a coordinated manner. If this result is generally applicable, then it may suggest that fully coordinated directed response requires the presence of an organelle like the Golgi complex, for example, or perhaps an intact centriole organizing center and its associated microtubular apparatus.

ACKNOWLEDGMENTS. The authors gratefully acknowledge Drs. Charles G. Cochrane and S. J. Singer for their helpful comments and discussions and thank Mrs. Monica Bartlett and Ms. Nancy McCarthy for typing the manuscript. Special thanks are extended to Dr. Manfred Schmitt for providing the scanning electron micrographs.

REFERENCES

Abercrombie, M., Heaysman, J. E. M., and Pegrum, S. M., 1970, The locomotion of fibroblasts in culture. III. Movements of particles on the dorsal surface of the leading lamella, *Exp. Cell Res.* **62**:389.

Adelstein, R. S., and Eisenberg, E., 1980, Regulation and kinetics of the actin-myosin-ATP interaction, *Ann. Rev. Biochem.* **49**:921.

Allan, R. B., and Wilkinson, P. C., 1978, A visual analysis of chemotactic and chemokinetic locomotion of human neutrophilic leucocytes. Use of a new chemotaxis assay with *candida albicans* as a gradient source, *Exp. Cell Res.* **111**:191.

Allen, R. A., Jesaitis, A. J., Sklar, L. A., Cochrane, C. G., and Painter, R. G., 1984, Physiochemical properties of the N-formyl peptide receptor of human neutrophils, *Fed. Proc.* **43**:1.

Amato, P. A., Unanue, E. R., and Taylor, D. L., 1983, Distribution of actin in spreading macrophages: A comparative study on living and fixed cells, *J. Cell. Biol.* **96**:750.

Anderson, D. C., Wible, L. J., Hughes, B. J., Smith, C. W., and Brinkley, B. R., 1982, Cytoplasmic microtubules in polymorphonuclear leukocytes: Effect of chemotactic stimulation and colchicine, *Cell* **31**:719.

Aswanikumar, S., Corcoran, B., Schiffmann, E., Day, A. R., Freer, R. J., Showell, H. J., Becker, E. L., and Pert, C. B., 1977, Demonstration of a receptor on rabbit neutrophils for chemotactic peptides, *Biochim. Biophys. Res. Commun.* **74**:810.

Autrum, H., 1981, Light and dark adaptation in invertebrates, in: *Handbook of Sensory Physiology,* Volume VIIC (H. Autrum, ed.), Springer, Berlin.

Ball, E. H., and Singer, S. J., 1982, Mitochondria are associated with microtubules and not with intermediate filaments in cultured fibroblasts, *Proc. Natl. Acad. Sci., USA* **79**:123.

Becker, E. L., 1979, A multifunctional receptor on the neutrophil for synthetic chemotactic oligopeptides, *J. Reticuloendothelial Soc.* **26**:701.

Bergman, K., Burke, P. V., Cerda–Olmedo, E., David, C. N., Delbruck, M., Foster, K. W., Goodell, E. W., Heisenberg, M., Meissner, G., Zalokar, M., Dennison, D. S., and Shropshire, W., Jr., 1969, Phycomyces, *Bacteriol. Rev.* **33**:99.

Bergmann, J. E., Kupfer, A., and Singer, S. J., 1984, Membrane insertion at the leading edge of motile fibroblasts, *Proc. Natl. Acad. Sci., USA* **80:**1367.

Berlin, R. D., and Oliver, J. M., 1978, Analogous ultrastructure and surface properties during capping and phagocytosis in leukocytes, *J. Cell Biol.* **77:**789.

Bourguignon, L. Y. W., and Singer, S. J., 1977, Transmembrane interactions and mechanism of capping of surface receptors by their specific ligands, *Proc. Natl. Acad. Sci, USA* **74:**5031.

Boyd, A., and Simon, M., 1982, Bacterial chemotaxis. *Ann. Rev. Physiol.* **44:**501.

Burgess, D. R., and Schroeder, T. E., 1977, Polarized bundles of actin filaments with microvilli of fertilized sea urchin eggs, *J. Cell Biol.* **74:**1032.

Burridge, K., and Feramisco, J., 1980, Microinjection and localization of a 130K protein in living fibroblasts: A relationship to actin and fibronectin, *Cell* **19:**587.

Burridge, K., and Feramisco, J. R., 1981, α-Actinin and vinculin from non-muscle cells: Calcium sensitive interactions with actin, *Cold Spring Harb. Symp. Quant. Biol.* **46:**587.

Byers, R. H., and Porter, K. R., 1977, Transformations in the structure of the cytoplasmic ground substance in erythropores during pigment aggregation and dispersion, *J. Cell Biol.* **75:**541.

Carlsson, L., Nystrom, L.–E., Sundkvist, I., Markey, F., and Lindberg, U., 1977, Actin polymerizability is influenced by profilin, a low molecular weight proten in nonmuscle cells, *J. Mol. Biol.* **115:**465.

Carlsson, L., Markey, F., Blikstad, I., Persson, T., and Lindberg, U., 1979, Reorganization of actin in platelets stimulated by thrombin as measured by DNAse I inhibition assay. *Proc. Natl. Acad. Sci., USA* **76:**6376.

Carp, H., 1982, Mitochondrial *N*-formyl methionyl proteins as chemoattractants for neutrophils, *J. Exp. Med.* **155:**264.

Carroll, R. C., and Gerrard, J. M., 1982, Phosphorylation of platelet actin binding protein during platelet activation, *Blood* **59:**466.

Cramer, E. B., and Gallin, J. I., 1979, Localization of submembraneous cations to the leading end of human neutrophils during chemotaxis, *J. Cell Biol.* **82:**369.

Davis, B. H., Walter, R. J., Pearson, C. B., Becker, E. L., and Oliver, J. M., 1982, Membrane activity and topography of F-Met-Leu-Phe-treated polymorphonuclear leukocytes, *Am. J. Pathol.* **108:**206.

Delbruck, M., and Reichhardt, W., 1956, System analysis for the light growth reactions of phycomyces, in: *Cellular Mechanisms in Differentiation and Growth,* Volume 14 (D. Rudnick, ed.), Princeton University Press, Princeton, New Jersey, p. 3.

Delbruck, M., and Varju D., 1961, Photoreactions in phycomyces. Responses to the stimulation of narrow test areas with ultraviolet light. *J. Gen. Physiol.* **44:**1177.

Dennison, D. S., and Bozof, R. P., 1973, Phototropism and local adaptation in *Phycomyces sporangiophores, J. Gen. Physiol.* **62:**157–168.

Dolmatch, B., and Niedel, J., 1983, Formyl peptide chemotactic receptor: Evidence for an active proteolytic fragment, *J. Biol. Chem.* **258:**7570.

Edds, K. T., 1977, Microfilament bundles. I. Formation with uniform polarity, *Exp. Cell Res.* **108:**452.

Elferink, J. G. R., Deierkauf, M., and Riemersma, J. C., 1982, Involvement of calmodulin in granulocyte chemotaxis: The effect of calmodulin inhibitors, *Res. Commun. Chem. Pathol. Pharmacol.* **38:**77.

Elgsater, A., and Branton, D., 1974, Intramembraneous particle aggregation in erythrocyte ghosts. I. The effects of protein removal. *J. Cell. Biol.* **63:**1018.

Fechheimer, M., and Zigmond, S. H., 1983, Changes in cytoskeletal proteins of polymorphonuclear leukocytes induced by chemotactic peptides, *Cell Motility* **3:**349.

Flanagan, J., and Koch, C. L. E., 1978, Cross-linked surface Ig attached to actin, *Nature (London)* **273**:278.

Fletcher, M. P., and Gallin, J. I., 1982, Human neutrophils contain an intracellular pool of putative receptors for the chemoattractant N-formyl methionylleucylphenylaline with a density of specific granules, *J. Cell Biol.* **95**:444a.

Fletcher, M. P., Seligmann, B. E., and Gallin, J. I., 1982, Correlation of human neutrophil secretion, chemoattractant receptor mobilization and enhanced functional capacity, *J. Immunol.* **128**:941.

Fox, J. E. B., and Phillips, D. R., 1982, Role of phosphorylation in mediating the association of myosin with the cytoskeletal structures of human platelets, *J. Biol. Chem.* **257**:4120.

Freer, R. J., 1981, Antagonists of the formylated peptide chemoattractants: structure-activity comparisons with formyl-methionyl-leucyl-phenyl-alanine-OH, *KROC Found. Ser.* **14**:161.

Freer, R. J., Day, A. R., Radding, J. A., Schiffman, E., Aswanikumar, S., Showell, H. J., and Becker, E. L., 1980, Further studies on the structural requirements for synthetic peptide chemoattractants, *Biochemistry* **19**:2404.

Freer, R. J., Day, A. R., Muthukumaraswamy, N., Pinon, D., Wu, A., Showell, H. J., and Becker, E. L., 1982, Formyl peptide chemoattractants: A model of the receptor on rabbit neutrophils, *Biochemistry* **21**:257.

Frey–Wyssling, A., 1957, *Macromolecules in Cell Structure*, Harvard University Press, Cambridge, Massachusetts.

Gallin, J. I., Wright, D. G., and Schiffmann, E., 1978, Role of secretory events in modulating human neutrophil chemotaxis, *J. Clin. Invest.* **62**:1364.

Gallin, J. I., Gallin, E. K., and Schiffmann, E., 1979, Mechanism of Leukocyte chemotaxis, in: *Advances in Inflammation Research* (G. Weissmann, ed.), Raven Press, New York, pp. 123.

Galvin, N. J., Stockhausen, D., Meyers–Hutchins, B. L., and Frazier, W. A., 1984, Association of the cyclic AMP chemotaxis receptor with the detergent-insoluble cytoskeleton of *Dictyostelium discoideum*, *J. Cell. Biol.* **98**:584.

Geiger, B., 1979, A 130K protein from chicken gizzard: Its localization at the termini of microfilament bundles in cultured chicken cells, *Cell* **18**:193.

Geiger, B., Tokuyasu, K. T., Dutton, A. H., and Singer, S. J., 1980, Vinculin, an intracellular protein localized at specialized sites where microfilament bundles terminate at cell membranes, *Proc. Natl. Acad. Sci., USA* **77**:4127.

Gerish, G., 1982, Chemotaxis in *Dictyostelium, Ann. Rev. Physiol.* **44**:535.

Gerish, G., and Keller, H. U., 1981, Chemotactic reorientation of granulocytes stimulated with micropipettes containing F-Met-Leu-Phe, *J. Cell Sci.* **52**:1.

Glenney, J. R., and Weber, K., 1981, Uncoupling of the Ca^{++}-dependent F-actin severing activity from the F-actin bundling activity of villin by mild *in vitro* proteolysis, *Proc. Natl. Acad. Sci., USA* **78**:2810.

Goetzl, E. J., Foster, D. W., and Goldman, D. W., 1981, Isolation and partial characterization of membrane protein constituents of human neutrophil receptors for chemotactic formyl methionyl peptides, *Biochemistry* **20**:5717.

Goldman, R. D., Schloss, J. A., and Starger, J. M., 1976, Organizational changes of actin-like microfilaments during animal cell movement, in: *Cell Motility, Part A*, Volume 3 (R. D. Goldman, T. Pollard, and J. Rosenbaum, eds.), Cold Spring Harbor Conferences on Cell Proliferation, p. 217.

Hartwig, J. H., and Stossel, T. P., 1975, Isolation and properties of actin, myosin and a new actin binding protein in rabbit alveolar macrophages, *J. Biol. Chem.* **250**:5696.

Hartwig, J. H., Davies, W. A., and Stossel, T. P., 1977, Evidence for contractile protein translocation in macrophage spreading, phagocytosis and phagolysosome formation, *J. Cell Biol.* **75**:956.

Hoffstein, S. T., 1979, Ultrastructural demonstration of calcium loss from local regions of the plasma membrane of surface-stimulated human granulocytes, J. Immunol. 123:1395.

Hoffstein, S., Goldstein, I. M., and Weissmann, G., 1977, Role of microtubule assembly in lysosomal enzyme secretion from human polymorphonuclear leukocytes. A reevaluation, J. Cell Biol. 73:242.

Hoffstein, S. T., Friedman, R. S., and Weissmann, G., 1982, Degranulation, membrane addition and shape change during chemotactic factor-induced aggregation of human neutrophils, J. Cell Biol. 95:234.

Jesaitis, A. J., and Yguerabide, J., 1980, Lateral mobility of plasma membrane lipids and Na^+-K^+-ATPase in cultured canine kidney cells, Fed. Proc. 39:2050.

Jesaitis, A. J., Naemura, J. R., Painter, R. G., Sklar, L. A., and Cochrane, C. G., 1982a, Intracellular localization of N-formyl chemotactic receptor and Mg^{+2} dependent ATPase in human granulocytes, Biophys. Biochim. Acta 719:556.

Jesaitis, A. J., Naemura, J. R., Painter, R. G., Schmitt, M., Sklar, L. A., and Cochrane, C. G., 1982b, The fate of the N-formylated chemotactic peptide receptor in stimulated human granulocytes: Subcellular fractionation studies, J. Cell. Biochem. 20:143.

Jesaitis, A. J., Naemura, J. R., Painter, R. G., Sklar, L. A., and Cochrane, C. G., 1983, The fate of the N-formylated chemotactic peptide in stimulated human granulocytes: Subcellular fractionation studies, J. Biol. Chem. 258:1968.

Jesaitis, A. J., Naemura, J. R., Sklar, L. A., Cochrane, C. G., and Painter, R. G., 1984, Rapid modulation of N-formyl chemotactic peptide receptors on the surface of human granulocytes: Formulation of high-affinity ligand-receptor complexes in transient association with the cell cytoskeleton, J. Cell. Biol., 98: 1378.

Keller, H. U., and Bessis, M., 1975, Migration and chemotaxis of anucleate cytoplasmic leukocyte fragments, Nature (London) 258:73.

Keller, H. U., Wissler, J. H., Hess, M. W., and Cottier, H., 1977, Relation between stimulus intensity and neutrophil chemotactic response, Experientia 33:534.

Keller, H. U., Wissler, J. H., Hess, M. W., and Cottier, H., 1978, Distinct chemokinetic and chemotactic responses in neutrophil granulocytes, Eur. J. Immunol. 8:1.

Koch, G. L. E., 1981, The anchorage of cell surface receptors to the cytoskeleton, in: Symposium of 2nd International Congress on Cell Biology (H. G. Schweiger, ed.), Berlin, Springer, p. 321.

Koo, C., Lefkowitz, R. J., and Snyderman, R., 1983, Guanine nucleotides modulate the binding affinity of oligopeptide chemoattractant receptor on human polymorphonuclear leukocytes, J. Clin. Invest. 72:748.

Koppel, K. E., Sheetz, M. P., and Schindler, M., 1981, Matrix control of protein diffusion in biological membranes, Proc. Natl. Acad. Sci., USA 78:3576.

Korchak, H. M., Roos, D., Giedd, K. N., Wynkoop, E. M., Vienne, K., Rutherford, L. E., Buyon, J. P., Rich, A. M., and Weissmann, G., 1983a, Granulocytes without degranulation: Neutrophil function in granule-depleted cytoplasts, Proc. Natl. Acad. Sci., USA 80:4968.

Korchak, H. M., Vienne, K., Wilkenfeld, C., Roberts, C. S., Rutherford, L. E., Haines, K. A., and Weissmann, G., 1983b, The role of calcium in neutrophil activation-mobilization of multiple calcium pools, J. Cell Biol. 97:605a.

Korn, E. D., 1978, Biochemistry of actomyosin-dependent cell motility, Proc. Natl. Acad. Sci., USA 75:588.

Korn, E. D., 1982, Actin polymerization and its regulation by proteins from non -muscle cells, Physiol. Rev. 62:672.

Koshland, D. E. Jr., 1981, Biochemistry of sensing and adaptation in a simple bacterial system, Ann. Rev. Biochem. 50:765.

Kupfer, A., Louvard, D., and Singer, S. J., 1982, Polarization of the Golgi apparatus and the microtubule-organizing center in cultured fibroblasts at the edge of an experimental wound, *Proc. Natl. Acad. Sci., USA* **79**:2603.

Lazarides, E., and Burridge, K., 1975, α-Actinin: Immunofluorescent localization of a muscle structural protein in nonmuscle cells, *Cell* **6**:289.

Lazarides, E., and Lindberg, U., 1974, Actin is the naturally occurring inhibitor of deoxyribonuclease. I. *Proc. Natl. Acad. Sci. USA* **71**:4742.

Luna, E. J., Fowler, V. M., Swanson, J., Branton, D., and Taylor, D. L., 1981, A membrane cytoskeleton from *Dictyostelium discoideum*. I. Identification and partial purification of an actin binding activity, *J. Cell Biol.* **88**:396.

Malawista, S. E., and Chevance, A. de B., 1982, The cytokineplast: Purified, stable and functional motile machinery from human blood polymorphonuclear leukocytes, *J. Cell. Biol.* **95**:960.

Malech, H. C., Root, R. K., and Gallin, J. I., 1976, Centriole, microtubule and microfilament orientation during human polymorphonuclear leukocyte chemotaxis, *Clin. Res.* **24**:314A.

Markey, F., Lindberg, U., and Eriksson, L., 1978, Human platelets contain profilin, a potential regulator of actin polymerizability, *FEBS Lett.* **88**:75.

Mescher, M. F., Jose, M. J. L., and Balk, S. P., 1981, Actin-containing matrix associated with the plasma membrane of murine tumor and lymphoid cells, *Nature (London)* **289**:139.

Mooseker, M. S. and Tilney, L. G., 1975, The organization of an actin filament-membrane complex: Filament polarity and membrane attachment in the microvilli of intestinal epithelial cells, *J. Cell Biol.* **67**:725.

Naccache, P. H., Showell, H. J., Becker, E. L., and Sha'afi, R. I., 1977a, Changes in ionic movements across rabbit leukocyte membranes during lysosomal enzyme release, *J. Cell Biol.* **76**:635.

Naccache, P. H., Showell, H. J., Becker, E. L., and Sha'afi, R. I., 1977b, Sodium, potassium, and calcium transport across rabbit polymorphonuclear leukocyte membranes. Effect of chemotactic factor, *J. Cell. Biol.* **73**:428.

Naccache, P. H., Volpi, M., Showell, H. J., Becker, E. L., and Sha'afi, R. I., 1979, Chemotactic factor-induced release of membrane calcium in rabbit neutrophils, *Science* **203**:461.

Nath, J., and Flavin, M., 1983, Tubulin heterogeneity revealed by tyrosinolation *in vivo*, *J. Cell Biol.* **91**:19004a.

Nath, J., Flavin, M., and Gallin, J. I., 1982, Tubulin tyrosinolation in human polymorphonuclear leukocytes: Studies in normal subjects and in patients with the Chediak–Higashi Syndrome, *J. Cell. Biol.* **95**:519.

Nicolson, G. L., 1973, Anionic sites of human erythrocyte membranes. I. Effects of trypsin, phospholipase C, and pH on the topography of bound positively charged colloidal particles, *J. Cell. Biol.* **57**:373.

Nicolson, G. L., and Painter, R. G., 1973, Anionic sites of human erythrocyte membranes. II. Antispectrin-induced transmembrane aggregation of the binding sites for positively charged colloidal particles, *J. Cell. Biol.* **59**:395.

Niedel, J. E., 1981, Detergent solubilization of the formyl peptide chemotactic receptor, *J. Biol. Chem.* **256**:9295.

Niedel, J. E., and Cuatrecasas, P., 1980, Formyl peptide chemotactic reception of leukocytes and macrophages, *Curr. Top. Cell Reg.* **17**:137.

Niedel, J., Wilkinson, and Cuatrecasas, P., 1979a, Receptor-mediated uptake and degradation of ^{125}I-chemotactic peptide by human neutrophils, *J. Biol. Chem.* **254**:10700.

Niedel, J. E., Kahane, I., and Cuatrecasas, P., 1979b, Receptor mediated internalization of fluorescent chemotactic peptide by human neutrophils, *Science* **205**:1412.

Niedel, J. E., Davis, J., and Cuatrecasas, P., 1980, Covalent affinity labelling of the formyl peptide chemotactic receptor, *J. Biol. Chem.* **255**:7063.

O'Farrell, P. H., 1975, High resolution two dimensional electrophoresis of proteins, *J. Biol. Chem.* **250**:4007.

Oliver, J. M., Krawiec, J. A., and Becker, E. L., 1978, The distribution of actin during chemotaxis in rabbit neutrophils, *J. Reticuloendothel. Soc.* **24**:697.

Painter, R. G., and Ginsberg, M. H., 1982, Concanavalin A induces interactions between surface glocoproteins and the platelet cytoskeleton, *J. Cell. Biol.* **92**:565.

Painter, R. G., and McIntosh, A. T., 1979, The regional association of actin and myosin with sites of particle phagocytosis, *J. Supramol. Struct.* **12**:369.

Painter, R. G., Whisenand, J., and McIntosh, A. T., 1981, Effects of cytochalasin B on actin and myosin association with particle binding sites in mouse macrophages: Implications with regard to the mechanism of action of the cytochalasins, *J. Cell. Biol.* **91**:373.

Painter, R. G., Schmitt, M., Jesaitis, A. J., Sklar, L. A., Preissner, K., and Cochrane, C. G., 1982, Photoaffinity labeling of the *N*-formyl peptide receptor of human polymorphonuclear leukocytes, *J. Cell. Biochem.* **20**:913.

Painter, R. G., Allen, R. A., Sklar, L. A., Schmitt, M., Cochrane, C. G., and Jesaitis, A. J., Intracellular processing of N-formylated chemotactic peptide receptors by human neutrophils, submitted.

Painter, R. G., Sklar, L. A., Jesaitis, A. J., Schmitt, M., and Cochrane, C. G., 1984, Activation of neutrophils by N-formyl chemotactic peptides, *Fed. Proc.,* in press.

Pilch, P. F., and Czech, M. P., 1979, Interaction of cross-linking agents with the insulin effector system of isolated fat cells: Covalent linkage of ^{125}I-insulin to a plasma membrane receptor protein of 140,000 daltons, *J. Biol. Chem.* **254**:3375.

Pilch, P. F., and Czech, M. P., 1980, The subunit structure of the high affinity insulin receptor: Evidence for a disulfide-linked receptor complex in fat cell and liver plasma membranes, *J. Biol. Chem.* **255**:1722.

Ramsey, W. S., 1972, Analysis of individual leukocyte behavior during chemotaxis, *Exp. Cell. Res.* **70**:129.

Rao, K. M. K., and Varani, J., 1982, Actin polymerization induced by chemotactic peptide and concanavalin A in rat neutrophils, *J. Immunol.* **129**:1605.

Roos, D., Voetman, A. A., and Meerhof, L. J., 1983, Functional activity of enucleated human polymorphonuclear leukocytes, *J. Cell. Biol.* **97**:368.

Rosenberg, S., Stracher, A., and Burridge, K., 1981, Isolation and characterization of a calcium-sensitive α-actinin-like protein from human platelet cytoskeletons, *J. Biol. Chem.* **256**:12986.

Schiffmann, E., 1982, Leukocyte chemotaxis, *Ann Rev. Physiol.* **44**:553.

Schiffmann, E., Corcoran, B. A., and Wahl, S. M., 1975, *N*-formyl methionyl peptides as chemoattractants for leukocytes, *Proc. Natl. Acad. Sci., USA* **72**:1059.

Schlessinger, J., Koppel, D. E., Axelrod, D., Jacobson, K., Webb, W. W., and Elson, E. L., 1976, Lateral transport on cell membranes: Mobility of Concanavalin A receptors on myoblasts, *Proc. Natl. Acad. Sci., USA* **73**:2409.

Schmitt, M., Painter, R. G., Jesaitis, A. J., Preissner, K., Sklar, L. A., and Cochrane, C. G., 1983, Photoaffinity labeling of the *N*-formyl peptide receptor binding site of intact human polymorphonuclear leukocytes. Evaluation of a label as suitable to follow the fate of the receptor-ligand complex, *J. Biol. Chem.* **258**:649.

Sefton, B. M., Hunter, T., Ball, E. H., and Singer, S. J., 1981, Vinculin: A cytoskeletal target of the transforming protein of Rous sarcoma virus, *Cell* **24**:165.

Seligmann, B. E., Fletcher, M. P., and Gallin, J. I., 1982, Adaptation of human neutrophil responsiveness to the chemoattractant N-formylmethionylleucylphenylalanine, *J. Biol. Chem.* **257**:6280.

Sellers, J. R., Pato, M. D., and Adelstein, R. S., 1981, Reversible phosphorylation of smooth muscle myosin, heavy meromyosin and platelet myosin, *J. Biol. Chem.* **256**:13137.

Sheetz, M. P., Schindler, M., and Koppel, D. G., 1980, Lateral mobility of integral membrane proteins is increased in spherocytic erythrocytes, *Nature* **285**:510.

Shelterline, P., and Hopkins, C. R., 1981, Transmembrane linkage between surface glycoproteins and components of the cytoplasm in neutrophil leukocytes, *J. Cell. Biol.* **90**:743.

Showell, H. J., Freer, R. J., Zigmond, S. H., Schiffmann, E., Aswanikumar, S., Corcoran, B., and Becker, E. L., 1976, The structure–activity relations of synthetic peptides as chemotactic factors and inducers of lysozomal enzyme secretion for neutrophils, *J. Exp. Med.* **143**: 1154.

Sklar, L. A., 1984, Sensory transduction and ligand-receptor dynamics in the human neutrophil, *Fed. Proc.* **43**:5.

Sklar, L. A., Jesaitis, A. J., Painter, R. G., and Cochrane, C. G., 1981a, The kinetics of neutrophil activation: The response to chemotactic peptides depends upon whether ligand-receptor interaction is rate-limiting, *J. Biol. Chem.* **256**:9909.

Sklar, L. A., Oades, Z. G., Jesaitis, A. J., Painter, R. G., and Cochrane, C. G., 1981b, Fluoresceinated chemotactic peptide and high affinity antibody to fluorescein as a probe of the temporal characteristics of neutrophil stimulation, *Proc. Natl. Acad. Sci., USA* **78**:7540.

Sklar, L. A., Finney, D. A., Oades, Z. G., Jesaitis, A. J., Painter, R. G., and Cochrane, C. G., 1984a, The dynamics of ligand-receptor interactions. Real-time analyses of association, dissociation and internalization of an *N*-formyl peptide and its receptors on the human neutrophil, *J. Biol. Chem.* **259**:5661.

Sklar, L. A., Jesaitis, A. J., and Painter, R. G., 1984b, The neutrophil N-formyl peptide receptor: The dynamics of ligand/receptor interactions and their relationship to cellular responses, in: *Contemporary Topics in Immunobiology*, Volume 14 (R. Snyderman, ed.,) Plenum Press, New York, p. 29.

Small, J. V., Isenberg, G., and Celis, J. E., 1978, Polarity of actin at the leading edge of cultured cells, *Nature (London)* **272**:638.

Snyderman, R., and Goetzl, E. J., 1981, Molecular and cellular mechanisms of leukocyte chemotaxis, *Science* **213**:830.

Southwick, F. S., and Stossel, T. P., 1983, Contractile proteins in leukocyte function, *Sem. Hematol.* **20**:305.

Stendahl, O. I., Hartwig, J. H., Brotschi, E. A., and Stossel, T. P., 1980, Distribution of actin binding protein and myosin in macrophages during spreading and phagocytosis, *J. Cell Biol.* **84**:215.

Stossel, T. P., 1978, The mechanism of leukocyte locomotion, in: *Leukocyte Chemotaxis: Methods, Physiology and Clinical Implications* (J. I. Gallin and P. G. Quie, eds.), Raven Press, New York, p. 143.

Stossel, T. P., Hartwig, J. H., Yin, H.–L., and Zaner, K. S., 1981, Structure of the cortical cytoplasm. *Cold Spring Harb. Symp. Quant. Biol.* **46**:569.

Sullivan, S. J., and Zigmond, S. H., 1982, Asymmetric receptor distribution on PMNs, *J. Cell. Biol.* **95**:418a.

Swanson, J. A., and Taylor, D. L., 1982, Local and spatially coordinated movements in Dictyostelium discoideum amoebae during chemotaxis, *Cell* **28**:225.

Taylor, D. L., and Condeelis, J. S., 1979, Cytoplasmic structure and contractibility in ameboid cells, *Int. Rev. Cytol.* **56**:57.

Taylor, D. L., and Fechheimer, M., 1982, Cytoplasmic structure and contractility: The solation–contraction coupling hypothesis, *Phil. Trans. R. Soc. London B.* **299**:185.

Taylor, D. L., and Wang, Y. L., 1978, Molecular cytochemistry: Incorporation of fluorescently labeled actin into living cells. *Proc. Natl. Acad. Sci., USA* **75**:857.

Taylor, D. L., Hellewell, S. B., Virgin, H. W., and Heiple, J. M., 1979, The solation-contraction coupling hypothesis of cell movements, in: *Cell Motility: Molecules and Organization* (S. Hatano, H. Ishikawa, and H. Sato, eds.), University of Tokyo Press, Tokyo, p. 363.

Taylor, D. L., Wang, Y. L., and Heiple, J., 1980a, The contractile basis of ameboid movement, VII. The distribution of fluorescently labeled actin in living amoebas, *J. Cell Biol.* **86:**590.

Taylor, D. L., Blinks, J. R., and Reynolds, G., 1980b, Contractile basis of ameboid movement. VIII. Aequorin luminescence during ameboid movement, endocytosis and capping, *J. Cell Biol.* **86:**599.

Taylor, D. L., Heiple, J., Wang, Y.-L., Luna, E. J., Tanasugarn, L., Brier, J., Swanson, J., Fechheimer, M., Amato, P., Rockwell, M., and Daley, G., 1981, Cellular and molecular aspects of ameboid movement, *Cold Spring Harb. Symp. Quant. Biol.* **46:**101.

Tsien, R. Y., Pozzan, T., and Rink, T. J., 1982, Calcium homeostasis in intact lymphocytes: Cytoplasmic free calcium monitored with a new, intracellularly trapped fluorescent indicator, *J. Cell. Biol.* **94:**325.

Vale, R. D., and Shooter, E. M., 1982, Alteration of binding properties and cytoskeletal attachment of nerve growth factor receptors in PC12 cells by wheat germ agglutinin, *J. Cell. Biol.* **94:**710.

Valerius, N. H., Stendahl, O. I., Hartwig, J. H., and Stossel, T. P., 1981, Distribution of actin-binding protein and myosin in polymorphonuclear leukocytes during locomotion and phagocytosis, *Cell* **24:**195.

Valerius, N. H., Stendahl, O. I., Hartwig, J. H., and Stossel, T. P., 1982, Distribution of actin binding protein and myosin in neutrophils during chemotaxis and phagocytosis, *Adv. Exp. Med. Biol.* **141:**19.

Wallach, D., Davies, P. J. A., and Pastan, I., 1978, Cyclic AMP dependent phosphorylation of filamin in smooth muscle, *J. Biol. Chem.* **253:**4739.

Walter, R. J., and Marasco, W. A., 1983, Localization of ^{125}I-NfNle-leu-phe-nle-tyr-lys on rabbit PMN, *J. Cell. Biol.* **97:**1590a.

Walter, R. J., Berlin, R. D., and Oliver, J. M., 1980, Asymmetric Fc receptor distribution of human PMN oriented in a chemotactic gradient, *Nature* **286:**724.

Wang, K., 1977, Filamin a new high molecular weight protein found in smooth muscle and non-muscle cells. Purification and properties of chicken gizzard filamin, *Biochemistry* **16:**1857.

Wang, K., Ash, J. F., and Singer, S. J., 1975, Filamin, a new high molecular weight protein found in smooth muscle cells, *Proc. Natl. Acad. Sci., USA* **72:**4483.

Wang, L.-L., and Bryan, J., 1981, Isolation of calcium-dependent platelet proteins that interact with actin, *Cell* **25:**637.

Ward, P. A., 1982, The chemotaxis system, *Monogr. Pathol.* **23:**54.

Webb, W. W., Barak, L. S., Tank, A. W., and Wu, E.-S., 1981, Molecular mobility on the cell surface, *Biochem. Soc. Symp.* **46:**191.

White, J. R., Naccache, P. H., and Sha'afi, R. I., 1982, The synthetic chemotactic peptide formyl-methionyl-leucyl-phenylalanine causes an increases in actin associated with the cytoskeleton in rabbit neutrophils, *Biochem. Biophys. Res. Commun.* **108:**1144.

Wilkinson, P. C., 1974, *Chemotaxis and Inflammation,* Churchill-Livingstone, Edinburgh.

Wilkinson, P. C., Michl, J., and Silverstein, S. C., 1980, Receptor distribution in locomoting neutrophils, *Cell Biol. Int. Rep.* **4:**736.

Williams, L. T., Snyderman, R., Pike, M. C., and Lefkowitz, R. J., 1977, Specific receptor sites for chemotactic peptides on human polymorphonuclear leukocytes, *Proc. Natl. Acad. Sci., USA* **74:**1204.

Woda, B. A., Yguerabide, J., and Feldman, J. D., 1980, The effect of local anesthetics on the lateral mobility of lymphocyte membrane proteins, *Exp. Cell Res.* **126:**327.

Wright, D. G., and Gallin, J. I., 1979, Secretory responses of human neutrophils: exocytosis of specific (secondary) granules by human neutrophils during adherence *in vitro* and during exudation *in vivo*, *J. Immunol.* **123**:285.

Yeltman, D. R., Jung, G., and Carraway, K. L., 1981, Isolation of α-actinin from sarcoma 180 ascites cell plasma membranes and comparison with smooth muscle α-actinin, *Biochim. Biophys. Acta* **688**:201.

Yin, H. L., Zaner, K. S., and Stossel, T. P., 1980, Ca^{2+} control of actin gelation, *J. Biol. Chem.* **255**:9494.

Yin, H. L., Albrecht, J. H., and Faltoum, A., 1981, Identification of gelsolin, a Ca^{2+} dependent regulatory protein of actin gel–sol transformation and its intracellular distribution in a variety of cells and tissues, *J. Cell. Biol.* **91**:901.

Yip, C. C., Yeung, C. W. T., and Moule, M. L., 1982, Subunit structure of insulin receptor of rat adipocytes as demonstrated by photoaffinity labeling, *Biochemistry* **21**:2940.

Yuli, I., and Snyderman, R., 1983, Rapid perpendicular light scattering (LS): a previously unrecognized response of human neutrophils (PMNS) to chemoattractants (CTX), *Clin. Res.* **31**:379A.

Zigmond, S. H., 1981, Consequences of chemotactic peptide receptor modulation for leukocyte orientation, *J. Cell. Biol.* **88**:644.

Zigmond, S. H., and Sullivan, S. J., 1979, Sensory adaptation of leukocytes to chemotactic peptides, *J. Cell Biol.* **82**:517.

Zigmond, S. H., Levitsky, H. J., and Kreel, B. J., 1981, Cell polarity: An examination of its behavioral expression and its consequences for polymorphonuclear leukocyte chemotaxis, *J. Cell. Biol.* **89**:585.

Zigmond, S. H., Sullivan, S. J., and Lauffenburger, D. A., 1982, Kinetic analysis of chemotactic receptor modulation, *J. Cell. Biol.* **92**:34.

3

THE INTERACTIONS OF CELLS WITH EXTRACELLULAR MATRIX COMPONENTS

Kenneth M. Yamada and Steven K. Akiyama

1. INTRODUCTION

The interactions of extracellular materials with the cell surface are important facets of the regulation of cell morphology and metabolism. Although it is not simple to unravel the complex actions and interrelationships of these molecules comprising the extracellular microenvironment of cells, their importance is now documented by a rapidly expanding literature (for recent reviews, see Hay, 1981; Chen, 1981; Kleinman et al., 1981; Aplin and Hughes, 1982; Grinnell, 1983; Yamada, 1983a,b). Many of the major molecules that constitute extracellular matrices have been purified. Some of the isolated molecules can have dramatic regulatory effects on cell morphology and polarity, organization of the cytoskeleton, and production of differentiated cell products. For example, the addition of purified cellular fibronectin to cells deficient in fibronectin can restore a more normal cell shape, adhesiveness, cell surface architecture, and organization of intracellular actin microfilaments (Yamada et al., 1976a,b; Willingham et al., 1977; Ali et al., 1977). If certain embryonic epithelial cells are stripped of their basal lamina, they cease morphogenesis and begin blebbing (Banerjee et al., 1977; Sugrue and Hay, 1981); treatment with purified fibronectin, laminin, or collagen can sometimes restore normal cell surface morphology and organization of microfilaments (Sugrue and Hay, 1981, 1982). Mammary epithelial cells

Kenneth M. Yamada and Steven K. Akiyama • Membrane Biochemistry Section, Laboratory of Molecular Biology, National Cancer Institute, National Institutes of Health, Bethesda, Maryland 20205.   Steven K. Akiyama was supported by Public Health Service grant number CA -06782 awarded by the National Cancer Institute, DHHS, and by Bethesda Research Laboratories, Inc., Gaithersburg, Maryland.

removed from their normal environment regain the ability to proliferate and to synthesize a set of differentiated products and a basal lamina if cultured within an artificial collagen gel (e.g., see Suard *et al.*, 1983, and references therein). These and many other examples of the regulatory effects of extracellular macromolecules on embryonic differentiation, cell motility, and growth demonstrate the importance of interactions of the cell surface with the extracellular milieu.

2. TYPES OF INTERACTIONS

Cells appear to interact with extracellular molecules by two general mechanisms. The simplest is by means of a direct, receptor-mediated interaction with an individual macromolecule. A second type of interaction is by an intermediary linkage molecule. For example, collagen is a major structural and scaffolding molecule with which cells appear to interact either directly at the plasma membrane or indirectly by means of specialized linkage molecules that span the gap between plasma membrane and collagen fibril (reviewed below). Cells show varying degrees of specificity for the type of extracellular molecule with which they interact, and different cells may or may not recognize the subtype of an extracellular molecule, e.g., in displaying specificity in binding to collagen types I, II, or IV. Examples of these types of interactions will be seen below.

3. EXPERIMENTAL APPROACHES

Five general approaches to analyzing the interactions of cells with extracellular molecules have been applied successfully by many investigators. Although other approaches do exist, these approaches will be described in the greatest detail because of their frequent application.

3.1. Cell Attachment to Substrates

In order to mimic the interactions of cells with the extracellular matrix, various purified extracellular molecules can be coated onto glass or plastic to function as substrates for cells in culture. The amounts of a protein adsorbed onto the substrate can be determined by measuring the binding of radioactive molecules or by immunoassays. The simplest cell-to-substrate interaction is attachment, which is measured as the percent of cells added to a dish that adhere in a given time and that remain attached after gentle washing. This type of approach has shown that cells can attach directly to native collagen, and that this attachment can be enhanced by a second molecule such as fibronectin (e.g., Linsenmayer *et al.*, 1978).

3.2. Cell Spreading on Substrates

Attachment of cells to a substrate is often accompanied by spreading of the cell body to form a flattened, fried-egg or pancake-shaped cell. Spreading of cells on a molecule adsorbed to substrates that would not otherwise support spreading is considered to indicate an interaction of the cell surface with that molecule. Cell spreading has been analyzed extensively, since it is simple to assay (reviewed by Grinnell, 1978).

3.3. Binding of Molecules to Cells

Direct determination of the interaction of an extracellular molecule with cells can sometimes be obtained by measuring the binding of a radiolabeled molecule to a monolayer of cells. For example, soluble type I collagen binds to fibroblasts, and the binding can be analyzed as for the binding of a hormone to its receptor to determine its specificity, saturability, and affinity (Goldberg, 1979). In other cases, however, binding cannot be measured easily, because the interaction requires multivalency or other factors. In this case, investigators have been forced to resort to binding the molecule to beads, and binding of the entire complex to cells can be analyzed (e.g., see Grinnell, 1980a).

3.4. Molecule–Molecule Interactions

The interactions of purified extracellular molecules with other molecules have been analyzed most extensively by immobilizing the molecules on inert supports and determining binding of other molecules to the complex. One simple approach is to trap one of the molecules on a nitrocellulose filter; its interaction with a second molecule that does not bind to the filter can be determined by measuring the amount of the second molecule which it binds on the filter. This approach has been used to analyze the binding of fibronectin to ligands (K. M. Yamada *et al.*, 1980) and of laminin to its cell surface receptor (Rao *et al.*, 1983; Malinoff and Wicha, 1983).

The most popular method for analyzing intermolecular interactions is by affinity chromatography. A molecule is coupled covalently to an inert support, e.g., agarose beads, and the binding of other molecules or fragments of molecules is then measured. This approach has been particularly successful in dissecting the organization of multifunctional molecules by determining which proteolytic fragments of the molecule bind to specific ligands attached to the column (reviewed in detail below). It has also been used to purify plasma membrane receptors for certain extracellular molecules.

3.5. Isolated Matrices

A fifth approach is to isolate intact extracellular matrices, to determine their composition and effects on cells *in vitro,* and to probe their functions by selectively removing certain constituents. Extracellular matrices isolated from dense cultures of cells (Chen *et al.,* 1978; Hedman *et al.,* 1979; Gospodarowicz *et al.,* 1982, 1983) or from tissues (Rojkind *et al.,* 1980; Weiss and Reddi, 1981; Enat *et al.,* 1983) are strikingly effective in promoting the growth and specialized secretory function of differentiated cells. One advantage of this approach is that the three-dimensional spatial organization of matrix molecules can be preserved, since spatial information may be important for certain cells (e.g., see Reddi, 1974, on the required geometry of matrix for bone induction). In addition, intact complexes of extracellular molecules are obtained to study synergistic interactions or activities of previously unidentified molecules present only in low concentrations.

4. CONCEPTS DERIVED FROM STUDIES OF CELL INTERACTIONS WITH GLASS AND PLASTIC SUBSTRATES

A major motivation for studying the function of extracellular molecules is to be able to maintain and to control differentiated cell function in tissue culture by mimicking the *in vivo* microenvironment *in vitro.* Cells plated directly onto artificial substrates such as tissue culture plastic in the absence of extracellular materials are often defective in growth or in the expression of differentiated cell functions (e.g., see Gospodarowicz *et al.,* 1982; Reid, 1982; Suard *et al.,* 1983). Although most cells can adhere directly to glass or plastic in protein-free culture medium, their attachment is often considered to be nonphysiological. Unless cells secrete an extracellular matrix, the interactions appear to involve abnormally tight apposition of the plasma membrane to the substrate, which cannot be disrupted by agents that normally disrupt cell adhesion, such as proteases and chelating agents (Pegrum and Maroudas, 1975; reviewed in Grinnell, 1978).

Cells normally attach to a proteinaceous layer adsorbed to the culture substrate, which is derived from serum proteins present in serum-containing culture media or from molecules synthesized by the cells themselves (reviewed by Grinnell, 1978). Successfully culturing cells in serum-free medium often requires the addition of "cell attachment factors" such as fibronectin to media (e.g., see Barnes and Sato, 1980). At least three factors can be isolated from serum that are involved in the adhesion and proliferation of cells in tissue culture. These proteins, fibronectin, serum-spreading factor, and chondronectin, will be described in more detail later. In addition, cellular fibronectin can be isolated from materials secreted by cultured cells.

4.1. Cell-Attachment Molecules *In Vitro*

Current concepts of cell-to-substrate adhesion often blur the distinction between cell attachment and spreading. Even though it is obvious that attachment and spreading are conceptually separable, many cell-attachment molecules also cause cell spreading on tissue culture substrates. Such spreading can involve the attachment of a large proportion of the plasma membrane to the substrate, and it may be related to the tendency of cells to bind, spread over, and metabolically respond to foreign bodies *in vivo* (Folkman and Greenspan, 1975). A role for fibronectin in regulating cell shape in normal physiological events has been suggested for endometrial fibroblasts (Grinnell *et al.*, 1982a). It is important to recall that many cells *in vivo* are not flattened like cells *in vitro*, but instead form columnar or cuboidal epithelia, or are elongated as fibroblastic or muscle cells, or remain rounded as chondrocytes.

The ability of a molecule to promote the spreading of a cell on a tissue culture dish is nevertheless useful to demonstrate its interaction with the cell surface. Not surprisingly, other molecules besides specific cell-attachment molecules can cause cell spreading. Lectins and antibodies directed against plasma membranes also cause cell spreading if coated onto plates (Grinnell and Hays, 1978b; Hughes *et al.*, 1979; Gjessing and Seglen, 1980). Even purified glycosidases and a glycosyl transferase maintained under ionic conditions restricting enzymatic activity can induce fibroblast spreading, apparently by binding to, but not cleaving, carbohydrate substrates on the cell surface (Rauvala *et al.*, 1981, 1983). On the other hand, even though hepatocytes can attach to asialoglycoproteins on dishes via their asialoglycoprotein receptors, the cells do not spread normally (Gjessing and Seglen, 1980). In addition, it has been reported that the binding of substrate-attached α_2-macroglobulin to its many receptors on fibroblasts does not increase cell attachment or spreading beyond the extent found in culture medium (Cassiman *et al.*, 1982).

Even though a number of molecules may mediate the attachment or spreading of cells, it should be emphasized that cells do not normally encounter molecules such as plant lectins, antibodies, or polylysine. The relevance of these particular experiments to physiological events is thus not clear, except to demonstrate general principles of adhesion.

4.2. Thresholds for Adhesion

Both cell attachment and cell spreading may be threshold phenomena. Increasing the density on a substrate of certain carbohydrate residues bound by hepatocytes has little effect on adhesion until the concentration reaches a critical density, beyond which cell adhesion increases rapidly (Weigel *et al.*, 1979). Similarly, the spreading of a fibroblast on the cell-attachment molecule fibronectin requires a minimum of 45,000 molecules/cell (Hughes *et al.*, 1979), and spreading on the lectin concanavalin A requires 6×10^5 molecules (Aplin and Hughes, 1981). Certain cellular mutants with defective adhesion can respond to unusually high quantities of added fibronectin (Klebe *et al.*, 1977; Pouysségur *et al.*, 1977; Norton and Izzard, 1982).

4.3. Immunological Identification of Cell-Attachment Molecules

The attachment of cells to culture substrates provides a convenient point of immunological attack. Antibodies against cell-to-substrate adhesion molecules would be expected to inhibit cell attachment to substrates. Antibodies capable of disrupting the adhesion of fibroblasts, epithelial cells, and myoblasts on plastic culture substrates have been produced, and the antigens they bind have been identified, but not yet characterized in molecular detail (Wylie *et al.*, 1979; Hsieh and Sueoka, 1980; Damsky *et al.*, 1981; Knudsen *et al.*, 1981; Greve and Gottlieb, 1982; Neff *et al.*, 1982). The plasma membrane antigens or receptors implicated in adhesion using this approach are generally found to be in the size range of 120,000–160,000 daltons. Each cell contains an average of 200,000–800,000 such molecules (Neff *et al.*, 1982). It is not yet certain whether these molecules are direct cell adhesion molecules, receptors for cell adhesion factors, or regulatory molecules. Two obvious approaches are (1) to examine whether these molecules are receptors for extracellular molecules such as fibronectin, and (2) to isolate these immunologically defined molecules and to insert them into cells defective in adhesion or into liposomes in order to determine whether they can directly mediate or modify attachment to specific substrates.

4.4. Interactions of Extracellular Molecules with the Cytoskeleton

Molecules interacting with the cell surface provide linkages between the extracellular environment and intracellular structural components. Although the existence of such connections is obvious, at present they remain relatively undefined at the molecular level. One approach has been to examine for morphological correlations, i.e., the codistribution of intracellular molecules and cell surface components. There is a clear, albeit variable, correlation between the pattern of actin microfilament bundles and extracellular fibronectin in cells cultured *in vitro,* especially in low concentrations of serum (Hynes and Destree, 1978; Singer, 1979, 1982; Chen and Singer, 1982). There is also a colocalization of intracellular vinculin, a component concentrated in adhesion plaques, and fibronectin, especially on the upper (ventral) surfaces of cells (Burridge and Feramisco, 1980; Singer and Paradiso, 1981). The addition of exogenous cellular fibronectin to transformed or mutant cells lacking the protein can promote the organization of microfilament bundles, with varying effects on focal contact formation (Willingham *et al.,* 1977; Ali *et al.,* 1977; Singer, 1982; Norton and Izzard, 1982).

Fibronectin does not appear to be present in most "focal" adhesions formed by cultured fibroblasts, which is significant in that this site is a major site of termination of microfilament bundles of cells; it is instead present in "close contacts" and thicker strands of extracellular material (Birchmeier *et al.,* 1980; Fox *et al.,* 1980; Chen and Singer, 1980, 1982; but see also Singer and Paradiso, 1981, and Hynes *et al.,* 1982). In fact, focal adhesive sites appear to be able to exclude other molecules, including fibronectin (Grinnell, 1980b; Avnur and Geiger, 1981), unless the molecule is covalently linked to the substrate (Chen and Singer, 1980). Certain apparent discrepancies in the literature concerning the presence of fibronectin in focal adhesions appear to be due to the existence of a previously unrecognized class of adhesion site, termed an "extracellular matrix (ECM)" contact, which involves linear fibrils containing fibronectin existing in very close contact with the substrate, coursing in parallel to intracellular microfilament bundles (Chen and Singer, 1982). The presence of fibronectin in close and ECM contacts may increase cell-to-substrate adhesion sufficiently to promote the formation of focal adhesions and actin microfilament organization.

There might be no specificity in the induction of intracellular microfilament bundle organization by extracellular molecules. For example, a stimulation of actin bundle formation similar to that induced by fibronectin is found by culturing cells on substrates of polylysine, native collagen, and laminin (Willingham *et al.*, 1977; Couchman *et al.*, 1983). In addition, the organization of actin microfilaments in the cortex of epithelial cells can also be produced by culturing the cells in the presence of laminin, collagen, or fibronectin (Sugrue and Hay, 1981, 1982). One interpretation of these findings is that the effects of extracellular attachment molecules on intracellular cytoskeletal organization may be indirect; i.e., any molecule that causes firm adherence of cells to a substrate will induce a similar reorganization of cytoskeletal proteins.

An alternative view is that extracellular molecules make specific interactions with specific transmembrane molecules, which make direct connection to cytoskeletal elements. The predominant intracellular molecules associated with actin microfilaments present in fibronectin-containing extracellular matrix contacts may differ from those in focal adhesions (alpha actinin versus vinculin, respectively), and the associations with actin may be end-on versus laterally at the two different sites (Chen and Singer, 1982). It is therefore likely that the molecular organization of these sites differs; it will therefore be of interest to determine whether different attachment proteins can induce the organization of different types of adhesion site. The mechanisms of the regulation of cytoskeletal organization by extracellular molecules should be an active area of research in the future.

5. COLLAGEN

5.1. Structure and Organization

Collagen is a major structural component of tissues, comprising an impressive 30% or more of the protein of adult animals. A number of excellent recent review articles have exhaustively covered the structure, biosynthesis, and functions of collagen (e.g., see Ramachandran and Reddi, 1976; Bornstein and Sage, 1980; Fessler and Fessler, 1978; Prockop *et al.*, 1979; Kleinman *et al.*, 1981; Linsenmayer, 1981; Aplin and Hughes, 1982; Burgeson, 1982; Miller, 1983). The following sections will briefly review the types and structures of the collagens and then will focus on the interactions of collagen with the cell surface.

Collagens appear to function as supporting elements at connective tissue sites throughout the body. There are more than 15 genetically distinct types of collagen subunits, each of which appears to be encoded by a separate gene. The number of unique collagen subunits that have been identified has increased at a rapid rate, and it is conceivable that even more will be discovered in the near

TABLE 1. Genetically Distinct Collagen Types

Collagen type	Subunits	Molecular composition	Major locations	Characteristic organization	Variations and other nomenclature	Selected recent references
Type I	$\alpha1(I)$ and $\alpha2(I)$	$[\alpha1(I)]_2\alpha2(I)$	Tendon, skin, bone	Large fibrils	Type I trimer $[\alpha1(I)]_3$ when lower $\alpha2(I)$	Ramachandran and Reddi, 1976 Linsenmayer, 1981
Type II	$\alpha1(II)$	$[\alpha1(II)]_3$	Cartilage	Smaller fibrils in a meshwork	—	Mayne and von der Mark, 1983
Type III	$\alpha1(III)$	$[\alpha1(III)]_3$	Skin, placenta, blood vessels	Fine fibrils in reticular networks	—	Lang et al., 1979 Fleischmajer et al., 1981
Type IV	$\alpha1(IV)$ and $\alpha2(IV)$	$[\alpha1(IV)]_3$ $[\alpha2(IV)]_3$	Basement membranes	"Chicken wire" networks	Also $[\alpha1(IV)]_2\alpha2(IV)$	Timpl et al., 1981 Foellmer et al., 1983 Duncan et al., 1983
Type V	$\alpha1(V)$, $\alpha2(V)$, and $\alpha3(V)$	$[\alpha1(V)]_2\alpha2(V)$	Pericellular and matrix	May be fibrillar	$\alpha1(V)_3$ is minor form, $\alpha1 = B$, $\alpha2 = A$, $\alpha3 = C$	Martinez–Hernandez et al., 1982 Bächinger et al., 1982 Linsenmayer et al., 1983
Type VI	(Perhaps 2)	?	Placenta	Tetramers disulfide bonded into microfibrils	Also termed "intima" collagen	Odermatt et al., 1983 Furthmayr et al., 1983
Type VII	$\alpha1(VII)$	$[\alpha1(VII)]_3$	Amnion	Fibrils under basement membranes	Also "LC" (long-chain) collagen	Bentz et al., 1983
Type VIII	?	?	Endothelial and other cells	?	Also "EC" (endothelial cell) collagen	Sage et al., 1980
Minor cartilage collagens	?	?	Cartilage	?	1α, 2α, 3α, HMW collagen (M1), LMW collagen (M2), SC or G collagen	Mayne and von der Mark, 1983

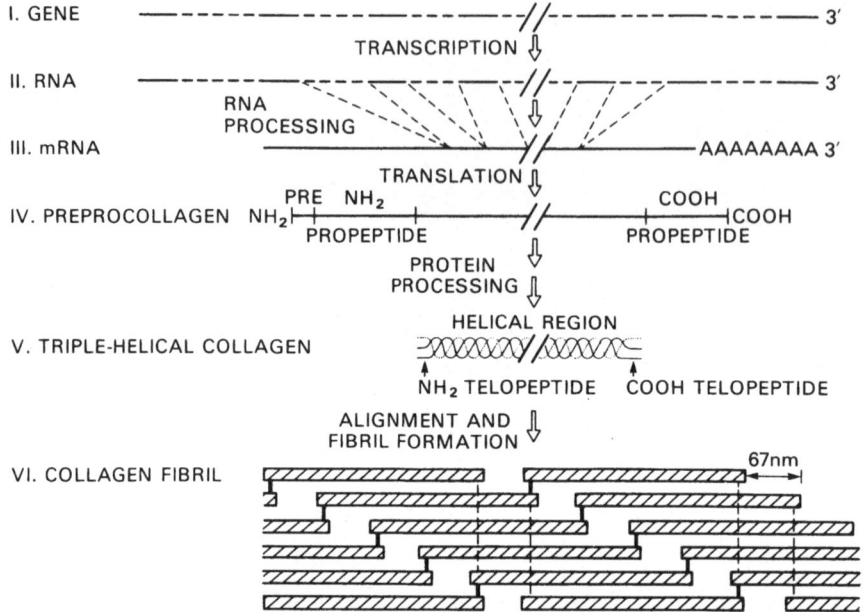

FIGURE 1. Complex biosynthetic pathways of a major extracellular molecule. The key steps in the biosynthesis of type I collagen are indicated schematically. I. The collagen gene contains at least 50 exons (solid bars) consisting of DNA sequences encoding contiguous portions of the collagen molecule; the exons are separated by noncoding introns (dashed line segments) of variable size. Note that only a small portion of the gene is shown (central portion is omitted). II. The gene is transcribed directly into RNA, which therefore contains sequences corresponding to both exons and introns. III. This precursor RNA is rapidly processed by "capping" of the 5′ end and by the addition of a series of adenosine residues to the 3′ end to form a poly A tail. In addition, a surprisingly accurate series of messenger-RNA-splicing events takes place in which all of the introns are excised from the precursor molecule to form the mature mRNA (dotted lines between lines II and III). IV. The collagen mRNA is translated on polyribosomes of the rough endoplasmic reticulum to synthesize the first collagen precursor, termed preprocollagen. At the NH_2 terminus of this precursor is located a hydrophobic sequence, termed the signal or leader sequence, which is important for the translocation of the protein into the lumen of the endoplasmic reticulum; this sequence is rapidly cleaved from the protein to yield procollagen. In addition, the protein is probably glycosylated during translation by the addition of N-linked oligosaccharides. The procollagen then undergoes intracellular processing in the Golgi apparatus, where some of the proline and lysine residues are hydroxylated; in addition, further glycosylation reactions occur there, including the addition of glucose and galactose residues to some of the hydroxylysine residues. The procollagen molecule is then processed proteolytically, probably by extracellular proteases that are specific for the removal of either NH_2 or COOH pro-peptides. V. The processed collagen has also been assembling spontaneously into a triple helix while being processed, and the final collagen molecule contains a long helical region of approximately 1000 amino acids. The ends of the collagen molecule are not helical; these regions are termed the telopeptide regions of the protein, which can serve as sites for the later formation of covalent cross-

future. Table I provides a current but tentative listing of the known collagen types and their subunit organization.

Depending on the genetic type, collagens can exist in the classical fibrillar forms or in a more amorphous or nonfibrillar organization. The first types of collagen to be well characterized are fibrillar and are termed collagen types I, II, and III (Table I). All these types form triple-helical fibrils *in vivo* and *in vitro;* the subunits comprising the helices can be identical or different depending on the collagen type; i.e., besides differences in subunits for each collagen type, there can be additional differences depending on whether the triple helices are homopolymers or heteropolymers. With interesting exceptions, collagens are usually synthesized and secreted in a transient form termed procollagen (the amino terminal secretory signal sequence of "preprocollagen" is removed before secretion), which is further processed by amino and carboxy propeptidases, leaving the more rigid, helical central part of the molecule as the functional unit. This triple-helical portion of the molecule then often self-associates laterally in a precise process of alignment to form the collagen fiber (fibril).

Type I collagen is the major collagen of connective tissue, and it consists of one $\alpha 1$ chain in a heteropolymeric helix with two $\alpha 2$ chains. It is a major component of tendons, fascia, skin, and bone and provides tensile strength to these tissues. Processed triple helices of type I collagen precipitate and align in parallel arrays to form striking, large fibers, which are readily recognized in electron microscopy by their banded fibrillar structure. Even though collagen fiber formation has been studied intensively by many laboratories, it is still not known for certain how this or other types of collagen are assembled into their distinctive patterns of fibril organization, e.g., into striking orthogonal arrays in the cornea (Trelstad and Silver, 1981).

Detailed studies of the specific steps in the formation of a single fibril of type I collagen reveal a complex process (Fig. 1). The messenger RNAs for collagen are produced from large and remarkably frequently interrupted genes. The genes for collagen types I, II, and III contain approximately 50 introns and exons (e.g., see Y. Yamada *et al.*, 1980). Their structures suggest that they may have evolved by many duplications of a single primordial gene only 54 bases

---

links. VI. The individual collagen molecules then spontaneously aggregate and precipitate into precisely aligned arrays; each molecule is offset from its neighbor by 234 amino acids (67 nm), a distance termed the "D" stagger or "quarter stagger." The protruding end of each molecule is available for the formation of covalent cross-links with the ends of other collagen molecules, so that a highly ordered, covalently stabilized, rigid fibril is formed. For further details, see Olsen (1981) and Miller (1983). Other extracellular molecules can also undergo a similarly complex series of biosynthetic steps.

in length. The noncoding regions between these small gene units must be processed out of the original RNA product of each gene before a functional messenger RNA is formed.

The original protein produced consists of a central region that will form the stiff triple helix by the intertwining of three chains, plus less rigidly structured ends. The central helical region contains a glycine at every third residue, which is often followed by a proline and a hydroxyproline residue to form the characteristic -Gly-Pro-Hyp- sequence, although there is considerable variability at the second and third positions (Ramachandran and Reddi, 1976; Linsenmayer, 1981). After the removal of the propeptides by proteases, the triple helices align in a staggered array, with each collagen triple helix offset from its neighbor by slightly less than a quarter of the helix. The parallel alignment of hydrophobic and hydrophilic polypeptide sequences accounts for the ultrastructural banding pattern characteristic of this collagen. The protruding sites of stagger at the ends of the helices provide sites for covalent cross-linking of the molecules by lysyl oxidase to form a tightly organized, rigid fiber (Fessler and Fessler, 1978; Olsen, 1981). Although it is clear that cells and their secreted enzymes continually interact with the collagen molecules as they are being assembled and processed, little is known about the exact sites and spatial relationships of the events.

Type II collagen is simpler in subunit organization, with only one type of subunit comprising the triple helix. This collagen is the characteristic, major collagen of cartilage, but it is also important in the vitreous humor of the eye and in the notochord and nucleus pulposus. Its fibrils are organized into fine fibrillar meshworks which enmesh proteoglycans; the collagen can be viewed as a structural framework holding within it the highly hydrated chondroitin sulfate proteoglycans and hyaluronic acid of cartilage. The latter molecules are thought to provide the shock-absorbing properties of cartilage, whereas the collagen provides coherence. Chondrocytes synthesize and are embedded within this collagenous meshwork. Cartilage also contains several "minor" collagen types of unknown function (Table I).

Type III collagen is the other major interstitial collagen besides type I. It is present in many connective tissues, especially skin and lung, but is not present in appreciable quantities in bone. The fibrils formed by type III collagen appear to be thinner than those of type I collagen (Fleischmajer *et al.*, 1981). These finer fibrils of type III collagen appear to be organized to allow distensibility in tissues such as lung.

Type IV collagen is the characteristic collagen of basement membranes. It consists of two types of subunit polypeptide organized into an unusual chicken-wire-type network (Timpl and Martin, 1982). In contrast to the first three collagen types, type IV contains a number of interruptions of the classical triple helix by noncollagenous polypeptide domains; these interruptions probably provide sites of flexibility which break up the rigid β structure of the triple helix. The final molecular form of type IV collagen is also unusual in that there is little evidence for processing of the chains beyond the procollagen stage; the chains are therefore large for collagen, approximately 175,000 and 185,000 daltons. Type IV collagen networks appear to be assembled precisely; four triple helices of type IV collagen overlap in head-to-head configurations at their amino termini to form a protease-resistant, disulfide-bonded structure termed the "7 S" region (Kühn et al., 1981; Duncan et al., 1983). The other ends of these molecules also appear to interact with yet another type IV collagen carboxy terminus, forming a network of interconnected molecules that may ultimately form a chicken wire or netlike pattern horizontally in the basal lamina (Timpl et al., 1981).

Collagen types V, VI, VII, and VIII are not as well characterized, and current data concerning their possible structure are summarized in Table I. Type V collagen is widespread in connective tissues and appears to be present in interstitial sites, and possibly also in basement membranes and closely associated with the cell surface (Martinez–Hernanadez et al., 1982; Linsenmayer et al., 1983). Since this chapter focuses on the interactions of these molecules with the cell surface, the references listed in Table I can be consulted for the details of organization and synthesis of other collagen types. It seems clear, however, that the interactions of each type of collagen with the cell surface may differ and each must be examined in detail in the future.

5.2. Cell Interactions with Three-Dimensional Collagen Matrices

Cells *in vivo* interact with collagenous matrices containing a number of other extracellular components. A simplified, *in vitro* model system is provided by collagen gels formed by the spontaneous assembly of isolated, purified collagen into three-dimensional gels, which are used to surround cells or explanted tissues (e.g., as described by Elsdale and Bard, 1972). Such gels have striking effects on the growth and cellular differentiation of a number of types of cells and tissues. Two striking examples include the stimulation of proliferation, maintenance of hormonal responsiveness, and synthesis of differentiated cell products by mammary explants embedded within collagen gels (Foster et al., 1983; Ormerod et al., 1983; Suard et al., 1983) and the conversion of the phenotype of nonmotile epithelial cells to a migratory, mesenchymal cell phenotype (Greenburg and Hay, 1982; see also Kleinman et al., 1981, and Hay, 1981 for reviews). Fibroblasts migrating in collagen gels have an appearance much closer to their *in vivo* morphology than when they are cultured on flat substrates of glass or plastic (Bard and Hay, 1975). Interestingly, the cells form adhesions to collagen fibrils that resemble the better characterized adhesion plaques on planar substrates of plastic (Grinnell and Bennett, 1981), although the molecular structure of these collagenous attachments remains to be characterized.

Cells appear to interact with collagen fibrils either directly or indirectly. Fibroblasts, hepatocytes, HeLa cells, and other cells can reportedly adhere directly to native collagen gels (Linsenmayer *et al.*, 1978; Grinnell and Minter, 1978; Schor and Court, 1979; Schor *et al.*, 1981a; Rubin *et al.*, 1981; Briles, 1982). The interaction of hepatocytes with collagen has been characterized, and it is relatively nonspecific in terms of the type of collagenous protein recognized. Hepatocytes adhere to collagen types I, II, III, IV, and V, and even to denatured collagens, albeit less rapidly. Even synthetic collagenlike peptides can serve as substrates, suggesting that the hepatocyte cell surface "receptor" for collagen simply recognizes general collagenous sequences, rather than specific collagen types (Rubin *et al.*, 1981).

Hepatocytes can also adhere to fibronectin-, laminin-, and asialoglycoprotein-coated substrates by other receptors (Rubin *et al.*, 1981; Johansson *et al.*, 1981; Briles, 1982). For example, the attachment of hepatocytes to fibronectin-coated substrates is inhibited by antifibronectin antibodies, which do not inhibit the attachment of these cells to laminin, and vice versa. Somatic cell mutants that cannot attach to native collagens, but readily attach to fibronectin and laminin, have also been isolated (Briles and Haskew, 1982; Briles, 1982). It is therefore clear that hepatocytes have the potential to interact with collagen by several distinct mechanisms, i.e., either directly or indirectly using fibronectin and possibly laminin as intermediaries.

Fibroblasts and certain other cells can attach and penetrate into collagen gels apparently without the need for any intermediary molecules (Schor and Court, 1979; Grinnell and Bennett, 1981). The fibroblast "receptor" for collagen is described in detail in Section 5.4. The adherence of fibroblasts to native collagen gels is strengthened by fibronectin (Linsenmayer *et al.*, 1978; Schor *et al.*, 1981a). On the other hand, fibronectin inhibits the penetration or locomotion of fibroblasts within collagenous gels (Schor *et al.*, 1981b; Couchman *et al.*, 1982). In contrast, fibronectin stimulates the migration of neural crest cells and melanoma cells in these gels, suggesting specificity in the effects of this intermediary molecule on different cell types (Schor *et al.*, 1981b; Rovasio *et al.*, 1983).

5.3. Cell Interactions with Two-Dimensional Collagen Substrates

A simpler model system for studying cell–collagen interactions involves collagen coated onto planar culture dishes. Collagen is spread onto these surfaces and dried to form a stable, easily stored substrate. Interestingly, cells now often require specific, intermediary attachment proteins to be able to attach to collagen in this form. Requirements for proteins such as fibronectin, laminin, or chondronectin have been identified for fibroblasts (Klebe, 1974; Kleinman et al., 1981), epithelial cells (Terranova et al., 1980), chondrocytes (Hewitt et al., 1980), and myoblasts (Chiquet et al., 1979). In some cases, there appears to be at least partial specificity for the type of protein that can function in cell attachment to collagen; e.g., laminin is more effective than fibronectin for attachment of certain malignant or transformed epithelial cells to native type IV collagen (Terranova et al., 1982), and chondronectin is reportedly active only for chondrocyte attachment to type II collagen (Hewitt et al., 1980). Nevertheless, clear specificity for only one attachment protein appears to be the exception rather than the rule at present. For example, a number of normal epithelial cells do not appear to have a specific requirement for laminin (see Section 7.2).

In addition, the seemingly differing requirement for intermediary linking molecules when cells attach to collagen existing as a two-dimensional substrate, but not as a three-dimensional gel, is puzzling. It is not obvious which situation more accurately reflects the *in vivo* interactions between cells and collagenous matrices, since extracellular matrices are more complex than either of these model systems, even if synthesized by a single type of cell *in vitro* (Hedman et al., 1979). It does appear likely, however, that cells can function opportunistically in interacting either directly or indirectly with collagenous structures depending on the presence or absence of additional proteins that can augment these interactions.

5.4. Fibroblast Receptor for Collagen

The binding of soluble collagen to fibroblasts has been characterized in detail (Goldberg, 1979, 1982; Goldberg and Burgeson, 1982). Mouse 3T3 fibroblasts bind type I collagen with a high apparent affinity of $K_D = 10^{-11}$ M, and there is the surprisingly large number of 5×10^5 such receptors per cell (Goldberg, 1979). This "receptor" recognizes the three fibrillar types of collagen, i.e., types I, II, and III. However, it does not bind native types IV or V or minor cartilage collagen types (Goldberg and Burgeson, 1982).

The binding of type I collagen requires a thermal activation step, e.g., heating to 37°C for 30 min; although the mechanism of this activation remains unknown, it may involve dissociation of larger collagen aggregates to a more readily bound form (Goldberg, 1982). Once bound, the collagen molecules dissociate relatively slowly from the receptor (half-life of 2.5 hr at room temperature and 0.7 hr at 37°C). They are apparently only minimally endocytosed.

There are reportedly at least several sites on collagen that are recognized by this receptor; unlike the hepatic cell receptor, the fibroblast receptor does not appear to recognize simple Gly-X-Y sequences but can instead bind several specific regions of denatured collagen chains (Goldberg, 1982). A possible function for this receptor is as an organizing center for the assembly of type I collagen fibrils (Goldberg, 1982).

The receptor activity is inhibited by proteases and concanavalin A, but not by a variety of other treatments including heparinase, chondroitinase, or hyaluronidase treatment; these results suggest that the receptor may be a glycoprotein (Goldberg, 1979, 1982). An important goal in the future will be the isolation and characterization of this receptor, e.g., by affinity chromatography. Another possible candidate for binding of collagen to the surface of some cells, however, may be heparan sulfate proteoglycan (Stamatoglou and Keller, 1982). For example, a form of heparan sulfate proteoglycan that can bind collagen but not fibronectin is present on a myeloma cell line; this molecule is reported to be involved in the attachment of this cell to three-dimensional collagen gels (Stamatoglou and Keller, 1982). Molecular identification of the receptor or receptors for collagen and their roles in fibrillogenesis and the interactions of cells with collagen should be possible, now that their binding parameters have been established.

5.5. Platelet Receptor for Collagen

Platelets interact with native collagens in the physiologically important processes of adhesion, aggregation, and activation. Platelet adhesion to the trimeric collagen molecule appears to require the native triple helix of collagen and does not occur to the individual $\alpha 1$ and $\alpha 2$ collagen chains of type I collagen (see Shadle and Barondes, 1982, and references therein). In contrast, platelet aggregation and activation to release the contents of platelet granules reportedly occurs with both native collagens and primary structural determinants on isolated α chains (e.g., see Chiang and Kang, 1976). It is not entirely clear whether the processes of adhesion and induction of aggregation are mediated by the same receptor mechanism.

A putative platelet receptor for type I collagen has been isolated and purified recently (Saito *et al.*, 1982; Chiang and Kang, 1982). The receptor binds to collagen fibrils and to preparations of the $\alpha 1(I)$ chain of collagen, and it can be purified by affinity chromatography on collagen affinity columns. It is a protein with subunits of apparent size 65,000 or 75,000 daltons that are reportedly linked by disulfide bonds to form a tetramer; the receptor remains active, however, even after the reduction of disulfide bonds (Chiang and Kang, 1982; Saito *et al.*, 1982).

The purified platelet collagen receptor activity is destroyed by proteases, and the binding of radioactively labeled collagen can be reversed completely in the presence of excess unlabeled α1 chains. The purified receptor binds to collagen with an affinity of $K_D = 2 \times 10^{-8}$ M, which is similar to affinities determined initially with purified platelet plasma membranes (Chiang and Kang, 1976, 1982). Surprisingly, the number of free α1 chains apparently bound to each receptor at saturation is three times the number of receptors (Chiang and Kang, 1982); one possible interpretation of this result is that the chains become self-associated at the time of binding, perhaps even into a triple-helical molecule, although there is no evidence as yet for our speculation.

In tests of the physiological significance of this receptor, it was examined for its ability to compete with the functional platelet receptor for collagen; the purified receptor was capable of inhibiting the binding of platelets to collagen (Saito *et al.*, 1982; Chiang and Kang, 1982). This result suggests that the purified receptor may be crucial to platelet interactions with collagen. It will be important to evaluate the role of this putative receptor further using monospecific antibodies to probe its function. Another valuable area of investigation will be in the mechanisms by which binding of collagen by the receptor is transduced into platelet-to-platelet aggregation and the activation of exocytosis during the platelet release reaction.

5.6. Chondrocyte Receptor for Collagen

Although the attachment of cartilage chondrocytes to type II collagen is enhanced by chondronectin (Hewitt *et al.*, 1980), these cells nevertheless appear to be capable of direct interactions with collagen. A putative receptor for collagen has been isolated by affinity chromatography from a crude plasma membrane fraction of chondrocytes (Mollenhauer and von der Mark, 1983). This protein receptor has an estimated size of 31,000 daltons, and it appears to be a hydrophobic protein that is reportedly inserted in functional form into liposomes. It has an unusually high content of fucose (20% plus only 10% content of other sugars). This putative collagen receptor also binds to other types of collagen, including types I, III, V, and M; relative affinities for each type of collagen remain to be determined. Important questions for the future include determinations of this molecule's specificity for the type of collagen, and for collagen in general (e.g., in comparison to binding to laminin or fibronectin), saturability, and affinity of binding. The possible existence of the same or unique receptors for other types of collagen on other types of cells should also be investigated.

6. FIBRONECTIN

The glycoprotein fibronectin (*fibre* = fiber + *nectere* = to bind, connect) is a major component of the cell surface, many extracellular matrices, and plasma (recent reviews include Mosesson and Amrani, 1980; Saba and Jaffe, 1980; Pearlstein *et al.*, 1980; Ruoslahti *et al.*, 1981a; Hynes, 1981; Hörmann, 1982; Furcht, 1983; Yamada, 1983a, 1984). The form of fibronectin in plasma is termed plasma fibronectin, which is a relatively soluble protein present at 0.3 g/liter (Mosesson and Amrani, 1980). Plasma fibronectin is probably synthesized in the liver by hepatocytes (Voss *et al.*, 1979; Foidart *et al.*, 1980; Owens and Cimino, 1982; Tamkun and Hynes, 1983), although production by endothelial cells has not been excluded (Saba and Jaffe, 1980). The cell surface, non-plasma-derived form of fibronectin is loosely termed cellular fibronectin, and it is characteristically located in striking fibrillar arrays surrounding cells (Fig. 2). Cellular fibronectin is synthesized by a large number of cell types including fibroblasts and certain epithelial cells (reviewed in Hynes and Yamada, 1982).

6.1. Biological Functions of Fibronectin

Cellular and plasma fibronectins appear to function in a variety of cellular events, especially those involving cell adhesion and migration (Table II). As for other extracellular constituents, it is often difficult to be certain that fibronectin is centrally involved in a particular event; the evidence for its function is usually based on a combination of *in vivo* localization data and *in vitro* functional data. For example, immunofluorescence studies show that fibronectin appears in association with migratory cells such as neural crest and possibly cerebellar cells at the time of migration (e.g., see Critchley *et al.*, 1979; Thiery *et al.*, 1982a; Hatten *et al.*, 1982), and tissue culture studies in some cases show that purified fibronectin permits cell migration. However, definitive proof that fibronectin or any other extracellular molecule is necessary for such events *in vivo* will require further experimentation, including examining for the effects of inhibiting its function *in vivo* with antibodies and searching for mutant cells and organisms in which it is absent or defective.

An unusual property of fibronectin is its dramatic effects on cell behavior when added to cells in tissue culture. Cells often adhere more tenaciously to substrates, spread out and flatten, lose microvilli, and migrate more rapidly (Yamada *et al.*, 1976a,b, 1978; Willingham *et al.*, 1977; Ali *et al.*, 1977; Ali and Hynes, 1978). More complex effects on embryonic development are re-

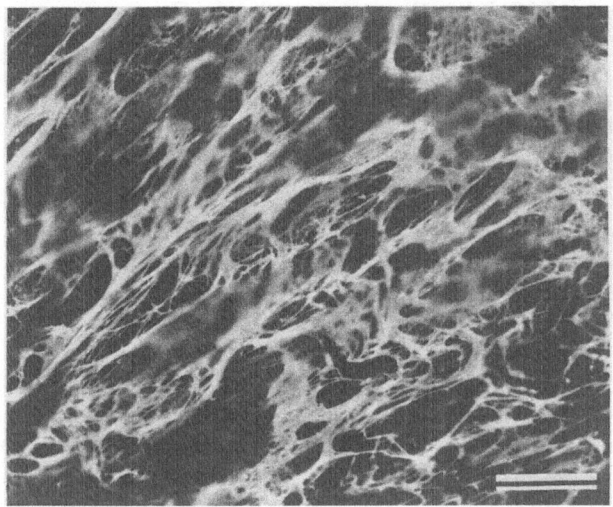

FIGURE 2. Extracellular matrix containing fibronectin. A confluent culture of chick embryo fibroblasts was stained for fibronectin with a rhodamine-labeled, affinity-purified antibody to chick cellular fibronectin. The matrix forms a dense network surrounding the cells (dark areas). Bar indicates 25 μm.

viewed in Section 10. One simplifying interpretation of these effects is that fibronectin is acting as an adhesive molecule (Yamada *et al.*, 1976b; Hynes and Yamada, 1982). An adhesive function for the molecule is simple to demonstrate in assays for cell-to-substrate adhesion *in vitro*, where fibronectins can mediate cell attachment and spreading on collagen, fibrin, and glass or plastic substrates (Klebe *et al.*, 1981; Grinnell, 1978; Grinnell and Feld, 1982). A surprising variety of cell types can be shown to utilize fibronectin in attaching to two-dimensional collagenous substrates, including fibroblasts and certain epithelial cells (reviewed in Kleinman *et al.*, 1981). In addition, simple fibronectin substrates alone can induce the attachment of many cells, including fibroblasts, myoblasts, hepatocytes, epidermal cells, and others (reviewed in Kleinman *et al.*, 1981; Hynes and Yamada, 1982; Furcht, 1983). Moreover, cellular fibronectin can agglutinate fixed erythrocytes (Yamada *et al.*, 1975; Vuento, 1979) and embryonic cells (Yamada *et al.*, 1978), although this cell–cell adhesive activity may be weak in comparison to other more specific mechanisms.

Fibronectin is also a multifunctional binding molecule. It binds to a variety of ligands including collagen, fibrin, heparin, hyaluronic acid, and some contractile proteins, often by means of specific binding sites (reviewed below).

TABLE II.   Biological Activities of Fibronectin

Activity[a]	Selected references[b]
Cellular adhesion	
Cell attachment to collagen and fibrin	Klebe, 1974; Grinnell et al., 1980; Kleinman et al., 1981
Cell attachment and spreading on culture substrates	Grinnell, 1978
Hemagglutination	Yamada et al., 1975
Cell–cell aggregation	Yamada et al., 1978
Cellular morphology	
Flattening and fibroblastic cell shape	Yamada et al., 1976a,b; Ali et al., 1977
Suppression of microvilli and lamellipodia	Yamada et al., 1976b
Cytoskeletal (microfilament) organization	Willingham et al., 1977; Ali et al., 1977; Singer, 1979
Cell migration	
Increased rate of migration	Ali and Hynes, 1978; Yamada et al., 1978; Rovasio et al., 1983
Chemotaxis or haptotaxis	Gauss–Müller et al., 1980
Wound healing	Nishida et al., 1983
Phagocytosis	
Stimulation of reticuloendothelial system function	Saba and Jaffe, 1980
Hemotasis and thrombosis	
Transglutaminase-mediated cross-linking to fibrin	Mosher, 1976
Platelet spreading and aggregation (not definitive)	Arneson et al., 1980; Lahav et al., 1982
Embryonic differentiation	
Stimulation of adrenergic phenotype of neural crest cells	Sieber–Blum et al., 1981
Inhibition of chondrogenic phenotype and myoblast fusion	Pennypacker et al., 1979; West et al., 1979; Podleski et al., 1979

[a] Functional activities determined in assays performed in vitro.
[b] Arbitrarily selected references; see text and reviews for more complete bibliography.

These binding activities of fibronectin may be involved in cell-to-substrate adhesion, e.g., to collagen or fibrin, but they may also be related to its promotion of phagocytosis by cells of the reticuloendothelial system (Saba and Jaffe, 1980), or even to a function in cross-linking between these molecules in extracellular spaces.

The effects of fibronectin on cell migration are not fully understood, but a striking correlation has been shown between cell migratory events and the presence of fibronectin in embryonic cell migratory pathways or in healing wounds (reviewed in Sections 7.2 and 10). *In vitro,* fibronectin will stimulate the migration of normal and oncogenically transformed fibroblastic cells (Ali and Hynes, 1978; Yamada *et al.*, 1978; Couchman *et al.*, 1982), adhesion mutants defective in cell surface glycoconjugates (Pouysségur *et al.*, 1977), and neural crest cells (Rovasio *et al.*, 1983).

Cells are also stimulated to migrate directionally toward fibronectin in assays using gradients of fibronectin in Boyden chambers; fragments of fibronectin specialized for binding to the cell surface are also active in these assays (Gauss-Müller *et al.*, 1980; Postlethwaite *et al.*, 1981; Seppä *et al.*, 1981; Greenburg *et al.*, 1981). One speculation is that fibronectin promotes the formation of "close contacts" of the cells with the substrate, which is the class of adhesions of cells with substrates thought to be important in cell migration as possible sites of motility-related interactions between intracellular actin microfilaments, the plasma membrane, and the substrate (Chen and Singer, 1982). It should be emphasized, however, that an essential role for fibronectin in cell migration remains to be proven definitively.

Fibronectin also appears to function as a growth factor for cells. In low concentrations (0.2–2 μg/ml), fibronectins can serve as "competence" factors for the stimulation of fibroblast proliferation (Hall and Ganguly, 1981; Bitterman *et al.*, 1983). In conjunction with "progression" factors such as insulin or fibroblast growth factor, fibronectin functions as a weak, but potentially physiologically important, competence factor as part of a two-step mechanism of growth control regulating the stimulation of cell division (Bitterman *et al.*, 1983). Fragments of fibronectin may be particularly effective in growth stimulation of fibroblastic cells cultured in serum-free medium (Humphries and Ayad, 1984).

Under other conditions, particularly in the presence of serum, cell-surface-derived fibronectin can stimulate replication, whereas plasma fibronectin does not (Yamada *et al.*, 1982; Couchman *et al.*, 1982). In another system in which DNA replication occurs, but cytokinesis is faulty in the absence of serum, fibronectin permits cells to undergo normal cytokinesis (Orly and Sato, 1979). It therefore appears likely that fibronectin, in probable concert with other extracellular matrix molecules (Greenburg and Gospodarowicz, 1982; Gospodarowicz *et al.*, 1983), can function to permit or to promote cell proliferation in place of other growth factors.

6.2. Functional Domains of Fibronectin

As found for other large, complex proteins, the most effective approach to understanding fibronectin's mechanism of action has been to dissect its functions into a series of more easily analyzed subsets of activities, i.e., into a series of distinct ligand-binding activities. The binding interactions listed in Table III have

TABLE III.   Interactions of Fibronectin

Binding interaction	Selected references[a]
Eukaryotic plasma membranes	(See text)
Collagen types I–V and gelatin	Klebe, 1974; Engvall and Ruoslahti, 1977
Fibrin and fibrinogen	Mosesson, 1978; Hörmann and Seidl, 1980
Heparin, hyaluronic acid, and proteoglycans	Stathakis and Mosesson, 1977; Perkins *et al.*, 1979; Yamada *et al.*, 1980; Isemura *et al.*, 1982b
Factor XIII$_a$ transglutaminase	Mosher, 1976; Mosher *et al.*, 1980
Actin	Keski–Oja *et al.*, 1980
Myosin	Koteliansky *et al.*, 1982
DNA	Zardi *et al.*, 1979
Gangliosides	Kleinman *et al.*, 1979; Yamada *et al.*, 1983
Bacteria	Kuusela, 1978; Proctor *et al.*, 1982
Clq component of complement	Menzel *et al.*, 1981; Pearlstein *et al.*, 1982; Bing *et al.*, 1982
Asymmetric acetylcholinesterase	Emmerling *et al.*, 1981; but see Grassi *et al.*, 1983
Thrombospondin	Lahav *et al.*, 1982

[a] Arbitrarily selected references; see text and reviews for more complete bibliography.

been analyzed by ligand-binding studies and in many cases by proteolytically cleaving the molecule followed by affinity chromatography to identify binding domains on fibronectin. There is an extensive literature concerning these studies, which involve the work of many laboratories. Since a lengthy review and bibliography solely concerning these studies was published recently (Yamada, 1984), this chapter will focus on the key findings and cite only selected references, with emphasis on the interactions of fibronectin with cells. The following sections will first summarize the functional domain structure of fibronectin, then will review each binding activity and the fibronectin domains involved, and finally will consider how the domains interact in biological processes.

6.3. Summary Map of Functional Domains

A variety of proteases have been used to cleave fibronectin into protease-resistant, structural domains. Nearly all these fragments retain specific binding activities for at least one of the ligands bound by fibronectin (reviewed in Mosher, 1984); even the broad-spectrum protease pronase will produce such domains under controlled conditions (K. M. Yamada *et al.*, 1980). The overall map of these structural and functional domains is shown in Fig. 3. Fibronectin is composed of subunits linked by interchain disulfide bonds at the carboxy terminus of the molecule (Petersen *et al.*, 1983). Plasma fibronectins contain two of these subunits, whereas fibroblast cellular fibronectins isolated from the cell surface contain dimers and multimers (cf., Mosesson and Umfleet, 1970; Yamada *et al.*, 1977; McConnell *et al.*, 1978).

Plasma fibronectins contain two subunits, termed the A and B chains (Fig. 3). These chains appear to contain similar functional domains, but their protease cleavage products differ at a site near the carboxy terminus (Richter and Hörmann, 1982; Hayashi and Yamada, 1983; Sekiguchi *et al.*, 1981; Sekiguchi and Hakomori, 1983). It is therefore conceivable that the chains are produced from different messenger RNAs. At present, it is thought that there is one site on each subunit for binding to eukaryotic cell surfaces, one for collagen, two for heparin, and two or three for fibrin (Fig. 3).

The structural domains of fibronectin are comprised of a striking series of repetitive loop structures, which are usually stabilized by intrachain disulfide bonds (Fig. 4). At present, slightly more than half of the bovine plasma fibronectin molecule has been sequenced (Petersen *et al.*, 1983). It is clear that protease-resistant domains of the molecule actually contain a substructure of several such disulfide-bonded loop units per domain, and the loop structures are not necessarily unique to each domain. For example, the Petersen "type I" homology unit is found in heparin-, collagen-, and probably fibrin-binding domains along the molecule (Fig. 4).

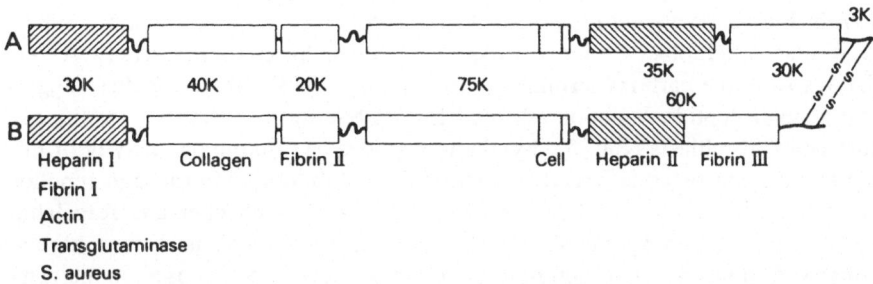

FIGURE 3. Structural and functional domains of fibronectin. Human plasma fibronectin is depicted as comprised of a series of protease-resistant domains (rectangles) separated by flexible, protease-susceptible regions of polypeptide (wavy lines). The A chain (top) and B chain are linked together at the carboxyl terminus (right end) by two disulfide bonds. The numbers indicate molecular weights of the domains (30K = 30,000), and the materials listed below the B chain are the interactions of each domain above.

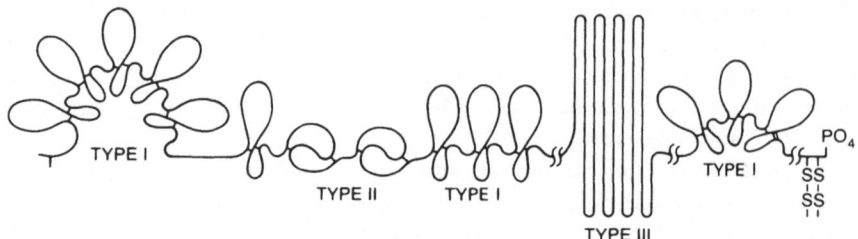

FIGURE 4. Repeating units in the amino acid sequence of fibronectin. Three types of repeating homology in the primary sequence of bovine plasma fibronectin are indicated as described by Petersen *et al.*, 1983, and Skorstengaard *et al.*, 1984. Approximately half of the molecule has been sequenced, and the nonsequenced portions are indicated by the wave breaks in the line. Type I contains disulfide-bonded loop structures, of which five are found in the first domain. The second, collagen-binding domain contains at least four type I units and two type II units. The type III unit, which is approximately twice the size of the other units (90 versus about 50 amino acids in length), is found in a heparin-binding domain. The terminal fibrin-binding domain appears to contain three more type I homology units. There is a phosphate group near the terminal interchain disulfide bonds.

6.4. Specific Binding Domains

6.4.1. Interaction Sites for Cells

Fibronectin interacts with eukaryotic and bacterial cell surfaces at different sites on the protein. The eukaryotic "cell-binding" site can be isolated free of other ligand-binding activities (Ruoslahti and Hayman, 1979; Hahn and Yamada, 1979b; Pierschbacher *et al.*, 1981, 1982, 1983; Hayashi and Yamada, 1983). A similar "chemotactic" domain has also been defined approximately 70,000 daltons from the carboxy terminus (Albini *et al.*, 1983). Recent studies suggest that a small peptide region is essential for the binding of fibronectin to the cell surface of fibroblasts (Fig. 5). Synthetic peptides can mimic the activity of this region and can serve as specific, competitive inhibitors of fibronectin function in adhesion assays (Pierschbacher *et al.*, 1983; Pierschbacher and Ruoslahti, 1984; Yamada and Kennedy, 1984). The active sequence is from a highly-conserved, unusually hydrophilic region that is likely to be exposed on the outer surface of the protein. The key amino acids are Gly-Arg-Gly-Asp-Ser; the first glycine residue is not essential for activity, but its presence substantially increases the activity of the peptide (Pierschbacher and Ruoslahti, 1984; Yamada and Kennedy, 1984; and unpublished results).

The presence of biological activity in such a short peptide sequence recalls the tiny peptide sequence from IgG called tuftsin, which can act as a site for binding to phagocytosis receptors on cells (Fridkin and Gottlieb, 1981; Nishioka *et al.*, 1981). Tuftsin, however, does not mimic the action of fibronectin peptides on the function of fibronectin (Yamada and Kennedy, 1984). Since the putative functional recognition site on fibronectin is so small, it appears most likely that binding occurs by the action of a cellular receptor binding to the fibronectin cell-recognition site, rather than by fibronectin actively binding to a cell surface molecule.

Staphylococcus aureus cells bind only to the amino-terminal domain of fibronectin under physiological salt conditions (Kuusela, 1978; Mosher and Proctor, 1980; Hayashi and Yamada, 1983). It has been suggested that this binding provides a selective advantage to the bacterium, which can bind to an extracellular matrix component such as fibronectin or laminin and establish infection sites (Speziale *et al.*, 1982). The site recognized by *S. aureus* is further away from the amino terminus than the heparin-binding site of this domain (Vartio, 1982). The bacterium can also be covalently cross-linked to this domain by factor XIII_a (plasma transglutaminase; Mosher and Proctor, 1980). This site for bacterial binding is entirely separate from the site for recognition of fibronectin by nucleated cells.

6.4.2. Binding Site for Collagen

Fibronectin may be important for mediating or strengthening cell interactions with collagen (Klebe, 1974; Kleinman *et al.*, 1981). Fibronectin can bind to collagen with moderately high affinity, with an estimated affinity to nonhelical collagen fragments of $K_D = 2$–5×10^{-9} M (Mosher, 1980). Fibronectin can bind to all of the collagen types tested to date, although its avidity for denatured collagens is highest (Jilek and Hörmann, 1978a, Engvall *et al.*, 1978). One explanation for this finding is that the site on collagen recognized best by fibronectin is an unusual region of decreased stability between residues 757 and 791 on the $\alpha 1$(I) chain (Kleinman *et al.*, 1978; Dessau *et al.*, 1978); this site might be exposed by a local unraveling of the collagen triple helix to a limited extent at 37°C, but it would be exposed most effectively after total separation of collagen chains after denaturation.

The region on fibronectin that binds to collagen has been isolated as a 30- to 40,000-dalton fragment, which can still bind to native type I collagen (Hahn and Yamada, 1979a; Ruoslahti *et al.*, 1979). The binding to collagen by fibronectin may occur by electrostatic mechanisms (Vuento *et al.*, 1982), and it is

CELL-BINDING REGION

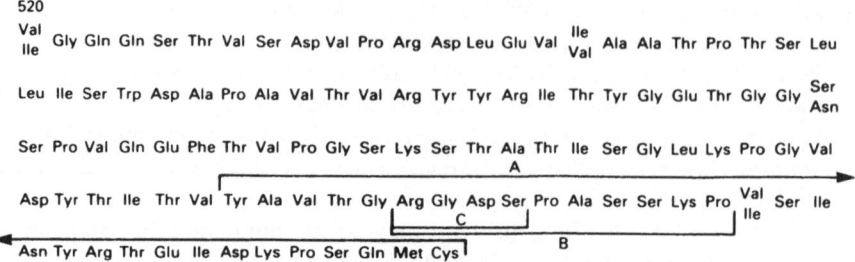

COLLAGEN-BINDING REGION

FIGURE 5. Functional and conserved amino acid sequences in fibronectins from different species. Sequences from near the cell-binding regions of bovine and human fibronectins are listed at the top. All the residues are identical except those where two amino acids are listed; the amino acid at the top is in the bovine plasma fibronectin and the amino acid at the bottom is in human plasma fibronectin. The bars indicate synthetic peptides described in the text. (A) A 30-amino-acid sequence with cell attachment activity (Pierschbacher *et al.*, 1983). (B) A ten-amino-acid sequence capable of blocking cell adhesion (Yamada and Kennedy, unpublished data). (C) A four-amino-acid sequence common to several weakly to moderately active peptides with cell attachment activity (Pierschbacher and Ruosalhti, personal communication). At the bottom are shown primary sequences from bovine compared to human fibronectins in the collagen-binding domain and then sequences from bovine compared to a chicken fibronectin sequence deduced from a DNA sequence from a recombinant DNA clone. Data are from Petersen *et al.*, 1983 (bovine cell-binding region); Pierschbacher *et al.*, 1982 (human cell-binding region); Skorstengaard *et al.*, 1984 (bovine collagen-binding domain); Pande and Shively, 1982 (human collagen-binding domain); and Hirano *et al.*, 1983 (chicken collagen-binding domain). The numbering is based on the residue number for bovine plasma fibronectin.

dependent on maintenance of the disulfide bonding of this cystine-rich region (Balian *et al.*, 1979a). The primary sequence of this domain contains a distinctive type of disulfide-bonded loop structure not yet found elsewhere in the molecule (Fig. 4, "type II"; Skorstengaard *et al.*, 1984); it is tempting to speculate that this structure is involved in the recognition of collagen by this domain.

6.4.3. Heparin-Binding Domains

The binding of fibronectin to heparin occurs with moderate affinity ($K_D = 10^{-7}$ to 4×10^{-9} M; K. M. Yamada *et al.*, 1980). The binding is specific for heparin or heparan sulfate, and other glycosaminoglycans such as hyaluronic acid do not compete for binding. Under physiological conditions, heparin is bound by two domains on each fibronectin subunit (Fig. 3). One binding domain is the 30,000-dalton amino-terminal domain (Sekiguchi and Hakomori, 1980; Richter *et al.*, 1981; Hayashi and Yamada, 1982). This binding is the only interaction of fibronectin with a ligand known to be regulated by a divalent cation; binding is modulated by physiological extracellular concentrations of calcium, but not by other divalent cations (Hayashi and Yamada, 1982). It is therefore possible that high local concentrations of calcium, e.g., near regions of high concentrations of negatively charged proteoglycans, might block the binding of fibronectin to heparin or heparan sulfate at this site.

The second binding site for heparin is located at the other end of the molecule, which has been isolated as a 50,000-dalton, pronase-resistant fragment (Hayashi *et al.*, 1980). This site is thought to be of higher affinity and is not regulated by divalent cations (K. M. Yamada *et al.*, 1980; Sekiguchi and Hakomori, 1980; Hayashi *et al.*, 1980; Ruoslahti *et al.*, 1981c; Richter *et al.*, 1981; Hayashi and Yamada, 1982). One speculation about the reason for the existence of two sites rather than one binding site for heparin, only one of which is regulated by calcium, is that this arrangement would allow fibronectin to bind to heparin at one end whether or not calcium is present, but would only permit cross-linking of heparin molecules in low concentrations of calcium (since binding of two heparin molecules so close to the carboxy terminus might be difficult, whereas such cross-linking by the two flexible ends of the molecule would be much more likely). Alternatively, the calcium effect might be related to some more general effect on the conformation and exposure of fibronectin; according to surface chemical studies, the molecular area of fibronectin expands dramatically in low concentrations of calcium (Holly *et al.*, 1983).

Fibronectin-mediated interactions of heparin with the cell surface by one or both domains may be important for stabilizing or strengthening the adhesive interactions of the cell with a substrate, although probably not as a primary adhesion mechanism (Laterra *et al.*, 1983). For example, although binding of heparin-related sites on the plasma membrane by platelet factor 4 attached to a substrate does not lead to normal cell spreading, fibronectin may bind to the cell surface and to heparan sulfate to mediate normal cell attachment and establishment of intracellular microfilament bundles (Laterra *et al.*, 1983). In addition, heparin can serve as a cofactor in the binding of gelatinized test particles to components of the reticuloendothelial system (Saba and Jaffe, 1980; Gudewicz *et al.*, 1980; Van De Water *et al.*, 1981). Heparan sulfate proteoglycan is an attractive candidate for mediating cell interactions with extracellular materials, since it may exist as a hydrophobic proteoglycan that inserts into membranes (Kjellén *et al.*, 1981; Rapraeger and Bernfield, 1983), although this hypothesis requires further testing.

6.4.4. Fibrin-Binding Domains

Fibronectin binds to fibrin and fibrinogen at a carboxy-terminal region of the fibrinogen Aα chain (Mosher, 1976; Stemberger and Hörmann, 1976; Stathakis and Mosesson, 1977). Fibronectin appears to bind to fibrin via three binding sites (Fig. 3). The domains that bind to fibrin under physiological ionic conditions are located at the amino-terminal end, at a more central site near the collagen-binding site, and at the carboxy-terminal end of the molecule (Hörmann and Seidl, 1980; Sekiguchi *et al.*, 1981; Hayashi and Yamada, 1983; Seidl and Hörmann, 1983). The interactions of fibronectin with fibrin or fibrinogen are relatively weak (Ruoslahti and Vaheri, 1975), and it can mediate the attachment of cells to planar fibrin-coated substrates effectively only after the two molecules are covalently cross-linked by transglutaminase (Grinnell *et al.*, 1980; Grinnell and Phan, 1983). Nevertheless, fibronectin may be crucial for cells to be able to interact with fibrin, since fibroblastic cells do not adhere to or penetrate clots in the absence of fibronectin (K. M. Yamada and D. W. Kennedy, unpublished results); this interaction may therefore be necessary for fibroblasts to enter the fibrin clot filling wounds during the process of wound healing (Grinnell, 1982).

6.4.5. Other Interactions of Fibronectin

Fibronectin interacts specifically with several other molecules but not with every one tested. Some of these binding sites have been identified, although the physiological importance of these interactions is as yet uncertain. More studies are clearly needed. Fibronectin binds to actin at the amino-terminal domain (Keski–Oja and Yamada, 1981; Hayashi and Yamada, 1983). Although fibronectin binds to DNA (Zardi *et al.*, 1979; Parsons *et al.*, 1979), the binding is very weak in physiological concentrations of NaCl and divalent cations, and no binding domains could be identified under these conditions (Hayashi and Yamada, 1983). Fibronectin binds to polyamines at several sites, especially within the collagen-binding domain (Isemura *et al.*, 1982a; Vartio, 1982). It will be interesting to determine whether these interactions, which can be relatively specific, are also physiologically important.

6.5. Functional Interrelationships of Domains

Various combinations of functional domains can account for different activities of fibronectin. Although the cell-binding or recognition site of fibronectin is sufficient to mediate tight adhesion and spreading of cells on plastic substrates as well as haptotaxis, it cannot mediate cell attachment to collagenous substrates (Ruoslahti and Hayman, 1979; Hahn and Yamada, 1979b; Postlethwaite *et al.*, 1981; Seppä *et al.*, 1981). This activity requires a combination of cell- and collagen-binding sites (Fig. 6). In fact, isolated cell- and collagen-binding sites can act as competitive inhibitors of fibronectin-mediated cell attachment to collagen (Hahn and Yamada, 1979b). Preliminary results indicate that artificial recombinants between cell- and collagen-binding domains regain the capacity to attach cells to collagen, even if the domains are isolated from different species (unpublished results).

Cell-to-cell adhesion, as measured by hemagglutination, requires a multimeric molecule; a monomeric fragment of fibronectin is inactive in this event (Hahn and Yamada, 1979b). This finding indicates that agglutination requires multivalency to link cells together, as is well known for antibodies and lectins.

Other important activities of fibronectin still remain to be examined, although it appears likely that certain combinations of domains can account for each activity. Activities that should be examined include cell attachment and migration in fibrin clots (cell + an unknown number of fibrin-binding domains), phagocytosis of gelatinized test particles by macrophages (cell + collagen + heparin domains), and restoration of normal morphology and alignment to transformed cells (cell + an unknown number of other domains). Another important question is whether the binding of one ligand induces cooperative effects on the binding of another ligand. One approach to these questions would be to create various recombinants between domains by cross-linking different purified fragments together and then testing them in various assays. Another more difficult approach would be to undertake a systematic series of *in vitro* mutagenesis experiments with fibronectin gene fragments in expression vectors, and to examine for alterations in the activity of mutated peptides isolated from *E. coli*.

6.6. Different Forms of Fibronectin

Although the plasma and fibroblast cellular forms of fibronectin are similar in structure and function, they are not identical (see a more detailed review by Yamada, 1983a). These proteins have very similar amino acid and carbohydrate compositions, as well as indistinguishable general secondary and tertiary struc-

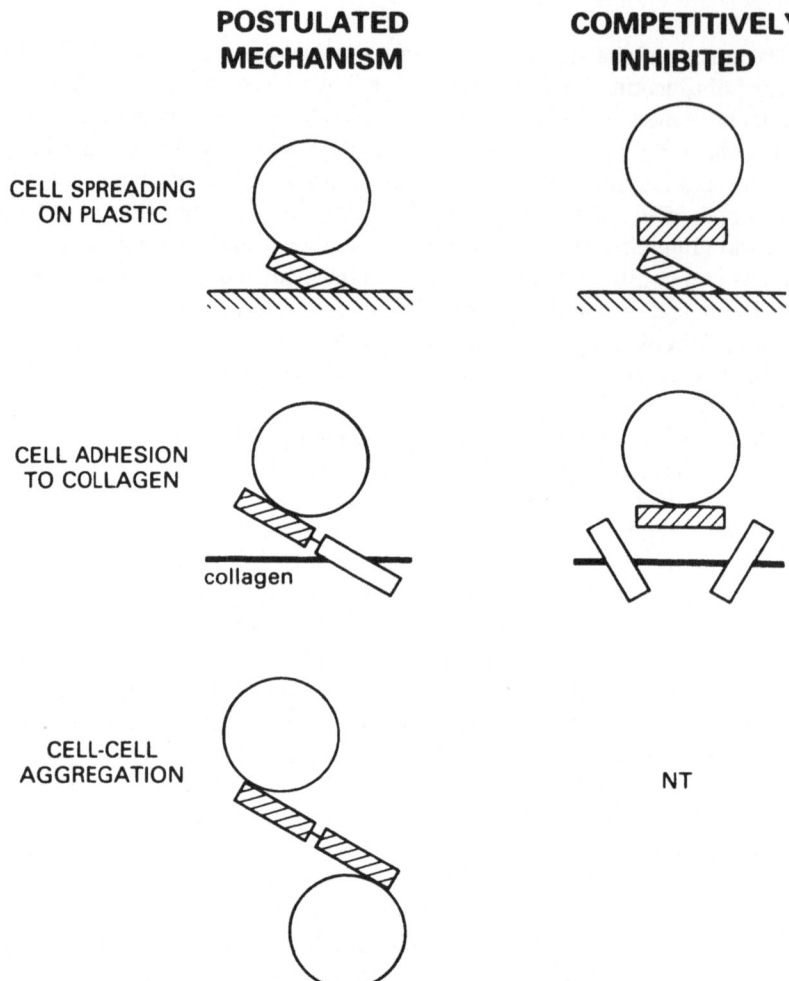

FIGURE 6. Current concepts of the mechanism of action of a modular adhesive molecule. Although based on data from fibronectin, these postulated mechanisms of action may be general for other matrix molecules. The left column shows the postulated mechanism of action of cell-binding (hatched bar) and collagen-binding (open bar) domains. Experimentally observed competitive inhibition by purified domains or the intact molecule is indicated at the right; sufficiently high concentrations of fibronectin sites can block cell or collagen sites and inhibit fibronectin-mediated adhesion. NT, not tested.

tures as determined by circular dichroism and tryptophanyl fluorescence spectroscopy (Mosesson *et al.*, 1975; Yamada *et al.*, 1977; Vuento *et al.*, 1977; Alexander *et al.*, 1978, 1979; Fukuda *et al.*, 1982). The two forms also appear to have the same series of functional domains (Hayashi and Yamada, 1981). Some of the differences noted between these two major forms of fibronectin include the following properties: cellular fibronectins are more active in some biological activities (e.g., agglutination), but identical in others (Yamada and Kennedy, 1979); cellular fibronectins are much less soluble at neutral pH (cf., Mosesson and Umfleet, 1970, and Yamada *et al.*, 1977); there are differences in apparent molecular weights of domains (Hayashi and Yamada, 1981) and in antigenicity (Atherton and Hynes, 1981); fibroblast cellular fibronectin exists primarily as multimers, whereas plasma fibronectin is dimeric (McConnell *et al.*, 1978); and some differences exist in their terminal sugars (Fukuda *et al.*, 1982).

6.6.1. Sources of Differences between Plasma and Cellular Fibronectins

The possibility that these two major forms of fibronectin are related by proteolytic processing of one form to another has been ruled out by mapping studies. Three differences in apparent molecular weights of domains have been mapped to the interior of the molecule, a result that would be highly improbable by simple proteolytic processing (Hayashi and Yamada, 1981). A monoclonal antibody also preferentially, though not absolutely, distinguishes fibroblast cellular from plasma fibronectin at a site apparently lacking carbohydrate (Atherton and Hynes, 1981). The most probable explanation of the major differences between these forms appears to be differences in messenger RNA splicing from a single gene, or major posttranslational modifications other than glycosylation.

A third possibility, that the proteins are the product of different genes, now appears unlikely from both DNA hybridization and protein evolutionary comparisons. Southern blot analysis using cDNA to human fibronectin shows no evidence for more than one gene, at least near the hybridization site of the probe (Kornblihtt *et al.*, 1983). Protein structural and immunological comparisons between the two types of fibronectin in three different species show that there are striking similarities in structure between plasma and cellular fibronectins within each species, but marked interspecies differences. This pattern of protein data is inconsistent with the production of fibronectin from two divergent genes (Akiyama and Yamada, 1984). This combination of independent DNA and protein data strongly suggests that there is only one gene for these two major forms of fibronectin.

6.6.2. Differences in Glycosylation

Fibronectins from some cells show major differences in extent and type of glycosylation. Amniotic fibronectin is more glycosylated than other types of fibronectin, with nearly twice the content of carbohydrate (Balian *et al.*, 1979b; Ruoslahti *et al.*, 1981b; Pande *et al.*, 1981). Fibronectin secreted by undifferentiated teratocarcinoma cells contains unusual oligosaccharides; the presence of lactosaminoglycan moieties as well as, surprisingly, covalently linked heparan sulfate results in a substantially higher-apparent-molecular-weight form of fibronectin. In contrast, fibronectin from a differentiated teratocarcinoma line showed no major differences from other fibronectins (Cossu and Warren, 1983; Cossu *et al.*, 1983). Finally, fibronectin secreted by chondrocytes *in vitro* contains substantial amounts of the high-mannose or hybrid type of oligosaccharide rather than only the complex form characteristic of plasma and fibroblast cellular fibronectins. Treatment of these cells by retinoic acid results in the production of a fibronectin molecule containing the usual complex oligosaccharide moiety, suggesting that there is control of glycosylation by this vitamin (Bernard *et al.*, 1984). Although the polypeptide moiety of chondrocyte fibronectin was not detectably different from fibroblast fibronectin (Bernard *et al.*, 1984), in most other cases, it is not clear whether there are any peptide differences accompanying the oligosaccharide modifications.

6.7. The Gene for Fibronectin

The fibronectin gene has been isolated as a set of overlapping recombinant DNA clones spanning the gene (Hirano *et al.*, 1983, Fig. 7). The clones were from a chicken "library" of genomic DNA fragments carried in a bacteriophage vector. The gene is one of the largest known, spanning 48,000 bases. It also contains at least 48 exons, which requires extensive processing of the original RNA transcript to form the final, processed messenger RNA (Hirano *et al.*, 1983). The exons themselves are unusual in their small and relatively similar size; with the exception of the two terminal exons, they show a montonous similarity of size, averaging 150 base pairs. There is no obvious evidence in the sizes of axons and introns of the present-day gene for past contributions by different genes that might have encoded proteins with different binding activities and which might have recombined to form the present gene.

The gene structure appears to be reflected in the structure of the protein (cf., Figs. 4 and 7; Hirano *et al.*, 1983, and Petersen *et al.*, 1983). The protein contains a series of homologous repeating units of the same equivalent size (45–50 amino acids) or of twice that size. The existence of small, repeating

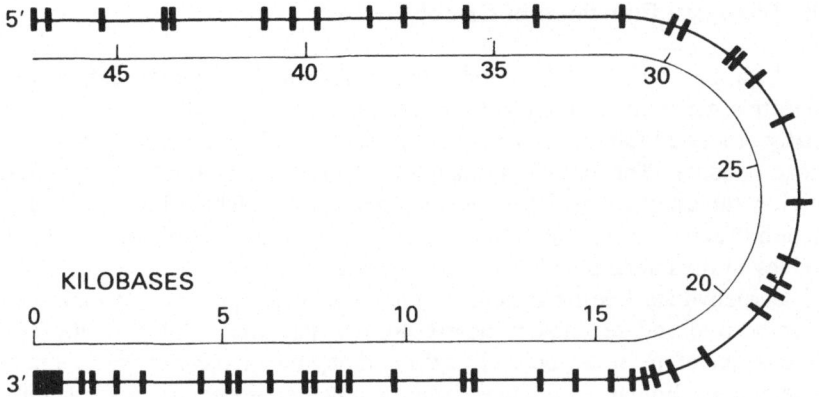

FIGURE 7. Schematic representation of the chicken fibronectin gene. The gene is depicted as the line with thick vertical bars; the gene is approximately 48 kilobases in length. The thick bars indicate coding regions (exons), and the line in between represents intronic material. Based on data by Hirano *et al.*, 1983.

units in both the gene and the final protein strongly suggests that the gene for fibronectin evolved from a series of gene duplications of one or only a few, small primordial genes. It appears likely that substantial evolutionary modification would have been required to obtain the present-day, highly complex, multifunctional protein from such a repetitive protein product. Although a similar mechanism is thought to have occurred for albumin, collagen, and other proteins (Brown, 1976; Y. Yamada *et al.*, 1980), it is somewhat surprising that a multifunctional protein would use a similar mechanism. It will be interesting to examine the structure of other genes for extracellular molecules to determine whether this mechanism was a general one for the generation of this class of molecules.

The existence of a large number of exons in the fibronectin gene would permit a limited amount of alteration in the final messenger RNA product dependent on the types of splicing. It will be particularly informative to compare the sequences of genomic DNA with the protein sequences of plasma and cellular fibronectins. One approach would be to compare fibronectin cDNAs from different tissues, based on recent cDNA cloning of chicken and human material (Fagan *et al.*, 1981; Kornblihtt *et al.*, 1983; Oldberg *et al.*, 1983). Splicing differences might also account for differences between the A and B chains of fibronectin, although posttranslational modifications are also possible explanations of the differences. The biological significance of these differences in the final product of the fibronectin gene are also important to elucidate.

6.8. Biological Functions of Carbohydrates

With the exception of teratocarcinoma fibronectin, cellular and plasma fibronectins generally contain only asparagine-linked carbohydrates, which are usually processed fully to the complex oligosaccharide containing sialic acid and fucose residues. The protein is therefore a good target for the glycosylation inhibitor tunicamycin, which relatively specifically inhibits this type of glycosylation (Tamura, 1982). It is therefore possible to test the role of carbohydrate residues in the interactions of this glycoprotein with cells.

Carbohydrate-free fibronectin synthesized in the presence of tunicamycin is synthesized and secreted in nearly normal quantities (Olden *et al.*, 1978). There is, therefore, no evidence that the carbohydrate moiety serves as a marker for secretion. Similar results have been obtained subsequently in most, but not all, glycoprotein secretory systems (Yamada and Olden, 1982). Nonglycosylated fibronectin is also fully active in a series of assays for biological activity of the molecule, including tests for its interaction with various cells (Olden *et al.*, 1979). There is, therefore, also no evidence for a role for the carbohydrate moiety in the interaction of fibronectin with the cell surface.

One major difference between nonglycosylated and normally glycosylated fibronectins, however, is in their stability against proteolytic degradation. Fibronectin molecules lacing carbohydrate are severalfold more sensitive to proteolytic attack, and the molecule undergoes a threefold higher rate of turnover on living cells (Olden *et al.*, 1978, 1979). The collagen-binding domain of fibronectin is the most heavily glycosylated. In the absence of carbohydrates, this domain is more than an order of magnitude more susceptible to proteolysis (Bernard *et al.*, 1982b). In the absence of carbohydrates, this domain is cleaved in a disulfide-stabilized loop; even the earliest cleavages cause a loss of the capacity of the domain to bind to collagen (Bernard *et al.*, 1982a; B. Bernard, M. Humphries, S. Newton, K. Yamada, and K. Olden, unpublished results). The carbohydrate moiety on fibronectin, therefore, appears to play no role in its synthesis, secretion, or yet known biological activities. It instead may act in a domain-specific fashion to protect a particularly important, protease-sensitive region of the molecule against proteolytic attack and abnormal rates of protein turnover.

6.9. Cell Surface Receptors for Fibronectin

6.9.1. Platelet Receptor for Fibronectin

Fibronectin can be bound to different cell types by means of at least three different types of receptor. The platelet receptor for fibronectin is expressed on the surface of platelets after their activation by thrombin (Ginsberg *et al.*, 1980; Plow and Ginsberg, 1981). This receptor binds fibronectin with moderately high avidity ($K_D = 3 \times 10^{-7}$ M), and there are approximately 100,000 binding sites per platelet. Although the involvement of fibrinogen in this binding has not been entirely excluded, the relatively high affinity suggests that some other, as yet unidentified platelet membrane component is involved; this receptor has not yet been isolated.

6.9.2. Bacterial Receptor for Fibronectin

Certain bacteria can bind to fibronectin in a functionally irreversible interaction (Proctor *et al.*, 1982; Ryden *et al.*, 1983). Although binding constants are not definitive with such a reaction, the estimated avidity is $K_D = 2 \times 10^{-10}$ M. There are an estimated 250–7500 receptors/bacterium, and the number and apparent affinity of receptor vary according to the type of growth medium (Proctor *et al.*, 1982). The receptor for fibronectin on *S. aureus* is a cell surface protein of 197,000 daltons, which apparently remains active even after degradation to a fragment of 18,000 daltons (Espersen and Clemmensen, 1982; Ryden *et al.*, 1983). It will be of considerable interest to determine whether this receptor is important to the pathogenicity of bacteria, and which amino acid sequence in fibronectin it recognizes.

6.9.3. Fibroblast "Receptor" for Fibronectin

The fibronectin "receptor" on fibroblasts has been surprisingly difficult to characterize. Under physiological conditions, plasma fibronectin was originally claimed to bind relatively poorly to these cells; activity was usually readily demonstrable only after the fibronectin has become bound to a substrate (Klebe, 1974; Pearlstein, 1978; Grinnell and Hayes, 1978a). In contrast, aggregated cellular fibronectin had been reported to bind weakly to cells (Hahn and Yamada, 1979b; Yamada, 1980). The "receptor" therefore appears to have low avidity for soluble fibronectin. Besides binding cellular fibronectin (albeit weakly), cells can also bind significant amounts of plasma fibronectin at 4°C or under certain monolayer culture conditions (Grinnell *et al.*, 1982b; McKeown–Longo and Mosher, 1983). The estimated affinities (4×10^{-8} M) are substantially higher than they appear to be for cells in suspension. The mechanisms of binding in these two situations are not known, although one speculation is that the plasma fibronectin forms aggregates at 4°C or, in the latter case, in the presence of unknown extracellular matrix components. The aggregates would then bind to cells with higher total avidity. It is also important to keep in mind the magnitude of binding required to demonstrate adhesive activity. Approximately 50,000 fibronectin molecules are reportedly required to mediate the spreading of each BHK cell (Hughes *et al.*, 1979); this number is much higher than required for hormone action.

More recently, the direct interaction of soluble fibronectin with fibroblastic cells in physiological medium, under assay conditions specifically designed to detect a low-affinity receptor, has been characterized (Akiyama and Yamada, 1984). Binding of tritiated fibronectin is specific and saturable with high (2 mg/ml) concentrations of fibronectin. There appears to be a single class of 500,000 sites per cell, suggesting that the fibronectin receptor is a major cell-surface component. The apparent dissociation constant of 8×10^{-7} M indicates that fibronectin binds with only moderate affinity. Results from both direct binding studies (Akiyama and Yamada, 1984) and biological assays for fibronectin receptor function (Oppenheimer-Marks and Grinnell, 1984) show that although the fibronectin receptor can be destroyed by trypsinization in the absence of divalent cations, it is resistant to trypsin in the presence of physiological concentrations of divalent cations. These results suggest that a protein stabilized by divalent cations is crucial for fibronectin receptor function.

On the other hand, it is well established that fibroblasts can bind or attach avidly to even low concentrations (1–10 μg/ml) of fibronectin adsorbed onto beads or tissue culture substrates, even at 37°C. Suggested explanations for this large apparent enhancement of activity by prior fibronectin binding to a substrate have included an allosteric activation of fibronectin by the substrate (Pearlstein, 1978) or, perhaps more likely, an increase in the apparent affinity of binding by permitting the multipoint attachment of a low-affinity receptor (discussed by Grinnell, 1980a, and analyzed theoretically by Brandts and Jacobson, 1983). A third possibility is that the substrate is required to concentrate fibronectin from solution into a thin zone of high fibronectin density. For example, if the fibronectin originally present in a 1.5-mm layer of culture medium containing 1 μg/ml fibronectin is concentrated by adsorption onto the substrate to form a layer 150 nm thick on the dish, the fibronectin would be concentrated 10,000-fold to 10 mg/ml (5×10^{-5} M monomer). None of these three mechanisms has been ruled out yet.

Using fibronectin-coated beads, a rough estimate of cell surface affinity to multivalent fibronectin on a substrate has been suggested by McAbee and Grinnell (1983) of 4×10^{-9} M involving a minimum of 60,000 fibronectin receptors/cell. If such a receptor recognizes even a four-amino-acid region of fibronectin (see Section 6.4.1), it may be a specialized binding molecule; simpler electrostatic binding mechanisms have not yet been excluded, however.

If a specific, saturable receptor exists for an adhesion protein, a theoretical prediction is that the protein or its active fragments could become inhibitory under certain conditions. This dualistic nature of adhesive protein function has been demonstrated for fibronectin (Yamada and Kennedy, 1984). An excess of plasma fibronectin in solution can block the function of a fixed amount of fibronectin adsorbed to a substrate, presumably by saturating all free cellular receptors and preventing their interaction with the substrate-bound fibronectin. The cell-binding fragment of fibronectin also inhibits fibronectin function, as do certain synthetic peptides from this region of the protein. From a biological viewpoint, these results may explain why the fibronectin receptor is of relatively low affinity: If the receptor were of high affinity, plasma fibronectin or fibronectin fragments would immediately saturate the receptors, preventing cell attachment to fibronectin bound to substrates. Adsorption of fibronectin to substrates increases the apparent affinity by its multivalency. In fact, substrate-bound fibronectin does out-compete inhibitory peptides as its concentration is increased, thus permitting cell adhesion (Yamada and Kennedy, 1984).

The molecular identity of the "receptor" on fibroblasts that recognizes fibronectin is still unknown. Candidates to date have included gangliosides (Kleinman et al., 1979; Yamada et al., 1981, 1983), a 47,000-dalton glycoprotein (Aplin et al., 1981; Hughes et al., 1981), or other protein molecules (Tarone et al., 1982; Oppenheimer–Marks and Grinnell, 1982). Gangliosides may be able to serve as, or at least to substitute for, a receptor function, since the addition of complex gangliosides such as G_{D1a} to a fibroblast mutant lacking gangliosides restores the ability of these cells to bind and organize fibronectin into a fibrillar matrix (Yamada et al., 1983). Gangliosides, however, cannot be the sole receptors for fibronectin, since some neuronal cells contain similar gangliosides and yet do not bind fibronectin (Raff et al., 1979), and the fibroblast "receptor" function is inhibited by proteases and certain lectins (Grinnell, 1980a; Oppenheimer–Marks and Grinnell, 1982; Tarone et al., 1982). Glycoproteins are therefore the more popular candidates for the fibronectin "receptor" at present.

Chemical cross-linking studies have implicated a 47,000-dalton glycoprotein as a possible receptor (Aplin et al., 1981). Antibodies raised against an impure preparation of this glycoprotein inhibit cell spreading on fibronectin substrates (Hughes et al., 1981). Although one reservation to these results is that the same glycoprotein is implicated in the lectin-mediated spreading of cells, it is possible that the same protein might be responsible for both interactions. As yet, however, the molecule has not been purified for more rigorous tests of its role in adhesion.

Other studies have implicated molecules of 80,000 and 120,000 daltons, since there are correlations between the presence of these molecules and the capacity of cells to undergo fibronectin-mediated cell spreading, as well as the converse correlation of complexing with antibodies to them and an inhibiton of spreading (Hsieh and Sueoka, 1980; Tarone et al., 1982). The relationship of these proteins to the general cell-to-substrate adhesion molecules described in Section 4.3 is not yet clear.

Monoclonal antibodies against one such set of molecules of 120,000–140,000 daltons inhibit myoblast adhesion to gelatin (Greve and Gottlieb, 1982; Neff *et al.*, 1982). Since myoblast attachment to gelatin requires fibronectin from the serum used in culture media (Chiquet *et al.*, 1979), it appears likely that these antibodies are inhibiting fibronectin-mediated adhesion. Preliminary results with chick fibroblasts further support the involvement of this antigen in adhesion to fibronectin-coated substrates, which is inhibited by this antibody (W. T. Chen, E. Hasegawa, and K. M. Yamada, unpublished results). Whether this protein is the fibronectin receptor or a regulatory molecule is not yet clear.

A final possibility is that the "receptor" for fibronectin can actually be several different molecules, depending on the type of cell. Experiments with monoclonal or monospecific polyclonal antibodies against each of the proposed receptor molecules, as well as the use of synthetic peptides to probe the binding site, may provide a more convincing identification of this important receptor.

7. LAMININ

7.1. Properties

The glycoprotein laminin (*lamina* = layer) is a characteristic component of basement membranes. In contrast to more generally distributed extracellular proteins such as fibronectin, laminin is generally present only in the basal lamina of a wide variety of tissues (Chung *et al.*, 1979; Timpl *et al.*, 1979; Ekblom *et al.*, 1980; Hogan *et al.*, 1980; Wewer *et al.*, 1981). The exact location of laminin within the basement membrane is not entirely clear, although it is generally felt to be present in substantial amounts in the lamina rara of the basal lamina, according to immunoelectron microscopic criteria (reviewed by Farquhar, 1981, but see also Laurie *et al.*, 1982). This location directly under cells suggests that it is important for the interactions of epithelia with the basement membrane.

In vitro adhesion assays with laminin purified from tumors or from parietal endoderm cell lines show that laminin can function as an adhesive protein (Terranova *et al.*, 1980; Vlodavsky and Gospodarowicz, 1981; Johansson *et al.*, 1981). Since type IV collagen is a major structural component of basement membranes, it is logical that laminin might mediate the attachment of certain epithelial cells to native type IV collagen (Terranova *et al.*, 1980). This hypothesis has been tested, with seemingly variable results (for example, compare Kleinman *et al.*, 1981 with Engvall and Ruoslahti, 1983). Such assays are usually conducted in the presence of cycloheximide to block the synthesis of endogenous laminin. Under these conditions, laminin is usually reported to function as an adhesive protein; the major uncertainty is the extent of cell-type specificity of laminin compared to other attachment proteins. Laminin has been reported to be necessary for the attachment of a transformed epidermal cell to type IV collagen, and fibronectin has no stimulatory effect, although direct attachment to type IV collagen by these cells is substantial (Terranova *et al.*, 1980). On the other hand, normal epidermal cells can attach and spread on fibronectin and epibolin, as well as laminin (Stenn *et al.*, 1983), even though they reportedly attach preferentially to type IV collagen compared to other collagen types (Murray *et al.*, 1979). Other transformed or tumor cells show varying degrees of preferential use of laminin as an attachment factor in comparison to fibronectin ranging from substantial (Vladovsky and Gospodarowicz, 1981) to modest (Terranova *et al.*, 1982). In contrast, a number of untransformed cells do not appear to show specificity for laminin, including hepatocytes (Johansson *et al.*, 1981; discussed previously), endothelial cells (Macarak and Howard, 1983; Palotie *et al.*, 1983), epidermal cells (Stenn *et al.*, 1983), and corneal epithelial cells (Sugrue and Hay, 1981; Scott *et al.*, 1983), One interpretation of these results is that epithelial cells can use any of several attachment proteins for adhesion. The specificity observed for some tumor cells may be related to the use of laminin by these cells for metastasis (Terranova *et al.*, 1982).

Since laminin is present in substantial quantities in basement membranes in apparent association with type IV collagen and a variety of epithelial cells, it has been suggested that laminin functions as a major adhesive protein for these cells *in vivo* (Kleinman *et al.*, 1981). However, it has been questioned whether laminin can actually bind to type IV collagen (Engvall and Ruoslahti, 1983), although evidence for binding has been obtained in certain assays (Kleinman *et al.*, 1983; Woodley *et al.*, 1983). One possible explanation of these discrepancies is that laminin may bind to type IV collagen with only modest affinity (N. C. Rao and L. A. Liotta, personal communication, 1983).

Although the nature of its binding to type IV collagen is still disputed, laminin clearly binds to heparin and heparan sulfate (Sakashita *et al.*, 1980; Del Rosso *et al.*, 1981; Woodley *et al.*, 1983). Binding to heparan sulfate proteoglycans may be important, since they are prominent constituents of the basal lamina; it has been suggested that the components of this structure can self-assemble (Kleinman *et al.*, 1983; Woodley *et al.*, 1983).

7.2. Specificity of Epithelial Cell Interactions

Although epithelial cells interact with laminin, fibroblastic cells may also interact with this attachment protein. Laminin can mediate the spreading of fibroblasts on artificial substrates, although it is an order of magnitude less active for one such cell line (Kennedy *et al.*, 1983; Couchman *et al.*, 1983). Laminin is substantially less effective in mediating attachment of a fibroblastic cell to type I collagen, suggesting that its greatest specificity may be in its recognition of native type IV collagen (Kennedy *et al.*, 1983; Kleinman *et al.*, 1983). On the other hand, epidermal cells interact readily with laminin, fibronectin, epibolin, and type IV collagen in simple assays for cell spreading on plastic substrates (Stenn *et al.*, 1983). These results can be viewed as showing that epidermal cells have "receptors" or interaction sites for a variety of such molecules, even though they might normally depend on laminin for interacting with type IV collagen and the basal lamina in intact tissues.

It is therefore possible that epithelial cells undergo different interactions with distinct attachment factors depending on their developmental state and local environment. For example, after wounding of the epithelium of the cornea, or epidermis, the wound surface becomes covered with fibronectin and fibrin within 8 hr (Fujikawa *et al.*, 1981; Clark *et al.*, 1982). Cell migration of epithelial cells to cover the wound then begins, and this migration is reportedly sensitive to inhibition by antifibronectin antibodies (Nishida *et al.*, 1983). Later, another protein termed bullous pemphigoid antigen appears in the newly formed basement membrane (Fujikawa *et al.*, 1981; Stanley *et al.*, 1981; Clark *et al.*, 1982). Finally, laminin and type IV collagen replace the fibronectin and fibrin as the wound heals (Fujikawa *et al.*, 1981; Clark *et al.*, 1982). Laminin may be the final attachment protein characteristic of certain mature epithelia, whereas fibronectin may be transiently involved only in the process of active epithelial cell migration.

Stage-specific requirements for attachment factors can also be seen in cell adhesion studies *in vitro*. Laminin *in vitro* can promote the attachment of both fetal liver cells (Hirata *et al.*, 1983) and regenerating liver cells (Carlsson *et al.*, 1983) from mice. However, laminin has little effect on normal adult liver cells which, instead, attach to fibronectin.

Cells of the peripheral nervous system can use both laminin and fibronectin as substrates for extension of neurites (Akers *et al.*, 1981; Baron–Van Evercooren *et al.*, 1982; Rogers *et al.*, 1983). In contrast, cells derived from the central nervous system may be able to interact only with laminin (Rogers *et al.*, 1983). This difference in specificity for attachment proteins may have implications for developmental regulation of axon extension, which may be regulated by both the capacity of cells to respond to adhesive proteins and the type of extracellular glycoproteins present in potential migratory regions.

7.3. Structure of Laminin

Laminin molecules are comprised of subunits of 200,000 and 400,000 daltons linked by disulfide bonds (Fig. 8; Chung *et al.*, 1979; Timpl *et al.*, 1979; Hogan *et al.*, 1980; Wewer *et al.*, 1981; Cooper *et al.*, 1981; Ott *et al.*, 1982;

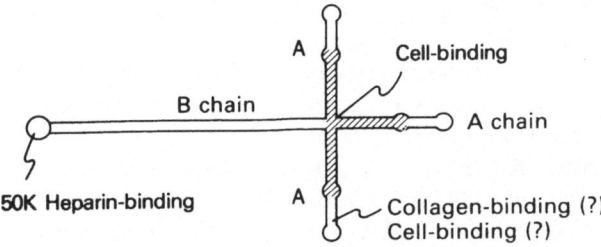

FIGURE 8. Structure of laminin. Laminin exists as a cruciform-shaped molecule according to rotary shadowing. The one B and possibly three A chains are linked together by disulfide bonds; each has distinctive numbers of globular domains. The cell-binding activity of the protein can be recovered in the central pepsin-resistant fragment and in one or more peripheral domains. Heparin-binding activity is present in the globular domain of the B chain, although other sites may exist. Binding activity for type IV collagen appears to require the peripheral domains, although the nature of the binding activity to collagen is not well understood. See text for details.

Rao *et al.*, 1982a,b; Engvall *et al.*, 1983). One of the large "B" subunits is probably linked by interchain disulfide bonds to three of the smaller "A" subunits to form a complete laminin molecule, which totals approximately 1,000,000 daltons. This stoichiometry and the possible existence of different forms of A chains require further examination. Laminin molecules examined by rotary shadowing electron microscopy display a highly distinctive cruciform shape (Fig. 8). The long arm is thought to be composed of the B subunit, and the short arms may each be a A subunit (Engel *et al.*, 1981; Ott *et al.*, 1982; Rao *et al.*, 1982b), although this conclusion should be verified by biochemical or immunological methods to evaluate the extent of interdigitation of the subunits.

Laminin produced by parietal endoderm cells from teratocarcinomas may contain a third polypeptide component of only 150,000 daltons that is not covalently attached, as well as only two B chains (Hogan *et al.*, 1980). Other types of laminin or laminin subunits may also exist, although they are not yet extensively characterized (Rohde *et al.*, 1980; Ohno *et al.*, 1983). Laminin purified from normal tissues has a similar structure to the material purified from tumors (Gospodarowicz *et al.*, 1981; Smith and Strickland, 1981; Clark and Kefalides, 1982; Ohno *et al.*, 1983).

7.3.1. Structural Domains of Laminin

A striking structural feature of laminin is its highly ordered arrangement of linear elements and more globular domains at the ends of the arms (Fig. 8; Engel *et al.*, 1981). Nearly identical structural features are found in laminin molecules isolated from insects, which suggests a marked evolutionary conservation of this overall configuration of structural units (L. Fessler, personal communication). This relatively highly structured shape of the molecule contrasts with the flexible and pleiomorphic morphology of the fibronectin molecule (e.g., see Erickson and Carrell, 1983; Rocco et al., 1983).

Laminin also contains distinct structural domains by biochemical criteria. Thermal denaturation studies combined with circular dichroism analyses of laminin undergoing digestion by elastase suggest that the protein is composed of domains that are resistant to elastase and heat denaturation, which are connected by regions of α helix that are destroyed by these treatments (Ott et al., 1982).

7.3.2. Functional Domains of Laminin

Proteases can cleave laminin into domains that still retain binding activities for interactions with the cell surface or with heparin (Timpl et al., 1979; Ott et al., 1982; Rao et al., 1982a,b; Timpl et al., 1983). The locations of these domains (Fig. 8) are still tentative, but it has been reported that the central region of the molecule, as well as one or more peripheral domains, contains sites for binding to the cell surface (Rao et al., 1982a,b; Timpl et al., 1983). For example, a pepsin-resistant, 300,000-dalton fragment from the center of the molecule can still bind to cells, although it can no longer promote the attachment of cells to type IV collagen (Rao et al., 1982b; Terranova et al., 1983). It has been suggested that the proteolysis removes binding sites for type IV collagen (Terranova et al., 1983). This central, pepsin-resistant fragment is rich in cystine residues and consists of a mixture of fragments held together by disulfide bonds (Rohde et al., 1980). A similar in vitro analysis has identified cell-binding activity in at least one other region of the molecule, apparently in one or more of the terminal globular domains (Timpl et al., 1983).

A globular domain of 50,000 daltons can be recovered after protease digestion which contains a heparin-binding site (Ott et al., 1982). This domain contains a substantial proportion of the β structure present in laminin, as indicated by circular dichroism studies. It appears to be located at the end of the β subunit. Other, weaker heparin-binding sites may also exist in laminin, however (Ott et al., 1982).

Figure 8 depicts a tentative functional map of laminin, and it is important to emphasize that its structure is not well understood. The existence of several possible cell-binding and heparin-binding sites, as well as the uncertainties about the existence and location of collagen-binding sites, requires clarification. One reason for the apparent difficulties in these analyses may be the high degree of structure in the laminin molecule. Unlike fibronectin, it may not possess con-

veniently located regions of flexible polypeptide chain that permit a rapid proteolytic dissection of the molecule into functional domains, and its cell-binding
regions may be more complex than the putative, single, non-disulfide-bonded
site on fibronectin. Further analyses of the functional organization of laminin
may therefore not be simple.

7.4. Cellular Receptors for Laminin

Putative plasma membrane receptors for laminin have been characterized
recently (Terranova *et al.*, 1983; Malinoff and Wicha, 1983; Rao *et al.*, 1983;
Lesot *et al.*, 1983). In contrast to fibronectin, but similar to collagen, laminin
binds to the cell surface in a saturable interaction with high apparent affinity
($K_D = 2 \times 10^{-9}$ M). There are approximately 100,000 copies of this receptor
per BL6 melanoma cell, and 50,000 on a fibrosarcoma cell. The high affinity
of this interaction has permitted the isolation of the receptor by affinity chromatography (Rao *et al.*, 1983; Malinoff and Wicha, 1983; Lesot *et al.*, 1983).
When plasma membranes from tumor cells or muscle are solubilized in nonionic
detergents and chromatographed on immobilized laminin, only the putative receptor binds (as well as type IV collagen in one study). The receptor can be
eluted by low pH; after neutralization, the receptor can be adsorbed to nitrocellulose filters, where it remarkably displays the same binding parameters as
the original cellular receptor. The receptor is a disulfide-linked complex containing a glycoprotein of approximately 67,000–69,000 daltons (Rao *et al.*, 1983;
Malinoff and Wicha, 1983).

The putative laminin receptor from muscle cells may differ in some of its
properties from the receptor from tumor cells. After solubilization, this receptor
could be eluted from laminin affinity columns with physiological concentrations
of salt. Reconstitution of the molecule into liposomes was reported to restore
its capacity to bind to laminin even at elevated ionic strength (Lesot *et al.*, 1983).
These possible differences between laminin receptors from different sources
require examination; it is important to determine whether there are different
classes of receptors for laminin with differing properties, or whether the observed
differences in salt sensitivity are simply the result of experimental differences,
such as differing detergent concentrations. Another important question involves
the specificity of these putative receptors, since one study could find only a
twofold greater extent of binding of the detergent-solubilized "receptor" to laminin than to fibronectin or to type I collagen (Lesot *et al.*, 1983).

The isolation of these putative receptors for laminin should provide valuable
opportunities to determine its distribution on different cell types and its mechanism of action. Important questions for the future include whether it functions

by means of a single subunit or as a complex with other molecules, whether this receptor can provide a link to intracellular molecules, such as cytoskeletal proteins, and whether the molecule is truly specific for binding to laminin.

8. SERUM SPREADING FACTOR(S)

Besides fibronectin and chondronectin (reviewed later), serum contains at least one additional glycoprotein attachment factor involved in the adhesion of cells to two-dimensional collagen substrates (Klebe *et al.*, 1978) and to plastic tissue culture dishes (Knox and Griffiths, 1980, 1982). Although fibronectin provides a well-documented cell-spreading activity (reviewed previously), it is not essential for the attachment and spreading of at least several cell types in tissue culture (Knox and Griffiths, 1980; Grinnell and Minter, 1978; Virtanen *et al.*, 1982). Although it is generally accepted that many cells require attachment factors of some type to attach, spread, and proliferate adequately in culture in a process termed "anchorage dependence" (Stoker *et al.*, 1968; Grinnell, 1978; Barnes and Sato, 1980), the *in vivo* physiological equivalent of this *in vitro* requirement for spreading factors is not yet clear. One hypothesis is that such factors cause alterations in cell shape, which alters expression of genes involved in growth regulation (Folkman and Greenspan, 1975; O'Neill *et al.*, 1979; Farmer *et al.*, 1978). It remains to be determined how the factors described later function *in vivo*, since it is not yet known what biological molecule or substrate would be the *in vivo* equivalent of the plastic substrate used to identify these molecules.

Many tissue culture media contain 5–15% serum to provide essential nutrients and growth factors. Serum proteins rapidly adsorb to artificial cell culture substrates, providing a layer of nonspecifically adsorbed serum proteins (e.g., see Pegrum and Maroudas, 1975). At this concentration of serum, only minimal amounts of serum fibronectin can attach, apparently because other proteins saturate the culture surface too rapidly (Knox and Griffiths, 1980; Grinnell and Minter, 1978). In contrast, another serum protein termed "serum spreading factor" is still capable of interacting with the cells; whether this factor is fully active because even minimal amounts are sufficient, or because it is more effective than fibronectin in competing for attachment to the substrate, is not yet clear. Serum spreading factors appear to be much more important for the *in vitro* spreading of some cells than fibronectin (Knox and Griffiths, 1982). Nevertheless, fibronectin may be necessary for the establishment of extensive arrays of actin microfilament bundles, although this point has not been examined extensively (Couchman *et al.*, 1982; Virtanen *et al.*, 1982; Laterra *et al.*, 1983).

Fibronectin itself is therefore not necessary for cell spreading if replaced by other factors. It is instructive to consider other cases in which fibronectin is not required. Plastic Petri dishes are modified by manufacturers to create tissue culture dishes, e.g., by a glow discharge. A key physical factor involved in this modification is the production of hydroxyl groups on the surface (Curtis *et al.*, 1983). Surfaces containing a high degree of hydroxylation are reported to permit cell spreading even in the absence of factors such as fibronectin; in this case, hydroxylation may be replacing the function of an attachment factor (Curtis *et al.*, 1983). Some cell lines can be cultured in defined medium lacking any protein supplements. Mouse L-cell lines that grow in serum-free medium produce substantial amounts of fibronectin themselves, which adsorbs to the culture substrate (Hynes, 1973; Yamada *et al.*, 1983).

Recently, several laboratories have independently described spreading factors that have apparent molecular weights of about 70,000. All these factors may actually be the same protein, although no direct comparisons have been published to date. The first protein reported to be purified to homogeneity was a spreading factor for both fibroblastic and epithelial cells, which has been isolated from fetal calf serum (Whateley and Knox, 1980). It is a monomer of 62–70,000 daltons containing 12% carbohydrate and with an acidic isoelectric point. This factor is reported to differ from fibronectin and laminin in its requirement for concurrent cellular protein synthesis in order to obtain cell attachment (Knox and Griffiths, 1980). It also mediates adhesion substantially more slowly than the attachment factor fibronectin. If these properties are confirmed with highly purified material, they will suggest a complex mechanism for promoting cell spreading, perhaps involving the new synthesis of a second cellular protein necessary for spreading.

Another factor of 70,000 daltons originally described by Holmes (1967) can be purified on glass bead columns and is required for the spreading and growth of a number of types of cell cultures in serum-free medium (Barnes *et al.*, 1980, 1983; Barnes and Silnutzer, 1983). This factor migrates as a doublet band on SDS gels, and it appears to act directly as a cell-attachment factor adsorbed to a substrate. For example, it can be subjected to disulfide bond reduction and denaturation in SDS, yet still displays activity after electroblotting onto nitrocellulose. Cells plated onto nitrocellulose sheets attach only at the sites where this protein or fibronectin was transferred (Hayman *et al.*, 1982). This protein has also been purified by affinity chromatography with a monoclonal antibody, and it has been found in tissues *in vivo* in a fibrillar pattern (Hayman *et al.*, 1983). It has been renamed "vitronectin" (*vitrum* = glass + *nectere* = to bind, connect) because of its affinity for glass substrates (Hayman *et al.*, 1983).

A third glycoprotein of 65,000 daltons termed "epibolin" has been isolated from human plasma as a protein required for the migration of epidermal cells (Stenn, 1981). This protein resembles "vitronectin" and the "serum spreading factor" of Whateley and Knox. These three proteins may simply be the same protein, a possibility that is important to examine in the near future with monoclonal antibodies and biological assay. It will be especially informative to obtain molecular comparisons of the mechanisms of action of this class of serum factor with fibronectin and laminin.

9. OTHER CELL-SUBSTRATE ADHESION FACTORS

9.1. Lectins, Antibodies, and Glycosidases

A number of other molecules that bind cell surface molecules can produce cell spreading on plastic substrates. For example, lectins such as concanavalin A, cationized ferritin, antiplasma membrane antibodies, and various glycosidases can mediate the spreading of fibroblasts (see Section 4.1 and Grinnell, 1978). On the other hand, although hepatocytes attach to asialoglycoproteins, they do not spread (Gjessing and Seglen, 1980). It is therefore possible that attachment and spreading are distinct phenomena, and that binding of some types of membrane proteins to the substrate is not sufficient to mediate cell spreading, e.g., because of poor coupling to cytoskeletal proteins. It will thus be important to establish whether seemingly nonspecific spreading factors such as lectins act simply by binding the plasma membrane as a whole to the substrate, or whether they act by binding to a specific receptor for a cell-attachment molecule, and only incidentally to the many other membrane molecules that share the carbohydrate determinant bound by the lectin (see also Rees et al., 1977).

9.2. Chondronectin

Chondronectin (chondros = cartilage + nectere = to bind, connect) is a glycoprotein purified from chicken and human sera which is reported to stimulate or mediate the attachment of chondrocytes to cartilage type II collagen in vitro (Hewitt et al., 1980, 1982a,b). The protein from chicken serum consists of subunits of approximately 70,000 daltons linked by disulfide bonds to form a covalent complex of 180,000 daltons (Hewitt et al., 1982a). At even low concentrations, e.g., 50 ng/ml, this protein is reported to stimulate the attachment of chondrocytes, but not of fibroblasts or epithelial cells, to type II collagen. In vivo, this protein is found at the interface between cells and the surrounding extracellular matrix (Hewitt et al., 1982a,b). Chondronectin may act by forming a complex with both cartilage proteoglycan and type II collagen, rather than by binding to type II collagen alone. It has been shown to have specificity for binding to chondroitin sulfate but not to certain other glycosaminoglycans (Hewitt *et al.*, 1982b). Although chondronectin is relatively specific for chondrocyte attachment to type II collagen, it would be of interest to evaluate its specificity in interactions with chondrocytes compared to other cells in the absence of collagen, as well as its specificity with other types of collagen as assayed by cells other than chondrocytes. It should also be noted that chondrocytes can also interact directly with type II collagen by means of a putative membrane receptor distinct from chondronectin (Mollenhauer and von der Mark, 1983; see Section 5.6 on collagen receptors). A receptor for chondronectin has not been identified as yet on the cell surface. It is possible that chondronectin may be involved more in strengthening a cell-to-collagen interaction than in acting as a direct connecting molecule between chondrocytes and collagen. Further studies with purified chondronectin should resolve these questions.

9.3. Interactions with Fibrin and Fibrinogen

When platelets are stimulated by thrombin or other agonists, they express receptors for the binding of fibrinogen and then aggregate. There are 4×10^4 fibrinogen receptors with an affinity of $K_D = 10^{-7}$ M per activated platelet. Chemical cross-linking and other methods indicate that this receptor contains a 105,000-dalton platelet membrane protein known as glycoprotein IIIa (e.g., see Bennett *et al.*, 1982).

Fibroblasts can bind both fibrin and fibrinogen, perhaps by a mechanism involving fibronectin (Colvin *et al.*, 1979). This interaction may be important for fibroblast migration into wounds (see Section 7.2). Macrophages can also bind fibrin, presumably as a prelude to endocytosis. Although binding can occur by a nonsaturable mechanism involving fibronectin (Jilek and Hörmann, 1978b), macrophages also bind fibrin monomer in a saturable, high-affinity interaction. The latter binding occurs at the amino terminus of the fibrin α chain (Gonda and Shainoff, 1982).

9.4. Other Platelet Interactions

Besides interactions with fibrinogen and fibronectin, platelets also interact with von Willebrand factor (factor VIII) and thrombospondin. Factor VIII von Willebrand factor mediates the adhesion of platelets to subendothelium exposed after injury (Weiss *et al.*, 1978; structure and function reviewed in Hoyer, 1980). Thrombospondin appears to function as a platelet lectin in promoting the aggregation of activated platelets (Gartner *et al.*, 1978; Jaffe *et al.*, 1982). This glycoprotein is comprised of three disulfide-linked subunits combined to form a 450,000-dalton protein (Margossian *et al.*, 1981). On the platelet plasma membrane, thrombospondin apparently binds to fibrinogen (Gartner *et al.*, 1981; Leung and Nachman, 1982) but may also interact with fibronectin; however, the latter interaction may not be required for function (Lahav *et al.*, 1982; Ginsberg *et al.*, 1983). Since thrombospondin is also present on fibroblasts in an extracellular matrix, as well as on endothelial cells, it may have more general functions than in the aggregation of platelets (McPherson *et al.*, 1981; Mosher *et al.*, 1982; Jaffe *et.al.*, 1983). Further evaluation of its functions and cellular receptors will be of interest in the future.

10. EXTRACELLULAR MOLECULES IN DEVELOPMENT

The cell surface and extracellular matrix are thought to have crucial functions in the processes of embryonic induction, organ morphogenesis, cell migration, and regulation of differentiated cell products. The possible involvement of various extracellular molecules in these events has been discussed in a number of publications; this chapter will therefore focus on certain key concepts and findings (for further discussion, see Wartiovaara *et al.*, 1980; Bernfield, 1980; Hay, 1981; Trinkaus, 1984; Yamada, 1983b).

Extracellular molecules are presumed to provide both structural and regulatory actions during development. For example, collagens are crucial structural molecules, and the disruption of type I collagen gene expression by the insertion of a retrovirus into a $5'$ site near the start of transcription of the $\alpha 1(I)$ gene is lethal to mouse embryos by days 12 to 13 of gestation (Schnieke *et al.*, 1983). This approach of using tumor virus integration into germ line genes as insertion mutagens may provide a general method for obtaining embryonic lethal mutations to study other genes required for embryonic development. Embryonic inducers probably involve extracellular molecules, although no convincing candidates have been identified to date (e.g., see Kratochwil, 1983). Two general approaches for beginning to determine the function of specific extracellular molecules in development have been to localize them immunologically *in vivo* at various stages of embryogenesis and to determine their effects on expression of differentiated function *in vitro*.

10.1. Time of Appearance of Development

Collagen does not appear to be synthesized in the developing mouse embryo until surprisingly late in the early phases of development; i.e., type I and type III collagens could not be detected for the first 7 days of development and were first detected at the 4-somite stage of mesenchymal tissues (Leivo *et al.*, 1980; Wartiovaara *et al.*, 1980). A substantial increase in mRNA for type I collagen is not seen until day 11 (Schnieke *et al.*, 1983). Type IV collagen of basement membranes appears earlier, appearing prior to implantation at day 3 in association with the blastocyst inner cell mass as it begins to differentiate into ectoderm and endoderm (Adamson and Ayers, 1979; Leivo *et al.*, 1980).

Laminin is first detected at a time earlier than the collagens and fibronectin, i.e., in the day 3 morula stage in which the embryos contain only 16–32 cells. In contrast, fibronectin appears slightly later in the mouse, when the inner cell mass is formed (reviewed in Wartiovaara *et al.*, 1980). In the chick embryo, fibronectin is first detected under the upper layer of the blastoderm immediately prior to the appearance of the primitive streak (Duband and Thiery, 1982b). Although it appears likely that these carefully choreographed patterns of appearance of specific extracellular proteins play some role in the organization or function of these embryos, little is known beyond these localization data.

10.2. Effects on Function—Migration

10.2.1. Laminin and Conditioned Medium Factors

Laminin is characteristically confined to basement membranes, where it is thought to function as an adhesive molecule (Kleinman *et al.*, 1981). It can also act as a potent promotor of neuronal outgrowth (Baron–Van Evercooren *et al.*, 1982; Rogers *et al.*, 1983).

At least two types of factors affecting neuronal development can be recovered from conditioned medium of explanted target cells (Collins, 1980; Coughlin *et al.*, 1981; Adler *et al.*, 1981; reviewed by Coughlin, 1984). Primary explants of a variety of tissues can produce these factors, which stimulate axon production and elongation, enhance the survival of dissociated neurons in culture, and increase levels of enzymes for the synthesis of differentiation-specific neurotransmitters. The factors appear to contain protein components, which are sensitive to denaturation by proteases and heat. These neuronotrophic factors exist as large complexes with apparent sizes of approximately 5,000,000 daltons.

They appear to function after binding to substrates, and an antiserum against the factor inhibits its *in vitro* activity (reviewed by Coughlin, 1984). A factor of this type, isolated from the culture media of endothelial cells, has been reported to be sensitive to heparinase, suggesting that it may contain an essential heparan sulfate proteoglycan component (Lander *et al.*, 1982). Since these factors may have important functions in directing and regulating neuronal outgrowth and differentiation, it will be of considerable interest to determine whether they are a class of related molecules as well as to determine their structure.

10.2.2. Fibronectin and Cell Migration

Fibronectin is present in connective tissue stroma as well as in association with basement membranes; it appears during gastrulation and is associated with a variety of other cell migratory events. For example, it is found in striking patterns along pathways along pathways of cell migration in the early avian embryo (Critchley *et al.*, 1979). It is present in large amounts in the basement membrane of the upper layer of gastrulating embryos, where it may form the substrate for migration of presumptive mesodermal cells; as cell migration ceases, the cells become enmeshed in a fibronectin matrix (Duband and Thiery, 1982b). Fibronectinlike antigenic material is also present in sea urchin embryos (Spiegel *et al.*, 1980); in fact, fibronectin antigen may be expressed in stage-specific fashion during the inward migration of sea urchin primary mesenchyme cells (Katow *et al.*, 1982). Fibronectin has also been detected along routes of cell movement during amphibian gastrulation (Boucaut and Darribere, 1983), and treatment of amphibian embryos with antifibronectin antibodies blocks gastrulation (J. P. Thiery, personal communication).

Fibronectin has also been found in association with cell migratory routes later in development, including along the neural crest pathway (Newgreen and Thiery, 1980; Mayer *et al.*, 1981; Duband and Thiery, 1982a; Thiery *et al.*, 1982a). There is a striking correlation between the presence of fibronectin during crest cell migration and its loss and subsequent appearance of the neuronal cell adhesion molecule N-CAM at the time of cessation of migration and formation of ganglia (Thiery *et al.*, 1982b). Fibronectin has also been reported to be closely associated with amphibian primordial germ cell migration (Heasman *et al.*, 1981), as well as with mouse cerebellar cell migration in the brain (Hatten *et al.*, 1982).

All these studies have suggested a role for fibronectin in promoting or guiding embryonic cell migration. Roles of fibronectin, laminin, and collagen have been compared in the adhesion and migration of explanted neural crest cells (Newgreen *et al.*, 1982; Rovasio *et al.*, 1983). The presence of fibronectin and a maintenance of a high local cell density were found to be required to mimic the conditions permitting optimal migration of these highly migratory cells (Rovasio *et al.*, 1983; Fig. 9). The regulation of neural crest cell migration *in vivo* is probably complex, since it is dependent on stage-specific tissue movements for the existence of transitory pathways for migration (Thiery *et al.*, 1982a). There may also be a surprising degree of bulk tissue shifting and movement, so that even inert latex beads can be translocated substantial distances during development (Bronner–Fraser, 1982); coating such beads with fibronectin can inhibit their translocation, suggesting that this molecule might inhibit or stimulate migration depending on the situation. All these concepts remain to be tested definitively, e.g., by immunological inhibition experiments and by examining for mutants in fibronectin and other extracellular molecules.

Tissue morphogenetic movements involved in the formation of feathers are also closely correlated with local accumulations of fibronectin; the rise in fibronectin is accompanied by a loss of type I and type III collagens at the same sites of morphogenetic activity (Mauger *et al.*, 1982). Another classical model system for analyzing cell–cell interactions during development involves experimentally

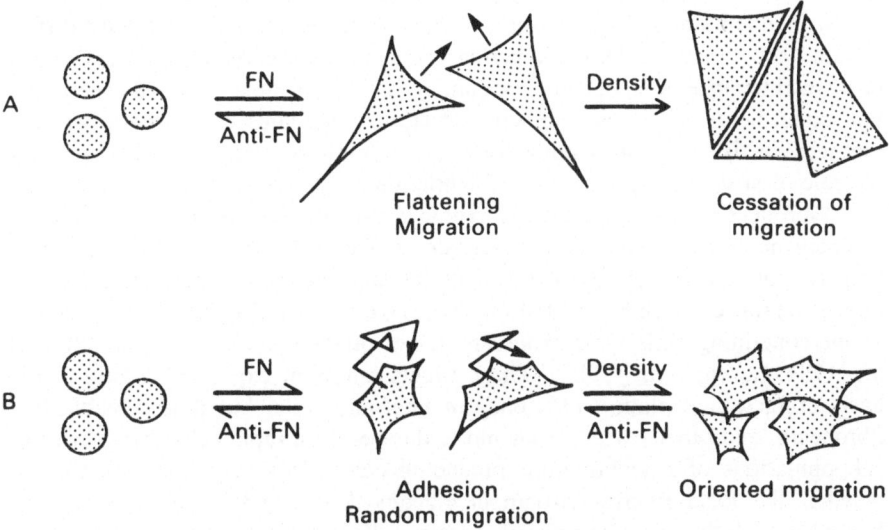

FIGURE 9. Comparison of effects of fibronectin on mesenchymal or fibroblastic cells and neural crest cells. Fibronectin promotes the migration and spreading of fibroblastic cells (A), but high density or growth in an extracellular matrix containing fibronectin inhibits locomotion. Neural crest cells appear to require fibronectin for adhesion, and it permits random migration. Rapid, oriented migration requires high cell density and also occurs in fibronectin-containing extracellular matrices. Antibodies to fibronectin inhibit its effects. See text and Rovasio *et al.*, 1983, for details.

confronting two tissues or cell types and then determining their movements relative to one another. The outcome of such interactions is apparently critically dependent on the relative amounts of fibronectin in apposed aggregates (Armstrong and Armstrong, 1981, 1984). An aggregate of heart myocyte tissue will actively spread over and enclose a second aggregate if the latter contains larger amounts of endogenously or exogenously added fibronectin (Armstrong and Armstrong, 1981). Moreover, if heart cells are dissociated and mixed to form randomized aggregates, they undergo a "sorting out" of mesenchyme cells from myocytes, as long as they are maintained under tissue culture conditions that stimulate fibronectin deposition (Armstrong and Armstrong, 1984).

10.3. Effects on Differentiation

Collagen and fibronectin can also influence the expression of differentiated cell products. As described in Section 5, this structural molecule can have dramatic effects on the maintenance of tissue architecture and function (e.g., see Suard *et al.*, 1983, and the review by Kleinman *et al.*, 1981). It is not yet clear whether these effects of collagen reflect a regulatory function or a structural action in maintaining local tissue architecture necessary for continued differentiated cell function. This distinction between function as a regulatory signal or by maintenance of an optimal local microenvironment may be partly semantic, but experiments in which synthetic structural molecules can provide gel-like matrices for tissues as substitutes for collagen might provide useful insights into the role of structural support in differentiation.

Fibronectin can modulate embryonic differentiation *in vitro*, although its mechanisms of action are not yet understood (Table II). Fibronectin or less pure preparations of extracellular matrix material can promote adrenergic differentiation of neural crest cells; treated cultures have increased numbers of catecholamine-containing cells with axons and fewer pigmented cells (Sieber–Blum *et al.*, 1981; Loring *et al.*, 1982). These effects cannot be explained simply on the basis of increased plating efficiency of neuronal cells, but it is possible that fibronectin and other extracellular molecules permit a rapid selective growth of subpopulations of catecholamine-producing cells. This important question of whether the local microenvironment dictates differentiation by selective enhancement of the proliferation of certain cells, as opposed to switching the biosynthesis of cell products within a single cell, is important to resolve in each of the many such differentiative events occurring during development (see also LeDouarin, 1980; Weston, 1983; Kratochwil, 1983).

Regulatory effects of fibronectin independent of any effect on cell proliferation can be demonstrated with cartilage chondrocytes *in vitro*. Chondrocytes treated with cellular fibronectin switch from the synthesis of sulfated proteogly-cans and type II collagen to the synthesis of type I collagen (Pennypacker *et al.*, 1979; West *et al.*, 1979). One possible mechanism for the action of fibronectin might be via its effects on cell shape, which appears to be an important regulator of gene expression in this particular cell type (Solursh *et al.*, 1982; Benya and Shaffer, 1982). Fibronectin-induced flattening of these cells might indirectly produce the switch in biosynthesis of differentiated cell products, although direct effects of this glycoprotein on gene expression while cells are maintained in a spherical morphology in collagen gels still remain to be explored.

Fibronectin is required for muscle development *in vitro*, apparently as an attachment molecule. Myoblasts will not attach to the gelatin substrates used for muscle cell culture unless fibronectin is present; without this early phase of attachment, cells do not grow (Chiquet *et al.*, 1979). On the other hand, later treatment of cellular fibronectin of a myoblast cell line inhibits its fusion into myotubes (Podleski *et al.*, 1979). Instead of fusing, cells continue to proliferate as mononucleated myoblasts; the converse effect of stimulation of the rate of myoblast fusion can be produced by gentle protease treatment of cells, which removes fibronectin and perhaps other cell surface proteins (Podleski *et al.*, 1979). Similar effects are seen with chick myoblasts, which can nevertheless proceed with the induction of differentiated muscle cell products even in the absence of fusion (K. Olden, personal communication). The biological impor-tance of the inhibitions of myoblast fusion and chondrocyte function is as yet uncertain, although they demonstrate the potential regulation of these events by unusually high amounts of cellular fibronectin. Since chicken myoblasts syn-thesize only low amounts of fibronectin (Chiquet *et al.*, 1981), perhaps only pathological regulatory disorders might cause such a large accumulation of fi-bronectin *in vivo*.

10.4. Glycosaminoglycans and Development

Alterations in the concentrations of hyaluronic acid and other glycosami-noglycans may have important effects on embryonic cell migration and differ-entiation. Rising concentrations of hyaluronic acid are correlated with cell mi-gration in the embryonic cornea and neural crest, and there is a good correlation between the production of hyaluronidase, disappearance of hyaluronic acid, cessation of migration, and differentiation of cells (reviewed by Toole, 1981). Migration of neural crest cells is also accompanied by alterations in glycosa-minoglycan content along migratory pathways (Weston, 1983). One major func-tion of hyaluronic acid may be to provide a water-filled gel that expands suffi-ciently to provide the pathways for migration; cells would migrate in spaces opened by hyaluronate, and removal of this glycosaminoglycan by hyaluronidase might lead to a collapse of these spaces and more effective cell–cell interactions in tissue formation (Toole, 1982).

A plasma membrane receptor for hyaluronic acid has been isolated from tumor cells recently (Underhill *et al.*, 1983). This receptor requires a minimum of six sugar residues of hyaluronate for binding. In detergent, the receptor displays an eightfold lower affinity and substantially less specificity in terms of the type of glycosaminoglycan it recognizes. These differing affinities may be the result of simple physical differences between the binding of individual receptors and the multivalent interactions of multiple receptors side by side in a membrane (Underhill *et al.*, 1983).

As reviewed previously, fibronectin, hyaluronic acid, and other extracellular molecules may play important roles in embryonic cell migration and possibly in the expression of differentiated phenotypes. It is obvious that much more experimentation will be required to determine how their plasma membrane receptors function, and how initial binding is transduced into their many biological effects.

ACKNOWLEDGMENTS.   We thank many colleagues for valuable discussions and for communication of results prior to publication.

REFERENCES

Adamson, E. D., and Ayers, S. E., 1979, The localization and synthesis of some collagen types in developing mouse embryos, *Cell* **16:**953–965.

Adler, R., Manthorpe, M., Skaper, S., and Varon, S., 1981, Polyornithine-bound neurite-promoting factors. Culture sources and responsive neurons, *Brain Res.* **206:**127–144.

Akers, R. M., Mosher, D. F., and Lilien, J. E., 1981, Promotion of retinal neurite outgrowth by substratum-bound fibronectin, *Dev. Biol.* **86:**179–188.

Akiyama, S. K., and Yamada, K. M., 1984, Comparisons of evolutionarily distinct fibronectins: Evidence for the origins of plasma and fibroblast cellular fibronectins from a single gene, submitted for publication.

Albini, A., Richter, H., and Pontz, B. F., 1983, Localization of the chemotactic domain in fibronectin, *FEBS Lett.* **156:**222–226.

Alexander, S. S., Jr., Colonna, G., Yamada, K. M., Pastan, I., and Edelhoch, H., 1978, Molecular properties of a major cell surface protein from chick embryo fibroblasts, *J. Biol. Chem.* **253:**5820–5824.

Alexander, S. S., Jr., Colonna, G., and Edelhoch, H., 1979, The structure and stability of human plasma cold-insoluble globulin, *J. Biol. Chem.* **254:**1501–1505.

Ali, I. U., and Hynes, R. O., 1978, Effects of LETS glycoprotein on cell motility, *Cell* **14:**439–446.

Ali, I. U., Mautner, V., Lanza, R., and Hynes, R. O., 1977, Restoration of normal morphology, adhesion and cytoskeleton in transformed cells by addition of a transformation-sensitive surface protein, *Cell* **11:**115–126.

Aplin, J. D., and Hughes, R. C., 1981, Cell adhesion on model substrata: Threshold effects and receptor modulation, *J. Cell Sci.* **50:**89–103.

Aplin, J. D., and Hughes, R. C., 1982, Complex carbohydrates of the extracellular matrix. Structure, interactions and biological roles, *Biochim. Biophys. Acta* **694:**375–418.

Aplin, J. D., Hughes, R. C., Jaffe, C. L., and Sharon, N., 1981, Reversible cross-linking of cellular components of adherent fibroblasts to fibronectin and lectin-coated substrata, *Exp. Cell Res.* **134:**488–494.

Armstrong, P. B., and Armstrong, M. T., 1981, Immunofluorescent histological studies of the role of fibronectin in the expression of the associative preferences of embryonic tissues, *J. Cell Sci.* **50:**121–133.

Armstrong, P. B., and Armstrong, M. T., 1984, A role for fibronectin in cell sorting, *J. Cell Sci.,* in press.

Arneson, M. A., Hammerschmidt, D. C., Furcht, L. T., and King, R. A., 1980, A new form of Ehlers–Danlos syndrome. Fibronectin corrects defective platelet function, *JAMA* **244:**144–147.

Atherton, B. T., and Hynes, R. O., 1981, A difference between plasma and cellular fibronectins located with monoclonal antibodies, *Cell* **25:**133–141.

Avnur, Z., and Geiger, B., 1981, The removal of extracellular fibronectin from areas of cell-substrate contact, *Cell* **25:**121–132.

Bächinger, H. P., Doege, K. J., Petschek, J. P., Fessler, L. I., and Fessler, J. H., 1982, Structural implications from an electron microscopic comparison of procollagen I, pC-collagen I, procollagen IV, and a *Drosophila* procollagen, *J. Biol. Chem.* **257:**14590–14592.

Balian, G., Click, E. M., Crouch, E., Davidson, J. M., and Bornstein, P., 1979a, Isolation of a collagen-binding fragment from fibronectin and cold-insoluble globulin, *J. Biol. Chem.* **254:**1429–1432.

Balian, G., Crouch, E., Click, E. M., Carter, W. G., and Bornstein, P., 1979b, Comparison of the structures of human fibronectin and plasma cold-insoluble globulin, *J. Supramol. Struct.* **12:**505–516.

Banerjee, S. D., Cohn, R. H., and Bernfield, M. R., 1977, Basal lamina of embryonic salivary epithelia: Production by the epithelium and role in maintaining lobular morphology, *J. Cell Biol.* **73:**445–463.

Bard, J. B. L., and Hay, E. D., 1975, The behavior of fibroblasts from the developing avian cornea, *J. Cell Biol.* **67:**400–418.

Barnes, D., and Sato, G., 1980, Serum-free medium. A review, *Cell* **22:**649–655.

Barnes, D. W., and Silnutzer, J., 1983, *J. Biol. Chem.* **258:**12548–12552.

Barnes, D., Wolfe, R., Serrero, G., McClure, D., and Sato, G., 1980, Effects of a serum spreading factor on growth and morphology of cells in serum-free medium, *J. Supramol. Struct.* **14:**47–63.

Barnes, D. W., Silnutzer, J., See, C., and Shaffer, M., 1983, Characterization of human serum spreading factor with monoclonal antibody, *Proc. Natl. Acad. Sci. USA* **80:**1362–1366.

Baron–Van Evercooren, A., Kleinman, H. K., Ohno, S., Marangos, P., Schwartz, J. P., and Dubois–Dalcq, M. E., 1982, Nerve growth factor, laminin, and fibronectin promote neurite growth in human fetal sensory ganglia cultures, *J. Neurosci. Res.* **8:**179–193.

Bennett, J. S., Vilaire, G., and Cines, D. B., 1982, Identification of the fibrinogen receptor on human platelets by photoaffinity labeling, *J. Biol. Chem.* **257:**8049–8054.

Bentz, H., Morris, N. P., Murray, L. W., Sakai, L. Y., Hollister, D. W., and Burgeson, R. E., 1983, Isolation and partial characterization of a new human collagen with an extended triple-helical structural domain, *Proc. Natl. Acad. Sci. USA* **80:**3168.

Benya, P. D., and Shaffer, J. D., 1982, Dedifferentiated chondrocytes reexpress the differentiated collagen phenotype when cultured in agarose gels, *Cell* **30:**215–224.

Bernard, B. A., Olden, K., and Yamada, K. M., 1982a, Carbohydrates protect the collagen binding domain of fibronectin against proteolytic degradation, in: *Extracellular Matrix* (S. Hawkes and J. L. Wang, eds.), Academic Press, New York, pp. 225–229.

Bernard, B. A., Yamada, K. M., and Olden, K., 1982b, Carbohydrates selectively protect a specific domain of fibronectin against proteases, *J. Biol. Chem.* **257**:8549–8554.

Bernard, B. A., De Luca, L. M., Hassell, J. R., Yamada, K. M., and Olden, K., 1984, Retinoic acid alters the proportion of high mannose to complex type oligosaccharides on fibronectin secreted by cultured chondrocytes, *J. Biol. Chem.*, **259**:5310–5315.

Bernfield, M., 1980, Organization and remodeling of the extra-cellular matrix in morphogenesis, in: *Morphogenesis and Pattern Formation: Implications for Normal and Abnormal Development* (L. L. Brinkley, B. M. Carlson, and T. G. Connelly, eds.), Raven Press, New York.

Bing, D. H., Almeda, S., Isliker, H., Lahav, J., and Hynes, R. O., 1982, Fibronectin binds to the Clq component of complement, *Proc. Natl. Acad. Sci. USA* **79**:4918–4201.

Birchmeier, C., Kreis, T. E., Eppenberger, H. M., Winterhalter, K. H., and Birchmeier, W., 1980, Corrugated attachment membrane in WI-38 fibroblasts: Alternating fibronectin fibers and actin-containing focal contacts, *Proc. Natl. Acad. Sci. USA* **77**:4108–4112.

Bitterman, P. B., Rennard, S. I., Adelberg, S., and Crystal, R. G., 1983, Role of fibronectin as a growth factor for fibroblasts, *J. Cell Biol.* **97**:1925–1927.

Bornstein, P., and Sage, H., 1980, Structurally distinct collagen types, *Ann. Rev. Biochem.* **49**:957–1003.

Boucaut, J.–C., and Darribere, T., 1983, Presence of fibronectin during early embryogenesis in amphibian *Pleurodeles waltlii*, *Cell Diff.* **12**:77–83.

Brandts, J. F., and Jacobson, B. S., 1983, A general mechanism for transmembrane signalling, based on clustering, in: *Survey and Synthesis of Pathology Research*, Volume 2, S. Carter AG, Basel, pp. 107–114.

Briles, E. I. B., 1982, Selection of a substratum-specific adhesion variants of rat hepatoma (HTC) cells, in: *Extracellular Matrix* (S. Hawkes, and J. L. Wang, eds.), Academic Press, New York, pp. 115–119.

Briles, E. B., and Haskew, N. B., 1982, Isolation of cloned variants of a rat hepatoma cell line with altered attachment to collagen, but normal attachment to fibronectin, *Exp. Cell Res.* **138**:436–441.

Bronner–Fraser, M. E., 1982, Distribution of latex beads and retinal pigment epithelial cells along the ventral neural crest pathway, *Dev. Biol.* **91**:50–63.

Brown, J. R., 1976, *Atlas of Protein Sequence and Structure*, Volume 5 (M. O. Dayhoff, ed.), National Biomedical Research Foundation, Washington, D. C., pp. 509–524.

Burgeson, R. E., 1982, Genetic heterogeneity of collagens, *J. Invest. Dermatol.* **79**:25s–30s.

Burridge, K., and Feramisco, J. R., 1980, Microinjection and localization of a 130K protein in living fibroblasts: A relationship to actin and fibronectin, *Cell* **19**:587–595.

Carlsson, R., Engvall, E., Freeman, A., and Ruoslahti, E., 1981, Laminin and fibronectin in cell adhesion: Enhanced adhesion of cells from regenerating liver to laminin, *Proc. Natl. Acad. Sci. USA*, **78**:2403–2406.

Cassiman, J. J., Van Der Schueren, B., Van Leuven, F., and Van Den Berghe, H., 1982, Qualitative and quantitative differences in spreading of human fibroblasts on various protein coats. Modulation of treatment of the cells with amines, *J. Cell Sci.* **54**:79–95.

Chen, L. B. (ed.), 1981, *Oncology Overview: Selected Abstracts on Fibronectin and Related Transformation-Sensitive Cell Surface Proteins*, International Cancer Research Data Bank, National Cancer Institute, Bethesda, Maryland.

Chen, L. B., Murray, A., Segal, R. A., Bushnell, A., and Walsh, M. L., 1978, Studies on intercellular LETS glycoprotein matrices, *Cell* **14**:377–391.

Chen, W. T., and Singer, S. J., 1980, Fibronectin is not present in the focal adhesions formed between normal cultured fibroblasts and their substrata, *Proc. Natl. Acad. Sci. USA* **77**:7318–7322.

Chen, W. T., and Singer, S. J., 1982, Immunoelectron microscopic studies of the sites of cell–substratum and cell–cell contacts in cultured fibroblasts, *J. Cell Biol.* **95:**205–222.

Chiang, T. M., and Kang, A. H., 1976, Binding of chick skin collagen α1 chain by isolated membranes from human platelets, *J. Biol. Chem.* **251:**6347–6351.

Chiang, T. M., and Kang, A. H., 1982, Isolation and purification of collagen α1(I) receptor from human platelet membrane, *J. Biol. Chem.* **257:**7581–7586.

Chiquet, M., Puri, E. C., and Turner, D. C., 1979, Fibronectin mediates attachment of chicken myoblasts to a gelatin-coated substratum, *J. Biol. Chem.* **254:**5475–5482.

Chiquet, M., Eppenberger, H. M., and Turner, D. C., 1981, Muscle morphogenesis: Evidence for an organizing function of exogenous fibronectin, *Dev. Biol.* **88:**220–235.

Chung, A. E., Jaffe, R., Freeman, I. L., Vergnes, J. P., Braginski, J. E., and Carlin, B., 1979, Properties of a basement membrane-related glycoprotein synthesized in culture by a mouse embryonal carcinoma-derived cell line, *Cell* **16:**277–287.

Clark, C. C., and Kefalides, N. A., 1982, Partial characterization of collagenous and noncollagenous basement membrane proteins synthesized by the 14.5-day rat embryo parietal yolk sac *in vitro, Conn. Tiss. Res.* **10:**303–318.

Clark, R. A., Lanigan, J. M., Dellapelle, P., Manseau, E., Dvorak, H. F., and Colvin, R. B., 1982, Fibronectin and fibrin provide a provisional matrix for epidermal cell migration during wound reepithelialization, *J. Invest. Dermatol.* **79:**264–269.

Collins, F., 1980, Neurite outgrowth induced by the substrate associated material from nonneuronal cells, *Dev. Biol.* **79:**247–252.

Colvin, R. B., Gardner, P. I., Roblin, R. O., Verderber, E. L., Lanigan, J. M., and Mosesson, M. W., 1979, Cell surface fibrinogen-fibrin receptors on cultured human fibroblasts. Association with fibronectin (cold insoluble globulin, LETS protein) and loss in SV40 transformed cells, *Lab. Invest.* **41:**464–473.

Cooper, A. R., Kurkinen, M., Taylor, A., and Hogan, B. L. M., 1981, Studies on the biosynthesis of laminin by murine parietal endoderm cells, *Eur. J. Biochem.* **119:**189–197.

Cossu, G., and Warren, L., 1983, Lactosaminoglycans and heparan sulfate are covalently bound to fibronectins synthesized by mouse stem teratocarcinoma cells, *J. Biol. Chem.* **258:**5603–5607.

Cossu, G., Andrews, P. W., and Warren, L., 1983, Covalent binding of lactosaminoglycans and heparan sulfate to fibronectin synthesized by a human teratocarcinoma cell line, *Biochem. Biophys. Res. Comm.* **111:**952–957.

Couchman, J. R., Rees, D. A., Green, M. R., and Smith, C. G., 1982, Fibronectin has a dual role in locomotion and anchorage of primary chick fibroblasts and can promote entry into the division cycle, *J. Cell Biol.* **93:**402–410.

Couchman, J. R., Höök, M., Rees, D. A., and Timpl, R., 1983, Adhesion, growth, and matrix production of fibroblasts on laminin substrates, *J. Cell Biol.* **96:**177–183.

Coughlin, M. D., 1984, Growth factors regulating autonomic nerve development, *Adv. Cell. Neurobiol.*, in press.

Coughlin, M. D., Bloom, E. M., and Black, I. B., 1981, Characterization of a neuronal growth factor from mouse heart-cell-conditioned medium, *Dev. Biol.* **82:**56–68.

Critchley, D. R., England, M. A., Wakely, J., and Hynes, R. O., 1979, Distribution of fibronectin in the ectoderm of gastrulating chick embryos, *Nature* **280:**498–500.

Curtis, A. S. G., Forrester, J. V., McInnes, C., and Lawrie, F., 1983, The adhesion of cells to polystyrene surfaces, *J. Cell Biol.* **97:**1500–1506.

Damsky, C. H., Knudsen, K. A., Dorio, R. J., and Buck, C. A., 1981, Manipulation of cell–cell and cell–substrate interactions in mouse mammary tumor epithelial cells using broad spectrum antisera, *J. Cell Biol.* **89:**173–184.

Del Rosso, M., Cappelletti, R., Viti, M., Vannucchi, S., and Chiarugi, V., 1981, Binding of the basement-membrane glycoprotein laminin to glycosaminoglycans. An affinity-chromatography study, *Biochem. J.* **199**:699–704.

Dessau, W., Adelmann, B. C., Timpl, R., and Martin, G. R., 1978, Identification of the sites in collagen alpha-chains that bind serum anti-gelatin factor (cold-insoluble globulin), *Biochem. J.* **169**:55–59.

Duband, J. L., and Thiery, J. P., 1982a, Distribution of fibronectin in the early phase of avian cephalic neural crest cell migration, *Dev. Biol.* **93**:308–323.

Duband, J. L., and Thiery, J. P., 1982b, Appearance and distribution of fibronectin during chick embryo gastrulation and neurulation, *Dev. Biol.* **94**:337–350.

Duncan, K. G., Fessler, L. I., Bächinger, H. P., and Fessler, J. H., 1983, Procollagen IV association to tetramers, *J. Biol. Chem.* **258**:5869–5877.

Ekblom, P., Alitalo, K., Vaheri, A., Timpl, R., and Saxen, L., 1980, Induction of a basement membrane glycoprotein in embryonic kidney: Possible role of laminin in morphogenesis, *Proc. Natl. Acad. Sci. USA* **77**:485–489.

Elsdale, T. R., and Bard, J. B. L., 1972, Collagen substrata for studies on cell behavior, *J. Cell Biol.* **41**:298–311.

Emmerling, M. R., Johnson, C. D., Mosher, D. F., Lipton, B. H., and Lilien, J. E., 1981, Cross-linking and binding of fibronectin with asymmetric acetylcholinesterase, *Biochemistry* **20**:3242–3247.

Enat, R., Jefferson, D. M., Ruiz–Opazo, N., Gatmaitan, Z., Leinwand, L., and Reid, L. M., 1983, Hepatocyte proliferation in vitro: Its dependence on the use of serum-free, hormonally defined medium and substrata of extracellular matrix, *Proc. Natl. Acad. Sci. USA* **81**:1411–1415.

Engel, J., Odermatt, E., Engel, A., Madri, J. A., Furthmayr, H., Rohde, H., and Timpl, R., 1981, Shapes, domain organizations and flexibility of laminin and fibronectin, two multifunctional proteins of the extracellular matrix, *J. Mol. Biol.* **150**:97–120.

Engvall, E., and Ruoslahti, E., 1977, Binding of soluble form of fibroblast surface protein, fibronectin, to collagen, *Int. J. Cancer* **20**:1–5.

Engvall, E., and Ruoslahti, E., 1983, Cell adhesion, protein binding, and antigenic properties of laminin, *Coll. Rel. Res.* **3**:359–369.

Engvall, E., Ruoslahti, E., and Miller, E. J., 1978, Affinity of fibronectin to collagens of different genetic types and to fibrinogen, *J. Exp. Med.* **147**:1584–1595.

Engvall, E., Krusius, T., Wewer, U., and Ruoslahti, E., 1983, Laminin from rat yolk sac tumor: Isolation, partial characterization, and comparison with mouse laminin, *Arch. Biochem. Biophys.* **222**:649–656.

Erickson, H. P., and Carrell, N. A., 1983, Fibronectin in extended and compact conformations: electron microscopy and sedimentation analysis, *J. Biol. Chem.* **257**:14539–14544.

Espersen, F., and Clemmensen, I., 1982, Isolation of fibronectin-binding protein from *Staphylococcus aureus, Infect. Immun.* **37**:526–531.

Fagan, J. B., Sobel, M. E., Yamada, K. M., DeCrombrugghe, B., and Pastan, I., 1981, Effects of transformation on fibronectin gene expression using cloned fibronectin cDNA, *J. Biol. Chem.* **526**:520–525.

Farmer, S. R., Ben–Ze'ev, A., Benecke, B.–J., and Penman, S., 1978, Altered Translatability of messenger RNA from suspended anchorage-dependent fibroblasts: Reversal upon cell attachment to a surface, *Cell* **15**:627–637.

Farquhar, M. G., 1981, The glomerular basement membrane: A selective macromolecular filter, in: *Cell Biology of Extracellular Matrix* (E. D. Hay, ed.), Plenum Press, New York, pp. 335–378.

Fessler, J. H., and Fessler, L. I., 1978, Biosynthesis of procollagen, *Ann. Rev. Biochem.* **47**:129–162.

Fleischmajer, R., Timpl, R., Tuderman, L., Raisher, L., Wiestner, M., Perlisch, J., and Graves,

P. N., 1981, Ultrastructural identification of extension aminopropeptides of type I and III collagens in human skin. *Proc. Natl. Acad. Sci. USA* **78**:7360–7364.

Foellmer, H. G., Madri, J. A., and Furthmayr, H., 1983, Monoclonal antibodies to Type IV collagen: Probes for the study of structure and function of basement membranes, *Lab. Invest.* **48**: 639–649.

Foidart, J. M., Berman, J. J., Paglia, L., Rennard, S., Abe, S., Perantoni, A., and Martin, G. R., 1980, Synthesis of fibronectin, laminin, and several collagens by a liver-derived epithelial line, *Lab. Invest.* **42**:525–532.

Folkman, J., and Greenspan, H. P., 1975, Influence of geometry on control of cell growth, *Biochim. Biophys. Acta* **417**:211–236.

Foster, C. S., Smith, C. A., Dinsdale, E. A., Monaghan, P., and Neville, A. M., 1983, Human mammary gland morphogenesis in vitro: The growth and differentiation of normal breast epithelium in collagen gel cultures defined by electron microscopy, monoclonal antibodies and autoradiography, *Dev. Biol.* **96**:197–216.

Fox, C. H., Cottler–Fox, M. H., and Yamada, K. M., 1980, The distribution of fibronectin in attachment sites of chick fibroblasts, *Exp. Cell Res.* **130**:477–481.

Fridkin, M., and Gottlieb, P., 1981, Tuftsin, Thr-Lys-Pro-Arg. Anatomy of an immunologically active peptide, *Mol. Cell Biochem.* **41**:73–97.

Fujikawa, L. S., Foster, C. S., Harrist, T. J., Lanigan, J. M., and Colvin, R. B., 1981, Fibronectin in healing rabbit corneal wounds, *Lab. Invest.* **45**:120–129.

Fukuda, M., Levery, S. B., and Hakomori, S., 1982, Carbohydrate structure of hamster plasma fibronectin. Evidence for chemical diversity between cellular and plasma fibronectins, *J. Biol. Chem.* **257**:6856–6860.

Furcht, L. T., 1983, Structure and function of the adhesive glycoprotein fibronectin, in: *Modern Cell Biology* (B. Satir, ed.), Liss, New York, pp. 53–117.

Furthmayr, H., Wiedemann, H., Timpl, R., Odermatt, E., and Engel, J., 1983, Electron-microscopical approach to a structural model of intima collagen, *Biochem. J.* **211**:303–311.

Gartner, T. K., Williams, D. C., Minion, F. C., and Phillips, D. R., 1978, Thrombin-induced platelet aggregation is mediated by a platelet plasma membrane-bound lectin, *Science* **200**:1281–1283.

Gartner, T. K., Gerrard, J. M., White, J. G., and Williams, D. C., 1981, Fibrinogen is the receptor for the endogenous lectin of human platelets, *Nature (London)* **289**:688–700.

Gauss–Müller, V., Kleinman, H. K., Martin, G. R., and Schiffmann, E., 1980, Role of attachment factors and attractants in fibroblast chemotaxis, *J. Lab. Clin. Med.* **96**:1071–1080.

Ginsberg, M. H., Painter, R. G., Forsyth, J., Birdwell, C., and Plow, E. F., 1980, Thrombin increases expression of fibronectin antigen on the platelet surface, *Proc. Natl. Acad. Sci. USA* **77**:1049–1053.

Ginsberg, M. H., Wencel, J. D., White, J. G., and Plow, E. F., 1983, Binding of fibronectin to α-granule-deficient platelets, *J. Cell Biol.* **97**:571–573.

Gjessing, R., and Seglen, P. O., 1980, Adsorption, simple binding and complex binding of rat hepatocytes to various in vitro substrata, *Exp. Cell Res.* **129**:239–249.

Goldberg, B., 1979, Binding of soluble type I collagen molecules to the fibroblast plasma membrane, *Cell* **16**:265–275.

Goldberg, B., 1982, Binding of soluble type I collagen to fibroblasts: Effects of thermal activation of ligand, ligand concentration, pinocytosis, and cytoskeletal modifiers, *J. Cell Biol.* **95**:747–751.

Goldberg, B., and Burgeson, R. E., 1982, Binding of soluble type I collagen to fibroblasts: Specificities for native collagen types, triple helical structure, telopeptides, propeptides, and cyanogen bromide-derived peptides, *J. Cell Biol.* **95**:752–756.

Gonda, S. R., and Shainoff, J. R., 1982, Adsorptive endocytosis of fibrin monomer by macrophages:

Evidence of a receptor for the amino terminus of the fibrin α-chain, *Proc. Natl. Acad. Sci. USA* **79**:4565–4569.

Gospodarowicz, D., Greenburg, G., Foidart, J. M., and Savion, N., 1981, The production and localization of laminin in cultured vascular and corneal endothelial cells, *J. Cell. Physiol.* **107**:171–183.

Gospodarowicz, D., Cohen, D. C., and Fujii, D. K., 1982, Regulation of cell growth by the basal lamina and plasma factors: Relevance to embryonic control of cell proliferation and differentiation, in: *Growth of Cells in Hormonally Defined Media: Hormones and Cell Culture*, Volume IX, Cold Spring Harbor Conferences on Cell Proliferation, Cold Spring Harbor, New York, pp. 95–124.

Gospodarowicz, D., Gonzalez, R., and Fujii, D. K., 1983, Are factors originating from serum, plasma, or cultured cells involved in the growth-promoting effect of the extracellular matrix produced by cultured bovine corneal endothelial cells, *J. Cell. Physiol.* **114**:191–202.

Grassi, J., Massoulié, J., and Timpl, R., 1983, Relationship of collagen-tailed acetylcholinesterase with basal lamina components. Absence of binding with laminin, fibronectin, and collagen types IV and V and lack of reactivity with different anti-collagen sera, *Eur. J. Biochem.* **133**:31–38.

Greenberg, J. H., Seppä, S., Seppä, H., and Hewitt, A. T., 1981, Role of collagen and fibronectin in neural crest cell adhesion and migration, *Dev. Biol.* **87**:259–266.

Greenburg, G., and Gospodarowicz, D., 1982, Inactivation of a basement membrane component responsible for cell proliferation but not for cell attachment, *Exp. Cell Res.* **140**:1–14.

Greenburg, G., and Hay, E. D., 1982, Epithelia suspended in collagen gels can lose polarity and express characteristics of migrating mesenchymal cells, *J. Cell Biol.* **95**:333–339.

Greve, J. M., and Gottlieb, D. I., 1982, Monoclonal antibodies which alter the morphology of cultured chick myogenic cells, *J. Supramol. Struct.* **18**:221–230.

Grinnell, F., 1978, Cellular adhesiveness and extracellular substrata, *Int. Rev. Cytol.* **53**:65–144.

Grinnell, F., 1980a, Fibroblast receptor for cell-substratum adhesion: Studies on the interaction of baby hamster kidney cells with latex beads coated by cold insoluble globulin (plasma fibronectin), *J. Cell Biol.* **86**:104–112.

Grinnell, F., 1980b, Visualization of cell–substratum adhesion plaques by antibody exclusion, *Cell Biol. Int. Rep.* **4**:1031–1036.

Grinnell, F., 1982, Fibronectin and wound healing, *Am. J. Dermatopathol.* **4**:185–188.

Grinnell, F., 1983, Cell attachment and spreading factors, in: *Growth and Maturation Factor* (G. Guroff, ed.), Wiley, New York, pp. 267–292.

Grinnell, F., and Bennett, M. H., 1981, Fibroblast adhesion on collagen substrate in the presence and absence of plasma fibronectin, *J. Cell Sci.* **48**:19–34.

Grinnell, F., and Feld, M. K., 1982, Fibronectin adsorption on hydrophilic and hydrophobic surfaces detected by antibody binding and analyzed during cell adhesion in serum-containing medium, *J. Biol. Chem.* **257**:4888–4893.

Grinnell, F., and Hays, D. G., 1978a, Cell adhesion and spreading factor. Similarity to cold insoluble globulin in human serum, *Exp. Cell Res.* **115**:221–229.

Grinnell, F., and Hays, D. G., 1978b, Induction of cell spreading by substratum-adsorbed ligands directed against the cell surface, *Exp. Cell Res.* **116**:275–284.

Grinnell, F., and Minter, D., 1978, Attachment and spreading of baby hamster kidney cells to collagen substrata: Effects of cold-insoluble globulin, *Proc. Natl. Acad. Sci. USA* **75**:4408–4412.

Grinnell, F., and Phan, T. V., 1983, Deposition of fibronectin on material surfaces exposed to plasma: Quantitative and biological studies, *J. Cell. Physiol.* **116**:289–296.

Grinnell, F., Feld, M., and Minter, D., 1980, Fibroblast adhesion to fibrinogen and fibrin substrata: Requirement for cold-insoluble globulin (plasma fibronectin), *Cell* **19**:517–525.

Grinnell, F., Head, J. R., and Hoffpauir, J., 1982a, Fibronectin and cell shape in vivo: Studies on the endometrium during pregnancy, *J. Cell Biol.* **94:**597–606.

Grinnell, F., Lang, B. R., and Phan, T. V., 1982b, Binding of plasma fibronectin to the surfaces of BHK cells in suspension at 4°C, *Exp. Cell Res.* **142:**499–504.

Gudewicz, P. W., Molnar, J., Lai, M. Z., Beezhold, D. W., Siefring, G. E., Jr., Credo, R. B., and Lorand, L., 1980, Fibronectin-mediated uptake of gelatin-coated latex particles by peritoneal macrophages, *J. Cell Biol.* **87:**427–433.

Hahn, L. H. E., and Yamada, K. M., 1979a, Identification and isolation of a collagen-binding fragment of the adhesive glycoprotein fibronectin, *Proc. Natl. Acad. Sci. USA* **76:** 1160–1163.

Hahn, L. H. E., and Yamada, K. M., 1979b, Isolation and biological characterization of active fragments of the adhesive glycoprotein fibronectin, *Cell* **18:**1043–1051.

Hall, W. M., and Ganguly, P., 1981, The relationship of serum fibronectin and cell shape to thrombin-induced inhibition of DNA synthesis in human fibroblasts, *J. Cell. Physiol.* **109:**271–280.

Hatten, M. E., Furie, M. B., and Rifkin, D. B., 1982, Binding of developing mouse cerebellar cells to fibronectin: A possible mechanism for the formation of the external granular layer, *J. Neurosci.* **2:**1195–1206.

Hay, E. D. (ed.), 1981, *Cell Biology of Extracellular Matrix*, Plenum Press, New York, 417 pp.

Hayashi, M., and Yamada, K. M., 1981, Differences in domain structures between plasma and cellular fibronectins, *J. Biol. Chem.* **256:**11292–11300.

Hayashi, M., and Yamada, K. M., 1982, Divalent cation modulcation of fibronectin binding to heparin and to DNA, *J. Biol. Chem.* **257:**5263–5267.

Hayashi, M., and Yamada, K. M., 1983, Domain structure of the carboxyl-terminal half of human plasma fibronectin, *J. Biol. Chem.* **258:**3332–3340.

Hayashi, M., Schlesinger, D. H., Kennedy, D. W., and Yamada, K. M., 1980, Isolation and characterization of a heparin-binding domain of cellular fibronectin, *J. Biol. Chem.* **255:**10017–10020.

Hayman, E. G., Engvall, E., A'Hearn, E., Barnes, D., Pierschbacher, M., and Ruoslahti, E., 1982, Cell attachment on replicas of SDS polyacrylamide gels reveals two adhesive plasma proteins, *J. Cell Biol.* **95:**20–23.

Hayman, E. G., Pierschbacher, M. D., Öhgren, Y., and Ruoslahti, E., 1983, Serum spreading factor (vitronectin) is present at the cell surface and in tissues, *Proc. Natl. Acad. Sci. USA* **80:**4003–4007.

Heasman, J., Hynes, R. O., Swan, A. P., Thomas, V., and Wylie, C. C., 1981, Primordial germ cells of *Xenopus* embryos: The role of fibronectin in their adhesion during migration, *Cell* **27:**437–447.

Hedman, K., Kurkinen, M., Alitalo, K., Vaheri, A., Johansson, S., and Höök, M., 1979, Isolation of the pericellular matrix of human fibroblast cultures, *J. Cell Biol.* **81:**83–91.

Hewitt, A. T., Kleinman, H. K., Pennypacker, J. P., and Martin, G. R., 1980, Identification of an adhesion factor for chondrocytes, *Proc. Natl. Acad. Sci. USA* **77:**385–388.

Hewitt, A. T., Varner, H. H., Silver, M. H., Dessau, W., Wilkes, C. M., and Martin, G. R., 1982a, The isolation and partial characterization of chondronectin, an attachment factor for chondrocytes, *J. Biol. Chem.* **257:**2330–2334.

Hewitt, A. T., Varner, H. H., Silver, M. H., and Martin, G. R., 1982b, The role of chondronectin and cartilage proteoglycan in the attachment of chondrocytes to collagen, in: *Limb Development and Regeneration* (R. O. Kelly, P. F. Goetinck, and J. A. MacCabe, eds.), Liss, New York, pp. 25–33.

Hirano, H., Yamada, Y., Sullivan, M., de Crombrugghe, B., Pastan, I., and Yamada, K. M.,

1983, Isolation of genomic DNA clones spanning the entire fibronectin gene, *Proc. Natl. Acad. Sci. USA* **80**:46–50.

Hirata, K., Yoshida, Y., Shiramatsu, K., Freeman, A. E., and Hayasaki, H., 1983, Effects of laminin, fibronectin and type IV collagen on liver cell cultures, *Exp. Cell Biol.* **51**:121–129.

Hogan, B. L. M., Cooper, A. R., and Kurkinen, M., 1980, Incorporation into Richert's membrane of laminin-like extracellular proteins synthesized by parietal endoderm cells of mouse embryo, *Dev. Biol.* **80**:289–300.

Holly, F. J., Dolowy, K., and Yamada, K. M., 1984, Comparative surface chemical studies of cellular fibronectin and submaxillary mucin monolayers: Effects of pH, ionic strength, and presence of calcium ions, *J. Colloid Interfac. Sci.*, in press.

Holmes, R., 1967, Preparation from human serum of an alpha-one protein which induces the immediate growth of unadapted cells in vitro, *J. Cell Biol.* **32**:297–308.

Hörmann, H., 1982, Fibronectin—mediator between cells and connective tissue, *Klin Wochenschr.* **60**:1265–1277.

Hörmann, H., and Seidl, M., 1980, Affinity chromatography on immobilized fibrin monomer, III. The fibrin affinity center of fibronectin, *Hoppe–Seylers Z. Physiol. Chem.* **361**:1449–1452.

Hoyer, L. W., 1980, The factor VIII complex: Structure and function, *Blood* **58**:1–13.

Hsieh, P., and Sueoka, N., 1980, Antisera inhibiting mammalian cell spreading and possible cell surface antigens involved, *J. Cell Biol.* **86**:866–873.

Hughes, R. C., Pena, S. D. J., Clark, J., and Dourmashkin, R. R., 1979, Molecular requirements for the adhesion and spreading of hamster fibroblasts, *Exp. Cell Res.* **121**:307–314.

Hughes, R. C., Butters, T. D., and Aplin, J. D., 1981, Cell surface molecules involved in fibronectin-mediated adhesion. A study using specific antisera, *Eur. J. Cell Biol.* **26**:198–207.

Humphries, M. J., and Ayad, S. R., 1984, Stimulation of DNA synthesis by cathepsin D digests of fibronectin, *Nature* **305**:811–813.

Hynes, R. O., 1973, Alteration of cell-surface proteins by viral transformation and by proteolysis, *Proc. Natl. Acad. Sci. USA* **70**:3170–3174.

Hynes, R. O., 1981, Fibronectin and its relation to cellular structure and behavior, in: *Cell Biology of Extracellular Matrix*, (E. D. Hay, ed.), Plenum Press, New York, pp. 295–333.

Hynes, R. O., and Destree, A. T., 1978, Relationships between fibronectin (LETS protein) and actin, *Cell* **15**:875–886.

Hynes, R. O., and Yamada, K. M., 1982, Fibronectins: Multifunctional modular glycoproteins, *J. Cell Biol.* **95**:369–377.

Hynes, R. O., Destree, A. T., and Wagner, D. D., 1982, Relationships between microfilaments, cell–substratum adhesion, and fibronectin, *Cold Spring Harbor Symp. Quant. Biol.* **46**:659–670.

Isemura, M., Hsu, C.–C., Yosizawa, Z., Odani, S., and Ono, T., 1982a, Interaction of fibronectin with arginine-agarose, *FEBS Lett.* **150**:243–246.

Isemura, M., Yosizawa, Z., Koide, T., and Ono, T., 1982b, Interaction of fibronectin and its proteolytic fragments with hyaluronic acid, *J. Biochem.* **91**:731–734.

Jaffe, E. A., Leung, L. L. K., Nachman, R. L., Levin, R. I., and Mosher, D. F., 1982, Thrombospondin is the endogenous lectin of human platelets, *Nature* **295**:246–248.

Jaffe, E. A., Ruggiero, J. T., Leung, L. L. K., Doyle, M. J., McKeown–Longo, P. J., and Mosher, D. F., 1983, Cultured human fibroblasts synthesize and secrete thrombospondin and incorporate it into extracellular matrix, *Proc. Natl. Acad. Sci. USA* **80**:998–1002.

Jilek, F., and Hörmann, H., 1978a, Cold-insoluble globulin (Fibronectin): Affinity to soluble collagen of various types, *Hoppe–Seylers Z. Physiol. Chem.* **359**:247–250.

Jilek, F., and Hörmann, H., 1978b, Fibronectin (cold-insoluble globulin): Mediation of fibrin-monomer binding to macrophages, *Hoppe–Seylers Z. Physiol. Chem.* **359**:1603–1605.

Johansson, S., Kjellén, L., Höök, M., and Timpl, R., 1981, Substrate adhesion of rat hepatocytes: A comparison of laminin and fibronectin as attachment proteins, *J. Cell Biol.* **90:**260–264.

Katow, H., Yamada, K. M., and Solursh, M., 1982, Occurrence of fibronectin on the primary mesenchyme cell surface during migration in the sea urchin embryo, *Differentiation* **22:**120–124.

Kennedy, D. W., Rohrbach, D. H., Martin, G. R., Momoi, T., and Yamada, K. M., 1983, The adhesive glycoprotein laminin is an agglutinin, *J. Cell. Physiol.* **114:**257–262.

Keski-Oja, J., and Yamada, K. M., 1981, Isolation of an actin-binding fragment of fibronectin, *Biochem. J.* **193:**615–620.

Keski-Oja, J., Sen, A., and Todaro, G. J., 1980, Direct association of fibronectin and actin molecules in vitro, *J. Cell Biol.* **85:**527–533.

Kjellén, L., Pettersson, I., and Höök, M., 1981, Cell-surface heparan sulfate: An intercalated membrane proteoglycan, *Proc. Natl. Acad. Sci. USA* **78:**5371–5375.

Klebe, R. J., 1974, Isolation of a collagen-dependent cell attachment factor, *Nature* **250:**248–251.

Klebe, R. J., Rosenberger, P. G., Naylor, S. L., Burns, R. L., Novak, R., and Kleinman, H., 1977, Cell attachment to collagen. Isolation of a cell attachment mutant, *Exp. Cell Res.* **104:**119–125.

Klebe, R. J., Hall, J. R., Naylor, S. L., and Dickey, W. D., 1978, Bioautography of cell attachment proteins, *Exp. Cell Res.* **115:**73–78.

Klebe, R. J., Bentley, K. L., and Schoen, R. C., 1981, Adhesive substrates for fibronectin. *J. Cell. Physiol.* **109:**481–488.

Kleinman, H. K., McGoodwin, E. B., Martin, G. R., Klebe, R. J., Fietzek, P. P., and Woolley D. E., 1978, Localization of the binding site for cell attachment in the alpha-1(I) chain of collagen, *J. Biol. Chem.* **253:**5642–5646.

Kleinman, H. K., Martin, G. R., and Fishman, P. H., 1979, Ganglioside inhibition of fibronectin-mediated cell adhesion to collagen, *Proc. Natl. Acad. Sci. USA* **76:**3367–3371.

Kleinman, H. K., Klebe, R. J., and Martin, G. R., 1981, Role of collagenous matrices in the adhesion and growth of cells. *J. Cell Biol.* **88:**473–485.

Kleinman, H. K., McGarvey, M. L., Hassell, J. R., and Martin, G. R., 1983, Formation of a supramolecular complex is involved in the reconstitution of basement membrane components, *Biochemistry* **22:**4969–4974.

Knox, P., and Griffiths, S., 1980, The distribution of cell-spreading activities in sera: A quantitative approach, *J. Cell Sci.* **46:**97–112.

Knox, P., and Griffiths, S., 1982, The abnormal morphology of polyoma-transformed baby hamster kidney cells is due to a failure to respond to 70K spreading factor, *J. Cell Sci.* **55:**301–316.

Knudsen, K., Rao, P. E., Damsky, C. H., and Buck, C. A., 1981, Membrane glycoproteins involved in cell substratum adhesion, *Proc. Natl. Acad. Sci. USA* **78:**6071–6075.

Kornblihtt, A. R., Vibe-Pedersen, K., and Baralle, F. E., 1983, Isolation and characterization of cDNA clones for human and bovine fibronectins, *Proc. Natl. Acad. Sci. USA* **80:**3218–3222.

Koteliansky, V. E., Gneushev, H. N., Glukhova, M. A., Shartava, A. S., and Smirnov, V. N., 1982, Fibronectin has an affinity to vinculin, α-actinin, tropomyosin and myosin, *FEBS Lett.* **143:**168–169.

Kratochwil, K., 1983, Embryonic induction, in: *Cell Interactions and Development* (K. M. Yamada, ed.), Wiley, New York, pp. 99–122.

Kühn, K., Wiedemann, H., Timple, R., Ristelli, J., Dieringer, H., Voss, T., and Glanville, R. W., 1981, Macromolecular structure of basement membrane components. Identification of 7S collagen as a crosslinking domain of type IV collagen, *FEBS Lett.* **125:**123–128.

Kuusela, P., 1978, Fibronectin binds to *Staphylococcus aureus*, *Nature* **276:**718–720.

Lahav, J., Schwartz, M. A., and Hynes, R. O., 1982, Analysis of platelet adhesion with a radioactive chemical crosslinking reagent: Interaction of thrombospondin with fibronectin and collagen, *Cell* **31**:253–262.

Lander, A. D., Fujii, D. K., Gospodarowicz, D., and Reichardt, L. F., 1982, Characterization of a factor that promotes neurite outgrowth: Evidence linking activity to heparin sulfate proteoglycan, *J. Cell Biol.* **94**:574–585.

Lang, H., Glanville, R. W., Fietzek, P. P., and Kühn, K., 1979, The covalent structure of calf skin type III collagen. IV. The amino acid sequence of the cyanogen bromide peptide α1(III)CB5 (positions 552–788), *Hoppe-Seyler's Z. Physiol. Chem.* **360**:841–850.

Laterra, J., Silbert, J. E., and Culp, L. A., 1983, Cell surface heparan sulfate mediates some adhesive responses to glycosaminoglycan-binding matrices, including fibronectin, *J. Cell Biol.* **96**:112–123.

Laurie, G. W., Leblond, C. P., and Martin, G. R., 1982, Localization of type IV collagen, laminin, heparan sulfate, proteoglycan, and fibronectin to the basal lamina of basement membranes, *J. Cell Biol.* **95**:340–344.

LeDouarin, N., 1980, The ontogeny of the neural crest in avian embryo chimaeras, *Nature* **286**:663–669.

Leivo, I., Vaheri, A., Timpl, R., and Wartiovaara, J., 1980, Appearance and distribution of collagens and laminin in the early mouse embryo, *Dev. Biol.* **76**:100–114.

Lesot, H., Kühl, U., and von der Mark, K., 1983, Isolation of a laminin-binding protein from muscle cell membrane, *EMBO J.* **2**(6):861–865.

Leung, L., and Nachman, R., 1982, Complex formation of platelet thrombospondin with fibrinogen, *J. Clin. Invest.* **70**:542–549.

Linsenmayer, T. F., 1981, Collagen, in: *Cell Biology of Extracellular Matrix* (E. D. Hay, ed.), Plenum Press, New York, pp. 5–37.

Linsenmayer, T. F., Gibney, E., Toole, B. P., and Gross, J., 1978, Cellular adhesion to collagen, *Exp. Cell Res.* **116**:470–474.

Linsenmayer, T. F., Fitch, J. M., Schmid, T. M., Zak, N. B., Gibney, E., Sanderson, R. D., and Mayne, R., 1983, Monoclonal antibodies against chicken type V collagen: Production, specificity, and use for immunocytochemical localization in embryonic cornea and other organs, *J. Cell Biol.* **96**:124–132.

Loring, J., Glimelius, B., and Weston, J. A., 1982, Extracellular matrix materials influence quail neural crest cell differentiation in vitro, *Dev. Biol.* **90**:165–174.

Macarak, E. J., and Howard, P. S., 1983, Adhesion of endothelial cells to extracellular matrix proteins, *J. Cell. Physiol.* **116**:76–86.

Malinoff, H. L., and Wicha, M. S., 1983, Isolation of a cell surface receptor protein for laminin from murine fibrosarcoma cells, *J. Cell Biol.* **96**:1475–1479.

Margossian, S. S., Lawler, J. W., and Slayter, H. S., 1981, Physical characterization of platelet thrombospondin, *J. Biol. Chem.* **256**:7495–7500.

Martinez–Hernandez, A., Gay, S., and Miller, E. J., 1982, Ultrastructural localization of type V collagen in rat kidney, *J. Cell Biol.* **92**:343–349.

Mauger, A., Demarchez, M., Herbage, D., Grimaud, J.–A., Druguet, M., Hartmann, D., and Sengel, P., 1982, Immunofluorescent localization of collagen types I and II, and of fibronectin during feather morphogenesis in the chick embryo, *Dev. Biol.* **94**:93–105.

Mayer, B. W., Jr., Hay, E. D., and Hynes, R. D., 1981, Immunocytochemical localization of fibronectin in embryonic chick trunk and area vasculosa, *Dev. Biol.* **82**:267–286.

Mayne, R., and von der Mark, K., 1983, Collagens of cartilage, in: *Cartilage,* Volume 1 (B. K. Hall, ed.), Academic Press, New York, pp. 181–214.

McAbee, D. D., and Grinnell, F., 1983, Fibronectin-mediated binding and phagocytosis of polystyrene latex beads by baby hamster kidney cells, *J. Cell Biol.* **97**:1515–1523.

McConnell, M. R., Blumberg, P. M., and Rossow, P. W., 1978, Dimeric and high molecular weight forms of the large external transformation-sensitive protein on the surface of chick embryo fibroblasts, *J. Biol. Chem.* **253**:7522–7530.

McKeown–Longo, P. J., and Mosher, D. F., 1983, Binding of plasma fibronectin to cell layers of human skin fibroblasts, *J. Cell Biol.* **97**:466–472.

McPherson, J., Sage, H., and Bornstein, P., 1981, Isolation and characterization of a glycoprotein secreted by aortic endothelial cells in culture. Apparent identity with platelet thrombospondin, *J. Biol. Chem.* **256**:11330–11336.

Menzel, E. J., Smolen, J. S., Liotta, L., and Reid, K. B., 1981, Interaction of fibronectin with Clq and its collagen-like fragment (CLF), *FEBS Lett.* **129**:188–192.

Miller, E. J., 1983, The structure of collagen, in: *Connective Tissue Diseases* (B. M. Wagner, R. Fleischmajer, and N. Kaufman, eds.), Williams & Wilkens, Baltimore, Maryland, pp. 4– 15.

Mollenhauer, J., and von der Mark, K., 1983, Isolation and characterization of collagen-binding glycoprotein from chondrocyte membranes, *EMBO J.* **2**:45–50.

Mosesson, M. W., 1978, Structure of human plasma cold-insoluble globulin and the mechanism of its precipitation in the cold with heparin or fibrin–fibrinogen complexes, *Ann. N.Y. Acad. Sci.* **312**:11–30.

Mosesson, M. W., and Amrani, D. L., 1980, The structure and biologic activities of plasma fibronectin, *Blood* **56**:145–158.

Mosesson, M. W., and Umfleet, R. A., 1970, The cold-insoluble globulin of human plasma, *J. Biol. Chem.* **245**:5728–5736.

Mosesson, M. W., Chen, A. B., and Huseby, R. M., 1975, The cold-insoluble globulin of human plasma: Studies of its essential structural features, *Biochim. Biophys. Acta* **386**:509–524.

Mosher, D. F., 1976, Action of fibrin-stabilizing factor on cold-insoluble globulin and alpha 2-macroglobulin in clotting plasma, *J. Biol. Chem.* **251**:1639–1645.

Mosher, D. F., 1980, Fibronectin, *Prog. Hemost. Thromb.* **5**:111–151.

Mosher, D. F. (ed.), 1984, *Fibronectin,* Academic Press, New York, in press.

Mosher, D. F., and Proctor, R. A., 1980, Binding and factor XIIIa-mediated cross-linking of a 27-kilodalton fragment of fibronectin to *Staphylococcus aureus, Science* **209**:927–929.

Mosher, D. F., Schad, P. E., and Vann, J. M., 1980, Cross-linking of collagen and fibronectin by factor XIIIa. Localization of participating glutaminyl residues to a tryptic fragment of fibronectin, *J. Biol. Chem.* **255**:1181–1188.

Mosher, D. F., Doyle, J. J., and Jaffe, E. A., 1982, Synthesis and secretion of thrombospondin by cultured human endothelial cells, *J. Cell Biol.* **93**:343–348.

Murray, J. C., Stingl, G., Kleinman, H. K., Martin, G. R., and Katz, S. I., 1979, Epidermal cells adhere preferentially to type IV (basement membrane) collagen, *J. Cell Biol.* **80**:197–202.

Neff, N. T., Lowrey, C., Decker, C., Tovar, A., Damsky, C., Buck, C., and Horwitz, A. F., 1982, A monoclonal antibody detaches embryonic skeletal muscle from extracellular matrices, *J. Cell Biol.* **95**:654–666.

Newgreen, D., and Thiery, J. P., 1980, Fibronectin in early avian embryos: Synthesis and distribution along the migration pathways of neural crest cells, *Cell Tissue Res.* **211**:269–291.

Newgreen, D. F., Gibbins, I. L., Sauter, J., Wallenfelds, B., and Wutz, R., 1982, Ultrastructural and tissue-culture studies on the role of fibronectin, collagen and glycosaminoglycans in the migration of neural crest cells in the fowl embryo, *Cell Tissue Res.* **221**:521–549.

Nishida, T., Nakagawa, S., Awata, T., Ohashi, Y., Watanabe, K., and Manabe, R., 1983, Fibronectin promotes epithelial migration of cultured rabbit cornea in situ, *J. Cell Biol.* **97**:1653–1657.

Nishioka, K., Amoscato, A. A., and Babcack, G. F., 1981, Tuftsin: A hormone-like tetrapeptide with antimicrobial and antitumor activities, *Life Sci.* **28**:1081–1090.

Norton, E. K., and Izzard, C. S., 1982, Fibronectin promotes formation of the close cell-to-substrate contact in cultured cells, *Exp. Cell Res.* **139**:463–467.

Odermatt, E., Risteli, J., van Delden, V., and Timpl, R., 1983, Structural diversity and domain composition of a unique collagenous fragment (intima collagen) obtained from human placenta, *Biochem. J.* **211**:295–302.

Ohno, M., Martinez–Hernandez, A., Ohno, N., and Kelfalides, N., 1983, Isolation of laminin from human placental basement membranes: Amnion, chorion and chorionic microvessels, *Biochem. Biophys. Res. Comm.* **112**:1091–1098.

Oldberg, Å., Linney, E., and Ruoslahti, E., 1983, Molecular cloning and nucleotide sequence of cDNA clone coding for the cell attachment domain in human fibronectin, *J. Biol. Chem.* **258**:10193–10196.

Olden, K., Pratt, R. M., and Yamada, K. M., 1978, Role of carbohydrates in protein secretion and turnover: Effects of tunicamycin on the major cell surface glycoprotein of chick embryo fibroblasts, *Cell* **13**:461–473.

Olden, K., Pratt, R. M., and Yamada, K. M., 1979, Role of carbohydrate in biological function of the adhesive glycoprotein fibronectin, *Proc. Natl. Acad. Sci. USA* **76**:3343–3347.

Olsen, B. R., 1981, Collagen biosynthesis, in: *Cell Biology of Extracellular Matrix*, (E. D. Hay, ed.), Plenum Press, New York, pp. 139–177.

O'Neill, C. H., Riddle, P. N., and Jordan, P. W., 1979, The relation between surface area and anchorage dependence of growth in hamster and mouse fibroblasts, *Cell* **16**:909–918.

Oppenheimer–Marks, N., and Grinnell, F., 1982, Inhibition of fibronectin receptor function by antibodies against baby hamster kidney cell wheat germ agglutinin receptors, *J. Cell Biol.* **95**:876–884.

Orly, J., and Sato, G., 1979, Fibronectin mediates cytokinesis and growth of rat follicular cells in serum-free medium, *Cell* **17**:295–305.

Ormerod, E. J., Warburton, M. J., Hughes, C., and Rudland, P. S., 1983, Synthesis of basement membrane proteins by rat mammary epithelial cells, *Dev. Biol.* **96**:269–275.

Ott, U., Odermatt, E., Engel, J., Furthmayr, H., and Timpl, R., 1982, Protease resistance and conformation of laminin, *Eur. J. Biochem.* **123**:63–72.

Owens, M. R., and Cimino, C. D., 1982, Synthesis of fibronectin by the isolated perfused rat liver, *Blood* **59**:1305–1309.

Palotie, A., Peltonen, L., Risteli, L., and Risteli, J., 1983, Effect of the structural components of basement membranes on the attachment of teratocarcinoma-derived endodermal cells, *Exp. Cell Res.* **144**:31–37.

Pande, H., and Shively, J. E., 1982, NH$_2$-terminal sequences of DNA-, heparin-, and gelatin-binding tryptic fragments from human plasma fibronectin, *Arch. Biochem. Biophys.* **213**:258–265.

Pande, H., Corkill, J., Sailor, R., and Shively, J. E., 1981, Comparative structural studies of human plasma and amniotic fluid fibronectins, *Biochem. Biophys. Res. Commun.* **101**:265–272.

Parsons, R. G., Todd, H. D., and Kowal, R., 1979, Isolation and identification of a human serum fibronectin-like protein elevated during malignant disease, *Cancer Res.* **39**:4341–4345.

Pearlstein, E., 1978, Substrate activation of cell adhesion factor as a prerequisite for cell attachment, *Int. J. Cancer* **22**:32–35.

Pearlstein E., Gold, L. I., and Garcia-Pardo, A., 1980, Fibronectin: A review of its structure and biological activity, *Mol. Cell. Biochem.* **29**:103–128.

Pearlstein, E., Sorvillo, J., and Gigli, I., 1982, The interaction of human plasma fibronectin with a subunit of the first component of complement, Clq, *J. Immunol.* **128**:2036–2039.

Pegrum, S. M., and Maroudas, N. G., 1975, Early events in fibroblast adhesion to glass. An electron microscopic study, *Exp. Cell Res.* **95**:416–422.

Pennypacker, J. P., Hassell, J. R., Yamada, K. M., and Pratt, R. M., 1979, The influence of an adhesive cell surface protein on chondrogenic expression in vitro, *Exp. Cell Res.* **121**:411–415.

Perkins, M. E., Ji, T. H., and Hynes, R. O., 1979, Cross-linking of fibronectin to sulfated proteoglycans at the cell surface, *Cell* **16**:941–952.

Petersen, T. E., Thøgersen, H. C., Skorstengaard, K., Vibe–Pedersen, K., Sahl, P., Sottrup–Jensen, L., and Magnusson, S., 1983, Partial primary structure of bovine plasma fibronectin: Three types of internal homology, *Proc. Natl. Acad. Sci. USA* **80**:137–141.

Pierschbacher, M. D., Hayman, E. G., and Ruoslahti, E., 1981, Location of the cell-attachment site in fibronectin with monoclonal antibodies and proteolytic fragments of the molecule, *Cell* **26**:259–267.

Pierschbacher, M. D., Ruoslahti, E., Sundelin, J., Lind, P., and Peterson, P. A., 1982, The cell attachment domain of fibronectin. Determination of the primary structure, *J. Biol. Chem.* **257**:9593–9597.

Pierschbacher, M. D., Hayman, E. G., and Ruoslahti, E., 1983, Synthetic peptide with cell attachment activity of fibronectin, *Proc. Natl. Acad. Sci. USA* **80**:1224–1227.

Plow, E. F., and Ginsberg, M. H., 1981, Specific and saturable binding of plasma fibronectin to thrombin-stimulated human platelets, *J. Biol. Chem.* **256**:9477–9482.

Podleski, T. R., Greenberg, I., Schlessinger, J., and Yamada, K., 1979, Fibronectin delays the fusion of L6 myoblasts, *Exp. Cell Res.* **123**:104–126.

Postlethwaite, A. E., Keski–Oja, J., Balian, G., and Kang, A. H., 1981, Induction of fibroblast chemotaxis by fibronectin. Localization of the chemotactic region to a 140,000-molecular weight non-gelatin-binding fragment, *J. Exp. Med.* **153**:494–499.

Pouysségur, J., Willingham, M., and Pastan, I., 1977, Role of cell surface carbohydrates and proteins in cell behavior: Studies on the biochemical reversion of an N-acetylglucosamine-deficient fibroblast mutant, *Proc. Natl. Acad. Sci. USA* **74**:243–247.

Prockop, D. J., Kivirikko, K. I., Tuderman, L., and Guzman, N. A., 1979, The biosynthesis of collagen and its disorders, *N. Engl. J. Med.* **301**:13–23, 77–85.

Proctor, R. A., Mosher, D. F., and Olbrantz, P. J., 1982, Fibronectin binding to *Staphylococcus aureus, J. Biol. Chem.* **257**:14788–14794.

Raff, M. C., Fields, K. L., Hakomori, S.-I., Mirsky, R., Pruss, R. M., and Winter, J., 1979, Cell-type-specific markers for distinguishing and studying neurons and the major classes of glial cells in culture, *Brain Res.* **174**:283–308.

Ramachandran, G. N., and Reddi, A. H. (eds.), 1976, *Biochemistry of Collagen,* Plenum Press, New York, 536 pp.

Rao, C. N., Margulies, I. M. K., Goldfarb, R. H., Madri, J. A., Woodley, D. T., and Liotta, L. A., 1982a, Differential proteolytic susceptibility of laminin alpha and beta subunits, *Arch. Biochem. Biophys.* **219**:65–70.

Rao, C. N., Inger, M. K., Margulies, T. S. T., Terranova, V. P., Madri, J. A., and Liotta, L. A., 1982b, Isolation of a subunit of laminin and its role in molecular structure and tumor cell attachment, *J. Biol. Chem.* **257**:9740–9744.

Rao, C. N., Barsky, S. H., Terranova, V. P., and Liotta, L. A., 1983, Isolation of a tumor cell laminen receptor, *Biochem. Biophys. Res. Comm.* **111**:804–808.

Rapraeger, A. C., and Bernfield, 1983, Heparan sulfate proteoglycans from mouse mammary epithelial cells, *J. Biol. Chem.* **258**, 3632–3636.

Rauvala, H., Carter, W. G., and Hakomori, S.-I., 1981, Studies on cell adhesion and recognition I. Extent and specificity of cell adhesion triggered by carbohydrate-reactive proteins (glycosidases and lectins) and by fibronectin, *J. Cell Biol.* **88**:127–137.

Rauvala, H., Prieels, J.-P., and Finne, J., 1983, Cell adhesion mediated by a purified fucosyltransferase, *Proc. Natl. Acad. Sci. USA* **80**:3991–3995.

Reddi, A. H., 1974, Bone matrix in the solid state: Geometric influence on differentiation of fibroblasts, in: *Advances in Biological and Medical Physics,* Volume 15, (J. H. Lawrence and J. W. Gofman, eds.), Academic Press, New York, pp. 1–18.

Rees, D. A., Lloyd, C. W., and Thom, D., 1977, Control of grip and stick in cell adhesion through lateral relationships of membrane glycoproteins, *Nature* **267**:124–128.

Reid, L., 1982, Regulation of growth and differentiation of mammalian cells by hormones and extracellular matrix, in: *From Gene to Protein: Translation into Biotechnology,* Fourteenth Miami Winter Symposium (W. J. Whelan, ed.), Academic Press, New York, pp. 53–73.

Richter, H., and Hörmann, H., 1982, Early and late cathepsin D-derived fragments of fibronectin containing the C-terminal interchain disulfide cross-link, *Hoppe–Seylers Z. Physiol. Chem.* **363**:351–364.

Richter, H., Seidl, M., and Hörmann, H., 1981, Location of heparin-binding sites of fibronectin. Detection of hitherto unrecognized transamidase sensitive site, *Hoppe–Seylers Z. Physiol. Chem.* **362**:399–408.

Rocco, M., Carson, M., Hantgan, R., McDonagh, J., and Hermans, J., 1983, Dependent of the shape of the plasma fibronectin molecule on solvent composition: Ionic strength and glycerol content, *J. Biol. Chem.* **258**:14545–14549.

Rogers, S. L., Letourneau, P. C., Palm, S. L., McCarthy, J., and Furcht, L., 1983, Neurite extension by peripheral and central nervous system neurons in response to substratum bound fibronectin and laminin *Dev. Biol.* **98**:212–220.

Rohde, H., Bächinger, H. P., and Timpl, R., 1980, Characterization of pepsin fragments of laminin in a tumor basement membrane. Evidence for the existence of related proteins, *Hoppe–Seylers Z. Physiol. Chem.* **361**:1651–1660.

Rojkind, M., Gatmaitan, Z., MacKensen, S., Giambrone, M.–A., Ponce, P., and Reid, L. M., 1980, Connective tissue biomatrix: Its isolation and utilization for long-term cultures of normal rate hepatocytes, *J. Cell Biol.* **87**:255–263.

Rovasio, R. A., Delouvée, A., Yamada, K. M., Timpl, R., and Thiery, J. P., 1983, Neural crest cell migration: Requirements for exogenous fibronectin and high cell density, *J. Cell Biol.* **96**:462–473.

Rubin, K., Höök, M., Öbrink, B., and Timpl, R., 1981, Substrate adhesion of rat hepatocytes: Mechanism of attachment to collagen substrates, *Cell* **24**:463–470.

Ruoslahti, E., and Hayman, E. G., 1979, Two active sites with different characteristics in fibronectin, *FEBS Lett.* **97**:221–224.

Ruoslahti, E., and Vaheri, A., 1975, Interaction of soluble fibroblast surface antigen with fibrinogen and fibrin. Identity with cold insoluble globulin of human plasma, *J. Exp. Med.* **141**:497–501.

Ruoslahti, E., Hayman, E. G., Kuusela, P., Shively, J. E., and Engvall, E., 1979, Isolation of a tryptic fragment containing the collagen-binding site of plasma fibronectin, *J. Biol. Chem.* **254**:6054–6059.

Ruoslahti, E., Engvall, E., and Hayman, E. G., 1981a, Fibronectin: Current concepts of its structure and functions, *Coll. Rel. Res.* **1**:95–128.

Ruoslahti, E., Engvall, E., Hayman, E. G., and Spiro, R. G., 1981b, Comparative studies on amniotic fluid and plasma fibronectins, *Biochem. J.* **193**:295–299.

Ruoslahti, E., Hayman, E. G., Engvall, E., Cothran, W. C., and Butler, W. T., 1981c, Alignment of biologically active domains in the fibronectin molecule, *J. Biol. Chem.* **256**:7277–7281.

Ryden, C., Rubin, K., Speziale, P., Höök, M., Lindberg, M., and Wadstrom, T., 1983, Fibronectin receptors from *Staphylococcus aureus, J. Biol. Chem.* **258**:3396–3401.

Saba, T. M., and Jaffe, E., 1980, Plasma fibronectin (opsonic glycoprotein): Its synthesis by vascular endothelial cells and role in cardiopulmonary integrity after trauma as related to reticuloendothelial function, *Am. J. Med.* **68**:577–594.

Sage, H., Pritzl, P., and Bornstein, P., 1980, A unique, pepsin-sensitive collagen synthesized by aortic endothelial cells in culture, *Biochemistry* **19**:5747–5755.

Saito, Y., Imada, T., and Inada, Y., 1982, Platelet's adhering protein: Isolation of a new non-fibronectin protein from bovine platelet membrane, *Thromb. Res.* **25**:143–147.

Sakashita, S., Engvall, E., and Ruoslahti, E., 1980, Basement membrane glycoprotein laminin binds to heparin, *FEBS Lett.* **116**:243–246.

Schnieke, A., Harbers, K., and Jaenisch, R., 1983, Embryonic lethal mutation in mice induced by retrovirus insertion into the $\alpha 1$(I)collagen gene, *Nature (London)* **304**:315–320.

Schor, S. L., and Court, J., 1979, Different mechanisms in the attachment of cells to native and denatured collagen, *J. Cell Sci.* **38**:267–281.

Schor, S. L., Schor, A. M., and Bazill, G. W., 1981a, The effects of fibronectin on the migration of human foreskin fibroblasts and Syrian hamster melanoma cells into three dimensional gels of native collagen fibres, *J. Cell. Sci.* **48**:301–314.

Schor, S. L., Schor, A. M., and Bazill, G. W., 1981b, The effects of fibronectin on the adhesion and migration of chinese hamster ovary cells on collagen substrata, *J. Cell. Sci.* **49**:299–310.

Scott, D. M., Murray, J. C., and Barnes, M. J., 1983, Investigation of the attachment of bovine corneal endothelial cells to collagens and other components of the subendothelium role of fibronectin, *Exp. Cell Res.* **144**:472–478.

Seidl, M., and Hörmann, H., 1983, Affinity chromatography on immobilized fibrin monomer, IV. Two fibrin-binding peptides of a chymotryptic digest of human plasma fibronectin, *Hoppe–Seylers Z. Physiol. Chem.* **364**:83–92.

Sekiguchi, K., and Hakomori, S., 1980, Functional domain structure of fibronectin, *Proc. Natl. Acad. Sci. USA* **77**:2662–2665.

Sekiguchi, K., and Hakomori, S., 1983, Domain structure of human plasma fibronectin. Differences and similarities between human and hamster fibronectins, *J. Biol. Chem.* **256**:3967–3973.

Sekiguchi, K., Fukuda, M., and Hakomori, S., 1981, Domain structure of hamster plasma fibronectin. Isolation and characterization of four functionally distinct domains and their unequal distribution between two subunit polypeptides, *J. Biol. Chem.* **256**:6452–6462.

Seppä, H. E., Yamada, K. M., Seppä, S. T., Silver, M. H., Kleinman, H. K., and Schiffman, E., 1981, The cell binding fragment of fibronectin is chemotactic for fibroblasts, *Cell Biol. Int. Rep.* **5**:813–819.

Shadle, P. J., and Barondes, S. H., 1982, Adhesion of human platelets to immobilized trimeric collagen, *J. Cell Biol.* **95**:361–365.

Sieber–Blum, M., Sieber, F., and Yamada, K. M., 1981, Cellular fibronectin promotes adrenergic differentiation of quail neural crest cells in vitro, *Exp. Cell Res.* **133**:285–295.

Singer, I. I., 1979, The fibronexus: A transmembrane association of fibronectin-containing fibers and bundles of 5 nm microfilaments in hamster and human fibroblasts, *Cell* **16**:675–685.

Singer, I. I., 1982, Association of fibronectin and vinculin with focal contacts and stress fibers in stationary hamster fibroblasts, *J. Cell Biol.* **92**:398–408.

Singer, I. I., and Paradiso, P. R., 1981, A transmembrane relationship between fibronectin and vinculin (130 kd protein): Serum modulation in normal and transformed hamster fibroblasts, *Cell* **24**:481–492.

Skorstengaard, K., Thøgersen, H. C., and Petersen, T. E., 1984, Complete primary sequence of the collagen-binding domain of bovine fibronectin, *Eur. J. Biochem.* **140**:235–243.

Smith, K. K., and Strickland, S., 1981, Structural components and characteristics of Reichert's membrane, an extra-embryonic basement membrane, *J. Biol. Chem.* **256**:4654–4661.

Solursh, M., Linsenmayer, T. F., and Jensen, K. L., 1982, Chondrogenesis from single limb mesenchyme cells, *Dev. Biol.* **94**:259–264.

Speziale, P., Höök, M., Wadstrom, T., and Timpl, R., 1982, Binding of the basement membrane protein laminin to *Escherichia coli*, *FEBS Lett.* **146**:55–58.

Spiegel, E., Burger, M., and Spiegel, M., 1980, Fibronectin in the developing sea urchin embryo, *J. Cell Biol.* **87**:309–313.

Stamatoglou, S. C., and Keller, J. M., 1982, Interactions of cellular glycosaminoglycans with plasma fibronectin and collagen, *Biochim. Biophys. Acta* **719:**90–97.

Stanley, J. R., Alvarez, O. M., Berke, E. W., Jr., Eaglstein, W. H., and Katz, S. I., 1981, Detection of basement membrane zone antigens during epidermal wound healing, *J. Invest. Dermatol.* **77:**240–243.

Stathakis, N. E., and Mosesson, M. W., 1977, Interactions among heparin, cold-insoluble globulin, and fibrinogen in formation of the heparin-precipitable fraction of plasma, *J. Clin. Invest.* **60:**855–856.

Stemberger, A., and Hörmann, H., 1976, Affinity chromatography on immobilized fibrinogen and fibrin monomer. The behavior of cold-insoluble globulin, *Hoppe–Seylers Z. Physiol. Chem.* **357:**1003–1005.

Stenn, K. S., 1981, Epibolin: A protein of human plasma that supports epithelial cell movement, *Proc. Natl. Acad. Sci. USA* **78:**6907–6911.

Stenn, K. S., Madri, J. A., Tinghitella, T., and Terranova, V. P., 1983, Multiple mechanisms of dissociated epidermal cell spreading, *J. Cell Biol.* **96:**63–67.

Stoker, M., O'Neill, C., Berryman, S., and Waxman, V., 1968, Anchorage and growth regulation in normal and virus-transformed cells, *Int. J. Cancer* **3:**383–393.

Suard, Y. M. L., Haeuptle, M.–T., Farinon, E., and Kraehenbuhl, J.–P., 1983, Cell proliferation and milk protein gene expression in rabbit mammary cell cultures, *J. Cell Biol.* **96:**1435–1442.

Sugrue, S. P., and Hay, E. D., 1981, Response of basal epithelial cell surface and cytoskeleton to solubilized extracellular matrix molecules, *J. Cell Biol.* **91:**45–54.

Sugrue, S. P., and Hay, E. D., 1982, Interaction of embryonic corneal epithelium with exogeneous collagen, laminin, and fibronectin: Role of endogenous protein synthesis, *Dev. Biol.* **92:**97–106.

Tamkun, J. W., and Hynes, R. O., 1983, Plasma fibronectin is synthesized and secreted by hepatocytes, *J. Biol. Chem.* **258:**4641–4647.

Tamura, G. (ed.), 1982, *Tunicamycin,* Japan Scientific Press, Tokyo.

Tarone, G., Galetto, G., Prat, M., and Comoglio, P. M., 1982, Cell surface molecules and fibronectin-mediated cell adhesion: Effect of proteolytic digestion of membrane proteins, *J. Cell Biol.* **94:**179–186.

Terranova, V. P., Rohrbach, D. H., and Martin, G. R., 1980, Role of laminin in the attachment of PAM 212 (epithelial) cells to basement membrane collagen, *Cell* **22:**719–726.

Terranova, V. P., Liotta, L. A., Russo, R. G., and Martin, G. R., 1982, Role of laminin in the attachment and metastasis of murine tumor cells, *Cancer Res.* **42:**2265–2269.

Terranova, V. P., Rao, C. N., Kalebic, T., Margulies, I. M., and Liotta, L. A., 1983, Laminin receptor on human breast carcinoma cells, *Proc. Natl. Acad. Sci. USA* **80:**444–448.

Thiery, J. P., Duband, J.–L., and Delouvée, A., 1982a, Pathways and mechanism of trunk avian neural crest cell migration and localization, *Dev. Biol.* **93:**324–343.

Thiery, J. P., Duband, J.–L., Rutishauser, U., and Edelman, G. M., 1982b, Cell adhesion molecules in early chicken embryogenesis, *Proc. Natl. Acad. Sci. USA* **79:**6737–6741.

Timpl, R., and Martin, G. R., 1982, Components of basement membranes, in: *Immunochemistry of the Extracellular Matrix,* Volume 2 (H. Furthmayr, ed.), CRC Press, Boca Raton, Florida, pp. 119–150.

Timpl, R., Rohde, H., Robey, P. G., Rennard, S. I., Foidart, J.–M., and Martin, G. R., 1979, Laminin-A glycoprotein from basement membranes, *J. Biol. Chem.* **254:**9933–9937.

Timpl, R., Wiedemann, H., van Delden, V., Furthmayr, H., and Kühn, K., 1981, A network model for the organization of type IV collagen molecules in basement membranes, *Eur. J. Biochem.* **120:**203–211.

Timpl, R., Johansson, S., van Delden, V., Oberbäumer, I., and Höök, M., 1983, Characterization of protease-resistant fragments of laminin mediating attachment and spreading of rat hepatocytes, *J. Biol. Chem.* **258**:8922–8927.

Toole, B. P., 1981, Glycosaminglycans in morphogenesis, in: *Cell Biology of Extracellular Matrix* (E. D. Hay, ed.), Plenum Press, New York, pp. 259–294.

Toole, B. P., 1982, Developmental Role of hyaluronate, *Conn. Tiss. Res.* **10**:93–100.

Trelstad, R. L., and Silver, F. M., 1981, Matrix assembly, in: *Cell Biology of Extracellular Matrix* (E. D. Hay, ed.), Plenum Press, New York, pp. 179–215.

Trinkaus, J. P., 1984, *Cells into Organs: The Forces That Shape the Embryo,* Prentice–Hall, Englewood Cliffs, New Jersey.

Underhill, C. B., Chi–Rosso, G., and Toole, B. P., 1983, Effects of detergent solubilization on the hyaluronate-binding protein from membranes of simian virus 40-transformed 3T3 cells, *J. Biol. Chem.* **258**:8086–8091.

van De Water, L., III, Schroeder, S., Crenshaw, E. B., III, and Hynes, R. O., 1981, Phagocytosis of gelatin-latex particles by a murine macrophage line is dependent on fibronectin and heparin, *J. Cell Biol.* **90**:32–39.

Vartio, T., 1982, Characterization of the binding domains in the fragments cleaved by cathepsin G from human plasma fibronectin, *Eur. J. Biochem.* **123**:223–233.

Virtannen, I., Vartio, T., Badley, R. A., and Lehto, V. P., 1982, Fibronectin in adhesion, spreading and cytoskeletal organization of cultured fibroblasts, *Nature* **298**:660–663.

Vlodavsky, I., and Gospodarowicz, D., 1981, Respective role of laminin and fibronectin in adhesion of human carcinoma and sarcoma cells, *Nature* **289**:304–306.

Voss, B., Allam, S., Rauterberg, J., Ullrich, K., Gieselmann, V., and Von Figura, K., 1979, Primary cultures of rat hepatocytes synthesize fibronectin, *Biochem. Biophys. Res. Comm.* **90**:1348–1354.

Vuento, M., 1979, Hemagglutinin activity of human plasma fibronectin, *Hoppe–Seylers Z. Physiol. Chem.* **360**:1327–1333.

Vuento, M., Wrann, M., and Ruoslahti, E., 1977, Similarity of fibronectins isolated from human plasma and spent fibroblast culture medium. *FEBS Lett.* **82**:227–231.

Vuento, M., Salonen, E., Osterlund, K., and Stenman, U. H., 1982, Essential charged amino acids in the binding of fibronectin to gelatin, *Biochem. J.* **201**:1–8.

Wartiovaara, J., Leivo, I., and Vaheri, A., 1980, Matrix glycoproteins in early mouse development and in differentiation of teratocarcinoma cells, in: *The Cell Surface: Mediator of Developmental Processes* (S. Subtelny and N. K. Wessels, eds.), Academic Press, New York, pp. 305–324.

Weigel, P. H., Schnaar, R. L., Kuhlenschmidt, M. S., Schmell, E., Lee, R. T., Lee, Y. C., and Roseman, S., 1979, Adhesion of hepatocytes to immobilized sugars: A threshold phenomenon, *J. Biol. Chem.* **254**:10830–10838.

Weiss, H. F., Baumgartner, H. R., Tschopp, T. B., Turitto, V. T., and Cohen, D., 1978, Correction by factor VIII of the impaired platelet adhesion to subendothelium in von Willebrand disease, *Blood* **51**:267–279.

Weiss, R. E., and Reddi, A. H., 1981, Appearance of fibronectin during the differentiation of cartilage, bone, and bone marrow, *J. Cell Biol.* **88**:630–636.

West, C. M., Lanza, R., Rosenbloom, J., Lowe, M., Holtzer, H., and Avdalovic, N., 1979, Fibronectin alters the phenotypic properties of cultured chick embryo chondroblasts, *Cell* **17**:491–501.

Weston, J. A., 1983, Regulation of neural crest cell migration and differentiation, in: *Cell Interactions and Development: Molecular Mechanims* (K. M. Yamada, ed.), Wiley, New York, pp. 153–184.

Wewer, U., Albrechtsen, R., and Ruoslahti, E., 1981, Laminin, a noncollagenous component of epithelial basement membranes synthesized by a rat yolk sac tumor, *Cancer Res.* **41**:1518–1524.

Whateley, J. G., and Knox, P. L., 1980, Isolation of a serum component that stimulates the spreading of cells in culture, *Biochem. J.* **185**:349–354.

Willingham, M. C., Yamada, K. M., Yamada, S. S., Pouysségur, J., and Pastan, I., 1977, Microfilament bundles and cell shape are related to adhesiveness to substratum and are dissociable from growth control in cultured fibroblasts, *Cell* **10**:375–380.

Woodley, D. T., Rao, C. N., Hassell, J. R., Liotta, L. A., Martin, G. R., and Kleinman, H. K., 1983, Interactions of basement membrane components, *Biochim. Biophys. Acta* **761**:278–283.

Wylie, D. E., Damsky, C. H., and Buck, C. A., 1979, Studies on the function of cell surface glycoproteins. I. Use of antisera to surface membranes in the identification of membrane components relevant to cell-substrate adhesion, *J. Cell Biol.* **80**:385–402.

Yamada, K. M., 1980, Fibronectin: Transformation-sensitive cell surface protein, *Lymphokine Reports* **1**:231–254.

Yamada, K. M., 1983a, Cell surface interactions with extracellular materials, *Ann. Rev. Biochem.* **52**:761–799.

Yamada, K. M., 1983b, *Cell Interactions and Development: Molecular Mechanisms*, Wiley, New York.

Yamada, K. M., 1984, Structural and functional domains of fibronectin, in: *Fibronectin* (D. F. Mosher, ed.), Academic Press, New York, in press.

Yamada, K. M., and Kennedy, D. W., 1979, Fibroblast cellular and plasma fibronectins are similar but not identical, *J. Cell Biol.* **80**:492–498.

Yamada, K. M., and Olden, K., 1982, Actions of tunicamycin on vertebrate cells, in: *Tunicamycin* (G. Tamura, ed.), Japan Scientific Societies Press, Tokyo, pp. 119–144.

Yamada, K. M., Yamada, S. S., and Pastan, I., 1975, The major cell surface glycoprotein of chick embryo fibroblasts is an agglutinin, *Proc. Natl. Acad. Sci. USA* **72**:3158–3162.

Yamada, K. M., Yamada, S. S., and Pastan, I., 1976a, Cell surface protein partially restores morphology, adhesiveness, and contact inhibition of movement to transformed fibroblasts, *Proc. Natl. Acad. Sci. USA* **73**:1217–1221.

Yamada, K. M., Ohanian, S. H., and Pastan, I., 1976b, Cell surface protein decreases microvilli and ruffles on transformed mouse and chick cells, *Cell* **9**:241–245.

Yamada, K. M., Schlesinger, D. H., Kennedy, D. W., and Pastan, I., 1977, Characterization of a major fibroblast cell surface glycoprotein, *Biochemistry* **16**:5552–5559.

Yamada, K. M., Olden, K., and Pastan, I., 1978, Transformation-sensitive cell surface protein: Isolation, characterization, and role in cellular morphology and adhesion, *Ann. N.Y. Acad. Sci.* **312**:256–277.

Yamada, K. M., Kennedy, D. W., Kimata, K., and Pratt, R. M., 1980, Characterization of fibronectin interactions with glycosaminoglycans and identification of active proteolytic fragments, *J. Biol. Chem.* **255**:6055–6063.

Yamada, K. M., Kennedy, D. W., Grotendorst, G. R., and Momoi, T., 1981, Glycolipids: Receptors for fibronectin? *J. Cell. Physiol.* **109**:343–351.

Yamada, K. M., Kennedy, D. W., and Hayashi, M., 1982, Fibronectin in cell adhesion, differentiation and growth, *Cold Spring Harbor Conf. Cell Proliferation* **9**:131–143.

Yamada, K. M., Critchley, D. R., Fishman, P. H., and Moss, J., 1983, Exogenous gangliosides enhance the interaction of fibronectin with ganglioside-deficient cells, *Exp. Cell Res.* **143**:295–302.

Yamada, Y., Avvedimento, V. E., Mudryj, M., Ohkubo, H., Vogeli, G., Irani, M., Pastan, I., and de Crombrugghe, B., 1980, The collagen gene: Evidence for its evolutionary assembly by amplification of a DNA segment containing an exon of 54 bp, *Cell* **22**:887–892.

Zardi, L., Siri, A., Carnemolla, B., Santi, L., Gardner, W. D., and Hoch, S. O., 1979, Fibronectin: A chromatin-associated protein? *Cell* **18**:649–657.

4

THE HUMAN ERYTHROCYTE AS A MODEL SYSTEM FOR UNDERSTANDING MEMBRANE CYTOSKELETON INTERACTIONS

Vann Bennett

1. INTRODUCTION

The prevailing view of membrane structure presented in many undergraduate textbooks is that of proteins floating in the plane of a fluid phospholipid bilayer (Singer and Nicolson, 1972). Many observations in the last 10 years indicate, however, that biological membranes do not behave as simple two-dimensional solutions of proteins. Measurements of the rates of lateral diffusion of membrane proteins in a variety of systems have revealed populations of proteins that are not mobile. The proteins that are mobile move at rates 10- to 100-fold slower than predicted on the basis of membrane viscosity (reviewed by Cherry, 1979). The diffusion constants of these slowly diffusing proteins are increased 200-fold in areas of the cell where the membrane is separated from the cytoplasm (Tank et al., 1982). Furthermore, diffusion of proteins may be nonrandom in some cells (Smith et al., 1979) and can require metabolic energy as in formation of caps of surface-labeled proteins in lymphocytes (Unanue and Karnovsky, 1973). These examples indicate possibilities for long-range interactions and organization in cell membranes and have led to proposals that at least some membrane proteins

Vann Bennett ● Department of Cell Biology and Anatomy, Johns Hopkins University School of Medicine, Baltimore, Maryland 21205. Work funded in part by grants from the National Institutes of Health (Research Career Development Award and research grant RO1 AM29808) and the Muscular Dystrophy Association.

have direct interactions with underlying cytoplasmic proteins (Singer, 1974; Nicolson, 1974; Edelman, 1976; Bourguignon and Singer, 1977).

Association between membrane proteins and structural proteins on the cytoplasmic membrane surface has been demonstrated in human erythrocytes, and the details of this linkage have been elucidated. This chapter will cover briefly the organization of membrane proteins in the erythrocyte, which has been reviewed elsewhere (Lux, 1979; Branton et al., 1981), and will focus on membrane–protein associations. Observations also will be reviewed indicating that erythrocyte membrane proteins have closely related forms in other tissues, and that the structure of the erythrocyte membrane will have general relevance for other cells.

2. OVERVIEW OF THE HUMAN ERYTHROCYTE MEMBRANE

The erythrocyte is an unusually durable cell, surviving thousands of passes through the circulation during its 120-day life. The basis for the resilience of this cell lies in the mechanical properties of the plasma membrane, since the usual cytoplasmic structural proteins and organelles have been lost in differentiation. Direct measurements of deformation of erythrocytes have demonstrated that the membrane behaves like a semisolid with elastic properties that are not observed with simple lipid vesicles (Rand and Burton, 1964; Evans and Hochmuth, 1978; Kwok and Evans, 1981). It is this ability of erythrocyte membranes to store energy during brief periods of deformation, such as occur in the circulation, and then return to their original shape that provides these cells with their longevity.

The structural component of the erythrocyte responsible for the elastic properties is a membrane-associated assembly of proteins commonly (and somewhat inaccurately) referred to as the cytoskeleton or membrane skeleton. This structure was discovered by Steck and co-workers (Yu et al., 1973) by the simple maneuver of extracting erythrocyte ghosts with a nonionic detergent that removed phospholipids and integral proteins. A meshwork of proteins remained after solubilization of the traditional membrane components, and this meshwork retained the same shape as the original ghosts. The cytoskeleton has been visualized as a two-dimensional meshwork of filaments about 100–140 nm in length (Hainfield and Steck, 1977; Nermut, 1981). The principal protein components of the cytoskeleton are spectrin, band 4.1, erythrocyte actin, and band 4.9 (nomenclature of Steck, 1974, based on mobility of proteins on SDS-gels; see Table I for a list of membrane proteins).

Several lines of evidence indicate the importance of the cytoskeleton for

TABLE I. Major Proteins in Human Red Cell Membranes

Protein	Subunit (M_r)	Probable assembly state	Approximate copies/cell
Peripheral proteins			
Spectrin	$\alpha = 260,000$	$(\alpha,\beta)_2$ tetramer	10^5 tetramers
	$\beta = 225,000$		
Ankyrin	215,000	Monomer	10^5
Band 4.1[a]	78,000	?	2×10^5
Band 4.2[a]	72,000	?	2×10^5
Band 4.9[a]	45,000	?	5×10^4
Actin[b]	43,000	Oligomer of 12–17 subunits	5×10^5
Glyceraldehyde 3-phosphodehydrogenase	35,000	Tetramer	5×10^5
Band 7[a]	29,000	?	5×10^5
Band 8[a]	23,000	?	10^5
Tropomyosin	29,000	Dimer	7×10^4 dimers
	27,000		
Integral proteins			
Band 3[a]	89,000	Dimer/tetramer	10^6
Glycophorin A[b]	31,000	Dimer	4×10^5
Glycophorin B[c]	23,000	?	$\sim 10^5$
Glycophorin C[c] (glycoconnectin)	29,000	?	$\sim 10^5$

[a] Nomenclature of Steck (1974).
[b] Data from Pinder and Gratzer (1983).
[c] Nomenclature of Furthmayr (1979); estimated by Lux (1983) assuming 60% carbohydrate for all three glycophorins.

maintaining erythrocyte shape. Extraction of cytoskeletal components spectrin and actin from ghosts at low ionic strength is accompanied by disintegration of ghosts into small vesicles. Strains of anemic mice have been developed by Bernstein and co-workers (Bernstein, 1980; Russell, 1979) that have substantial reduction in the amount of spectrin in their erythrocytes (Greenquist *et al.*, 1978). Erythrocytes from these animals are extremely fragile, and in the severely affected strains the erythrocytes lyse soon after entering the circulation. These spectrin-deficient cells have been experimentally reconstituted with spectrin, with some improvement in their structural stability (Shohet, 1979). Finally, human beings with hereditary hemolytic anemias have abnormally shaped or fragile erythrocytes and have been discovered to have defects in amounts or function of cytoskeletal proteins (see section below).

Spectrin was discovered by Marchesi and Steers (1968) in extracts of erythrocyte ghosts and is the major component of the cytoskeleton. Spectrin, as viewed by shadowing with platinum, is a flexible, rod-shaped molecule about 100 nm in length composed of two parallel polypeptide chains of $M_r = 260,000$ (alpha chain) and 225,000 (beta chain) Shotton *et al.*, 1979). Spectrin heterodimers self-associate at one end of the molecule to form tetramers of 200 nm in length. It has been proposed (Shotton *et al.*, 1979; Tyler *et al.*, 1980b) that in the tetramer, each alpha chain is associated by head–head linkage with a beta chain, so that no alpha–alpha or beta–beta contacts occur. The tetramer is most likely the major form of spectrin in ghosts based on chemical cross-linking experiments (Ji *et al.*, 1980), presence of spectrin tetramer in membrane extracts prepared under mild conditions (Ralston, 1978), and characterization of the dimer–tetramer equilibrium (Ungewickell and Gratzer, 1978). It has also been proposed that higher oligomers of spectrin such as hexamers and octomers may also be present in membranes and account for the polymerization of spectrin into a meshwork (Morrow and Marchesi, 1981).

Spectrin has specific binding sites for a number of proteins. Spectrin binds to F-actin by lateral association (Brenner and Korn, 1979; Cohen *et al.*, 1980; Fowler *et al.*, 1981), and this association is promoted by band 4.1 (4.1a and 4.1b) (Ungewickell *et al.*, 1979; Fowler and Taylor, 1980; Cohen *et al.*, 1980; Cohen and Foley, 1981). Spectrin also binds directly to band 4.1 in the absence of actin (Tyler *et al.*, 1980b) with a K_D of 10^{-7}M and a stoichiometry of 2 mols 4.1 per spectrin dimer. The measurements of binding of spectrin and band 4.1 have been reproduced in other laboratories (Wolfe *et al.*, 1982; Goodman *et al.*, 1982) although binding of spectrin to band 4.1 in the absence of actin could not be detected by sedimentation on sucrose gradients (Ungewickell *et al.*, 1979). The binding sites for actin and band 4.1 have been localized to the same region at the tails of spectrin tetramers by rotary shadowing (Tyler *et al.*, 1980b; Cohen *et al.*, 1980). It is possible that all three proteins are associated in a ternary complex at this site. Band 4.1 has recently been reported to cross-react with acumentin, an actin-capping protein from macrophages (Spiegel *et al.*, 1982), and might be expected to also bind directly to actin. Spectrin also binds to calmodulin, although with low affinity ($K_D \sim 5$ uM) Sobue *et al.*, 1981). The association of spectrin and calmodulin was dependent on calcium but was measured under unusual conditions of 6 M urea. The physiological significance of this weak interaction is not clear, but it should be kept in mind that calmodulin is present in cytosol at micromolar levels (Jarrett and Penniston, 1978). Furthermore, spectrin from other tissues (see below) also binds calmodulin, although with higher affinity.

The domain structure of spectrin has been resolved by isolation of proteolytic fragments of individual subunits corresponding to the sites for formation of tetramer, binding to ankyrin (see below), as well as sites of interchain cross-linking between the subunits (Morrow *et al.*, 1980; Speicher *et al.*, 1980, 1982; Marchesi, 1983). The binding of spectrin to actin/band 4.1 apparently requires both subunits, since isolated subunits were inactive (Calvert *et al.*, 1980). The amino acid sequence of spectrin alpha and beta chains is currently being determined by Marchesi, Speicher, and their colleagues.

Erythrocyte actin was first visualized as filaments in spectrin extracts (Marchesi and Steers, 1968), although the identity of this protein as actin was not established until later (Tilney and Detmers, 1975; Sheetz *et al.*, 1976). Purified erythrocyte actin is capable of polymerizing to form 7-nm filaments, activates myosin ATPase, and has all the properties of other cellular actins. In spite of these similarities to other actins, erythrocyte actin has not been visualized as filaments in cells or ghosts even though the concentration of 200 µg/ml is well above its critical concentration for polymerization in solution. Actin is thought, instead, to exist as oligomers of 12–17 subunits (Brenner and Korn, 1980; Pinder and Gratzer, 1983) on the basis of indirect studies. The basis for stable actin oligomers probably involves accessory actin-binding proteins. Band 4.9, for example, which is another cytoskeletal protein, binds to actin and may fragment and/or bundle existing actin filaments (Siegel and Branton, 1982). Tropomyosin has also been discovered in erythrocyte membranes (Fowler and Bennett, 1984), and the association of this protein could stabilize short filaments or regulate association of actin with proteins such as spectrin or band 4.9.

Another unexplained feature of erythrocyte actin is that it is present as the beta isoform instead of a mixture of isoelectric variants as in other cells (Pinder *et al.*, 1978. Since erythroblasts presumably have different actins before terminal differentiation, this suggests that some feature of beta-actin strongly favors its association with the membrane in erythrocytes relative to other isoforms. It will be important in future studies of actin interactions in erythrocytes to use erythrocyte actin. Erythrocyte actin, as a pure beta isoform, also will be useful for studies of beta-actin function in other cells.

It has not been established how spectrin, actin, and band 4.1 form the anastomosing meshwork of the cytoskeleton. One possibility is that actin oligomers may bind to more than one spectrin tetramer/band 4.1 complex, with each tetramer in turn linked to other oligomers (Brenner and Korn, 1980). It has also been proposed that spectrin may self-associate without requirement for actin and band 4.1 to form an extended polymer (Morrow and Marchesi, 1981). These possibilities are not mutually exclusive, and both types of polymerization mechanisms may occur. Resolution of the type of linkage would be facilitated by a method for reassembling a cytoskeletal meshwork *in vitro*. An obvious difficulty in such an approach is that criteria for a true "cytoskeleton" have not been established.

The erythrocyte membrane contains two major structural proteins, band 3 and glycophorin A, which span the membrane bilayer with distinct domains expressed on outer and cytoplasmic surfaces (Steck, 1974, 1978; Marchesi *et al.*, 1976). Band 3 is a glycoprotein present in about 10^6 copies/cell and contains an anion transport channel (Cabantchik *et al.*, 1978). An important feature of band 3, in terms of the cytoskeleton, is that this membrane protein contains a large domain of $M_r \sim 43,000$, which is localized on the inner surface of the membrane (Steck *et al.*, 1976; Bennett and Stenbuck, 1980b). This cytoplasmic domain of band 3 is associated with glycolytic enzymes aldolase, phosphofructokinase, and glyceraldehyde-3-phosphodehydrogenase as well as with a $M_r = 72,000$ polypeptide (band 4.2) of unknown function (Steck, 1978). The binding site for the glycolytic enzymes is localized within 75 residues of the NH_2-terminus (Tsai *et al.*, 1982; Murthy *et al.*, 1981). The cytoplasmic region of band 3 also interacts with hemoglobin (Salhany and Shaklai, 1979; Salhany *et al.*, 1980). Band 3 is a stable homodimer in detergent extracts as well as erythrocyte ghosts (Steck, 1972; Yu and Steck, 1975; Nigg and Cherry, 1979a). Band 3 dimers associate to form tetramers or higher oligomers in the membrane (Nigg and Cherry, 1979b) and in solution (Dorst and Schubert, 1979; Nakashima *et al.*, 1981).

Glycophorin A (PAS bands 1 and 2) is the major sialic-acid-containing glycoprotein and contains some blood group antigens and binding sites for lectins and viruses (Marchesi *et al.*, 1976). Glycophorin also may contain a recognition site that allows invasion of erythrocytes by malarial parasites (Pasval *et al.*, 1982; Perkins, 1981). Glycophorin A has been sequenced (Marchesi *et al.*, 1976), and the portion of the polypeptide chain exposed on the cytoplasmic membrane surface has been determined (Cotmore *et al.*, 1977). Unlike band 3, the cytoplasmic domain of glycophorin is small ($M_r = 3,000$). Glycophorin A is a stable dimer, even in the presence of sodium dodecylsulphate. Glycophorin is thought to be associated with band 3 in erythrocyte membranes since addition of antibody against glycophorin slows the rotational diffusion rate of band 3 (Nigg *et al.*, 1980). The linkage between band 3 and glycophorin is not maintained after these proteins have been solubilized in nonionic detergents (Yu and Steck, 1975). Band 3 and glycophorin are major constituents of the intramembrane particles visualized by freeze–fracture electron microscopy (Yu and Branton, 1976; Pinto da Silva and Nicolson, 1974; Marchesi *et al.*, 1976).

3. ASSOCIATION OF SPECTRIN WITH THE MEMBRANE

3.1. Evidence for a Spectrin–Membrane Interaction

The cytoskeleton of erythrocytes must be linked to the membrane in some way in order to account for the mechanical properties of erythrocytes. This logical inference is supported by the fact that the protein components of the cytoskeleton remain associated with membranes after repeated washes and can be extracted only at extremes of ionic strength. The next question is which protein(s) of the cytoskeleton are involved in the attachment of this structure to the membrane. A definite bias in answering this question has been that spectrin, as the major component of the skeleton, also is the linkage protein. As will be discussed, this idea clearly is at least partly correct, but additional associations were and still are theoretically possible.

Morphological experiments provided initial evidence that spectrin is linked in some way to integral membrane proteins. Incorporation of bivalent antispectrin Ig into ghosts caused aggregation in the plane of the membrane of the glyco-proteins (Nicolson and Painter, 1973). It was subsequently reported that intra-membrane particles (containing band 3 and glycophorin) could be aggregated by low ionic strength or acidic pH only after removal of spectrin and actin (Elgsaeter and Branton, 1974; Elgsaeter *et al.*, 1976). It was also observed that spectrin was localized on the inner surface of the membrane in patterns coincident with those of the intramembrane particles (Shotton *et al.*, 1978). These obser-vations led to the conclusion that the intramembrane particles and their constituent proteins were associated with spectrin. An alternative explanation to direct con-tact between integral proteins and spectrin is that the integral proteins were simply trapped in the underlying tangle of cytoskeletal proteins without specific protein linkages (Steck, 1974). The entrapment theory was supported by mea-surements of rotational diffusion of eosin-labeled band 3, which, to a first approximation, was unaffected by removal of spectrin and actin and was consistent with unrestricted motion in the membrane (Cherry *et al.*, 1976). Further indirect evidence against an integral protein–spectrin linkage were reports that spectrin can associate with artificial lipid membranes without additional proteins (Mombers *et al.*, 1979; Haest *et al.*, 1978).

The interaction of spectrin with erythrocyte membranes has been analyzed directly by measuring reassociation of purified, radiolabeled spectrin with inside-out vesicles depleted of spectrin and actin (Bennett and Branton, 1977; Bennett, 1977; Litman *et al.*, 1980; Baskin and Langdon, 1981). This approach is analogous to studies with ligands such as hormones that associate with receptors on the cell surface. Binding assays with radiolabeled hormones have permitted characterization of receptors and provided a way to monitor purification of receptor molecules. Binding studies with spectrin have demonstrated the presence of a protein attachment site and led to identification of this spectrin-binding protein. Before discussing these experiments, it is important to consider some potential difficulties in performing such binding measurements. A major problem in measuring reassociation of spectrin, as in all binding studies, is to distinguish specific binding from nonspecific or biologically irrelevant interactions. Spectrin is especially prone to aggregate and to adsorb to surfaces. Moreover, no activity of spectrin or vesicles is known to be altered after reassociation, and a functional assay is therefore not available to monitor specific binding. In view of the lack of functional correlates of spectrin binding to membranes, it was important to establish other criteria to distinguish specific interactions. The following features would be expected for a selective association resembling the attachment of spectrin in erythrocytes: (1) Binding should be limited to the inner surface of the membrane. (2) The ionic and pH dependence for binding should be the same as for association of spectrin with ghost membranes. (3) Binding should be abolished by treatments of spectrin, and possibly vesicles, which denature proteins. (4) Binding should be saturable at a level similar to the amount of spectrin present in ghost membranes. Another consideration in binding measurements with radioactive probes is to ensure that the labeling procedure does not alter or damage the protein. Spectrin initially was metabolically labeled with $^{32}P_i$ (label incorporates into the beta subunit in unstimulated cells) in order to maintain as closely as possible the native state of the protein (Bennett and Branton, 1977; Bennett, 1977). Identical measurements have also been obtained by radiolabeling with ^{125}I (Litman *et al.*, 1980), with reductive formylation (Baskin and Langdon, 1981), and with Bolton Hunter reagent (Tyler *et al.*, 1980b; Agre *et al.*, 1981).

Binding of radiolabeled spectrin heterodimer to erythrocyte membranes exhibited features consistent with a specific, protein–protein interaction. Quantitatively similar results have been reported using spectrin tetramer (Goodman and Weidner, 1980). Spectrin associated with inside-out vesicles in amounts

10–15 times larger than right-side-out vesicles, indicating a binding site localized on the inner membrane surface (Fig. 1). Binding to inverted vesicles exhibited an identical dependence on ionic strength, divalent metal ions, and pH as did association of spectrin with native membranes. Spectrin bound to inverted membranes with a K_D of 10^{-7} - 10^{-8} M to a single class of sites present in about 10^5 copies/cell, which is equivalent to the number of spectrin tetramers. These sites involved a protein since the binding capacity of membranes was destroyed by mild proteolysis (Fig. 2), by extraction with dilute acid (Fig. 1), or by reaction

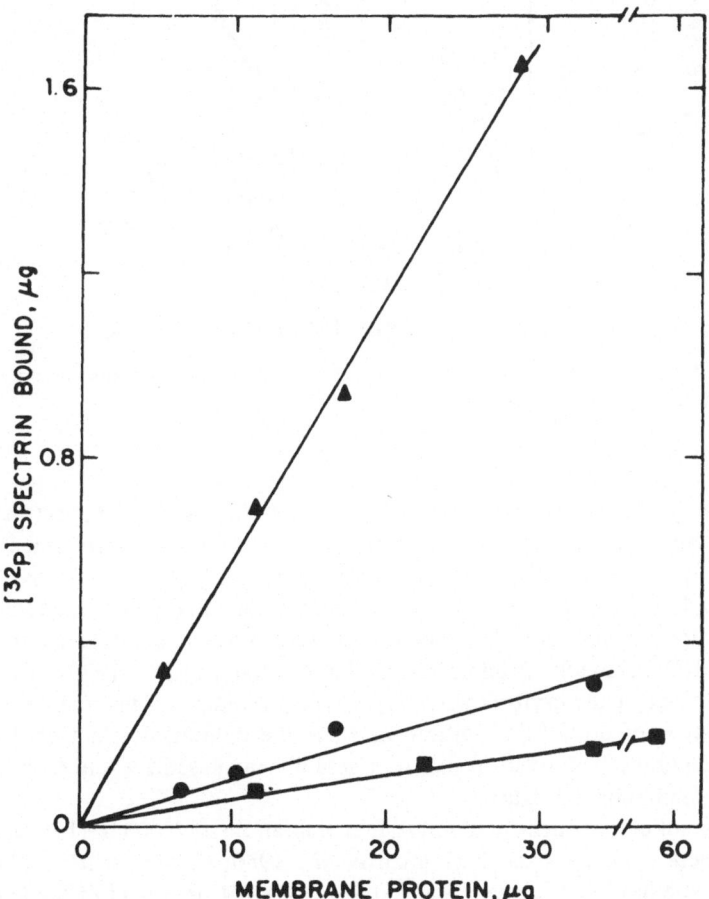

FIGURE 1. Binding of [^{32}P] spectrin heterodimer to spectrin-depleted inside-out vesicles (▲), right-side-out vesicles (■), and acetic-acid-extracted inside-out vesicles (●). From Bennett and Branton (1977).

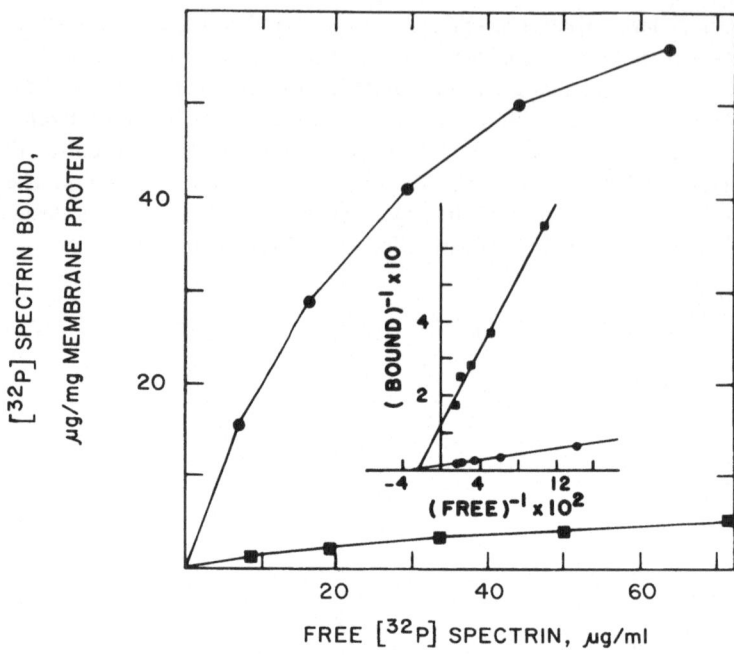

FIGURE 2. Binding of [^{32}P] spectrin heterodimer to control (●) and chymotrypsin-digested (■) inside-out vesicles. From Bennett (1978).

with sulfhydral-reactive compounds (Bennett, 1978). Ability of spectrin to reassociate with vesicles required some tertiary structure in spectrin since binding activity was destroyed by thermal denaturation in a highly cooperative manner between 49°C and 51°C (Fig. 3). It is of interest that this is the same temperature range where erythrocytes disintegrate and where spectrin exhibits major changes in its ORD behavior (Brandts *et al.*, 1977; Chang *et al.*, 1979). Binding of spectrin to extracted erythrocyte lipids was also evaluated. This intereaction does occur but is of doubtful significance, since the binding was not abolished by heat denaturation of spectrin and was actually increased by this treatment (V. Bennett, unpublished data).

The subunit of spectrin involved in membrane association has been identified as the beta chain or band 2 (Litman *et al.*, 1980; Calvert *et al.*, 1980). The binding site has been further localized to a 50,000M_r fragment of the beta chain produced by cleavage with 2-nitro-5-thiocyanobenzoic acid, and this binding fragment is located at the same end of the spectrin molecule involved in head-to-head association to form tetramers (Morrow *et al.*, 1980). The binding site

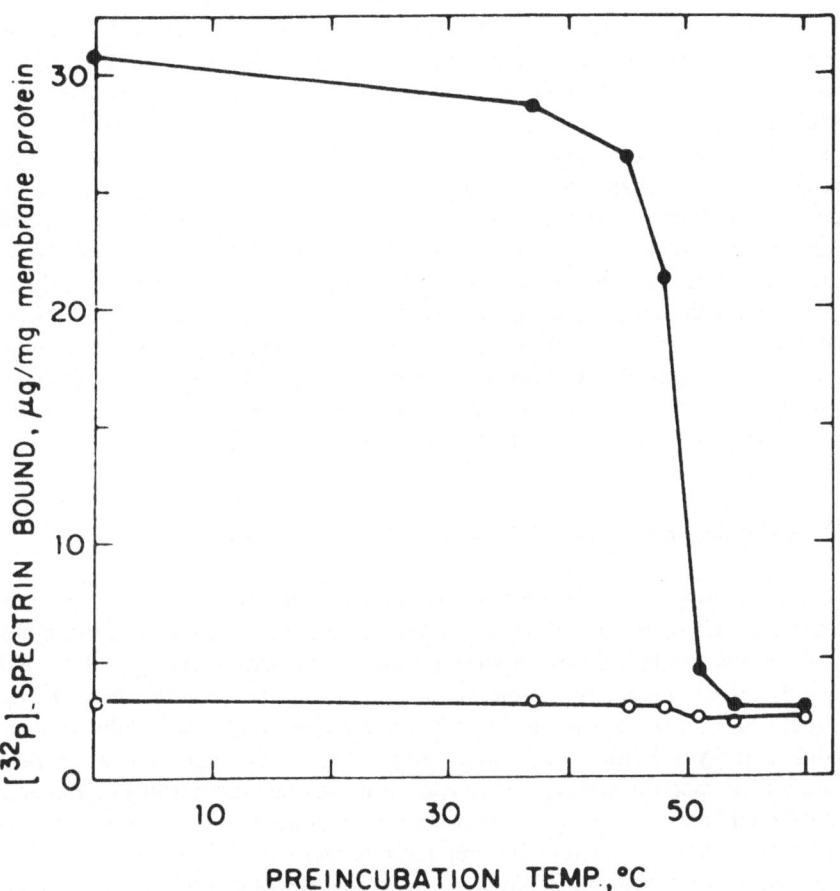

FIGURE 3.   Effect of preincubation temperature on binding of [^{32}P] spectrin heterodimer to inside-out (●) and right-side-out vesicles (○). From Bennett and Branton (1977).

on the spectrin molecule has been visualized by rotary shadowing of spectrin in the presence of the binding protein, and this site is localized about 20 nm from the head of spectrin (Tyler *et al.*, 1979).

These experiments indicated that a specific protein–protein association could be reconstituted between the beta subunit of spectrin and some protein localized on the cytoplasmic surface of the membrane. The affinity for binding of spectrin of 10^{-7}–10^{-8} M suggests a relatively rapid rate of dissociation from its membrane site; a half-life of about 15 min, for example, can be calculated from the rate of association for a spectrin–membrane complex with a K_D of 10^{-7} M. The apparent stability of attachment of spectrin in ghosts indicates that additional interactions of spectrin are not reconstituted in the binding assay. These additional linkages most likely include lateral associations of spectrin to form the meshwork visualized in detergent-extracted cytoskeletons. It also is conceivable that spectrin interacts with the membrane surface by low affinity binding. It should be emphasized, in this regard, that the actual local concentration of spectrin at the membrane surface in erythrocytes is 10^{-5} M or higher. Unfortunately, spectrin in the test tube at such concentrations becomes a gel, and measurement of low-affinity interactions is technically very difficult. Future experiments could be approached with defined domains of spectrin that retain activity but have better solubility properties. The initial placement and concentration of spectrin along the membrane probably depends on the high affinity interaction, and it has been proposed that this is an essential first event in assembly of the membrane–cytoskeleton complex (Morrow and Marchesi, 1981).

3.2. Identification of the High-Affinity Attachment Site for Spectrin

The identity of the membrane attachment protein was obviously the next question. In this section I will review first some strategies that were not successful in solving the problem. These studies are instructive since similar problems will probably arise in elucidating protein associations in other systems. In principle, it should have been possible to identify the binding protein by solubilizing the binding activity from membranes and then purifying the protein using as an assay inhibition of spectrin binding. This approach was not successful (V. Bennett, unpublished data), however, for reasons that now are known to be due in part to protease activity in membrane preparations and the tendency of the binding protein to aggregate. A more fundamental problem was that the binding protein itself has an attachment site on the membrane and can bind to membranes. Thus this protein either had no effect or actually increased binding of spectrin to membranes in competition assays. Another approach was to prepare detergent-extracted cytoskeletons from erythrocytes, which were assumed to retain the binding protein (Sheetz and Sawyer, 1978; Goodman and Branton, 1978). Spectrin was then extracted from the cytoskeletons, and the spectrin-depleted residues were used as a source of binding protein. A protein of $M_r = 95,000$ termed "band 3'," in these residues was proposed as the spectrin-binding site since it was present in about the right amount to account for spectrin binding (Goodman and Branton, 1978). It is now evident that the 95,000-M_r polypeptide was a proteolytic fragment of a larger protein, and that it is not the binding site for spectrin.

A third approach to identifying spectrin-binding proteins was to chemically cross-link ghost membranes by a reversible reaction, as was first performed by Wang and Richards (1974). Spectrin alpha subunit (band 1) was cross-linked to band 3 by oxidation, which presumably formed disulfide-linked complexes (Liu and Palek, 1979). Drawbacks to this type of experiment are that (1) only protein complexes with contiguous reactive groups will be cross-linked and therefore a negative result cannot rule out binding, and (2) cross-linking can occur between proteins owing to proximity and may not reflect a common binding or recognition site; the proteins of erythrocyte membranes are extremely concentrated and therefore especially susceptible to cross-linking between neighboring components. Currently available knowledge indicates that the observations of spectrin–band 3 complexes probably suffer from both problems. The spectrin-binding protein does not readily cross-link to spectrin and thus was missed. Furthermore, spectrin does not associate directly with band 3 although these proteins exist close enough to each other to be cross-linked.

An immunological strategy also was unsuccessful. Antisera were prepared against several erythrocyte membrane proteins (bands 3, 4.1, and 4.2), and the antibodies were then demonstrated to inhibit binding of spectrin to vesicles. It was incorrectly concluded that spectrin associated with bands 4.1 and 4.2 and possibly band 3 (Litman et al., 1978). These experiments have similar problems as the cross-linking and are much more laborious. The binding interaction will not necessarily be blocked by antibodies, and antibody against a neighboring protein could interfere with binding indirectly. An additional, technical problem is that antibody preparations are frequently contaminated by serum proteases that can inhibit binding simply by proteolysis.

The studies that led to identification of the binding protein were based on the observation that during mild proteolysis of inside-out vesicles a polypeptide was released from the membranes that competed for binding of spectrin to undigested vesicles (Fig. 4) (Bennett, 1978). A 72,000-M_r proteolytic fragment was purified using as an assay inhibition of spectrin binding. This polypeptide by a number of criteria had properties expected for the spectrin attachment site (Bennett, 1978). Release of the 72,000-M_r fragment during digestion of membranes paralleled closely loss of membrane-binding capacity for spectrin. The purified fragment was a potent competitive antagonist of spectrin binding, with a K_i approximately the same as the K_D for spectrin binding to vesicles (Fig. 5). The ability of the fragment to compete for binding was abolished by reaction with N-ethylmaleimide, which also destroyed spectrin-binding activity in vesicles. The 72,000-M_r fragment associated with spectrin in solution, and a complex of these proteins was isolated with approximately 1 mol of fragment per mol of spectrin heterodimer. The fragment was localized exclusively on the cytoplasmic

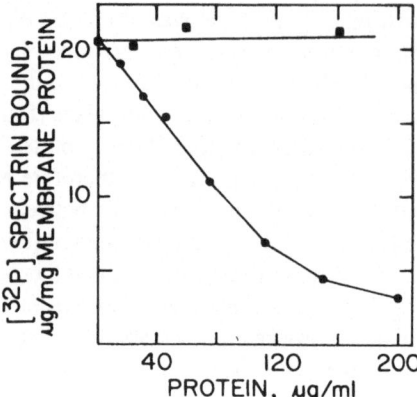

FIGURE 4. Inhibition of binding of [^{32}P] spectrin vesicles by protein solubilized during digestion of control (●) and acid-extracted (■) inside-out vesicles with chymotrypsin. From Bennett (1978).

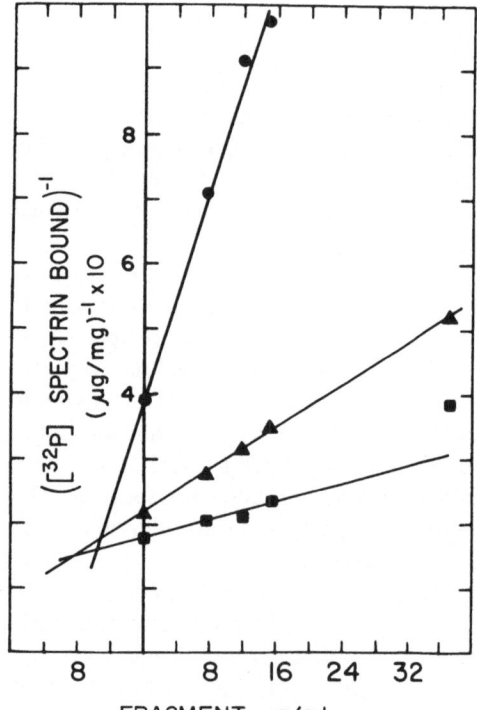

FIGURE 5. Dixon plot of inhibition of binding of [^{32}P] spectrin to inside-out vesicles in the presence of increasing concentrations of 72,000-M_r fragment with 16 (●), 32 (▲), and 47 (■) ug/ml [^{32}P] spectrin. From Bennett (1978).

surface of the membrane since it was released in equal amounts from vesicles prepared from erythrocytes that were untreated or extensively proteolyzed on the external membrane surface. The amount of 72,000-M_r fragment in ghost membranes and intact cells was measured at $\sim 10^5$ copies/cell by radioimmunoassay, which is the same as the number of binding sites for spectrin (Bennett, 1979). Finally, affinity-purified Ig directed against the 72,000-M_r fragment blocked binding of spectrin to inverted vesicles (Fig. 6) (Bennett and Stenbuck, 1979a).

A 215,000-M_r polypeptide (band 2.1) localized on the inner surface of the plasma membrane has been identified as the precursor of the 72,000-M_r fragment on the basis of cross-reaction with antifragment antibody (Bennett and Stenbuck, 1979a), by comparative peptide mapping (Luna *et al.*, 1979; Yu and Goodman, 1979), and by selective extraction of band 2.1 from membranes (Bennett and Stenbuck, 1979a). Affinity-purified Ig against the 72,000-M_r fragment cross-reacted only with band 2.1 and the closely migrating minor bands 2.2 and 2.3 (Bennett and Stenbuck, 1979a,b). Bands 2.2 and 2.3 are present in variable amounts in different preparation of ghosts and most likely are degradation products of band 2.1. Peptide maps were prepared from the 72,000-M_r fragment which corresponded only with band 2.1, 2.2, and 2.3 among the membrane polypeptides with a molecular weight above 72,000M_r (Luna *et al.*, 1979; Yu

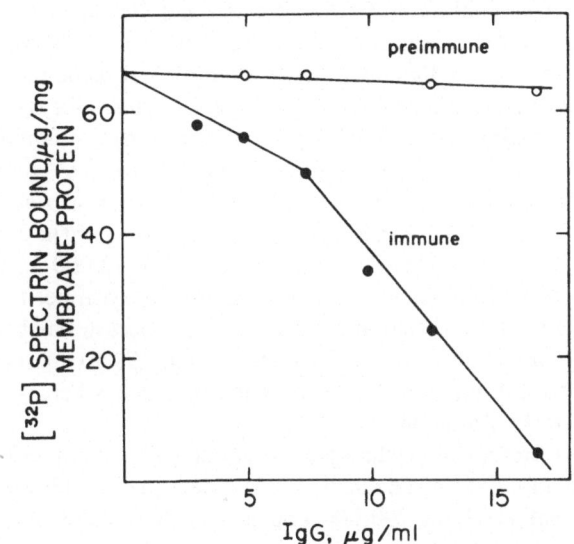

FIGURE 6. Effect of preimmune and affinity-purified anti-72,000-M_r-fragment Ig on binding of [^{32}P] spectrin to inside-out vesicles. From Bennett and Stenbuck (1979a).

and Goodman, 1979). Further evidence that the 72,000-M_r fragment is derived from band 2.1 was obtained by selective removal of band 2.1, which abolished production of the fragment and destroyed spectrin-binding activity in the extracted membranes (Bennett and Stenbuck, 1979a). In contrast, band 4.1, which has been identified as a spectrin-binding protein in the cytoskeleton (see previous discussion), was extracted completely with no effect on binding of spectrin or production of the 72,000-M_r fragment (Bennett and Stenbuck, 1979a; Cohen and Foley, 1982).

Direct evidence that spectrin associated with band 2.1 in membranes was provided by isolation with antifragment IgG of a specific complex of spectrin and 2.1 in a 1 : 1 molar ratio formed during reassociation of spectrin with inverted vesicles (Bennett and Stenbuck, 1979a). Furthermore, partly purified 2.1 was a competitive antagonist of spectrin binding to inverted vesicles. It was concluded from these experiments and from the evidence that 2.1 was the precursor of the 72,000-M_r fragment that 2.1 was the high-affinity membrane attachment site for spectrin. In recognition of this binding function the protein has been named ankyrin, from the Greek "ankyra," which means anchor.

Ankyrin has been purified in milligram amounts as a water-soluble protein in the absence of detergent (Tyler *et al.*, 1980b; Bennett and Stenbuck, 1980a). The isolated protein contains a single polypeptide chain and is somewhat asymmetrical with a frictional ratio of 1.46 (Bennett and Stenbuck, 1980a). Ankyrin has some hydrophobic character based on the behavior in charge-shift electrophoresis but is much less hydrophobic than an integral membrane protein such as band 3. Ankyrin as isolated from the membrane is the same protein identified as the spectrin-binding site since this protein can be almost completely immunoprecipitated by anti-72,000-M_r-fragment IgG. Furthermore, controlled digestion of purified ankyrin produces a 72,000-M_r fragment as well as fragments corresponding to bands 2.2 and 2.3. Ankyrin binds to spectrin in solution with high affinity (K_D 5 × 10^{-8}M) at the same site as the 72,000-M_r fragment and forms a complex with a stoichiometry of 1 mol of ankyrin per mol of spectrin heterodimer (Fig. 7). Association of ankyrin with spectrin has been visualized by low-angle rotary shadowing (Tyler *et al.*, 1979, 1980b) and occurs at a site on spectrin about 20 nm from the head of the molecule where spectrin heterodimers join to form a tetramer.

Ankyrin is present in erythrocytes in about 100,000 copies/cell, as determined by radioimmunoassay (Bennett, 1979). Spectrin heterodimers, on the other hand, are present in about 200,000 copies/cell. It is likely that the spectrin molecules present in excess over ankyrin-binding sites are associated with membrane-bound spectrin to form tetramers. Two lines of evidence support this conclusion. Spectrin most likely is a tetramer or higher-order oligomer in erythrocyte membranes (see Section 1). Furthermore, spectrin tetramer associates

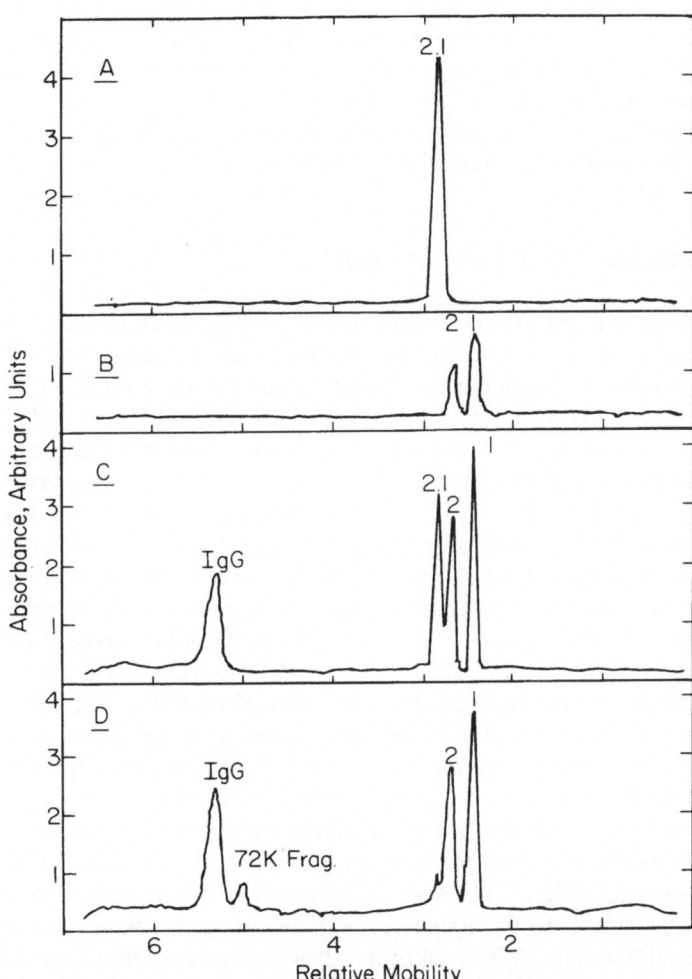

FIGURE 7. Association of ankyrin with spectrin at the same site as the 72,000-M_r fragment. Spectrin heterodimer (83 nM) and ankyrin (372 nM) were incubated in the presence and absence of 72,000-M_r fragment (4.9 μm). Spectrin and associated proteins were isolated by immunoprecipitation with antispectrin Ig, and the precipitates were analyzed by SDS electrophoresis. The gels were stained with Coomassie blue and scanned in a densitometer. 2.1, Ankyrin; 1, spectrin alpha chain; 2, spectrin beta chain. (A) Ankyrin alone; (B) spectrin alone; (C) immunoprecipitate from incubation of ankyrin and spectrin alone; (D) immunoprecipitate from incubation of ankyrin, spectrin, and 72,000-M_r fragment. From Bennett and Stenbuck (1980a).

with vesicles with the same affinity and with the same number of binding sites as spectrin heterodimers (Goodman and Weidner, 1980). It thus seems likely that each spectrin tetramer binds to the membrane at only one of its two potential ankyrin sites. The reason that spectrin tetramers bind monovalently may be that the distance between ankyrin sites on the membrane is greater than the distance between binding sites on the tetramers, or that occupation of one ankyrin site in some way inhibits binding at the second site.

3.3. Association of Ankyrin and Band 3

The fact that ankyrin is a water-soluble protein raised new questions about the linkage between spectrin and integral membrane proteins. If it is correct that spectrin is associated with intramembrane particles and that ankyrin is the binding protein, then ankyrin should provide a linkage between spectrin and some component of the complex of proteins that form the intramembrane particles. Several different types of experiments have demonstrated that, in fact, ankyrin is directly associated with band 3 at its 43,000-M_r cytoplasmic domain. Antiankyrin IgG coprecipitates band 3 with ankyrin in a 1 : 1 molar ratio from detergent extracts of spectrin-depleted vesicles (Fig. 8) (Bennett and Stenbuck, 1979b). The association of band 3 with ankyrin was specific since the sialoglycoproteins (glycophorins) were not present, and since band 3 was not immunoprecipitated after denaturation of the extract. The ankyrin-associated band 3 was nearly identical to the free population of band 3 on the basis of its CNBr fragments and also was demonstrated to span the membrane. Furthermore, antibodies raised against the 43,000-M_r cytoplasmic domain of band 3 cross-reacted with the ankyrin-linked band 3 (Bennett and Stenbuck, 1980b).

Some band 3 remains firmly associated with cytoskeletal proteins and persists after repeated washes in detergent (Bennett and Stenbuck, 1979b; Sheetz, 1979; Bennett, 1982). This fraction of band 3, which represents about 10^5 copies/cell, is most likely associated with ankyrin rather than spectrin or other cytoskeletal proteins. This conclusion is based on several observations. Spectrin binds to ankyrin-linked band 3 but not free band 3 (Bennett and Stenbuck, 1979b). Band 3 will rebind to cytoskeletal assemblies that contain ankyrin but not to ankyrin-depleted cytoskeletons (V. Bennett, unpublished data). Furthermore, an oligomer containing ankyrin and band 3 in a 1 : 1 molar ratio has been isolated from cytoskeletons and purified (Bennett, 1982). The ankyrin and band 3 were associated in a complex, since the molecular weight of the oligomer, calculated from hydrodynamic values, was the sum of that of ankyrin and band 3, and since antibody against ankyrin precipitated ankyrin as well as band 3. Ankyrin-associated band 3 is very similar to the major population of band 3 since two-

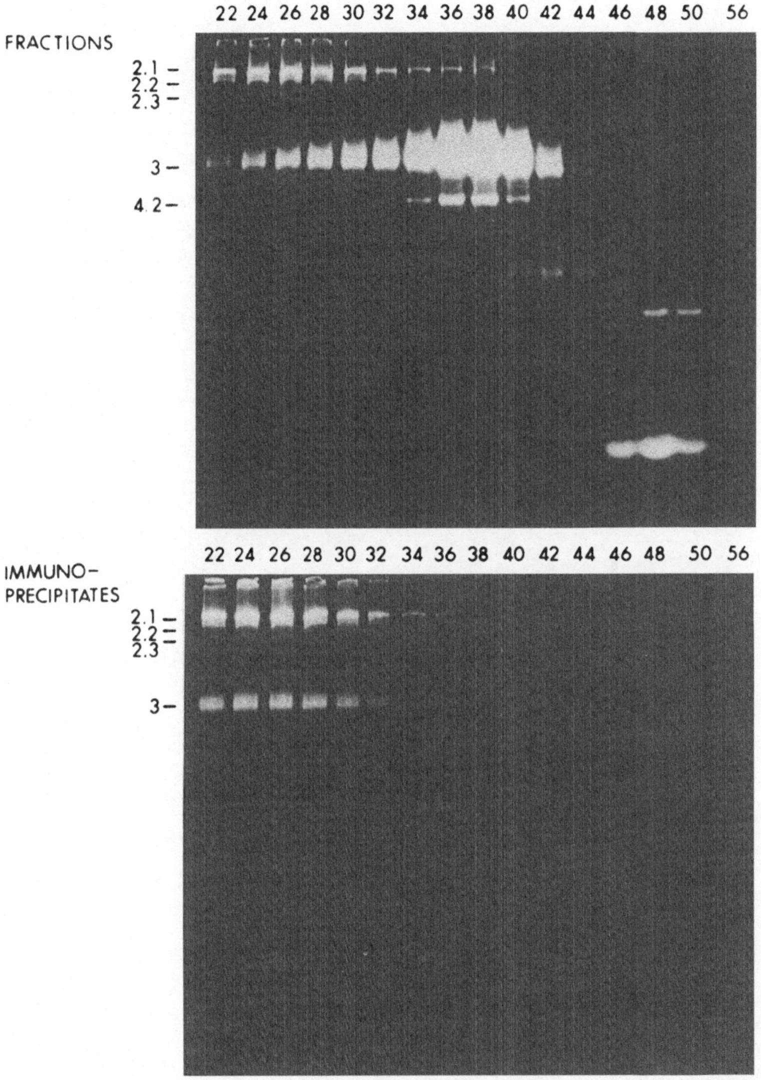

FIGURE 8. Immunoprecipitation of an ^{125}I-labeled ankyrin–band 3 complex with antiankyrin Ig from TX-100 solubilized protein chromatographed on an Ultrogel AcA22 column. Note that band 3 is immunoprecipitated only in the fractions containing ankyrin (bands 2.1, 2.2). Spectrin-depleted vesicles were radiolabeled with 125I–Bolton–Hunter reagent, solubilized in TX-100, and fractionated by gel filtration. The fractions (top panel) were immunoprecipitated with antiankyrin IgG. From Bennett and Stenbuck (1979b).

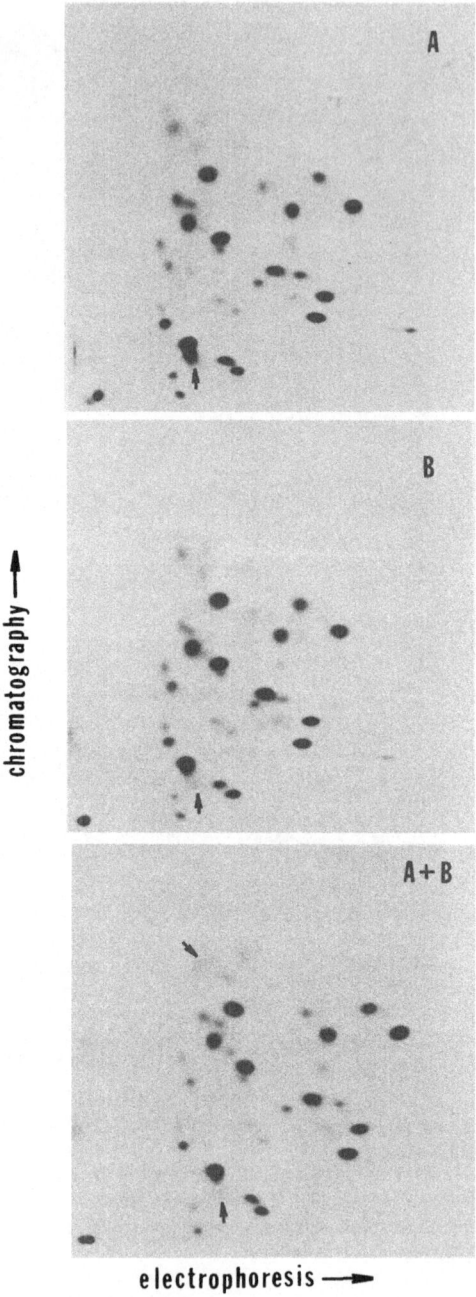

FIGURE 9. Two-dimensional peptide maps of ^{125}I-labeled free band 3 (A), ankyrin-associated band 3 (B), and 1:1 mixture of free band 3 and ankyrin-associated band 3 (A + B). Free band 3 and ankyrin-associated band 3 were purified as described (Bennett, 1982). From Bennett (1982).

dimensional peptide maps of these proteins are nearly identical (Fig. 9). The ankyrin band 3 also contains a DIDS-reactive site and thus presumably has an anion transport channel.

A 43,000-M_r cytoplasmic domain of band 3 is released by mild proteolysis and has been purified (Bennett and Stenbuck, 1980b). Ankyrin binds to this band 3 fragment in solution with a K_D of 5 $\times$ 10^{-9} M and in a 1 : 1 molar ratio (Fig. 10). This observation demonstrates that ankyrin can bind directly to a specific domain of band 3 without assistance of intermediary proteins. Ankyrin radiolabeled with ^{125}I–Bolton–Hunter reagent will reassociate with inverted vesicles that have been depleted of ankyrin as well as most of the peripheral membrane proteins (Bennett and Stenbuck, 1980b; Hargreaves *et al.*, 1980). Band 3 is the binding site for ankyrin since the reassociation is abolished by anti-band-3-fragment IgG, by selective proteolytic cleavage of band 3, and by the cytoplasmic fragment of band 3. Furthermore, ankyrin binds to vesicles and to liposomes reconstituted with band 3 in a nearly identical manner.

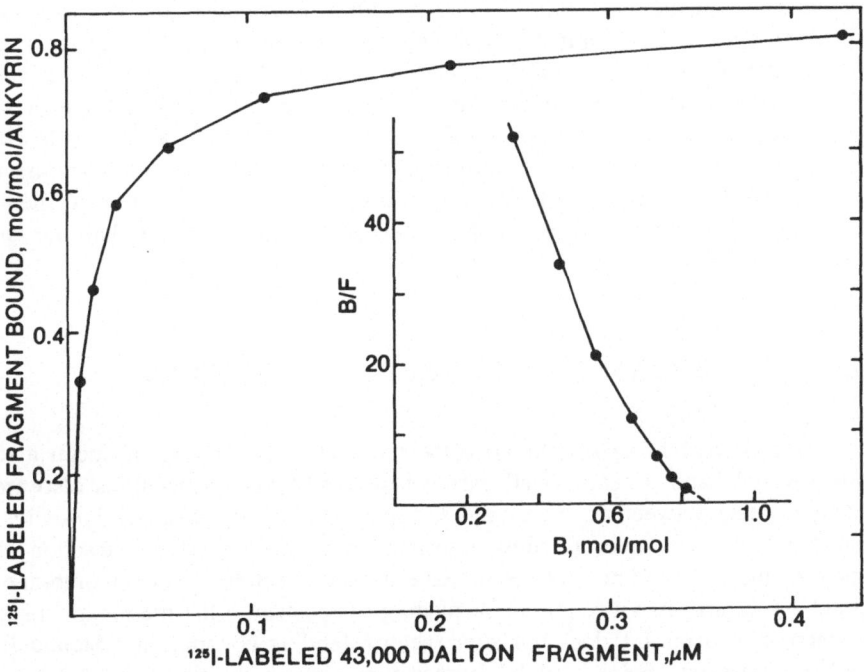

FIGURE 10. Binding of ^{125}I-labeled band 3 43,000-M_r fragment to pure ankyrin. Ankyrin was incubated with purified ^{125}I-labeled band 3 fragment, and ankyrin and associated fragment were then immunoprecipitated with antiankyrin Ig. From Bennett and Stenbuck (1980b).

Measurements of ankyrin reassociation with membranes and liposomes have demonstrated convincingly that ankyrin binds directly to band 3. However, these studies have also raised some new questions which remain unanswered. The maximal number of binding sites for ankyrin is about 200,000 copies/cell whereas band 3 is present in about 10^6 copies/cell. It is unlikely that ankyrin is binding to a subpopulation of band 3 since the free population of band 3 isolated by extraction of cytoskeletons (which retain ankyrin-linked band 3) can still bind to ankyrin either when reconstituted into liposomes (Hargreaves *et al.*, 1980) or in solution (Bennett and Stenbuck, 1980b). Furthermore, ankyrin-linked band 3 has been purified as an oligomeric complex with ankyrin and found to be nearly identical to the free population of band 3 by two-dimensional peptide mapping (Fig. 9). Another possibility that is consistent with available data is that ankyrin binds to band 3 only in certain oligomeric states, such as a tetramer but not a monomer or dimer. It is pertinent, in this regard, that band 3 may exist as a tetramer in erythrocyte ghosts and in this case would be present in near equivalence to the number of ankyrin sites as about 250,000 assemblies/cell.

A further complexity of ankyrin rebinding is that the 43,000-M_r fragment of band 3 competes for ankyrin binding with a much lower affinity than this fragment binds to ankyrin in solution (Bennett and Stenbuck, 1980b; Hargreaves *et al.*, 1980). Moreover, Scatchard plots of binding data were curvilinear and thus consistent with either two distinct types of site, one site which can have either high or low affinity depending on unknown factors, or a negatively cooperative type of interaction of ankyrin with a single class of sites. The conclusion from these experiments is that details of ankyrin–band 3 association in membranes are not understood and that this linkage is more complex than the ankyrin–spectrin interaction.

4. OTHER MEMBRANE–CYTOSKELETAL ASSOCIATIONS

One linkage of cytoskeleton to the membrane by binding of spectrin to ankyrin and band 3 seems well established. Additional associations also are likely to occur between the membrane and spectrin or other cytoskeletal proteins. Band 4.1, for example, remains associated with vesicles after extraction of spectrin and actin and possibly could have some interaction with the membrane. Band 4.1 clearly is not required for binding of spectrin to the membrane since extraction of band 4.1 does not alter spectrin binding (Bennett and Stenbuck, 1979a; Cohen and Foley, 1982). Furthermore, patients lacking band 4.1 have unaltered amounts of membrane-associated spectrin (Tchernia *et al.*, 1981). Measurements of reassociation of ^{125}I-labeled band 4.1 with vesicles and liposomes have demonstrated binding to glycophorin A (Anderson and Lovrien, 1984). Band 4.1 has also been proposed as the binding site for a minor sialoglycoprotein termed "glyconnectin" on the basis of indirect evidence (Mueller and Morrison, 1981). Glyconnectin remains in detergent-extracted cytoskeletons, and it has been inferred that glyconnectin is associated with some protein component of the cytoskeletons. Evidence that glyconnectin is associated with band 4.1 is based on the observation that membranes deficient in band 4.1 (from a patient with hereditary elliptocytosis) do not retain glyconnectin in the cytoskeleton. However, glyconnectin was not demonstrated to be present in the intact membranes either and may have been proteolyzed. Proof of a band 4.1–glyconnectin linkage will involve measurements of reassociation of pure proteins and demonstration by direct experiments that a complex exists in membranes. It would also be important to establish that glyconnectin is a membrane-spanning polypeptide and, if so, to isolate its cytoplasmic domain as has been done with band 3.

Actin is another cytoskeletal component that conceivably could associate directly with the membrane. However, the conclusion of several studies is that actin does not associate with membranes depleted of spectrin and band 4.1 (Cohen and Branton, 1979; Fowler *et al.*, 1981; Cohen and Foley, 1980, 1982). Band 4.9 is another potential candidate for mediating a membrane linkage, but its membrane association has not yet been examined.

5. IMPLICATIONS FOR THE ERYTHROCYTE MEMBRANE

The spectrin–ankyrin–band 3 complex provides a mechanism for linking the membrane bilayer, through integral membrane proteins, to the underlying cytoskeleton (Fig. 11). Ankyrin is present in about 10^5 copies/cell and is thus associated with 10^5 band 3 molecules which are components of intramembrane particles. The ankyrin–particle stoichiometry is not known, but would be 1 : 1 if ankyrin is distributed randomly, corresponding to 10^5 cytoskeleton-linked particles. The average nearest neighbor distance between ankyrin-linked particles would be about 35 nm, which is comparable to the apparent size of spectrin tetramer in solution (Ralston, 1978) and considerably less than the 200-nm length of rotary-shadowed spectrin tetramers (Shotton *et al.*, 1979). Linkage of the membrane to the spectrin assembly at 35-nm intervals would only allow evagination of vesicles with diameters 600 Å or less, which approaches the minimal limit for membranes observed with artificial bilayer vesicles. During maturation of erythrocytes, membrane vesicles are released from spectrin-free areas (Tokuyasu *et al.*, 1979), and this process presumably continues until the distance

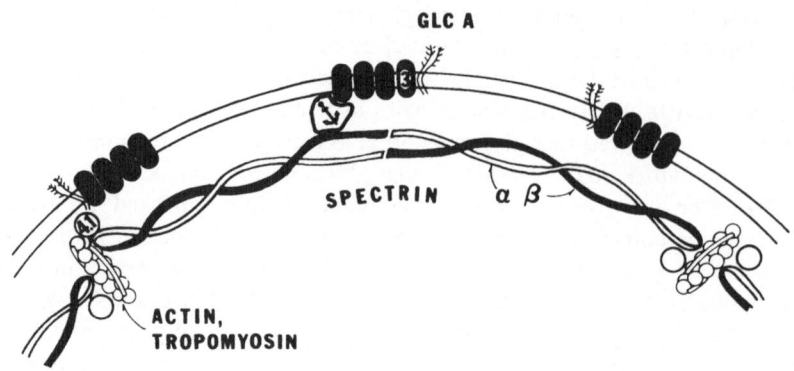

FIGURE 11. Schematic model of arrangement of proteins in human erythrocyte membranes.

between membrane anchoring sites decreases to a point where no further vesiculation occurs.

The fact that only a fraction of band 3 molecules is linked to the cytoskeleton is in agreement with measurements of the rotational diffusion of band 3 which indicated an unhindered local environment and little interaction with spectrin (Cherry *et al.*, 1976). More detailed measurements of band 3 rotational diffusion have resolved a population of molecules with decreased mobility which can be freed from restraints by cleavage of the cytoplasmic domain or by extraction of ankyrin and band 4.1 with high salt (Nigg and Cherry, 1980). About 40% of band 3 is immobilized in this way, which is consistent with ankyrin linkage to 10^5 band 3 dimers which are in equilibrium with another 10^5 dimers to form band 3 tetramers.

Association of integral proteins to the cytoskeleton would be expected to have a substantial effect on their lateral mobility. Measurements of lateral mobility of band 3 and glycophorin confirm that these proteins are restricted in their lateral movement (Peters *et al.*, 1974; Fowler and Branton, 1977). Several types of experiments indicate that linkage to ankyrin and spectrin restricts mobility of band 3 and glycophorin. One approach has been to examine mobility of band 3 in membranes from erythrocytes deficient in spectrin (Sheetz *et al.*, 1980). Band 3 (covalently labeled with a fluorescent probe) exhibited a 50-fold higher rate of diffusion in spectrin-deficient membranes, as measured by the technique of fluorescence photobleach–recovery. Another approach has been to expose erythrocyte ghosts to conditions known to promote dissociation of spectrin from ankyrin (e.g., low ionic strength). Such incubations increased the diffusion constant for band 3 up to 50-fold and also increased substantially the mobile fraction of band 3 from 10% to 80% (Golan and Veatch, 1980, 1982). Selective dissociation of spectrin from ankyrin can be achieved by use of the 72,000-M_r ankyrin fragment to compete with membrane-bound ankyrin for spectrin sites in ghosts. Exposure of ghosts to this spectrin-binding fragment increased the rate of diffusion of band 3 as well as the proportion of band 3 molecules that were mobile (Fowler and Bennett, 1978; Golan and Veatch, 1982).

Only a fraction of band 3 proteins interact with high affinity with spectrin, and yet more than 90% of these proteins are restricted in lateral movement (Golan and Veatch, 1980). It is possible that immobilization results from lower-affinity interaction of band 3 with spectrin or other cytoskeletal proteins, or that band 3 is simply trapped in the cytoskeletal meshwork. Another possibility which may occur in parallel is that band 3 tetramers associate with each other with low affinity, which would decrease the diffusion constant owing to increased effective size of the aggregate and would immobilize those proteins associated with an ankyrin-linked tetramer. These alternatives could be resolved by measurements of diffusion as a function of integral protein density, since interaction with cytoskeletal elements would be independent of this parameter, and pro-tein–protein interactions would be sensitive to protein concentration.

Regulation of membrane properties by modulation of protein–protein as-sociations is an intriguing possibility, especially in view of the fact that spectrin (Palmer and Verpoorte, 1971; Harris and Lux, 1980), ankyrin, and the cyto-plasmic domain of band 3 (Bennett and Stenbuck, 1979a) are all phosphorylated in erythrocytes. Crenation of cells accompanies dephosphorylation of spectrin during metabolic depletion of ATP, and conversely, ATP-dependent phospho-rylation of spectrin has been correlated with restoration of normal disc mor-phology (Sheetz and Singer, 1977; Birchmeier and Singer, 1977). However, phosphorylation of spectrin has no effect on association with ankyrin since cleavage of phosphate groups with phosphatase or proteolytic removal of the phosphorylated peptides has no effect on spectrin binding to membranes (An-derson and Tyler, 1980). Furthermore, dephosphorylation of ankyrin and band 3 by phosphatase has no effect on ankyrin–spectrin or ankyrin–band 3 association (V. Bennett, unpublished data). The phosphorylated state of spectrin also has no influence on binding to F-actin (Brenner and Korn, 1979) or on self-association to form tetramers (Ungewickell and Gratzer, 1978). The role of phosphorylation of spectrin and ATP-dependent shape changes in erythrocytes has been reeval-uated, with the conclusion that spectrin phosphorylation is not directly involved in shape change (Patel and Fairbanks, 1981). The function of phosphorylation in the erythrocyte membrane thus remains obscure.

An important future area will be to understand the basis for Ca^{2+} effects on erythrocyte shape and mechanical properties. Calcium at micromolar levels causes erythrocytes to become spherical and decreases the deformability of the membrane (Weed *et al.*, 1969). Erythrocytes contain calmodulin in about 5×10^5 copies/cell (Jarrett and Penniston, 1978), and it is likely that calmodulin mediates at least some of the effects of calcium on membrane properties (Nelson *et al.*, 1983). Spectrin binds calmodulin directly in a calcium-dependent manner (Sobue *et al.*, 1981), although functional consequences of this interaction are not known. Two other calmodulin-binding proteins present in ~ 1000 copies/cell have been identified that are distinct from the calcium transporter (Agre *et al.*, 1983).

6. MEMBRANE PROTEINS IN ABNORMAL ERYTHROCYTES

A series of protein–protein linkages have been outlined that are involved in the two-dimensional meshwork of the cytoskeleton and that connect the cytoskeleton to the membrane. How important are these proteins and their associations for maintaining a normal cell, and do they have any clinical relevance? Recent studies with hereditary hemolytic anemias in human beings and mice have introduced the beginnings of genetics in this area and already have demonstrated clearly that abnormal or deficient membrane proteins lead to major defects in erythrocyte shape and mechanical stability.

Hereditary spherocytosis (HS) and elliptocytosis (HE) are relatively common ($\sim 1 : 5000$) disorders of the erythrocyte membrane leading to fragile and/or abnormally shaped cells (reviewed by Lux, 1983). Erythrocytes are accumulated in the spleen in both cases, and these diseases can be essentially cured by splenectomy. Hereditary spherocytosis is in about 80% of the cases transmitted as an autosomal dominant trait, whereas hereditary elliptocytosis is almost entirely autosomal dominant. These diseases were recognized initially on the basis of abnormally appearing erythrocytes associated with varying degrees of anemia that could not be explained by altered hemoglobin, glycolytic enzymes, or autoimmune features. It is likely in both disorders that the molecular lesions will be heterogeneous. A distinctive feature of the spherocytosis group is that these cells are more susceptible to hypotonic lysis, presumably owing to a decreased surface area/volume ratio.

Several families with the dominant form of HS have defective binding of spectrin and band 4.1 (Wolfe *et al.*, 1982; Goodman *et al.*, 1982). The capacity of spectrin for band 4.1 binding is reduced by about 50%, with an unaltered affinity in the remaining sites. Pyropoikilocytosis is a rare recessively inherited variant of hereditary elliptocytosis and is characterized by erythrocytes that fragment and assume bizarre shapes after warming to 45–46°, which is about 4° less than the disintegration temperature for normal erythrocyte (Chang *et al.*, 1979). Spectrin isolated from these cells denatures at the same temperature of 45–46°, based on change in the ORD spectrum, whereas normal spectrin denatures at 49–51°. Spectrin from patients with pyropoikilocytosis also has a lowered affinity for the dimer–tetramer equilibrium (Liu *et al.*, 1981; Knowles *et al.*, 1983). It is likely that a change in spectrin amino acid sequence is responsible for these features since an altered tryptic cleavage has been detected in an 80,000-M_r domain of the alpha chain (the portion of the molecule involved in dimer self-association with the adjacent beta chain). A similar defect in dimer self-association has also been observed in some patients with mild hereditary elliptocytosis (Coetzer and Zail, 1982; Liu *et al.*, 1982). It has not been established that the altered affinity for tetramer formation measured *in vitro* actually causes the disease. It is possible that the abnormal spectrin is less stable in erythrocytes and is degraded or precipitated.

High-affinity-binding ankyrin-binding sites are reduced by 50% in erythrocyte membranes in two families with an elliptocytosis-type anemia (Agre *et al.*, 1981). The cytoplasmic domain of band 3 purified from the defective membranes binds to ankyrin normally, however. These erythrocytes may have an altered arrangement of band 3 molecules rather than a defect in the binding site itself.

Spectrin-binding sites and ability of spectrin to bind to normal membranes were unaltered in the erythrocytes deficient in high-affinity ankyrin binding and also were normal in several cases of spherocytosis (Agre *et al.*, 1981, 1982). These results do not exclude defects in spectrin binding in some families, however. It is also possible that spectrin-binding defects play a role in hemolysis in other forms of anemia. The binding of ankyrin to spectrin is sensitive to alkylation of a sulfhydryl group on ankyrin (Bennett, 1978). It is possible that metabolic disorders, such as glucose-6-phosphate dehydrogenase deficiency, which lead to lowered levels of reduced glutathione could cause oxidation of ankyrin sulfhydryl groups. Such oxidation could result in inter- and intramolecular cross-linking by disulfide formation or lead to oxidation of -SH groups to sulfones and sulfoxides. In either case, the ability of ankyrin to bind to spectrin would be compromised and possibly result in increased erythrocyte fragility.

The defects discussed up to this point have involved altered function of structural proteins. Another mechanism for disease is the absence or decreased quantity of these proteins owing to abnormal synthesis or instability of the product. The first examples of deficient proteins came from strains of mutant mice that were developed at the Jackson Laboratory by Bernstein and his co-workers (reviewed by Bernstein, 1980; Russell, 1979). The mice were selected for hemolytic disease, and four distinct strains are now available. These mice have been also selected for enhanced viability of the homozygotes and have been produced in highly inbred strains to provide a common genetic background. SDS-gels of ghost membranes from these strains revealed a striking decrease in the quantity of spectrin, and the extent of deficiency correlated with the severity of the anemia (Greenquist et al., 1978). The mechanism for decreased amounts of spectrin is under investigation and appears to involve decreased synthesis of either alpha- or beta-spectrin subunits depending on the strain (Bodine et al., 1983). This system together with cDNA probes and other techniques of molecular biology should provide an understanding of spectrin synthesis as the thalassemias have with hemoglobin.

Two examples of deficient proteins have also been reported for human hemolytic anemias. A form of elliptocytosis with abnormal erythrocyte morphology and increased fragility has been associated with lack of band 4.1 (Tchernia et al., 1981; Mueller and Morrison, 1981). The parents had 50% of the normal complement of band 4.1 and were clinically normal, whereas the affected individuals had a complete absence of band 4.1. It is not clear whether band 4.1 is lacking owing to ineffective synthesis, abnormal degradation, or decreased association with membranes, or whether band 4.1 is present but at a different molecular weight. An unresolved discrepancy is that the parents lacking 50% of band 4.1 are essentially normal whereas patients lacking 40% of spectrin-binding sites for band 4.1 have a significant anemia.

Two siblings with a severe recessive form of spherocytosis have a 50% reduction in spectrin polypeptides on SDS-gels (Agre et al., 1982). Spectrin was not degraded to a lower-molecular-weight form, since no immunoreactive bands were detected in erythrocytes or ghosts by blot transfer. Furthermore, spectrin was not lost during preparation of membranes, since analysis by radioimmunoassay demonstrated an equivalent reduction of spectrin in intact erythrocytes and ghost membranes. The spectrin remaining in the membranes was normal in membrane binding and in formation of tetramers. The parents (established by HLA typing) were clinically normal and had normal amounts of spectrin. The fact that the patients, who apparently are homozygous for a gene, have only a 50% decrease in spectrin raises some interesting questions. The mutant spectrin could be unstable and subject to proteolysis during the life-span of affected erythrocytes. It is possible that spectrin polypeptides are encoded by more than one gene, as is the case with hemoglobin, and a lack of one gene leads to a partial loss of spectrin. It also is conceivable that a single defect causes the mRNA to be unstable or not processed efficiently leading to partial synthesis. As a first step in answering these questions it will be important to measure levels of mRNA and rates of biosynthesis of spectrin subunits in normal and affected individuals. Preliminary experiments in this laboratory indicate that reticulocyte preparations from peripheral blood do not have active synthesis of spectrin, however, and it will be necessary to obtain more primitive cells from bone marrow.

In summary, studies with patients and mice with abnormally fragile erythrocytes have revealed defects in function and amounts of cytoskeletal proteins. This work is relatively recent, and future studies will be required to clearly establish the primary cause for abnormal membranes. It also is evident, in the case of deficient spectrin and band 4.1, that cDNA probes and techniques for measuring biosynthesis will be necessary for further understanding.

7. ERYTHROCYTE PROTEINS IN OTHER CELLS

The erythrocyte membrane is a highly specialized structure that enables erythrocytes to survive without cytoplasmic support. The erythrocyte membrane is not a unique system, however, and recent studies indicate that the major structural proteins will be present and organized in a similar way in many other types of cells. The idea that the organization of the erythrocyte membrane is directly applicable to other cells has only gradually been accepted. It was suggested soon after the discovery of spectrin in an influential review (Guidotti, 1972) that spectrin might be related to myosin. The analogy between spectrin and myosin was supported indirectly by observations that erythrocyte membranes also contained actin (Tilney and Detmers, 1975; Sheetz et al., 1976). Unfortunately, a direct assay was negative for spectrin in cultured fibroblasts by radioimmunoassay (Hiller and Weber, 1977).

These results led to the widely held conclusion that spectrin was not present in other cells, and that the other erythrocyte membrane proteins were also unique. In retrospect, this point of view was not likely since many proteins and enzymes in erythrocytes are found elsewhere. For example, all the glycolytic enzymes are present in erythrocytes, as are two commonly occurring isoforms of carbonic anhydrase. Calmodulin (Jarrett and Penniston, 1978), actin (Tilney and Detmers, 1975), and tropomyosin (Fowler and Bennett, 1984) also are present in erythrocytes. Immunoreactive forms of ankyrin (Bennett, 1979; Bennett and Davis, 1981, 1983; Bennett et al., 1982a), spectrin (Goodman et al., 1981; Repasky et al., 1982; Davis and Bennett, 1982; Bennett et al., 1982b; Nelson and Lazarides, 1983; Pollard, 1983), band 4.1 (Cohen et al., 1982), and band 3 (Bennett et al., 1982a) have recently been detected in a variety of cells and tissues. The immunoreactive forms of spectrin and ankyrin belong to two general families: (1) a membrane-associated group that contains proteins quite similar to the erythrocyte membrane proteins and (2) microtubule-associated forms of spectrin and ankyrin that share some function and antigenic determinants but are more distantly related.

7.1. Membrane-Associated Analogs of Spectrin and Ankyrin

Within a period of about a year many laboratories independently discovered a protein that now is known to be closely related to spectrin. A high-molecular-weight protein with subunits of $\sim$ 250,000 and 220,000 M_r that associated with actin filaments was isolated from brain (Davies and Klee, 1981; Shimo–Oka and Watanabe, 1981; Levine and Willard, 1981; Glenney et al., 1982a,b,c). This protein also binds to calmodulin in a Ca^{2+}-dependent manner as determined by overlays of SDS-gels and adsorption to calmodulin affinity columns (Davies and Klee, 1981; Kakiuchi et al., 1981; Sobue et al., 1982; Palfrey et al., 1982; Glenney et al., 1982a,b). The protein was localized by immunofluorescence to the plasma membrane of many types of cells (Levine and Willard, 1981; Repasky et al., 1982; Burridge et al., 1982). The brain protein and a similar actin-binding protein from intestinal epithelial cells have been visualized by low-angle rotary shadowing as flexible rods about 200 nM in length which are similar to erythrocyte spectrin (Glenney et al., 1982a,b,c; Bennett et al., 1982a,b; Davis and Bennett, 1983a) (Fig. 12).

The brain protein has been identified as a form of spectrin based on the following properties shared with mammalian erythrocyte spectrin: (1) ability to bind to ankyrin sites on erythrocyte membranes; (2) similar structure of a tetramer with the morphology of a 200-nM flexible rod; (3) common antigenic sites in both alpha and beta subunits (Burridge et al., 1982; Bennett et al., 1982b). An additional important feature in common between the brain protein and erythrocyte spectrin is the fact that functional hybrid molecules can be formed with the alpha subunit of brain and the beta subunit from erythrocyte spectrin (Davis and Bennett, 1983a). The subunits of brain spectrin are most likely arranged the same way as those of erythrocyte spectrin, with laterally associated alpha, beta dimers attached by head-to-head linkage of each alpha chain with a beta chain (Davis and Bennett, 1983a). Furthermore, the amino acid compositions of brain and erythrocyte spectrin are remarkably similar (Glenney et al., 1982c; Davis and Bennett, 1983a). The brain protein also binds to erythrocyte band 4.1, and its association with actin is promoted by band 4.1 (Burns et al., 1983; Lin et al., 1983). The brain protein will be referred to here as brain spectrin on the basis of these similarities to erythrocyte spectrin. It has also been named calmodulin-

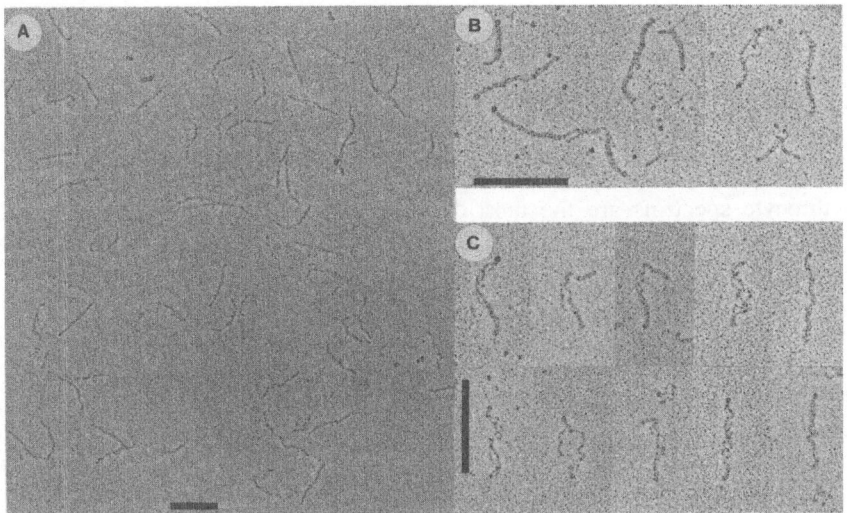

FIGURE 12. Electron micrographs of rotary-shadowed brain spectrin (A, B, and top row of C) and pig erythrocyte spectrin (bottom row of C). Scale bar = 200 nM. From Bennett *et al.* (1982b).

binding protein I (Davies and Klee, 1981), brain actin-binding protein (Shimo–Oka and Watanabe, 1981), and fodrin (Levine and Willard, 1981) before the close relationship to erythrocyte spectrin was recognized.

Brain spectrin comprises about 3% by weight of the total membrane protein (30 picomoles/mg), and spectrin polypeptides in other tissues also are significant bands on SDS-gels. It is surprising, in retrospect, that these proteins present in such quantities were not recognized previously. One significant difficulty is the limited degree of cross-reactivity between mammalian erythrocyte spectrin and tissue spectrin. These proteins also have distinct peptide maps (Bennett *et al.*, 1982b) and thus are products of different genes. Experience with highly conserved proteins such as actin and calmodulin has led to the expectation that closely related proteins in different tissues and species will have nearly the same amino acid sequence and identical antigenic sites. However, it appears to be more generally the case that proteins occur as members of multigene families with different sequences and variable cross-reactivity. Examples of families are the intermediate filament proteins that include five noncross-reacting members, the different forms of myosin, and multiple variants of tubulin.

The spectrin family is not as easy to define as the myosins, for example, which have an actin-activated ATPase activity. The criteria for a spectrin that seems appropriate at this point are (1) two distinct subunits of $M_r \sim$ 220,000–260,000 arranged as (alpha, beta) tetramers, (2) binding sites for actin, ankyrin, calmodulin, and band 4.1, (3) shape of a flexible rod 200 nM in length, and (4) ability of isolated subunits to hybridize with subunits of erythrocyte spectrin. It should be emphasized that cross-reactivity with mammalian erythrocyte spectrin is not an essential feature. It is likely that mammalian erythrocyte spectrins are the most divergent members of the spectrin family. Avian erythrocyte spectrin appears much closer to tissue spectrins in terms of cross-reactivity, even when compared with mammalian tissues (Repasky *et al.*, 1982). Thus the divergence of mammalian erythrocyte spectrin occurred relatively recently, most likely during evolution of nonnucleated erythrocytes.

An area of future interest will be the functional differences between members of the spectrin family and regulation of the expression of these spectrins during development. It is already known that tissue spectrins and avian erythrocyte spectrin bind calmodulin in overlays of SDS-gels, whereas mammalian erythrocyte spectrin does not bind calmodulin under these conditions (Palfrey *et al.*, 1982; Bartelt *et al.*, 1982), presumably owing to a lower affinity of this spectrin for calmodulin. The beta chains (defined here as the ankyrin-binding subunit) of brain spectrin, erythrocyte spectrin, and a spectrin from intestinal epithelial cells all have differences in M_r and are less cross-reactive than the corresponding alpha chains (Glenney *et al.*, 1982b; Nelson and Lazarides, 1983). Another difference between spectrins is the dimer–tetramer equilibrium. Mammalian erythrocyte-spectrin tetramers are relatively unstable above 30° in dilute solution (Ungewickell and Gratzer, 1978) whereas brain-spectrin tetramers are quite stable and dissociate to dimers only after partial denaturation.

Ankyrin has been detected in membrane fractions of various tissues by radioimmunoassay using ^{125}I-labeled 72,000-M_r fragment and antibody against the 72,000-M_r fragment (Bennett, 1979). The amounts of ankyrin estimated by this assay were 25 pmoles/mg in brain, 5 pmoles/mg in liver, kidney, and fat cells, and 15 pmoles/mg in testes. These values are of course rough estimates and assume equivalent cross-reactivity between tissue and erythrocyte ankyrin. It is of interest that the amounts of ankyrin, at least in brain, are approximately the same as the numbers of copies as spectrin. The cross-reacting polypeptide in liver plasma membranes is a band of M_r = 190,000 (Bennett *et al.*, 1982a), and in brain the M_r is $\sim$ 210,000. In both cases, other cross-reacting bands of lower M_r are present in variable amounts depending on use of protease inhibitors, and these bands probably are proteolytic fragments.

It has not been possible in preliminary studies to solubilize the 210,000-M_r brain polypeptide in high yield or without substantial proteolysis (Bennett and Davis, unpublished data). However, if brain membranes were digested with chymotrypsin, soluble fragments of 72,000 and 95,000 M_r were produced that cross-react with ankyrin (Davis and Bennett, 1983b). The 72,000-M_r fragment from brain membranes has been purified about 500-fold by affinity chromatography on erythrocyte spectrin–Sepharose (Bennett and Davis, 1983; Davis and Bennett, 1984). The brain fragment was not the result of contamination with erythrocyte ankyrin since two-dimensional peptide maps revealed no common peptides from pig brain and pig erythrocytes. The brain fragment binds to brain spectrin at a site about 80 nM from the closest end (Davis and Bennett, 1984), which is the same region where erythrocyte ankyrin binds on erythrocyte spectrin (Tyler *et al.*, 1980b).

These initial experiments demonstrate the presence of a protein in brain membranes that has properties very similar to erythrocyte ankyrin: (1) shared antigenic sites; (2) similar domain structure as evidenced by cleavage to fragments of nearly identical M_r to erythrocyte ankyrin fragments; and (3) spectrin-binding activity present in a 72,000-M_r fragment. The brain ankyrin is present in about 30 pmoles/mg or in a 1 : 1 ratio to brain spectrin, based on radioimmunoassay with displacement of binding of ^{125}I-labeled brain fragment to antierythrocyte ankyrin Ig (Davis and Bennett, 1984). Thus it is likely that brain spectrin and other tissue spectrin are attached to membranes by an ankyrin linkage very similar to the arrangement in erythrocytes.

It is not known how tissue ankyrins are attached to membranes, but a reasonable guess is that analog(s) of band 3 will be involved. Liver plasma membranes contain several polypeptides that cross-react with antibody against the cytoplasmic domain of band 3 (Bennett *et al.*, 1982a). An interesting possibility is that the regions of band 3 involved in association with ankyrin may be a component of diverse membrane proteins. If such a common domain is present, then linkage to ankyrin would provide a mechanism for localization of different membrane proteins to the same region of the cell surface.

7.2. Possible Functions of Spectrin and Ankyrin in Nonerythroid Cells

The primary function of spectrin and ankyrin in erythrocytes is to provide structural support for the lipid bilayer. It is likely that spectrin and ankyrin in other cells will have a similar supportive activity. Examples of cells without substantial cytoplasm and extended plasma membrane surfaces that probably require such a submembrane lattice are endothelial cells, alveolar cells in lung, and lens epithelial cells. It is pertinent that lung and lens membranes are enriched in spectrin (Davis and Bennett, 1983a). Another important function of spectrin system will be to mediate association of actin filaments with plasma membranes.

Erythroid complexes of spectrin–actin–band 4.1 are active as nuclei for initiation of actin polymerization *in vitro* (Lin and Lin, 1979), and a similar complex containing brain spectrin has a similar activity (Lin *et al.*, 1983). The linkage of actin filaments to plasma membranes, which are in turn attached to external surfaces, could be essential to convert contractile activity of actin–myosin into useful work. Thus the attachment of actin to spectrin on membranes may be required for various types of cell motility including movement of macrophages and migrating cells during development and muscle contraction. Spectrin is not localized by immunofluorescence to adhesion plaques in cultured cells (Burridge *et al.*, 1982) and thus is not the only site of actin attachment to membranes.

The spectrin–ankyrin–band 3 type of linkage also provides a mechanism for association of integral proteins with cytoplasmic structural proteins. It is conceivable that this system in other cells could permit localization of ion channels and receptors in specific regions of the cell surface. In brain, for example, the postsynaptic densities are enriched in receptors for appropriate neurotransmitters. Spectrin is a major component of postsynaptic densities (Carlin *et al.*, 1983) and is at least in the right location to participate in localization of receptors. Spectrin may be associated in some way with cell surface proteins in lymphocytes, since capping of surface components results in redistribution of spectrin underneath the clustered membrane proteins (Levine and Willard, 1983). Many other proteins also are localized under caps including actin, myosin, myosin light-chain kinase, a-actinin, and calmodulin, and it is not known how these proteins are clustered. Spectrin may be a common linking protein with binding sites for actin, and thus other actin-binding proteins, as well as cell surface proteins via ankyrin.

Spectrin may also be involved in active intracellular movement of proteins and organelles. In brain, Willard and his colleagues have studied anterograde axonal transport of spectrin and found it moves at different rates with several different groups of proteins (Lorenz and Willard, 1978; Levine and Willard, 1980). Other proteins, in contrast, appear to move in discrete groupings. These workers have suggested that spectrin may link at least some proteins to an actomyosin system and have also suggested that capping of membrane proteins also is mediated by spectrin (Levine and Willard, 1983).

Spectrin is a major binding site for calmodulin, but the function of this association is not known. One possibility is that spectrin simply provides a reservoir of calmodulin close to the membrane so that Ca-transport activity can be rapidly regulated. It is also conceivable that calmodulin may regulate some behavior of spectrin. Calmodulin and calcium do not modulate association of ankyrin and spectrin from brain or erythrocytes (V. Bennett, unpublished data), but other activities have not been examined.

7.3. Microtubule-Associated Analogs of Spectrin and Ankyrin

Two high-molecular-mass microtubule-associated proteins (MAPs) have been demonstrated to cross-react with affinity-purified antibodies against erythrocyte ankyrin and spectrin, respectively (Bennett and Davis, 1981; Davis and Bennett, 1982). Both preparations of antibodies cross-reacted with multiple polypeptides in crude extracts of brain. In each case, microtubule protein isolated by repeated cycles of polymerization was greatly enriched in only one of the cross-reacting polypeptides. Antibody against ankyrin cross-reacted with a polypeptide $M_r = 370,000$ in the region of MAP1, and antibody against spectrin cross-reacted with a polypeptide of $M_r = 300,000$ that comigrated with MAP2. Relatively little is known about MAP1, which has not been purified and contains more than one polypeptide. MAP2, however, has been isolated (Herzog and Weber, 1978; Kim et al., 1979) and is known to form projections extending from microtubules polymerized in vitro (Murphy and Borisy, 1975; Kim et al., 1979).

The cross-reactivity of MAP2 and antispectrin antibody has been characterized in more detail (Davis and Bennett, 1982). Pure MAP2 cross-reacted with antibody after electrophoretic transfer from SDS-polyacrylamide gels to nitrocellulose, and MAP 2 also was immunoprecipitated by anti-spectrin antibody. The immunoreactivity involved the major component of MAP2 rather than a minor contaminant since peptide maps of MAP2 and the polypeptides immunoprecipitated by antibody were nearly identical. The antigenic determinants shared by MAP2 and spectrin reside in the alpha subunit of spectrin, because binding of ^{125}I-labeled MAP2 to antispectrin IgG was displaced by the alpha subunit and not the beta subunit. The homology between MAP2 and the alpha subunit is only partial because MAP2 displaced binding of ^{125}I-labeled a subunit to antispectrin antibody by a maximum of 20%. Furthermore, peptide maps of MAP2 and the α subunit were dissimilar.

The distribution of ankyrin immunoreactivity in cultured HeLa cells has been visualized by immunofluorescence microscopy (Bennett and Davis, 1981). The major cross-reacting bands in HeLa cells were polypeptides of 220,000 M_r and 250,000 M_r, which have the same M_r as HeLa cell microtubule-associated proteins (Bulinski and Borisy, 1980a; Weatherbee et al., 1980). Ankyrin immunofluorescence was localized in a filamentous meshwork, surrounding the nucleus and extending through the cytoplasm, and in an intensely staining punctate pattern over nuclei. The cytoplasmic meshwork resembled patterns obtained with tubulin and tubulin-associated proteins and was sensitive to colchicine. The punctate nuclear pattern has not been observed with tubulin but has been reported with a monoclonal antibody against HeLa MAP (Izant et al., 1982). The distribution of cytoplasmic staining with antiankyrin Ig changed dramatically during mitosis with rearrangements to the spindle pole and later to the cleavage furrow that paralleled closely the behavior of microtubule-associated proteins (Bulinsky and Borisy, 1980; Izant et al., 1982).

The surprising finding of shared antigenic sites between erythrocyte membrane proteins and microtubule-associated proteins suggested the possibility of functional homology as well. Partly purified MAP1 and pure MAP2 did not displace binding of spectrin or ankyrin to erythrocyte membranes. Erythrocyte ankyrin did associate with microtubules polymerized from pure brain tubulin. The binding of ankyrin was saturable at a ratio of 1 mol ankyrin per 4 mol tubulin dimer, and binding was displaced by brain MAPs. The apparent K_D for ankyrin was 2–4 μm with 10μm tubulin (Bennett and Davis, 1981) (Fig. 13). Ankyrin also promoted the polymerization of pure tubulin, provided that the tubulin was near its critical concentration (Bennett and Davis, 1982).

The fact that ankyrin associates with microtubules suggests that ankyrin may have evolved from a MAP. Mature human erythrocytes lack tubulin, but tubulin is present as a membrane-associated marginal band in mammalian erythroblasts and primitive circulating fetal erythrocytes (VanDeurs and Behnke, 1973). Furthermore, a protein similar to MAP2 has been identified in the marginal band of nucleated erythrocytes (Sloboda and Dickersin, 1980). Thus ankyrin in erythrocyte precursor cells and nucleated erythrocytes may bind to tubulin and maintain the position of the marginal band. Conversely, MAP1 may have some of the properties of ankyrin, such as binding sites for a membrane-bound analog

$\longrightarrow$

FIGURE 13. Association of erythrocyte ankyrin with brain tubulin. (A) Brain microtubules after two cycles of assembly and disassembly were depolymerized and applied to DE-53 cellulose in a buffer containing 50 nM PIPES, 1 mM NaEGTA, 0.1 mM MgCl$_2$, 0.1 mM GTP, and 0.1 mM dithiothreitol, pH 7.0. The column was washed with 10 volumes of buffer, and a fraction enriched in MAPS was eluted with 0.1 M KCl; the column was washed with 10 volumes of 0.2 M KCl, and tubulin was eluted with 0.4 M KCl. Erythrocyte ankyrin and purified tubulin dimer (dialyzed against assay buffer) were centrifuged for 20 min at 200,000 g and then combined at 2 μm and 10 μm, respectively, in 5% sucrose, 100 mM PIPES, 1 mM NaEGTA, 1 mM GTP, 0.5 mM MgCl$_2$, and 0.2 mM dithiothreitol. After 30 min at 37°C, the solutions were layered over 20% sucrose in assay buffer, and microtubules were collected by centrifugation for 20 min at 200,000 g at 25°C. The pellets were analyzed on an SDS-polyacrylamide gel. Control lanes: a, erythrocyte ankyrin; b, pure tubulin; c, MAP fraction. Pellet lanes: 1, ankyrin alone; 2, tubulin alone; 3, ankyrin and tubulin chilled to 0°C after polymerization; 4, ankyrin and tubulin; 5, ankyrin and tubulin plus MAP fraction (400 μg ml^{-1}.) (B) Various concentrations of ankyrin were incubated alone or with 10 μM tubulin for 30 min at 37°C, and microtubules were collected and analyzed on an SDS-polyacrylamide gel. The relative peak areas were estimated by densitometry. The data are corrected for the amount of ankyrin sedimented in the absence of tubulin, which was less than 5% of the value in the presence of tubulin. From Bennett and Davis (1981).

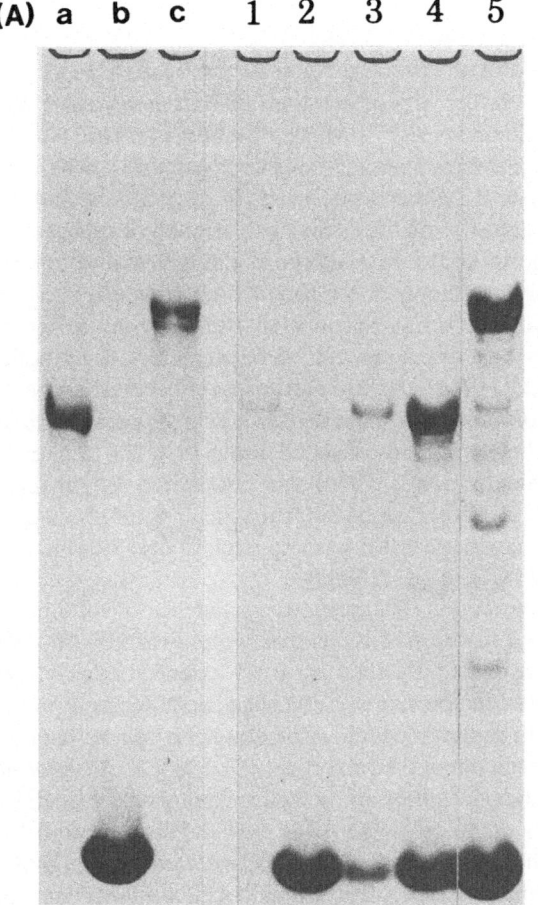

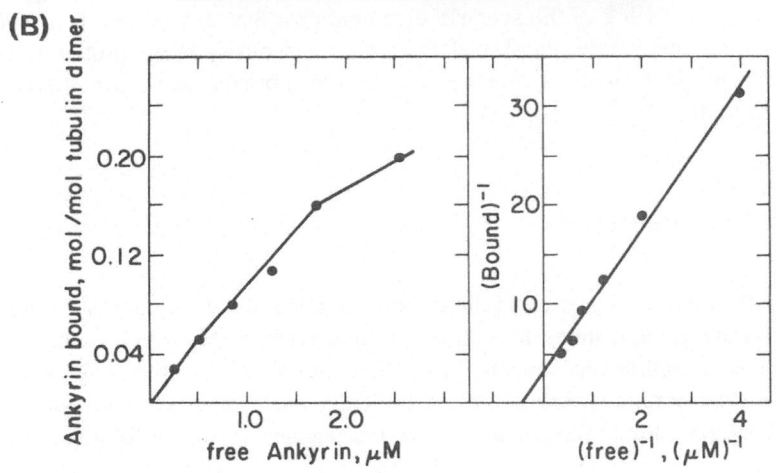

of band 3 or for a spectrin. Such an activity of MAP1 could explain how microtubules interact with membranes (Bhattacharyya and Wolff, 1975; Sherline *et al.*, 1977; Stephens, 1977; Moskalewski *et al.*, 1977). Ankyrin analogs could also mediate membrane associations during mitosis, where a spindle-associated membrane system has been observed (Hepler, 1980), and a membrane linkage presumably also is required during constriction of the cleavage furrow.

Spectrin, in contrast to ankyrin, did not associate with microtubules or affect the polymerization of tubulin. An activity shared by spectrin and MAP2 is association with actin. Spectrin associates with muscle actin (Brenner and Korn, 1979), and a direct association between MAP2 and actin has been reported (Sattilaro *et al.*, 1981). A further similarity between spectrin and MAP2 is the distinctive morphology of these proteins, which appear by rotary shadowing with platinum to be flexible rod-shaped molecules 200 nm in length for spectrin tetramers (Shotton *et al.*, 1979) and 180 nm in length for MAP2 (Voter and Erickson, 1982). Other high-molecular-mass actin-binding proteins such as filamin and macrophage actin-binding protein also exhibit a similar flexible rod morphology (Tyler *et al.*, 1980a).

These considerations suggest the existence of a family of related proteins with the shared features of molecular weights of 260,000–300,000, the ability to bind to actin, and the shape of an extended, flexible rod. An important difference between spectrin and the other actin-binding proteins is that spectrin contains a beta subunit, which associates with ankyrin and thus allows spectrin to bind to the membrane (Calvert *et al.*, 1980; Morrow *et al.*, 1980). Filamin and macrophage actin-binding protein are homodimers, did not associate with erythrocyte membranes (Tyler *et al.*, 1980a), and presumably lack a beta-type subunit. Furthermore, the homology between spectrin and MAP2 is localized to the alpha subunit. It is conceivable that MAP2 and the other actin-binding proteins also have an analog of the spectrin beta chain but that this subunit is dissociated during the procedures employed to extract and purify these proteins. Such a protein might provide a linkage between membranes and actin filaments or microtubules.

8. CONCLUSION

The studies discussed in the review have led to a detailed understanding of membrane–protein intractions in the human erythrocyte, which promises to be of clinical significance and to have direct relevance to other cell types. The plasma membrane of red cells and probably many other cells is composed of the traditional lipid bilayer and, in addition, an underlying scaffolding of spectrin molecules that are attached to integral membrane proteins by a specific protein linkage mediated by ankyrin. The strategy of sequential removal of one protein followed by its purification, reassociation, and identification of its binding site also may have application in elucidating details of other complex membrane structures. For example, desmosomes, postsynaptic densities, Z discs, and adhesion plaques most likely represent stable assemblies of protein components that are ultimately attached to the membrane and would be amenable to such an approach. It is probable, based on the erythrocyte analogy, that end-on attachment of filament-forming proteins such as actin or intermediate filament proteins to membranes will require several intermediate proteins that may themselves be associated to form an extensive, lateral structure.

ACKNOWLEDGMENTS.   Valuable assistance by Jonathan Davis is gratefully acknowledged. The manuscript was prepared by Arlene Daniel.

REFERENCES

Agre, P., Orringer, E., Chui, D., and Bennett, V., 1981, A molecular defect in two families with hemolytic poikilocytic anemia, *J. Clin. Invest.* **68**:1566–1576.

Agre, P., Orringer, E., and Bennett, V., 1982, Deficient red cell spectrin in severe recessively inherited spherocytosis, *N. Engl. J. Med.* **306**:1155–1161.

Agre, P., Gardner, K., and Bennett, V., 1983, Association between human erythrocyte calmodulin and the cytoplasmic surface of human erythrocyte membranes, *J. Biol. Chem.* **258**:6258–6265.

Anderson, J. M., and Tyler, J. M., 1980, State of spectrin phosphorylation does not affect erythrocyte shape or spectrin binding to erythrocyte membranes, *J. Biol. Chem.* **255**:1259–1265.

Anderson, R. A. and Lovrien, R. E., 1984, Glycophorin is linked by band 4.1 to the human erythrocyte membrane skeleton, *Nature* **307**:655–658.

Bartelt, D. C., Carlin, R. K., Scheele, G. A., and Cohen, W. D., 1982, The cytoskeletal system of nucleated erythrocytes II. Presence of a high molecular weight calmodulin-binding protein, *J. Cell Biol.* **95**:278–284.

Baskin, G. S., and Langdon, R. G., 1981, A spectrin-dependent ATPase of the human erythrocyte membrane, *J. Biol. Chem.* **256**:5428–5435.

Bennett, V., 1977, Human erythrocyte spectrin: phosphorylation in intact cells and purification of the ^{32}P-labeled protein in a nonaggregated state, *Life Sci.* **21**:433–440.

Bennett, V., 1978, Purification of an active proteolytic fragment of the membrane attachment site for human erythrocyte spectrin. *J. Biol. Chem.* **253**:2292–2299.

Bennett, V., 1979, Immunoreactive forms of human erythrocyte ankyrin are present in diverse cells and tissues, *Nature* **281**:597–599.

Bennett, V., 1982, Isolation of an ankyrin-band 3 oligomer from human erythrocyte membranes, *Biochim. Biophys. Acta* **689**:475–484.

Bennett, V., and Branton, D., 1977, Selective association of spectrin with the cytoplasmic surface of human erythrocyte plasma membranes, *J. Biol. Chem.* **252**:2753–2763.

Bennett, V., and Davis, J., 1981, Erythrocyte ankyrin: Immunoreactive analogues are associated with mitotic structures in cultured cells and with microtubules in brain, *Proc. Natl. Acad. Sci. USA* **78:**7550–7554.

Bennett, V., and Davis, J., 1982, Immunoreactive forms of human erythrocyte ankyrin are localized in mitotic structures in cultured cells and are associated with microtubules in brain, *Cold Spring Harbor Symp. Quant. Biol.* **46:**647–657.

Bennett, V., and Davis, J., 1983, Spectrin and ankyrin in brain, *Cell Motility,* **3:**623–633.

Bennett, V., and Stenbuck, P. J., 1979a, Identification and partial purification of ankyrin, the high affinity membrane attachment site for human erythrocyte spectrin, *J. Biol. Chem.* **254:**2533–2541.

Bennett, V., and Stenbuck, P. J., 1979b, The membrane attachment protein for spectrin is associated with band 3 in human erythrocyte membranes, *Nature* **280:**468–473.

Bennett, V., and Stenbuck, P. J., 1980a, Human erythrocyte ankyrin. Purification and properties, *J. Biol. Chem.* **255:**2540–2548.

Bennett, V., and Stenbuck, P. J., 1980b, Association between ankyrin and the cytoplasmic domain of band 3 isolated from the human erythrocyte membrane, *J. Biol. Chem.* **255:**6424–6432.

Bennett, V., Davis, J., and Fowler, W., 1982a, Immunoreactive forms of erythrocyte spectrin and ankyrin in brain, *Phil. Trans. R. Soc. Lond.* **B299:**301–312.

Bennett, V., Davis, J., and Fowler, W., 1982b, Brain spectrin, a membrane-associated protein related in structure and function to erythrocyte spectrin, *Nature* **299:**126–131.

Bernstein, S. E., 1980, Inherited hemolytic disease in mice: a review and update, *Lab. Animal Sci.* **30:**197–205.

Bhattacharyya, B., and Wolff, J., 1975, Membrane-bound tubulin in brain and thyroid tissue, *J. Biol. Chem.* **250:**7639–7646.

Birchmeier, W., and Singer, S. J., 1977, On the mechanism of ATP-induced shape changes in human erythrocyte membranes. II. The role of ATP, *J. Cell Biol.* **73:**647–659.

Bodine, D. M., Birkenmeier, C. S., and Barker, J. E., 1983, Genetic analysis of spectrin deficient mice, Proceedings of Meeting on Actin–Membrane Interaction, Chapel Hill, North Carolina.

Bourguignon, L. Y. W., and Singer, S. J., 1977, Transmembrane interactions and the mechanism of capping of surface receptors by their specific ligands, *Proc. Natl. Acad. Sci. USA* **74:**5031–5035.

Brandts, J. F., Erickson, L., Lysko, K., Schwartz, A. T., and Taverna, R. D., 1977, Calorimetric studies of the structural transitions of the human erythrocyte membrane. The involvement of spectrin in the A transition, *Biochemistry* **16:**3450–3454.

Branton, D., Cohen, C. M., and Tyler, J., 1981, Interaction of cytoskeletal proteins on the human erythrocyte membrane, *Cell* **24:**24–32.

Brenner, S. L., and Korn, E. D., 1979, Spectrin-actin interaction. Phosphorylated and dephosphorylated spectrin tetramer cross-link F-actin, *J. Biol. Chem.* **254:**8620–8627.

Brenner, S. L., and Korn, E. D., 1980, Spectrin/actin complex isolated from sheep erythrocytes accelerates actin polymerization by simple nucleation. Evidence for oligomeric actin in the erythrocyte cytoskeleton, *J. Biol. Chem.* **255:**1670–1676.

Bulinski, J. C., and Borisy, G., 1980a, Immunofluorescence localization of HeLa cell microtubule-associated proteins on microtubules *in vitro* and *in vivo*, *J. Cell Biol.* **87:**792–801.

Burns, N. R., Ohanian, V., and Gratzer, W. B., 1983, Properties of brain spectrin (fodrin), *FEBS Lett.* **153:**165–168.

Burridge, K., Kelly, T., and Mangeat, P., 1982, Nonerythrocyte spectrins: actin-membrane attachment proteins occurring in many cell types, *J. Cell Biol.* **95:**478–486.

Cabantchik, Z. I., Knauf, P. A., and Rothstein, A., 1978, The anion transport system of the red blood cell. The role of membrane protein evaluated by the use of 'probes,' *Biochim. Biophys. Acta* **515:**239–302.

Calvert, R., Bennett, P., and Gratzer, W., 1980, Properties and structural role of the subunits of human spectrin, *Eur. J. Biochem.* **107**:355–361.

Carlin, R. C., Bartelt, D. C., and Siekevitz, P., 1983, Identification of fodrin as a major calmodulin-binding protein in postsynaptic density preparations, *J. Cell Biol.* **96**:443–448.

Chang, K., Williamson, J. R., and Zarkowsky, H. S., 1979, Effect of heat on the circular dichroism of spectrin in hereditary pyropoikilocytosis, *J. Clin. Invest.* **64**:326–328.

Cherry, R. J., 1979, Rotational and lateral diffusion of membrane proteins, *Biochim. Biophys. Acta* **559**:289–327.

Cherry, R. J., Burkli, A, Busslinger, M., Schneider, G., and Parish, G. R., 1976, Rotational diffusion of band 3 proteins in the human erythrocyte membrane, *Nature* **263**:389–393.

Coetzer, T., and Zail, S., 1982, Spectrin tetramer-dimer equilibrium in hereditary elliptocytosis, *Blood* **59**:900–905.

Cohen, C. M., and Branton, D., 1979, The role of spectrin in erythrocyte membrane stimulated actin polymerization, *Nature* **279**:163–165.

Cohen, C. M., and Foley, S. F., 1980, Spectrin dependent and independent association of F-actin with the erythrocyte membrane, *J. Cell Biol.* **86**:694–698.

Cohen, C. M., and Foley, S. F., 1982, The role of band 4.1 in the association of actin with erythrocyte membranes, *Biochim. Biophys. Acta* **688**:691–701.

Cohen, C. M., Tyler, J. M., and Branton, D., 1980, Spectrin–actin associations studied by electron microscopy of shadowed preparations, *Cell* **21**:875–883.

Cohen, C. M., Foley, S. F., and Korsgren, C., 1982, A protein immunologically related to erythrocyte band 4.1 is found on stress fibers of non-erythroid cells, *Nature* **299**:648–650.

Cotmore, S. F., Furthmayr, H., and Marchesi, V. T., 1977, Immunocytochemical evidence for the transmembrane orientation of glycophorin A, *J. Molec. Biol.* **113**:539–553.

Davies, P., and Klee, C., 1981, Calmodulin-binding proteins: A high molecular weight calmodulin-binding protein from bovine brain. *Biochem. Inter.* **3**:203–212.

Davis, J., and Bennett, V., 1982, Microtubule-associated protein 2, a microtubule-associated protein from brain, is immunologically related to the subunit of erythrocyte spectrin, *J. Biol. Chem.* **257**:5816–5820.

Davis, J., and Bennett, V., 1983, Brain Spectrin: Isolation of subunits and formation of hybrids with erythrocyte spectrin subunits, *J. Biol. Chem.* **258**:7757–7766.

Davis, J., and Bennett, V., 1984, Brain ankyrin: Purification of a 72,000 Mr spectrin-binding domain, *J. Biol. Chem.* **259**:1874–1881.

Dorst, H., and Schubert, D., 1979, Self-association of band 3-protein from human erythrocyte membranes in aqueous solutions, *Hoppe–Seylers Z. Physiol. Chem.* **360**:1605–1618.

Edelman, G. M., 1976, Surface modulation in cell recognition and cell growth, *Science* **192**:218–226.

Elgsaeter, A., and Branton, D., 1974, Intramembrane particle aggregation in erythrocyte ghosts. I. The effects of protein removal, *J. Cell Biol.* **63**:1018–1030.

Elgsaeter, A., Shotton, D. M., and Branton, D., 1976, Intramembrane particle aggregation in erythrocyte ghosts II. The influence of spectrin aggregation, *Biochim. Biophys. Acta* **426**:101–122.

Evans, E. A., and Hochmuth, R. M., 1978, Mechanochemical properties of membranes, *Curr. Top. Membr. Transp.* **10**:1–64.

Fowler, V., and Bennett, V., 1978, Association of spectrin with its membrane attachment site restricts lateral mobility of human erythrocyte integral membrane proteins, *J. Supramol. Struct.* **8**:215–221.

Fowler, V. and Bennett, V., 1984, Erythrocyte tropomyosin: Purification and properties, *J. Biol. Chem.* **259**:5978–5989.

Fowler, V., and Branton, D., 1977, Lateral mobility of human erythrocyte integral membrane proteins, *Nature* **268**:23–26.

Fowler, V., and Taylor, D. L., 1980, Spectrin plus band 4.1 cross-link actin. Regulation by micromolar calcium, *J. Cell Biol.* **85**:361–376.

Fowler, V. M., Luna, E. J., Hargreaves, W. R., Taylor, D. L., and Branton, D., 1981, Spectrin promotes the association of F-actin with the cytoplasmic surface of the human erythrocyte membrane, *J. Cell Biol.* **88**:388–395.

Furthmayr, H., 1979, Glycophorins A, B and C: A family of sialoglycoproteins. Isolation and characterization of trypsin-derived peptides, in: *Normal and Abnormal Red Cell Membranes* (S. Lux, V. Marchesi, and F. Fox, eds.), Liss, New York, p. 195.

Glenney, J., Glenney, P., Osborn, M., and Weber, K., 1982a, An F-actin-and calmodulin-binding protein from isolated intestinal brush borders has a morphology related to spectrin, *Cell* **28**: 843–854.

Glenney, J., Glenney, P., and Weber, K., 1982b, Erythroid spectrin, brain fodrin, and intestinal brush border proteins (TW-260/240) are related molecules containing a common calmodulin-binding subunit bound to a variant cell type specific subunit, *Proc. Natl. Acad. Sci. USA* **79**:4002–4005.

Glenney, J., Glenney, P., and Weber, K., 1982c, F-actin-binding and cross-linking properties of porcine brain fodrin, a spectrin-related molecule, *J. Biol. Chem.* **257**:9781–9787.

Golan, D. E., and Veatch, W., 1980, Lateral mobility of band 3 in the human erythrocyte membrane studied by fluorescence photobleaching recovery: evidence for control by cytoskeletal interactions, *Proc. Natl. Acad. Sci. USA* **77**:2537–2541.

Golan, D. E., and Veatch, W. R., 1982, Lateral mobility of band 3 in the human erythrocyte membrane. Control by ankyrin-mediated interactions, *Biophys. J.* **37**:177a.

Goodman, S., and Branton, D., 1978, Spectrin binding and the control of membrane protein mobility, *J. Supramol. Struct.* **8**:455–463.

Goodman, S. R., and Weidner, S. A., 1980, Binding of spectrin tetramers to human erythrocyte membranes, *J. Biol. Chem.* **255**:8082–8086.

Goodman, S., Zagon, I., and Kulikowski, R., 1981, Identification of a spectrin-like protein in nonerythroid cells, *Proc. Natl. Acad. Sci. USA* **78**:7570–7574.

Goodman, S. R., Shiffer, K. A., Casoria, L. A., and Eyster, M. E., 1982, Identification of the molecular defect in the erythrocyte membrane skeleton of some kindreds with hereditary spherocytosis, *Blood* **60**:772–784.

Greenquist, K. C., Shohet, S. B., and Bernstein, S. E., 1978, Marked reduction in spectrin in hereditary spherocytosis in the common house mouse, *Blood* **51**:1149–1155.

Guidotti, G., 1972, Membrane proteins, *Ann. Rev. Biochem.* **41**:731–752.

Haest, C. W. M., Plasa, G., Kamp, D., and Deuticke, B., 1978, Spectrin as a stabilizer of the phospholipid asymmetry in the human erythrocyte membrane, *Biochim. Biophys. Acta* **509**:21–32.

Hainfield, J. F., and Steck, T. L., 1977, The sub-membrane reticulum of the human erythrocyte: a scanning electron microscope study, *J. Supramol. Struct.* **6**:301–317.

Hargreaves, W. R., Giedd, K. N., Verkleij, A., and Branton, D., 1980, Reassociation of ankyrin with band 3 in erythrocyte membranes and in lipid vesicles, *J. Biol. Chem.* **255**:11965–11972.

Harris, H. W., and Lux, S. E., 1980, Structural characterization of the phosphorylation sites of human erythrocyte spectrin, *J. Biol. Chem.* **255**:11512–11520.

Hepler, P., 1980, Membranes in the mitotic apparatus of barley cells, *J. Cell Biol.* **86**:490–499.

Herzog, W., and Weber, K., 1978, Fractionation of brain microtubule-associated proteins: Isolation of two different proteins which stimulate tubulin polymerization *in vitro, Eur. J. Biochem.* **92**:1–8.

Hiller, G., and Weber, K., 1977, Spectrin is absent in various tisse culture cells, *Nature* **266**:181–183.

Izant, J. G., Weatherbee, J. A., and McIntosh, J. R., 1982, A microtubule-associated protein in the mitotic spindle and the interphase nucleus, *Nature* **295**:248–250.

Jarrett, H. W., and Penniston, J. T., 1978, Purification of the Ca^{2+} -stimulated ATPase activator from human erythrocytes. Its membership in the class of Ca^{2+}-binding modulator proteins, *J. Biol. Chem.* **253**:4676–4682.

Ji, T. H., Kiehm, D. J., and Middaugh, G. R., 1980, Presence of spectrin tetramer on the erythrocyte membrane, *J. Biol. Chem.* **255**:2990–2993.

Kakiuchi, S., Sobue, K., and Fujita, M., 1981, Purification of a 240,000 M_r calmodulin-binding protein from a microsomal fraction of brain, *FEBS Lett.* **132**:144–148.

Kim, H., Binder, L. J., and Rosenbaum, J. L., 1979, The periodic association of MAP2 with brain microtubules *in vitro*, *J. Cell Biol.* **80**:266–276.

Knowles, W. J., Morrow, J. S., Speicher, D. W., Zarkowsky, H. S., Mohandas, N., Shohet, S. B., and Marchesi, V. T., 1983, Molecular and functional changes in spectrin from patients with hereditary poikilocytosis, *J. Clin. Invest.* **71**:1867–1877.

Kwok, R., and Evans, E., 1981, Thermoelasticity of large lecithin bilayer vesicles, *Biophys. J.* **35**:637–652.

Levine, J., and Willard, M., 1980, The composition and organization of axonally transported proteins in the retinal ganglion cells of the guinea pig, *Brain Res.* **194**:137–154.

Levine, J., and Willard, M., 1981, Fodrin: axonally transported polypeptides associated with the internal periphery of many cells, *J. Cell Biol.* **90**:631–643.

Levine, J., and Willard, M., 1983, Redistribution of fodrin (a component of the cortical cytoplasm) accompanying capping of cell surface molecules, *Proc. Natl. Acad. Sci. USA* **80**:191–195.

Lin, D. C., and Lin, S., 1979, Actin polymerization induced by motility-related high-affinity cytochalasin binding complex from human erythrocyte membranes, *Proc. Natl. Acad. Sci. USA* **76**:2345–2349.

Lin, D. C., Flanagan, M. D., and Lin, S., 1983, Complexes containing actin and spectrin from erythrocyte and brain, *Cell Motility* **3**:375–382.

Litman, P., Chen, J. H., and Marchesi, V. T., 1978, Spectrin binds to the inner surface of the human red cell membrane via associations with band 4.1-4.2 and 3, *J. Supramol. Struct.* **8**(suppl. 2):209.

Litman, D., Hsu, C. J., and Marchesi, V. T., 1980, Evidence that spectrin binds to macromolecular complexes on the inner surface of the red cell membrane, *J. Cell Sci.* **42**:1–22.

Liu, S. C., and Palek, J., 1979, Cross-linkings between spectrin and band 3 in human erythrocyte membranes, *J. Supramol. Struct.* **10**:97–109.

Liu, S. C., Palek, J., Prchal, J., and Castleberry, R. P., 1981, Altered spectrin dimer-dimer association and instability of membrane skeletons in hereditary pyropoikilocytosis, *J. Clin. Invest.* **68**:597–605.

Liu, S. C., Palek, J., and Prchal, J. T., 1982, Defective spectrin dimer-dimer association in hereditary elliptocytosis, *Proc. Natl. Acad. Sci. USA* **79**:2072–2076.

Lorenz, T., and Willard, M., 1978, Subcellular fractionation of intra-axonally transported polypeptides in the rabbit visual system, *Proc. Natl. Acad. Sci. USA* **75**:505–509.

Luna, E. J., Kidd, G. H., and Branton, D., 1979, Identification by peptide analysis of the spectrin-binding protein in human erythrocytes, *J. Biol. Chem.* **254**:2526–2532.

Lux, S. E., 1979, Spectrin–actin membrane skeleton of normal and abnormal blood cells, *Semin. Hematol.* **16**:21–51.

Lux, S. E., 1983, Disorders of the red cell membrane skeleton: Hereditary spherocytosis and herditary elliptocytosis, in: *The Metabolic Basis of Inherited Disease* (Stanbury, Wyngaarden, Fredrickson, Goldstein, and Brown, eds.) McGraw–Hill, New York. pp. 1573–1605.

Marchesi, V. T., 1983, The red cell membrane skeleton: recent progress, *Blood* **61**:1–11.

Marchesi, V. T., and Steers, E., 1968, Selective solubilization of a protein component of the red cell membrane, *Science* **159**:203–204.

Marchesi, V. T., Furthmayr, H., and Tomita, M., 1976, The red cell membrane, *Ann. Rev. Biochem.* **45**:667–697.

Mombers, C., Verkleij, A. J., DeGier, J., and Van Deenen, L. L. M., 1979, The interaction of spectrin-actin and synthetic phospholipids II. The interaction with phosphatidylserine, *Biochim. Biophys. Acta* **551**:271–281.

Morrow, J. S., and Marchesi, V. T., 1981, Self assembly of spectrin oligomers *in vitro:* A basis for a dynamic cytoskeleton, *J. Cell Biol.* **88**:463–468.

Morrow, J. S., Speicher, D. W., Knowles, W. J., Hsu, C. J., and Marchesi V. T., 1980, Identification of functional domains of human erythrocyte spectrin, *Proc. Natl. Acad. Sci. USA* **77**:6592–6596.

Moskalewski, S., Thyberg, J., Hinek, A., and Friberg, U., 1977, Fine structure of the Golgi complex during mitosis of cartilaginous cells *in vitro, Tissue Cell* **9**:185–196.

Mueller, T. J., and Morrison, M., 1981, Glycoconnectin (PAS2), a membrane attachment site for the human erythrocyte cytoskeleton, in: *Erythrocyte Membranes 2: Recent Clinical and Experimental Advances* (W. Kruckeberg, J. Eaton, and G. Brewer, eds.), Liss, New York, pp. 95–112.

Murphy, D. B., and Borisy, G. G., 1975, Association of high-molecular-weight proteins with microtubules and their role in microtubule assembly *in vitro, Proc. Natl. Acad. Sci. USA* **72**:2696–2700.

Murthy, S. N., Liu, T., Kaul, R. K., Kohler, H., and Steck, T. L., 1981, The aldolase-binding site of the human erythrocyte membrane is at the NH_2 terminus of band 3, *J. Biol. Chem.* **256**:11203–11208.

Nakashima, H., Nakagawa, Y., and Makino, S., 1981, Detection of the associated state of membrane proteins by polyacrylamide gradient gel electrophoresis with nondenaturing detergents—application to band 3, *Biochim. Biophys. Acta* **643**:509–518.

Nelson, G. A., Andrews, M. L., and Karnovsky, M. J., 1983, Control of erythrocyte shape by calmodulin, *J. Cell Biol.* **96**:730–735.

Nelson, W. J., and Lazarides, E., 1983, Expression of the subunit of spectrin in nonerythroid cells, *Proc. Natl. Acad. Sci. USA* **80**:363–367.

Nermut, M. L., 1981, Visualization of the "membrane skeleton" in human erythrocytes by freeze-etching, *Eur. J. Cell Biol.* **25**:265–271.

Nicolson, G. L., 1974, Transmembrane control of the receptors on normal and tumor cells I. Cytoplasmic influence over cell surface components, *Biochim. Biophys. Acta* **457**:57–108.

Nicolson, G. L., and Painter, R. G., 1973, Anionic sites of human erythrocyte membranes II. Antispectrin-induced transmembrane aggregation of the binding sites for positively charged colloidal particles. *J. Cell Biol.* **59**:395–406.

Nigg, E., and Cherry, R. J., 1979a, Dimeric association of band 3 in the erythrocyte membrane demonstrated by protein diffusion measurements, *Nature* **277**:493–494.

Nigg, E. A., and Cherry, R. J., 1979b, Influence of temperature and cholesterol on the rotational diffusion of band 3 in the human erythrocyte membrane, *Biochemistry* **18**:3457–3465.

Nigg, E. A., and Cherry, R. J., 1980, Anchorage of a band 3 population at the erythrocyte cytoplasmic membrane surface: protein rotational diffusion measurements, *Proc. Natl. Acad. Sci. USA* **77**:4702–4706.

Nigg, E. A., Bron, C., Giradet, M., and Cherry, R. J., 1980, Band 3–glycophorin A association in erythrocyte membranes demonstrated by combining protein diffusion measurements with antibody-induced cross-linking, *Biochemistry* **19:1887–1893**.

Palfrey, C., Schiebler, W., and Greengard, P., 1982, A major calmodulin-binding protein common to various vertebrate tissues, *Proc. Natl. Acad. Sci. USA* **79**:3780–3784.

Palmer, F. B., and Verpoorte, J., 1971, The phosphorus components of solubilized erythrocyte membrane protein, *Can. J. Biochem.* **49:**337–347.

Pasval, G., Wainscoat, J. S., and Weatherall, D. J., 1982, Erythrocytes deficient in glycophorin resist invasion by the malarial parasite Plasmodium falciparum, *Nature* **297:**64–66.

Patel, V. P., and Fairbanks, G., 1981, Spectrin phosphorylation and shape change of human erythrocyte ghosts, *J. Cell Biol.* **88:**430–440.

Perkins, M., 1981, Inhibitory effects of erythrocyte membrane proteins on the *in vitro* invasion of the human malarial parasite (Plasmodium falciparum) into its host cell, *J. Cell Biol.* **90:**563–567.

Peters, R., Peters, J., Tews, K. H., and Bahr, W., 1974, A microfluorimetric study of translational diffusion in erythrocyte membranes, *Biochim. Biophys. Acta* **367:**282–294.

Pinder, J. C., and Gratzer, W. B., 1983, Structural and dynamic states of actin in the erythrocyte, *J. Cell Biol.* **96:**768–775.

Pinder, J. C., Ungewickell, E., Bray, D., and Gratzer, W. B., 1978, The spectrin-actin complex and erythrocyte shape, *J. Supramol. Struct.* **8:**439–445.

Pinto da Silva, P., and Nicolson, G. L., 1974, Freeze–etch localization of concanavalin A receptors to the membrane intercalated particles of human erythrocyte ghost membranes, *Biochim. Biophys. Acta* **363:**311–319.

Pollard, T. D., 1983, Meeting Highlights, Critique and Perspectives *Cell Motility* **3:**693–697.

Ralston, G. B., 1978, Physical–chemical studies of spectrin, *J. Supramol. Struct.* **8:**361–373.

Rand, R. P., and Burton, A. C., 1964, Mechanical properties of the red cell membrane I. Membrane stiffness and intracellular pressure, *Biophys. J.* **4:**115–135.

Repasky, E., Granger, B., and Lazarides, E., 1982, Widespread occurrence of avian spectrin in nonerythroid cells, *Cell* **29:**821–833.

Russell, E. S., 1979, Hereditary anemia of the mouse: A review for geneticists, *Adv. Genetics* **20:**357–459.

Salhany, J. M., and Shaklai, N., 1979, Functional properties of human hemoglobin bound to the erythrocyte membrane, *Biochemistry* **18:**893–899.

Salhany, J. M., Cordes, K. A., and Gaines, E. D., 1980, Light-scattering measurements of hemoglobin binding to the erythrocyte membrane. Evidence for transmembrane effects related to a disulfonic stilbene binding to band 3, *Biochemistry* **19:**1447–1454.

Sattilaro, R. F., Dentler, W. L., and LeCluyse, E. L., 1981, Microtubule-associated proteins (MAPs) and the organization of actin filaments *in vitro*, *J. Cell Biol.* **90:**467–473.

Sheetz, M. P., 1979, Integral membrane protein interaction with triton cytoskeletons of erythrocytes, *Biochim. Biophys. Acta* **557:**122–134.

Sheetz, M. P., and Sawyer, D., 1978, Triton shells of intact erythrocytes, *J. Supramol. Struct.* **8:**399–412.

Sheetz, M. P., and Singer, S. J., 1977, On the mechanism of ATP-induced shape changes in human erythrocyte membranes I. The role of the spectrin complex, *J. Cell Biol.* **73:**638–646.

Sheetz, M. P., Painter, R. G., and Singer, S. J., 1976, Relationships of the spectrin complex of human erythrocyte membranes to the actomyosins of muscle cells, *Biochemistry* **15:**4486–4492.

Sheetz, M. P., Schindler, M., and Koppel, D., 1980, Lateral mobility of integral membrane proteins is increased in spherocytic erythrocytes, *Nature* **285:**510–512.

Sherline, P., Lee, Y. C., and Jacobs, L., 1977, Binding of microtubules to pituitary secretory granules and secretory granule membranes, *J. Cell Biol.* **72:**380–389.

Shimo–Oka, T., and Watanabe, Y., 1981, Stimulation of actomyosin Mg^{2+}-ATPase activity by a brain microtubule-associated protein fraction. High-molecular-weight actin-binding protein is the stimulating factor, *J. Biochem.* **90:**1297–1307.

Shohet, S. B., 1979, Reconstitution of spectrin deficient mouse erythrocyte membranes, *J. Clin. Invest.* **64:**483–493.

Shotton, D. M., Thompson, K., Wofsy, L., and Branton, D., 1978, Appearance and distribution of surface protein of the human erythrocyte membrane, *J. Cell Biol.* **76:**512–531.

Shotton, D. M., Burke, B. E., and Branton, D., 1979, The molecular structure of human erythrocyte spectrin, *J. Mol. Biol.* **131:**303–329.

Siegel, D. L., and Branton, D., 1982, Human erythrocyte band 4.9, *J. Cell Biol.* **95:**265a.

Singer, S. J., 1974, The molecular organization of membranes, *Ann. Rev. Biochem.* **43:**805–833.

Singer, S. J., and Nicolson, G. L., 1972, The fluid mosaic model of the structure of cell membranes, *Science* **175:**720–731.

Sloboda, R. D., and Dickersin, K., 1980, Structure and composition of the cytoskeleton of nucleated erythrocytes. I. The presence of microtubule-associated protein 2 in the marginal band, *J. Cell Biol.* **87:**170–179.

Smith, B. A., Clark, W. R., and McConnell, H. M., 1979, Anisotropic molecular motion on cell surfaces, *Proc. Natl. Acad. Sci. USA* **76:**5641–5644.

Sobue, K., Muramoto, Y., Fujita, M., and Kakiuchi, S., 1981, Calmodulin-binding protein of erythrocyte cytoskeleton, *Biochem. Biophys. Res. Comm.* **100:**1063–1070.

Sobue, K., Kanda, K., and Kakiuchi, S., 1982, Solubilization and partial purification of protein kinase systems from brain membranes that phosphorylate calspectin. A spectrin-like calmodulin-binding protein (fodrin), *FEBS Lett.* **150:**185–190.

Speicher, D. W., Morrow, J. C., Knowles, W. J., and Marchesi, V. T., 1980, Identification of proteolytically resistant domains of human erythrocyte spectrin, *Proc. Natl. Acad. Sci. USA* **77:**5673–5677.

Speicher, D. W., Morrow, J. S., Knowles, W. J., and Marchesi, V. T., 1982, A structural model of human erythrocyte spectrin. Alignment of chemical and functional domains, *J. Biol. Chem.* **257:**9093–9101.

Spiegel, J. E., Southwick, F. S., Lux, S. E., and Stossel, T. P., 1982, Erythrocyte protein 4.1 is immunologically related to acamentin from polymorphonuclear leukocytes, *Blood* **60:**25a.

Steck, T. L., 1972, Cross-linking the major proteins of the isolated erythrocyte membrane, *J. Mol. Biol.* **66:**295–305.

Steck, T. L., 1974, Organization of proteins in the human red blood cell membrane, *J. Cell Biol.* **62:**1–19.

Steck, T. L., 1978, The band 3 protein of the human red cell membrane: a review, *J. Supramol. Struct.* **8:**311–324.

Steck, T. L., Ramos, B., and Strapazon, E., 1976, Proteolytic dissection of band 3, the predominant transmembrane polypeptide of the human erythrocyte membrane, *Biochemistry* **15:**1154–1161.

Stephens, R. E., 1977, Major protein differences in cilia and flagella: Evidence for a membrane-associated tubulin, *Biochemistry* **16:**2047–2058.

Tank, D. W., Wu, E. S., and Webb, W. W., 1982, Enhanced molecular diffusibility in muscle membrane blebs: release of lateral constraints, *J. Cell Biol.* **92:**207–212.

Tchernia, G., Mohandas, N., and Shohet, S. B., 1981, Deficiency of skeletal membrane protein band 4.1 in homozygous hereditary elliptocytosis, *J. Clin. Invest.* **68:**454–460.

Tilney, L. G., and Detmers, P., 1975, Actin in erythrocyte ghosts and its association with spectrin, *J. Cell Biol.* **66:**508–520.

Tokuyasu, K. T., Schekman, R., and Singer, S. J., 1979, Domains of receptor mobility and endocytosis in the membranes of neonatal human erythrocytes and reticulocytes are deficient in spectrin, *J. Cell Biol.* **80:**481–486.

Tsai, T. H., Murthy, S. N., and Steck, T. L., 1982, Effect of red cell membrane binding on the catalytic activity of glyceraldehyde-3-phosphate dehydrogenase, *J. Biol. Chem.* **257:**1438–1442.

Tyler, J. M., Hargreaves, W. R., and Branton, D., 1979, Purification of two spectrin-binding proteins: biochemical and electron microscopic evidence for site-specific reassociation between spectrin and bands 2.1 and 4.1, *Proc. Natl. Acad. Sci. USA* **76:**5192–5196.

Tyler, J. M., Anderson, J. M., and Branton, D., 1980a, Structural comparison of several actin-binding macromolecules, *J. Cell Biol.* **85:**489–495.

Tyler, J. M., Reinhardt, B. N., and Branton, D., 1980b, Associations of erythrocyte membrane proteins—binding of purified bands 2.1 and 4.1 to spectrin, *J. Biol. Chem.* **255:**7034–7039.

Unanue, E. R., and Karnovsky, M. J., 1973, Redistribution and fate of Ig complexes on surface of B lymphocytes: Functional implications and mechanisms, *Transplant Rev.* **14:**184–210.

Ungewickell, E., and Gratzer, W., 1978, Self-association of human spectrin. A thermodynamic and kinetic study, *Eur. J. Biochem.* **88:**379–385.

Ungewickell, E., Bennett, P. M., Calvert, R., Ohanian, V., and Gratzer, W. B., 1979, *In vitro* formation of a complex between cytoskeletal proteins of the human erythrocyte, *Nature* **280:**811–814.

VanDeurs, B., and Behnke, O., 1973, The microtubule marginal band of mammalian red blood cells, *Z. Anat. Entw. Gesch.* **143:**43–47.

Voter, W. A., and Erickson, H. P., 1982, Electron microscopy of MAP2 (microtubule-associated protein 2), *J. Ultrastruct. Res.* **80:**374–382.

Wang, K., and Richards, F. M., 1974, An approach to nearest neighbor analysis of membrane proteins. Application to the human erythrocyte membrane of a method employing cleavable cross-linkages, *J. Biol. Chem.* **249:**8005–8018.

Weatherbee, J. A., Luftig, R. G., and Weihing, R. R., 1980, Purification and reconstitution of HeLa cell microtubules, *Biochemistry* **19:**4116–4123.

Weed, R. I., La Celle, P. L., Merrill, E., 1969, Metabolic dependence of red cell deformability, *J. Clin. Invest.* **48:**795–809.

Wolf, L. C., John, K. M., Falcone, J. C., Byrne, A. M., and Lux, S. E., 1982, A genetic defect in the binding of protein 4.1 to spectrin in a kindred with hereditary spherocytosis, *N. Engl. J. Med.* **307:**1367–1374.

Wu, E. S., Tank, D. W., and Webb, W. W., 1982, Unconstrained lateral diffusion of concanavilin A receptors on bulbous lymphocytes, *Proc. Natl. Acad. Sci. USA* **79:**4962–4966.

Yu, J., and Branton, D., 1976, Reconstitution of intramembrane particles in recombinants of erythrocyte protein band 3 and lipid; effects of spectrin-actin association, *Proc. Natl. Acad. Sci. USA* **73:**3891–3895.

Yu, J., and Goodman, S. R., 1979, Syndeins: The spectrin binding protein(s) of the human erythrocyte membrane, *Proc. Natl. Acad. Sci. USA* **76:**2340–2344.

Yu, J., and Steck, T. L., 1975, Associations of band 3, the predominant polypeptide of the human erythrocyte, *J. Biol. Chem.* **250:**9176–9184.

Yu, J., Fischman, D. A., and Steck, T. L., 1973, Selective solubilization of proteins and phospholipids from red blood cell membranes by nonionic detergents, *J. Supramol. Struct.* **1:**233–248.

5

REGULATION OF ASSEMBLY OF THE SPECTRIN-BASED MEMBRANE SKELETON IN CHICKEN EMBRYO ERYTHROID CELLS

Randall T. Moon, Ingrid Blikstad, and Elias Lazarides

1. INTRODUCTION: THE SORTING OF NEWLY SYNTHESIZED PROTEINS

How newly synthesized proteins are routed to their proper locations and assembled into higher-order structures within the cell remains a challenge to cell biology. The magnitude of the problem is considerable, given that some newly synthesized proteins such as secreted immunoglobulins (Dulis, 1983) are destined for secretion from cells, others such as many glycoproteins (Fitting and Kabat, 1982; Polonoff *et al.*, 1982) or hormone or neurotransmitter receptors (Jacobs *et al.*, 1983; Merlie *et al.*, 1982) become integrated into the plasma membrane, and the remaining proteins reside in intracellular vesicles, membranes, cytoskeletal structures, the nucleus, or the cytoplasm. As illustrated below, the selection of a protein's destination can be made either cotranslationally or posttranslationally.

Integral membrane proteins and secreted proteins are examples of proteins that reach their destination by virtue of a cotranslational selection process. Tre-

Randall T. Moon, Ingrid Blikstad, and Elias Lazarides ● Division of Biology, California Institute of Technology, Pasadena, California 91125. Work supported by grants from the National Institutes of Health (NIH), the National Science Foundation, and the Muscular Dystrophy Association. Randall T. Moon was supported by fellowships from the NIH and the American Cancer Society. Ingrid Blikstad was supported by a fellowship from the Swedish Natural Science Research Council. Elias Lazarides is the recipient of a Research Career Development Award from the NIH.

mendous advances have been made in our understanding of how these proteins
use a signal peptide in the growing nascent polypeptide chain and a signal
recognition particle to effect their cotranslational vectorial transport into or across
the endoplasmic reticulum (e.g., Meyer *et al.*, 1982; Walter and Blobel, 1982).

Posttranslational selection processes can also determine the subcellular distribution of proteins. Microinjection of labeled proteins into the cytoplasm of
frog oocytes has demonstrated that there are karyophilic proteins which migrate
into and accumulate in the nucleus, there are proteins such as actin which come
to reside in both the cytoplasm and the nucleus, and there are proteins such as
tubulin which remain primarily in the cytoplasm (reviewed by DeRobertis, 1983).
Although the molecular basis for these results remains unclear, in the oo-
cyte–nurse cell syncytium of the insect *Hyalophora cecropia,* it is apparent that
some soluble proteins migrate and their cytoplasmic distribution becomes nonrandom in response to endogenously generated electrical gradients (Woodruff
and Telfer, 1980). Some integral membrane proteins also reach their destination
by a posttranslational mechanism, as shown for mitochondrial membrane proteins
(Gasser and Schatz, 1983; reviewed by Wickner, 1980; Schatz and Butow, 1983).

Cytoskeletal proteins provide useful opportunities for the analysis of how
specific newly synthesized proteins attain a nonrandom distribution within cells,
and how proteins assemble into higher-ordered structures. That some cytoskeletal
elements have a nonrandom distribution within cells is clear from data showing
that the intermediate-filament proteins desmin and vimentin are present predominantly at the Z disc in late-stage myotubes and in adult skeletal muscle (Granger
and Lazarides, 1979; Gard and Lazarides, 1980). A second example of the
nonrandom distribution of a cytoskeletal protein is spectrin, which is a peripherally associated plasma membrane protein consisting of two subunits in erythroid
(reviewed in Lux, 1979; Branton *et al.*, 1981; Bennett, 1982; Marchesi, 1983;
Cohen, 1983; Goodman and Shiffer, 1983) and nonerythroid cells (Repasky *et
al.*, 1982; Goodman *et al.*, 1981; Burridge *et al.*, 1982; Glenney *et al.*, 1982;
Lazarides and Nelson, 1983; see also review by Nelson and Lazarides in this
text, Chapter 6).

It is easy to envision that microtubules and actin filaments are assembled
posttranslationally from precursors, since both these cytoskeletal filaments exist
in equilibrium with their constituent polypeptides and can be reversibly depolymerized. However, intermediate filaments and the erythroid spectrin–actin
network are not readily depolymerized and remain insoluble when cells are lysed
in physiological buffers. An important question, therefore, is whether or not
insoluble structures such as intermediate filaments and the spectrin–actin network
are assembled cotranslationally, and whether their polyribosomes have a nonrandom distribution within the cell. Alternatively, these structures may be as-

sembled posttranslationally from soluble precursors. Fulton and Wan (1983) have argued for the former mechanism of assembly. Our recent data (Blikstad and Lazarides, 1983a,b; Moon and Lazarides, 1983a,b,c), summarized below, are compatible with rapid posttranslational mechanisms of assembly for both the two spectrin subunits and the intermediate-filament proteins vimentin and synemin.

Intermediate filaments are cytoskeletal elements that are insoluble in the nonionic detergent Triton X-100. This property has been exploited experimentally to demonstrate that vimentin, the core intermediate-filament protein in erythroid cells (Blikstad and Lazarides, 1983a), and synemin, a high-molecular-weight protein peripherally associated with intermediate filaments (Moon and Lazarides, 1983a), are assembled posttranslationally from detergent-soluble precursors into detergent-insoluble structures, at least in chicken embryo erythroid cells. Two lines of evidence support the existence of a soluble precursor pool. First, pulse-chase experiments have demonstrated that newly synthesized detergent-soluble vimentin (Blikstad and Lazarides, 1983a; Moon and Lazarides, 1983b) and synemin (Moon and Lazarides, 1983a) can be chased into detergent-insoluble structures. Second, replacement of arginine with an analog, canavanine, blocks the assembly of newly synthesized vimentin, and the unassembled vimentin accumulates in the detergent-soluble compartment (Moon and Lazarides, 1983b). Although intermediate filaments in avian erythroid cells may be assembled posttranslationally from soluble precursors, these data do not preclude the possibility that the polyribosomes synthesizing intermediate-filament proteins are somehow associated with the cytoskeleton as has been suggested for polyribosomes in general (Lenk et al., 1977; Cervera et al., 1981).

A second example of a structure that is insoluble under physiological ionic conditions is the erythroid cell membrane skeleton. The most abundant protein of this membrane skeleton is spectrin, which associates with the membrane through the binding of its β subunit to the extrinsic membrane protein ankyrin. Ankyrin, in turn, associates with the membrane by binding to the integral membrane protein, the anion transporter (band 3). While this current model of protein arrangement in the human erythrocyte was developing (for reviews, see Lux, 1979; Branton et al., 1981; Bennett, 1982; Marchesi, 1983; Cohen, 1983; Goodman and Shiffer, 1983), several investigators perceived that data on the synthesis and assembly of the cytoskeletal proteins would contribute greatly to this specific model and to our general understanding of membrane biogenesis (Lodish, 1973; Koch et al., 1975a,b; Lodish and Small, 1975; Chang et al., 1976; Chan et al., 1978; Weise and Chan, 1978). Initial studies demonstrated that the anion transporter, spectrin, and ankyrin are synthesized simultaneously in mammals (Koch et al., 1975a,b; Chang et al., 1976), although termination of protein synthesis upon terminal differentiation is asynchronous (Chang et al., 1976; Chan et al., 1978).

In a related study, Weise and Chan (1978) have described in detail the appearance in the plasma membrane of newly synthesized membrane proteins in chicken embryo erythroid cells. Since the protein composition of the chicken erythrocyte membrane is similar to that of human red blood cells (Weise and Chan, 1978; Repasky et al., 1982), general principles of cytoskeletal assembly elucidated in chicken erythroid cells should also apply to mammalian erythroid cells. Like mammalian erythrocytes, chicken erythroid cells have a membrane-associated spectrin–actin network containing equimolar proportions of α-spectrin (240,000 mol.wt.) and β-spectrin (220,000 mol.wt.) (Repasky et al., 1982). Chicken erythroid cells also have an approximately 230,000-mol.wt. protein (260,000 in our gel system), globin, which resembles mammalian ankyrin in that it is associated with the plasma membrane (Beam et al., 1979), it is not extracted by Triton X-100 or extreme changes in ionic strength (Beam et al., 1979; Repasky et al., 1982), it is phosphorylated (Alper et al., 1980a,b), and it cross-reacts with antibodies raised against human ankyrin (see Blikstad et al., 1983; Moon and Lazarides, 1984). Finally, the band 3 and 3.1 proteins in chicken erythrocytes are integral membrane proteins which are glycosylated and largely soluble in Triton X-100 (Jackson, 1975; Weise and Ingram, 1976; Chan et al., 1978; Jay, 1983), similar to the mammalian anion transport protein (reviewed in Branton et al., 1981).

Weise and Chan (1978) found that although the synthesis of bands 3 and 3.1 in primitive erythroid cells (day 3 and 5 cells) was greater than in definitive erythroid cells (days 8 and 11), all major membrane proteins are nevertheless synthesized simultaneously in chicken embryo erythroid cells. They also found that inhibition of protein synthesis with cycloheximide resulted in a rapid block in the assembly of labeled membrane skeleton proteins (globin and spectrin), with the exception of bands 3 and 3.1 proteins (understandable in light of anion transport protein processing in the Golgi apparatus; see Braell and Lodish, 1982). Weise and Chan concluded that in light of the rapid, cycloheximide-sensitive appearance of newly synthesized membrane skeleton proteins in the membrane, synthesis and assembly of membrane skeleton proteins are tightly coupled events. The recognized limitation of their approach, however, was that they measured assembled membrane skeleton proteins and they would have missed any membrane protein precursors that were soluble following hypotonic lysis. Therefore, prior to our studies (Blikstad et al., 1983; Blikstad and Lazarides, 1983b; Moon and Lazarides, 1983c), summarized below, no studies have directly addressed the question of whether the erythroid cell membrane skeleton is assembled from soluble precursors, and what might be the major levels of regulation in its assembly.

2. SYNOPSIS

In pursuing the problem of the regulation of cytoskeletal assembly, our specific goal has been to elucidate the mechanisms by which newly synthesized chicken erythroid cell membrane skeleton proteins are assembled into a higher-order structure which is localized at the cytoplasmic surface of the plasma membrane. We have, therefore, been concerned with how the cell regulates synthesis and assembly in both space and time. The additional significance of this work is that it may further our general understanding of how the membranes of differentiating cells become specialized in different regions of the cell.

Our basic experimental approach has been to use the insolubility in Triton X-100 of the membrane skeleton of chicken embryo erythroid cells as an assay for whether skeletal proteins are assembled cotranslationally or posttranslationally (Blikstad *et al.*, 1983; Blikstad and Lazarides, 1983a,b; Moon and Lazarides, 1983a,b). We have also developed a cell-free system from chicken embryo erythroid cells to uncouple *in vitro* the synthesis and assembly of membrane skeleton proteins (Moon and Lazarides, 1983c). These experiments have suggested that the primary topographic cue for the assembly of the skeleton at the plasma membrane is provided by the membrane. On the basis of these observations we propose that the cotranslational insertion into the endoplasmic reticulum and transport to the plasma membrane of the anion transporter provides high-affinity binding sites at the plasma membrane for ankyrin, and hence spectrin, and thus dictates the spatial organization of the spectrin–actin network. The temporal coordination of assembly involves the simultaneous synthesis of both membrane-skeleton components studied thus far (spectrin and ankyrin), the presence of high-affinity membrane-binding sites to bind a substantial proportion of the newly synthesized spectrin and ankyrin, and a catabolic mechanism for scavenging unassembled reactants. On the basis of these concepts we propose the "receptor-mediated assembly and stabilization hypothesis" to describe the regulation of assembly of the spectrin-based membrane skeleton.

3. SYNTHESIS AND ASSEMBLY OF THE MEMBRANE SKELETON

3.1. Equimolar Assembly of α- and β-Spectrin Follows Their Unequal Synthesis

The equimolar proportions of α- and β-spectrin found in the membrane skeleton of erythroid cells could arise from the tightly and coordinately regulated synthesis of equimolar proportions of each subunit, as is the case in the globin subunits (e.g., Lodish, 1971). Alternatively, the synthesis of all or some membrane skeleton proteins could be under less stringent regulation, with the result being that the assembly of specific proportions of proteins would occur by only some of the newly synthesized molecules binding to receptors. The latter mechanism applies to the synthesis of α- and β-spectrin since, as we have observed, α-spectrin is synthesized in an approximately 3 : 1 ratio relative to β -spectrin, and the α-spectrin in molar excess of β-spectrin is not assembled into the cytoskeleton (Blikstad et al., 1983; Moon and Lazarides, 1983c).

Since α-spectrin is synthesized in excess of β-spectrin, it appears that the genes for these two proteins may not be transcribed into equimolar proportions of mRNA. To examine this point in more detail we have determined whether the 3 : 1 ratio of synthesis of α- to β-spectrin was due to their relative mRNA abundance or to translational control. Our approach was to use concentrations of emetine in vivo which slow, but do not stop, protein synthesis. mRNAs that may be inefficient at initiation would thus become fully loaded with ribosomes, and hence the net synthesis of different polypeptides would reflect the relative abundance of their mRNAs (Lodish, 1971). We found that chicken embryo erythroid cells preincubated with different concentrations of emetine before labeling in vivo with ^{35}S-methionine synthesized total proteins at 10–100% of the rate observed in cells not exposed to emetine, and that α-spectrin was always synthesized in excess relative to β-spectrin (Fig. 1a). The approximately 3 : 1 ratio of α- to β-spectrin was also always obtained by translation of avian erythroid mRNA in an mRNA-dependent rabbit reticulocyte lysate at either subsaturating (Fig. 1b) or saturating (data not shown) mRNA concentrations. Melting the RNA at 75°C or 95°C prior to translation, or substituting ^{3}H-leucine as the label, similarly resulted in the greater in vitro synthesis of α- over β-spectrin (Fig. 1b). Thus, the unequal synthesis of α- and β-spectrin is probably not due to differences in the secondary structure of the two mRNA populations, or to a difference in methionine content. Taken together these results suggest that the unequal synthesis of α- and β-spectrin is due to transcriptional rather than translational control, and that α-spectrin mRNA is present in excess of β-spectrin mRNA. Since spectrin subunits are nevertheless found in equimolar proportions in the membrane skeleton, the ratio of synthesis of membrane skeleton proteins does not determine the steady-state ratio of these proteins.

The threefold excess synthesis of α-spectrin relative to β-spectrin could be required to maintain the 1 : 1 stoichiometry observed at steady state in the membrane if assembled α-spectrin turned over threefold more rapidly than assembled β-spectrin. This is clearly not the case, however, since the α-spectrin in molar excess to β-spectrin does not assemble onto the cytoskeleton (Fig. 2), and since assembled subunits do not turn over (Fig. 3; Weise and Chan, 1978; see also Moon and Lazarides, 1984). Instead, the excess α-spectrin enters an

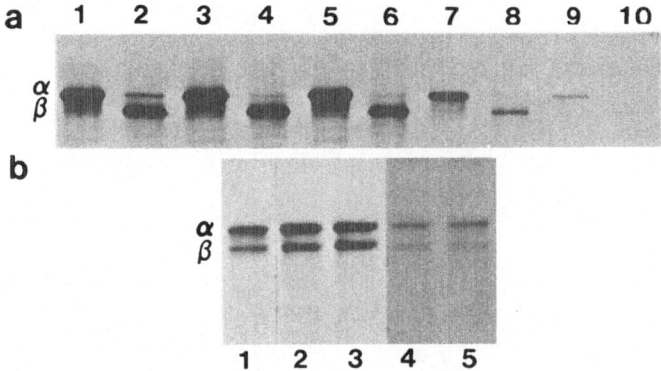

FIGURE 1. Unequal synthesis of α- and β-spectrin *in vivo* and *in vitro*. (a) Effect of emetine *in vivo* on the ratio of newly synthesized α- and β-spectrin. Erythroid cells from 10-day chicken embryos were cultured as described (Moon and Lazarides, 1983b). After 15 min at 35°C, emetine was added to the final concentrations given below, and 60 μCi of ^{35}S-methionione were added to each 250-μl cell suspension. After 30 min at 35°C with periodic agitation the cells were lysed (Moon and Lazarides, 1983b) and then processed for immunoprecipitation of α-spectrin (1,3,5,7,9) and β-spectrin (2,4,6,8,10) followed by SDS 10% polyacrylamide gel electrophoresis and fluorography. The incorporation of ^{35}S-methionine into total trichloracetic-acid-insoluble radioactivity is expressed as the percentage relative to the control for each emetine concentraiton. 1,2, 0 μM emetine control, 587,000 cpm in 10 μl of solubilized cells; 3,4, 0.02 μM emetine, 96% of control; 5,6, 0.1 μM emetine, 100% of control; 7,8, 0.5 μM emetine, 52% of control; 9,10, 2.5 μM emetine, 14% of control. The fluorograph was exposed for 72 hr. (b) Total RNA from 15-day chicken embryo erythroid cells was translated at subsaturating concentrations (approximately 100 μg/ml) in a mRNA-dependent rabbit reticulocyte lysate. The RNA was left on ice (1,4,5) or else heated for 1 min at 75°C (2) or 95°C (3) prior to rapid cooling and translation for 120 min at 30°C. The lysates were supplemented with ^{35}S-methionine (1,2,3), or ^{3}H-leucine (4,5). Aliquots of each lysate were used for immunoprecipitation in which the antibodies against α- and β-spectrin were mixed. Fluorographs were exposed 24 hr (1,2,3) and 24 days (4,5). Mammalian α-spectrin contains 1.2 mole % methionine and 11.9 mole % leucine, whereas β contains 1.8 mole % methionine and 11.6 mole % leucine (Anderson, 1979), and similar values have been obtained for porcine brain spectrin (Glenney *et al.*, 1982).

operationally defined soluble fraction (Triton X-100 soluble) (Fig. 2 and Blikstad *et al.*, 1983). Therefore, we have postulated (Blikstad *et al.*, 1983; Moon and Lazarides, 1983c) that the selective assembly of only some of the newly synthesized α-spectrin is due to the limited availability of β-spectrin, through which α-spectrin binds to the membrane skeleton.

Our finding that much of the newly synthesized β-spectrin enters the cytoskeleton (Blikstad *et al.*, 1983, and Fig. 3B) suggests that the extent of β-spectrin assembly is driven toward completion by the availability of β-spectrin binding sites (presumably ankyrin) and/or the high affinity of β-spectrin for these

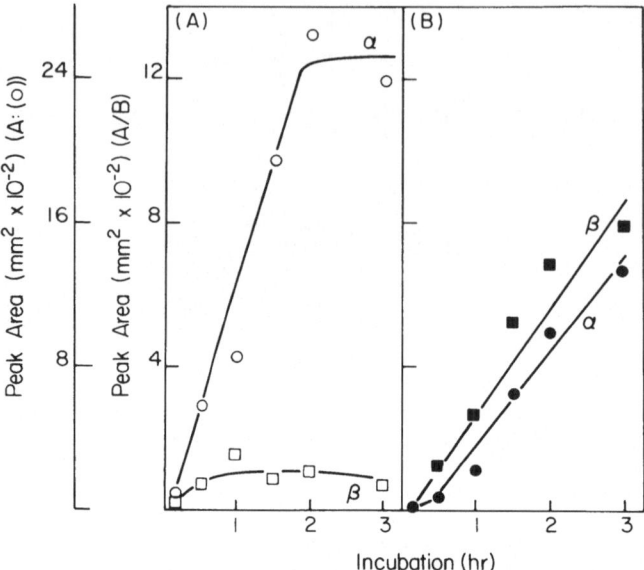

FIGURE 2.   Partitioning of newly synthesized α- and β-spectrin into soluble (A) and cytoskeletal (B) fractions of 10-day chicken embryo erythroid cells during continuous labeling in ^{35}S-methionine. O, α-spectrin, □, β-spectrin. See Moon and Lazarides (1983b,c) for experimental details on cell labeling, fractionation into soluble and cytoskeletal fraction, and immunoprecipitation. Data points represent quantification of labeled immunoprecipitates of spectrin at the indicated times.

binding sites. The data argue against the possibility that a large precursor pool of β-spectrin is necessary to favor assembly onto low-affinity binding sites. This is, however, speculative since the local concentrations of precursors in the cell are unknown, nor do we know whether there are critical concentrations of precursors required for assembly. In this content, Morrow and Marchesi (1981) have suggested that high local concentrations of spectrin are required at the membrane to promote spectrin oligomerization through low-affinity interactions.

3.2. Soluble Spectrin Is Selectively Catabolized

If only some of the newly synthesized spectrin subunits are actually assembled, one must postulate a mechanism for degrading unassembled subunits, lest they accumulate indefinitely in the cytoplasm. As shown in Fig. 3 (see also Blikstad *et al.*, 1983), soluble spectrin subunits do indeed turn over. Although some of these soluble subunits may serve as precursors to cytoskeletal spectrin (a possibility discussed in the Section 3.3), many of the soluble subunits are

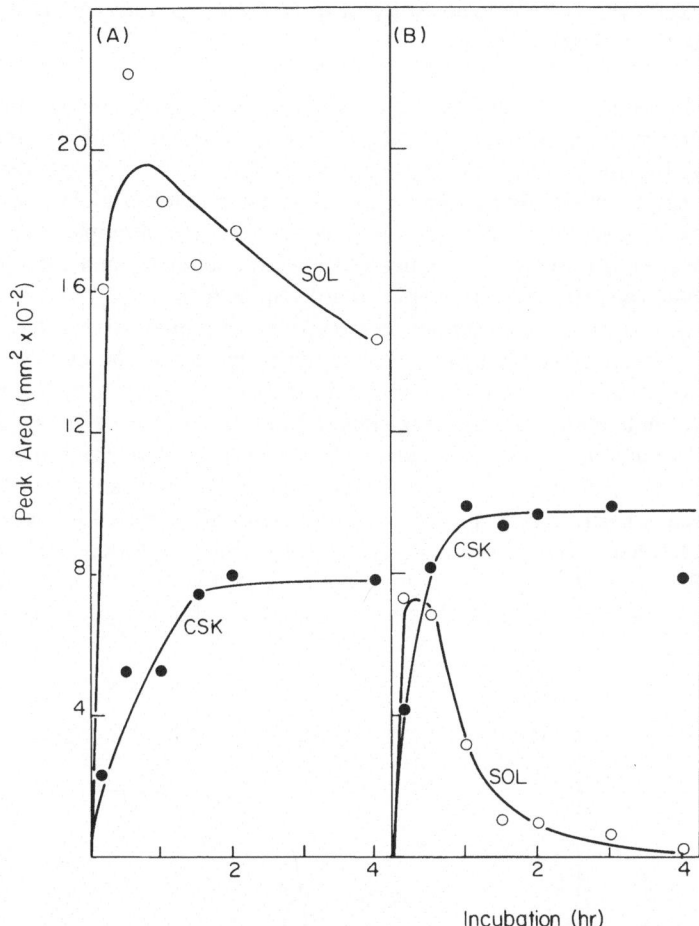

FIGURE 3. Stability of newly synthesized α-spectrin (A) and β-spectrin (B) in the soluble (SOL) and cytoskeletal (CSK) fractions of 10-day chicken embryo erythroid cells. Cells were pulse labeled for 10 min with ^{35}S-methionine and then chased with an excess of unlabeled methionine for 4 hr. Data represent quantification of immunoprecipitates. See Moon and Lazarides (1983b,c) for details on methods.

catabolized since there is not a proportional increase in cytoskeletal spectrin for each incremental decrease in soluble spectrin. The cytoskeletal spectrin is, however, stable and protected against degradation. The stability of cytoskeletal spectrin was demonstrated by the absence of spectrin turnover after a 4-hr (Fig. 3), 8-hr (data not shown), or 23-hr (Weise and Chan, 1978) chase following a brief pulse labeling.

3.3. Most Newly Synthesized Ankyrin Is Assembled, and Unassembled Ankyrin Is Catabolized

As discussed in the Introduction, available data suggest that globin is the avian counterpart to mammalian ankyrin. Our investigation of the synthesis, assembly, and turnover of globin, hereafter called ankyrin (though we do not rule out the possibility that mammalian ankyrin and chicken globin have some unique species-specific properties), demonstrates that the principles of receptor-mediated assembly of newly synthesized spectrin subunits, and catabolism of unassembled spectrin subunits, apply to ankyrin as well.

We investigated the assembly of ankyrin since a qualitative determination of the proportion of newly synthesized ankyrin that enters the cytoskeleton as opposed to the soluble fraction would, as in the case of spectrin, tell us whether there were enough available binding sites to bind most of the newly synthesized ankyrin. Expanding on our preliminary experiments on ankyrin (globin) assembly (Blikstad *et al.*, 1983), we found that in continuous labeling experiments, newly synthesized ankyrin, like β-spectrin, quickly saturated a small soluble pool and that most labeled ankyrin entered the cytoskeleton (Fig. 4A; Moon and Lazarides, 1984). This indicates that, as with β-spectrin, but not α-spectrin, there are sufficient cytoskeletal binding sites for binding most, but not all, newly synthe-

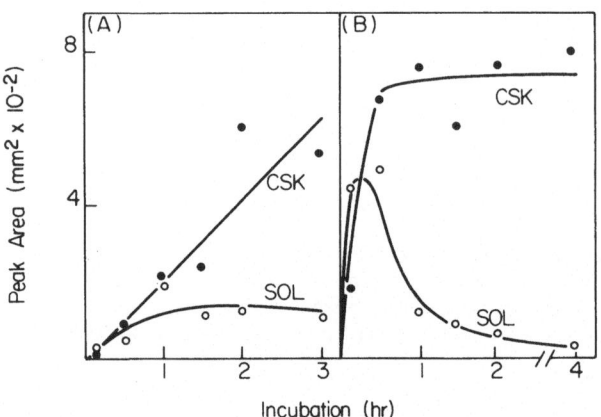

FIGURE 4. Distribution of newly synthesized globin between the soluble (SOL) and cytoskeletal (CSK) fractions of 10-day chicken embryo erythroid cells. Cells were labeled continuously with ^{35}S-methionine for 3 hr (A) or for 10 min then chased with unlabeled methionine (B) prior to separation into soluble and cytoskeletal fractions, immunoprecipitation with antibodies raised against mammalian ankyrin (kindly provided by V. Bennett), gel electrophoresis, and fluorography.

sized ankyrin. The efficient assembly of most ankyrin may involve the high affinity of the anion transporter for ankyrin (Kd 5–8 $\times$ 10^{-8} M from *in vitro* reconstitution studies, Hargreaves *et al.*, 1980).

We investigated the turnover of ankyrin in the soluble and cytoskeletal fractions to determine whether our observations on spectrin turnover (Section 2) could be generalized to include ankyrin. Pulse-chase experiments revealed that, like α- and β-spectrin, newly assembled cytoskeletal ankyrin does not turn over during a 4-hr chase period, whereas the soluble ankyrin was rapidly catabolized (Fig. 4B; Moon and Lazarides, 1984). Thus ankyrin, like spectrin, is resistant to turnover once assembled onto the membrane skeleton, and a mechanism exists for the selective catabolism of unassembled, soluble ankyrin.

3.4. Is the Assembly of Spectrin and Ankyrin Simultaneous Yet Independent?

Since spectrin, ankyrin, and the anion transporter are being synthesized and assembled simultaneously in differentiating erythroid cells (Weise and Chan, 1978; Blikstad *et al.*, 1983; Blikstad and Lazarides, 1983b; Moon and Lazarides, 1983b), is the extent of assembly of one membrane skeleton component such as spectrin tightly coupled to the simultaneous assembly of another membrane skeleton component such as ankyrin? If the assembly of membrane skeleton components were simultaneous but independent, i.e., not tightly coupled, then one might be able to experimentally interfere with assembly so that spectrin, but not ankyrin (or the reverse), is assembled. We have tried two approaches to selectively alter the ratio of assembly of spectrin to ankyrin. First, we have reversibly inhibited protein synthesis by inducing hypertonic initiation block (Saborio *et al.*, 1974; Mechler, 1981), which results in ribosomes "running off" from preexisting polyribosomes, and blocking reinitiation (Moon and Lazarides, unpublished observations). If, upon labeling with ^{35}S-methionine and reversal of the initiation block, there were large changes relative to controls in the ratio of newly assembled labeled spectrin to ankyrin, then this would argue against tightly coupled assembly. The result upon reversal of the initiation block was that α- and β-spectrin and ankyrin all entered the cytoskeleton simultaneously and in about the same relative proportions, irrespective of the extent of the prior inhibition of protein synthesis (Moon and Lazarides, unpublished observations). This further emphasizes that the assembly of spectrin and ankyrin is simultaneous but does not address the issue of whether assembly is tightly coupled.

Our second approach to investigate the possible tight coupling of the assembly of spectrin and ankyrin has been to synthesize these proteins *in vivo* in the presence of amino acid analogs. By altering the charge or conformation of the proteins, the analog might selectively prevent the assembly of either spectrin or ankyrin. This approach, based on our observation that the arginine analog canavanine blocks the assembly of vimentin into intermediate filaments (Moon and Lazarides, 1983b), has shown that the assembly of ankyrin can be blocked selectively with a proline analog (Moon and Lazarides, unpublished observations). Since some newly synthesized spectrin still assembles in the absence of ankyrin binding to the cytoskeleton, this suggests that ankyrin and spectrin can assemble independently yet simultaneously. The newly synthesized spectrin which binds the membrane skeleton independent of the binding of newly synthesized ankyrin may be binding both to preexisting membrane-bound ankyrin and to spectrin oligomers, or it may, in the absence of high affinity ankyrin sites, bind to a number of lower affinity sites.

3.5. Directing Membrane Skeleton Proteins to the Plasma Membrane: Cotranslational Membrane Insertion of the Anion Transporter and Posttranslational Binding of Spectrin

As discussed in the Introduction, part of the impetus for our studies was to determine if membrane skeleton proteins, like spectrin and ankyrin, are assembled cotranslationally, as suggested for HeLa cell cytoskeletal proteins (Fulton and Wan, 1983), or posttranslationally, like the erythroid cell intermediate-filament proteins synemin and vimentin (Blikstad and Lazarides, 1983a; Moon and Lazarides, 1983a). Data indicate that the anion transporter, the integral membrane protein that anchors the extrinsic membrane skeletal proteins to the plasma membrane, reaches the membrane cotranslationally, whereas the extrinsic membrane skeleton proteins are assembled posttranslationally (as hypothesized by Lodish and Small, 1975).

Like many other glycoproteins, the nascent chains of the anion transporter are cotranslationally inserted into the endoplasmic reticulum (Braell and Lodish, 1982). Unlike many proteins synthesized in this fashion, however, there is no detectable cleavage of an amino-terminal peptide, and the amino terminus is exposed to the cytoplasm after insertion of the completed polypeptide. After processing, the mature anion transporter appears at the plasma membrane 30–40 min after its synthesis (Braell and Lodish, 1981).

With regard to the membrane skeleton proteins, we examined whether spectrin nascent chains are cotranslationally bound to the cytoskeleton and found that they are not bound. By examining two-dimensional gels of erythroid cell soluble and cytoskeletal fractions pulse-labeled for very short periods of time, we observed labeled polypeptides in the cytoskeleton with a lower molecular weight than α-spectrin. This labeled cytoskeletal material was identified as α-spectrin nascent chains since they could be chased into full-sized α-spectrin with unlabeled methionione and immunoprecipitated with α-spectrin antibodies (Blikstad and Lazarides, 1983b). Since the nascent chains were found in the cytoskeleton, this agrees with numerous studies (e.g., Lenk et al., 1977; Cervera et al., 1981; VanVenrooij et al., 1981; Lemieux and Beaud, 1982; Moon et al., 1983) which have shown that polyribosomes are retained in the Triton X-100-extracted cytoskeleton. The α-spectrin nascent chains were not cotranslationally associated with the cytoskeleton, however, since they could readily be released quantitatively into the soluble fraction from the cytoskeletal fraction by puromycin, which causes premature termination of nascent chains (Blikstad and Lazarides, 1983b). Although β-spectrin nascent chains have not been identified on these gels owing to their poor isoelectric focusing, these data show that at least α-spectrin must associate with the cytoskeleton posttranslationally. This conclusion is supported by in vitro membrane-binding studies which show that newly synthesized α- and β-spectrin will bind posttranslationally to erythroid membranes (Moon and Lazarides, 1983c).

4. REGULATION OF ASSEMBLY OF SPECTRIN AND ANKYRIN: THE RECEPTOR-MEDIATED ASSEMBLY AND STABILIZATION HYPOTHESIS

The following data and principles form the basis for our hypothesis on the regulation of assembly of the membrane skeleton, described in the subsequent paragraph.

1. Proteins that comprise the membrane skeleton are synthesized simultaneously in developing erythroid cells. The presence of proteins capable of forming higher-order structures poses a problem for the cell in that assembly must be regulated so that assembled complexes accumulate only at the appropriate site.
2. An intrinsic membrane protein, which serves as a receptor for the membrane skeleton, the anion transporter, is cotranslationally inserted into the membrane in the endoplasmic reticulum and vectorially transported to the plasma membrane. The extrinsic membrane proteins, which bind to the anion transporter through noncovalent interactions, and which comprise the membrane skeleton, are assembled posttranslationally after their synthesis on membrane-free polyribosomes.
3. The assembly process is not coupled to synthesis. α-Spectrin is synthesized in severalfold excess relative to the availability of cytoskeletal

binding sites; similarly, β-spectrin and ankyrin are synthesized in excess of the amount of each which is assembled.

4. The assembly process is rapid, since spectrin and ankyrin are assembled into detergent-insoluble structures within minutes after synthesis.

5. Once extrinsic membrane proteins are assembled onto the detergent-insoluble membrane skeleton, they are resistant to catabolism. Soluble membrane skeleton proteins are, however, labile.

The preceding data on the synthesis, assembly, and turnover of spectrin and ankyrin enable us to propose a working hypothesis for the assembly of the erythroid cell membrane skeleton (Fig. 5). We formulate this hypothesis, the "receptor mediated assembly and stabilization hypothesis," as follows: "The assembly of the constituent polypeptides of the membrane skeleton requires both spatial and temporal regulation. The spatial localization of newly assembled spectrin and ankyrin is regulated predominantly by the vectorial flow of newly synthesized anion transporters to the plasma membrane, which provide localized high-affinity binding sites for ankyrin, and in turn, spectrin. The temporal regulation of assembly is achieved by the simultaneous synthesis of several polypeptides which sequentially and posttranslationally bind to each other. Ligands that are assembled are resistant to catabolism, whereas unassembled spectrin and ankyrin are catabolized, thus preventing the accumulation and assembly of reactants at inappropriate sites." The physiological significance of this hypothesis is that it explains how the cell assembles the membrane skeleton in the shortest possible period of time (all components are synthesized concurrently) while simultaneously ensuring that assembled components accumulate only at the plasma membrane (high-affinity receptors are localized only at the membrane and reactants not bound to the membrane are catabolized). The key feature of this hypothesis is that it explains how the relative proportions at steady state of membrane skeleton proteins are determined posttranslationally by the amount of each

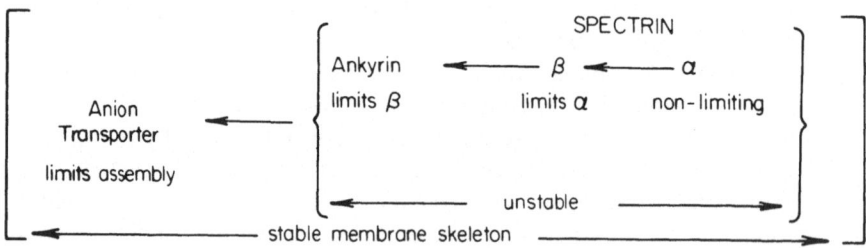

FIGURE 5. The receptor-mediated assembly and stabilization hypothesis. Details are given in Section 4 in the text.

protein binding to membrane receptors, and not by the amount of each protein synthesized.

We stress that the above hypothesis, although based on current data, is largely speculative since there are a great many uncertainties in our understanding of the structure of erythroid membrane skeletons, and in our understanding of the assembly process. With regard to membrane skeleton structure, we cannot rule out the presence of relatively unstudied membrane receptors for some proteins, including nonankyrin receptors for spectrin such as band 4.1 (Anderson and Lovrien, 1984). We do, however, propose that the general principles of assembly stated in the above hypothesis are compatible with multiple receptors for newly synthesized membrane skeleton proteins. With regard to the assembly process, the proportion of newly synthesized membrane skeleton proteins that bind to the skeleton will be influenced by the (largely unknown) local concentrations of ligands and receptors and their affinities. We, therefore, do not yet know why there is always some unassembled newly synthesized β-spectrin and ankyrin, although we propose that there are too few receptors on the membrane to bind all newly synthesized spectrin and ankyrin (Moon and Lazarides, 1984). Does the presence of some unassembled spectrin reflect the inefficiency of forming spectrin oligomers by high spectrin concentrations and low affinities (Morrow and Marchesi, 1981)? With regard to receptor availability, it is especially interesting that not all newly synthesized ankyrin becomes assembled, since there is an excess of anion transporters relative to ankyrin at steady state (Hargreaves et al., 1980; and reviewed in Branton et al., 1981; Bennett, 1982), and since the definitive chicken embryo erythroid cells in day 8 and 11 embryos synthesize enough anion transporters to bind all the ankyrin (Weise and Chan, 1978) if binding occurred in a 1 : 1 molar ratio. Why all newly synthesized anion transports do not, or cannot, bind ankyrin is not known (Moon and Lazarides, 1984). Despite these uncertainties, the above hypothesis serves as a framework for further thought and work on the assembly processes of membrane skeletons.

5. TESTING THE RECEPTOR-MEDIATED ASSEMBLY AND STABILIZATION HYPOTHESIS BY SEPARATING THE SYNTHESIS AND ASSEMBLY OF SPECTRIN *IN VITRO* IN A CELL-FREE SYSTEM

To test this hypothesis *in vivo* would be difficult since synthesis, assembly, and turnover cannot yet be readily uncoupled. We have, therefore, developed a cell-free system from chicken embryo erythroid cells where these events can be uncoupled (Moon and Lazarides, 1983c). The following aspects of the receptor-mediated assembly and stabilization hypothesis have been directly demonstrated *in vitro* using this chicken erythroid cell lysate (Moon and Lazarides, 1983c).

1. α-Spectrin is synthesized in excess of β-spectrin, and the α-spectrin in molar excess to β-spectrin cannot bind to membranes independently of β-spectrin. The assembly of α-spectrin is, therefore, limited by the availability of β-spectrin.
2. The binding of newly synthesized α- and β-spectrin to membranes can occur posttranslationally.

We are currently using this cell-free system to investigate our observation made *in vivo* that although assembled spectrin and ankyrin have long half-lives, a mechanism exists for the selective catabolism of unassembled, soluble spectrin and ankyrin. This cell-free system should also be useful for studying whether the β'- and γ-spectrin variants (Nelson and Lazarides, 1983; Lazarides and Nelson, 1983) are capable of binding to different erythroid and nonerythroid membranes. By supplementing the cell-free system with total mRNA containing mRNAs coding for different spectrin subunits and with membranes from different sources, it should be possible to establish whether particular spectrin subunits synthesized *in vitro* are competent to bind to receptors on different membranes, without having to first purify the spectrin subunits. Additionally, by adding antibodies or purified membrane components (e.g., spectrin and ankyrin), one could block specific assembly steps, or conversely, enable assembly by complementing the system with a missing component.

6. PROSPECTS FOR FUTURE INVESTIGATIONS

Recent advances in our understanding of the assembly of the erythroid membrane skeleton, summarized previously, have led us to propose here a hypothesis to describe the regulation of its assembly. The utility of this hypothesis would be tested, and our understanding of the assembly of cytoskeletal structures would be advanced, by data on the following issues:

1. What is the subcellular site of assembly of the cytoskeletal elements? It is possible that spectrin and ankyrin are assembled at the plasma membrane. If assembly occurs at the plasma membrane, then the very rapid assembly of newly synthesized spectrin and ankyrin would suggest that

spectrin and ankyrin polyribosomes are themselves localized near the plasma membrane. Although not discounting this possibility, we consider it more plausible that these proteins assemble onto the anion transporter at some point during the journey of the anion transporter from the rough endoplasmic reticulum to the plasma membrane. This speculation is based on available evidence that the region of the anion transporter to which ankyrin binds (Branton et al., 1981) may be exposed to the cytoplasm soon after synthesis on the rough endoplasmic reticulum (Braell and Lodish, 1981) and on data showing that assembly of spectrin and ankyrin occurs soon after synthesis (Weise and Chan, 1978; Blikstad et al., 1983; Moon and Lazarides, 1984). We further speculate that higher-order spectrin interactions (for review, see Branton et al., 1981; Marchesi, 1983) do not occur until the newly synthesized spectrin heterodimers or tetramers arrive at the plasma membrane. The basis for this speculation is that such spectrin–spectrin interactions are low affinity and hence may require the high spectrin concentrations found only at the membrane (Morrow and Marchesi, 1981).

2. Is there any physiological significance for the synthesis of excess α-spectrin relative to β-spectrin? Since $\alpha\beta$-spectrin assembly appears to be determined solely by the distribution and availability of the β-spectrin receptor (presumably ankyrin) and β-spectrin (Moon and Lazarides, 1983c), it may be that there is no selective pressure for regulating the synthesis of equimolar amounts of α-spectrin. It is also possible that the excess α-spectrin has cellular functions other than its cytoskeletal role, or that it enhances the rate of dimerization of $\alpha\beta$ spectrin.

3. What is the extent of coordination by transcriptional or translational control of the synthesis of any of the membrane skeleton proteins? Since the amount of newly synthesized spectrin and ankyrin that is assembled appears to be determined primarily by the availability and affinity of membrane-binding sites, there would appear to be a requirement only for the coexpression of these gene products. We postulate, therefore, that severalfold increases in spectrin or ankyrin mRNA abundance (e.g., by introducing extra gene copies) would not affect the level of expression of the other membrane skeleton proteins through feedback mechanisms.

4. Is the rapid and simultaneous assembly of most of the newly synthesized β-spectrin and ankyrin and one third of the α-spectrin (enough to be equimolar to β-spectrin) due to simultaneous binding to independent sites, or do any of these three polypeptides form a complex prior to binding to the anion transporter? Although our amino acid analog experiments suggest that spectrin and ankyrin need not form a complex

prior to assembly, we do not know the extent of oligomerization of newly synthesized spectrin that binds the membrane skeleton. The cell-free system from chicken erythroid cells (Moon and Lazarides, 1983c) may be useful for studying such complexes.

5. When in the assembly of spectrin and ankyrin do band 4.1 and actin assemble? Is most newly synthesized band 4.1 and actin driven to the membrane by the abundance of high-affinity binding sites, as we would postulate based on the precedent of β-spectrin and ankyrin? This would provide a mechanism by which actin can be segregated to several actin-binding sites in the cytoplasm.

6. What mechanisms underlie the apparent stabilization of assembled spectrin and ankyrin and the turnover of unassembled, soluble molecules? Can one block *in vivo* or *in vitro* the degradation of soluble proteins? We speculate that ubiquitin (Herschko, 1983) mediates this selective catabolism, but no data are presently available. Might spectrin and ankyrin phosphorylation, currently a phenomenon without a function (see review by Bennett, 1982), be involved in the protection from degradation of the assembled molecules? If so, does the presence of calmodulin or erythrocyte membranes (Agre *et al.*, 1983) play a role in spectrin and ankyrin phosphorylation? Alternatively, does phosphorylation of these molecules play a role in assembly by preventing dissociation from the membrane skeleton?

7. Can mutants or diseases be identified that result in defective membrane skeletons owing to defects in assembly rather than simply reduced synthesis of membrane skeleton components? Are there mutants or diseased states that affect the catabolism of soluble membrane skeleton proteins and hence accumulate these proteins in the cytoplasm?

7. IMPLICATIONS FOR THE ASSEMBLY AND LOCALIZATION OF NONERYTHROID CYTOSKELETAL STRUCTURES

Recent work on nonerythroid forms of spectrin (reviewed by Nelson and Lazarides in this volume, Chapter 6) has produced some interesting results. First, there are changes during differentiation in the subunit composition of spectrin (Nelson and Lazarides, 1983). Second, variants of β-spectrin may be nonrandomly localized within cells and thus contribute to the formation of membrane domains (Lazarides and Nelson, 1983) with potentially specialized properties

(Repasky *et al.*, 1982). Nonerythroid cells, therefore, regulate the temporal and spatial expression of spectrin, and we expect that this may be accomplished by mechanisms similar, though not necessarily identical, to those described for erythroid cells.

With regard to switches in the subunit composition of spectrin (e.g., during muscle differentiation; Nelson and Lazarides, 1983), our work (Moon and Lazarides, 1983c) predicts that the stable assembly and membrane location of spectrin variants is, as in erythroid cells, mediated posttranslationally by the availability and location in the membrane of β-spectrin receptors and the level of expression of β-spectrin like polypeptides. The distribution of α-spectrin would, therefore, passively follow the distribution of β-spectrin. By analogy to the selective catabolism of soluble erythroid spectrin and ankyrin, we also predict that similar catabolic pathways work with transcriptional controls in nonerythroid cells to regulate the amount of spectrin and its receptor.

Hypotheses for the mechanisms of restriction of spectrin variants to particular regions of the cells, such as the distribution of αγ-spectrin throughout neurons and the restriction to the cell body of α-β(β')-spectrin (Lazarides and Nelson, 1983), can be derived from work which suggests that the erythroid spectrin network restricts the lateral mobility of membrane proteins (reviewed in Branton *et al.*, 1981; Bennett, 1982; Goodman and Shiffer, 1983) and that mobile erythroid membrane domains lack spectrin (Tokuyasu *et al.*, 1979). We hypothesize that α- and β(β')-spectrin and ankyrin like β(β') receptor(s) are synthesized simultaneously in the cell body. The ankyrin like protein binds to the membrane (possibly to integral membrane proteins analogous to the anion transporter) soon after its synthesis, as does β(β')-spectrin and equimolar amounts of α-spectrin. Higher-order structures subsequently form, involving the membrane-bound spectrin subunits and additional proteins such as actin. The formation of higher-order structures restricts the lateral mobility of the β(β')-spectrin and its ankyrinlike receptor, and they become immobilized in the cell body (see further discussion in Chapter 6 by Nelson and Lazarides). How αγ-spectrin becomes localized throughout the neuron may be more complex if they are synthesized in the cell body and if the neuron degrades unassembled spectrins. Perhaps newly synthesized neuronal spectrin moves down axons by associating with organelles (Levine and Willard, 1981), which would protect them from catabolism, and which in effect would act as the receptor for αγ-spectrin.

ACKNOWLEDGMENTS.   We thank Dr. B. Granger for helpful discussions on the structure of erythroid cells and for his comments on this manuscript.

REFERENCES

Agre, P., Gardner, K., and Bennett, V., 1983, Association between human erythrocyte calmodulin and the cytoplasmic surface of human erythrocyte membranes, *J. Biol. Chem.* **258**:6258–6265.

Alper, S. L., Beam, K. G., and Greengard, P., 1980a, Hormonal control of Na^+K^- cotransport in turkey erythrocytes. Multiple site phosphorylation of goblin, a high molecular weight protein of the plasma membrane, *J. Biol. Chem.* **255**:4864–4871.

Alper, S. L., Palfrey, H. C., DeRiemer, S. A., and Greengard, P., 1980b, Hormonal control of protein phosphorylation in turkey erythrocytes. Phosphorylation by cAMP-dependent and Ca^{2+}-dependent protein kinases of distinct sites in goblin, a high molecular weight protein of the plasma membrane, *J. Biol. Chem.* **255**:11029–11039.

Anderson, J. M., 1979, Structural studies on human spectrin: Comparison of subunits and fragmentation of native spectrin, *J. Biol. Chem.* **254**:939–944.

Anderson, R. A. and Lovrien, R. E., 1984, Glycophorin is linked by band 4.1 to the human erythrocyte membrane skeleton, *Nature* **307**:655–658.

Beam, K. G., Alper, S. L., Palade, G. E., and Greengard, P., 1979, Hormonally regulated phosphoprotein of turkey erythrocytes: Localization to plasma membrane, *J. Cell Biol.* **83**:1–15.

Bennett, V., 1982, The molecular basis for membrane–cytoskeleton association in human erythrocytes, *J. Cell. Biochem.* **18**:49–65.

Blikstad, I., and Lazarides, E., 1983a, Vimentin filaments are assembled from a soluble precursor in avian erythroid cells, *J. Cell Biol.* **96**:1803–1808.

Blikstad, I., and Lazarides, E., 1983b, Synthesis of spectrin in avian erythroid cells: Association of nascent polypeptide chains with the cytoskeleton, *Proc. Natl. Acad. Sci. USA* **80**:2637–2641.

Blikstad, I., Nelson, W. J., Moon, R. T., and Lazarides, E., 1983, Synthesis and assembly of spectrin during avian erythropoiesis: Stoichiometric assembly but unequal synthesis of α- and β-spectrin, *Cell* **32**:1081–1091.

Braell, W. A., and Lodish, H. F., 1981, Biosynthesis of the erythrocyte anion transport protein, *J. Biol. Chem.* **256**:11337–11344.

Braell, W. A., and Lodish, H. F., 1982, The erythroid anion transport protein is cotranslationally inserted into microsomes, *Cell* **28**:23–31.

Branton, D., Cohen, C. M., and Tyler, J., 1981, Interaction of cytoskeletal proteins on the human erythrocyte membrane, *Cell* **24**:24–32.

Burridge, K., Kelley, T., and Mangeat, P., 1982, Nonerythrocyte spectrins: Actin–membrane attachment proteins occurring in many cell types, *J. Cell Biol.* **95**:478–486.

Cervera, M., Dreyfuss, G., and Penman, S., 1981, Messenger RNA is translated when associated with the cytoskeletal framework in normal and VSV-infected HeLa cells, *Cell* **23**:113–120.

Chan, L.-N. L., Mahoney, K. A., Wacholtz, M., and Sha'afi, R. I., 1978, Asynchronous termination of plasma membrane protein synthesis in erythroid cells, *Membrane Biochim.* **2**:47–61.

Chang, H., Langer, P. J., and Lodish, H. F., 1976, Asynchronous synthesis of erythrocyte membrane proteins, *Proc. Natl. Acad. Sci. USA* **73**:3206–3210.

Cohen, C. M., 1983, The molecular organization of the red cell membrane skeleton, *Semin. Hematol.* **20**:141–158.

DeRobertis, E. M., 1983, Nucleocytoplasmic segregation of proteins and RNAs, *Cell* **32**:1021–1025.

Dulis, B. H., 1983, Regulation of protein expression in differentiation by subunit assembly: Human membrane and secreted IgM, *J. Biol. Chem.* **258**:2181–2187.

Fitting, T., and Kabat, D., 1982, Evidence for a glycoprotein "signal" involved in transport between subcellular organelles: Two membrane glycoproteins encoded by murine leukemia virus reach the cell surface at different rates, *J. Biol. Chem.* **257**:14011–14017.

Fulton, A. B., and Wan, K. M., 1983, Many cytoskeletal proteins associate with the HeLa cytoskeleton during translation *in vitro, Cell* **32:**619–625.

Gard, D. L., and Lazarides, E., 1980, The synthesis and distribution of desmin and vimentin during myogenesis *in vitro, Cell* **19:**263–275.

Gasser, S. M., and Schatz, G., 1983, Import of proteins into mitochondria: *In vitro* studies on the biogenesis of the outer membrane, *J. Biol. Chem.* **258:**3427–3430.

Glenney, J. R., Jr., Glenney, P., and Weber, K., 1982, F-actin-binding and cross-linking properties of porcine brain fodrin, a spectrin-related molecule, *J. Biol. Chem.* **257:**9781–9787.

Goodman, S. R., and Shiffer, K., 1983, The spectrin membrane skeleton of normal and abnormal human erythrocytes: a review, *Am. J. Physiol.* **244:**C121–C141.

Goodman, S. R., Zagon, I. S., and Kulikowski, R. R., 1981, Identification of a spectrin-like protein in nonerythroid cells, *Proc. Natl. Acad. Sci. USA* **78:**7570–7574.

Granger, G. L., and Lazarides, E., 1979, Desmin and vimentin coexist at the periphery of the myofibril Z disc, *Cell* **18:**1053–1063.

Hargreaves, W. R., Giedd, K. N., Verkleij, A., and Branton, D., 1980, Reassociation of ankyrin with band 3 in erythrocyte membranes and in lipid vesicles, *J. Biol. Chem.* **255:**11965–11972.

Hershko, A., 1983, Ubiquitin: Roles in protein modification and breakdown, *Cell* **34:**11–12.

Jackson, R. C., 1975, The exterior surface of the chicken erythrocyte, *J. Biol. Chem.* **250:**617–622.

Jacobs, S., Frederick, F. C., Jr., and Cuatrecasas, P., 1983, Monensin blocks the maturation of receptors for insulin and somatomedin C: Identification of receptor precursors, *Proc. Natl. Acad. Sci. USA* **80:**1228–1231.

Jay, D. G., 1983, Characterization of the chicken erythrocyte anion exchange protein, *J. Biol. Chem.* **258:**9431–9436.

Koch, P. A., Gardner, F. H., Gartrell, J. E., Jr., and Carter, J. R., Jr., 1975a, Biogenesis of erythrocyte membrane proteins: *In vivo* studies in anemic rabbits, *Biochim. Biophys. Acta* **389:**177–187.

Koch, P. A., Gartrell, J. E., Jr., Gardner, F. H., and Carter, J. R., Jr., 1975b, Biogenesis of erythrocyte membrane proteins: *In vivo* studies in anemic rabbits, *Biochim. Biophys. Acta* **389:**162–176.

Lazarides, E., and Nelson, W. J., 1983, Erythrocyte and brain forms of spectrin in cerebellum: Distinct membrane-cytoskeletal domains in neurons, *Science* **220:**1296–1297.

Lemieux, R., and Beaud, G., 1982, Expression of vaccinia virus early mRNA in Ehrlich ascites tumor cells, *Eur. J. Biochem.* **129:**273–279.

Lenk, R., Ransom, L., Kaufmann, Y., and Penman, S., 1977, A cytoskeletal structure with associated polyribosomes obtained from HeLa cells, *Cell* **10:**67–78.

Levine, J., and Willard, M., 1981, Fodrin: Axonally transported polypeptides associated with the internal periphery of many cells, *J. Cell Biol.* **90:**631–643.

Lodish, H. F., 1971, Alpha and beta globin messenger ribonucleic acid: Different amounts and rates of initiation of translation, *J. Biol. Chem.* **246:**7131–7138.

Lodish, H. F., 1973, Biosynthesis of reticulocyte membrane proteins by membrane-free polyribosomes, *Proc. Natl. Acad. Sci. USA* **70:**1526–1530.

Lodish, H. F., and Small, B., 1975, Membrane proteins synthesized by rabbit reticulocytes, *J. Cell Biol.* **65:**51–64.

Lux, S. E., 1979, Spectrin–actin membrane skeleton of normal and abnormal red blood cells, *Semin. Hematol.* **16:**21–51.

Marchesi, V. T., 1983, The red cell membrane skeleton: Recent progress, *Blood* **61:**1–11.

Mechler, B., 1981, Membrane-bound ribosomes of myeloma cells. VI. Initiation of immunoglobulin mRNA translation occurs on free ribosomes, *J. Cell Biol.* **88:**42–50.

Merlie, J. P., Sebbane, R., Tzartos, S., and Lindstrom, J., 1982, Inhibition of glycosylation with tunicamycin blocks assembly of newly synthesized acetylcholine receptor subunits in muscle cells, *J. Biol. Chem.* **257**:2694–2701.

Meyer, D. I., Krause, E., and Dobberstein, B., 1982, Secretory protein translocation across membranes—the role of the "docking protein," *Nature* **297**:647–650.

Moon, R. T., and Lazarides, E., 1983a, Synthesis and post-translational assembly of intermediate filaments in avian erythroid cells: Vimentin assembly limits the rate of synemin assembly, *Proc. Natl. Acad. Sci. USA* **80**:5495–5499.

Moon, R. T., and Lazarides, E., 1983b, Canavanine inhibits vimentin assembly but not its synthesis in chicken embryo erythroid cells, *J. Cell Biol.* **97**:1309–1314.

Moon, R. T., and Lazarides, E., 1983c, β-spectrin limits α-spectrin assembly on membranes following synthesis in a chicken erythroid cell lysate, *Nature* **305**:62–65.

Moon, R. T., and Lazarides, E., 1984, Biogenesis of the avian erythroid membrane-skeleton: Receptor-mediated assembly and stabilization of ankyrin (globin) and spectrin, *J. Cell Biol.,* in press.

Moon, R. T., Nicosia, R. F., Olsen, C., Hille, M. B., and Jeffery, W. R., 1983, The cytoskeletal framework of sea urchin eggs and embryos: Developmental changes in the association of messenger RNA, *Develop. Biol.* **95**:447–458.

Morrow, J. S., and Marchesi, V. T., 1981, Self assembly of spectrin oligomers *in vitro:* A basis for a dynamic cytoskeleton, *J. Cell Biol.* **88**:463–468.

Nelson, W. J., and Lazarides, E., 1983, Switching of subunit composition of muscle spectrin during myogenesis *in vitro, Nature* **304**:364–368.

Polonoff, E., Machida, C. A., and Kabat, D., 1982, Glycosylation and intracellular transport of membrane glycoproteins encoded by murine leukemia viruses: Inhibition by amino acid analogues and by tunicamycin, *J. Biol. Chem.* **257**:14023–14028.

Repasky, E. A., Granger, B. L., and Lazarides, E., 1982, Widespread occurrence of avian spectrin in nonerythroid cells, *Cell* **29**:821–833.

Saborio, J. L., Pong, S. S., and Koch, D., 1974, Selective and reversible inhibition of initiation of protein synthesis in mammalian cells, *J. Mol. Biol.* **85**:195–211.

Schatz, G., and Butow, R. A., 1983, How are proteins imported into mitochondria? *Cell* **32**:316–318.

Tokuyasu, K. T., Schekman, R., and Singer, S. J., 1979, Domains of receptor mobility and endocytosis in the membranes of neonatal human erythrocytes and reticulocytes are deficient in spectrin, *J. Cell Biol.* **80**:481–486.

VanVenrooij, W. J., Sillekens, P. T. G., VanEekelen, C. A. G., and Reinders, R. J., 1981, On the association of mRNA with the cytoskeleton in uninfected and adenovirus-infected human KB cells, *Exp. Cell Res.* **135**:79–91.

Walter, P., and Blobel, G., 1982, Signal recognition particle contains a 7S RNA essential for protein translocation across the endoplasmic reticulum, *Nature* **299**:691–698.

Weise, M. J., and Chan, L. L., 1978, Membrane protein synthesis in embryonic chick eyrthroid cells, *J. Biol. Chem.* **253**:1892–1897.

Weise, M. J., and Ingram, V. M., 1976, Proteins and glycoproteins of membranes from developing chick red cells, *J. Biol. Chem.* **251**:6667–6673.

Wickner, W., 1980, Assembly of proteins into membranes, *Science* **210**:861–868.

Woodruff, R. I., and Telfer, W. H., 1980, Electrophoresis of proteins in intracellular bridges, *Nature* **286**:84–86.

6

ASSEMBLY AND ESTABLISHMENT OF MEMBRANE–CYTOSKELETON DOMAINS DURING DIFFERENTIATION
Spectrin as a Model System

W. James Nelson and Elias Lazarides

1. INTRODUCTION

The multifunctional capability of the eukaryotic cell is expressed and regulated, to a great extent, by the cell type-specific biophysical properties of the plasma membrane. The plasma membrane is essentially a barrier comprising a phospholipid bilayer structure which acts also as a matrix onto which and into which a variety of specific proteins are attached. It is these proteins which selectively modify the structure and properties of the plasma membrane to create a wide variety of domains of distinctive morphology and function.

An understanding of how the cell establishes and maintains these domains, given the high degree of lateral diffusion of proteins in the plane of the plasma membrane, requires the identification of the constituent membrane-associated proteins involved and the elucidation of their molecular organization. Early studies on membrane organization gave rise to the hypothesis that the plasma

W. James Nelson and Elias Lazarides • Division of Biology, California Institute of Technology, Pasadena, California 91125. Work supported by grants from the National Institutes of Health (NIH), National Science Foundation, and the Muscular Dystrophy Association. W. James Nelson was also supported by a Cancer Research Campaign International Fellowship awarded by the International Union Against Cancer and a Senior Postdoctoral Fellowship from the American Heart Association, Los Angeles Affiliate. Elias Lazarides is the recipient of a Research Career Development Award from the NIH.

membrane comprises a homogeneous, fluid mosaic of proteins and lipids in which the mobility of proteins in the plane of the membrane is determined by the fluidity of the surrounding lipid matrix (Singer and Nicolson, 1972). This implies that the plasma membrane exhibits little or no long-range molecular or functional order other than that determined by the lipid matrix. Although this is probably true for the erythrocyte, upon which this hypothesis was originally based, it is clearly an oversimplification for most other cells which express a variety of plasma membrane-associated functions. These functions, in many cases, are known to be segregated into distinct domains, which suggests that the distribution of components of the plasma membrane is nonuniform. Consequently, it has been proposed that long-range molecular interactions between proteins embedded in the lipid bilayer (intrinsic membrane proteins) and proteins peripherally associated with the membrane (extrinsic membrane proteins) may be involved in establishing this nonuniformity (Nicolson, 1976).

What is the identity of the proteins involved in these interactions? In an attempt to answer this question, considerable attention has been focused on molecular events in the subcortical cytoplasm, where many cellular functions associated with the plasma membrane, such as cell motility, endocytosis and exocytosis, mobility of cell surface receptors, and overall changes in cell shape, have been shown to be directly regulated by changes in the distribution and protein composition of the subcortical cytoskeleton (reviewed by Weihing, 1979). The mechanical basis for most, if not all, of these changes is provided by actin, which is a ubiquitous component of the subcortical cytoplasm in a wide variety of cell types. Actin, together with myosin, can form a contractile complex involved in many dynamic membrane functions (Pollard and Weihing, 1974) and, in association with a variety of regulatory proteins, can become cross-linked (reviewed by Schliwa, 1981) to provide a more static function, for example in the structural support of the plasma membrane. However, in order to be involved in these functions, the cytoskeleton must be attached, in some way, to the plasma membrane. Indeed this is indicated from the results of many studies which have shown that the cytoskeleton cannot be dissociated from the plasma membrane with the nonionic detergent Triton X-100 under physiological conditions (Ben–Ze'ev et al., 1979; Koch and Smith, 1978; Mescher et al., 1981). That the cytoskeleton is attached to, and modulates the function of, the plasma membrane suggests that a molecular continuity between integral membrane proteins embedded in the lipid bilayer and the cytoskeleton may provide a basis for establishing and maintaining distinct functional domains on the plasma membrane. Furthermore, since the subcortical cytoskeleton is continuous with elements of the cytoskeleton distributed throughout the cytoplasm, the interaction with the plasma membrane may provide an anchorage site for the cytoskeleton involved in integrating cellular functions not directly associated with the plasma membrane.

This review brings together the results of several recent studies that have identified spectrin as a common component of the subcortical cytoskeleton in a wide variety of cells, and which, by virtue of the fact that it interacts with membrane proteins, is involved in mediating attachment of the cytoskeleton to the plasma membrane. The analysis of the expression and assembly of spectrin during terminal differentiation provides, therefore, a model system to understand the mechanism(s) for establishing and maintaining membrane–cytoskeleton domains in general.

2. THE RED CELL: A WELL-DEFINED EXAMPLE OF MOLECULAR CONTINUITY BETWEEN THE PLASMA MEMBRANE AND THE CYTOSKELETON

The mammalian red cell has a distinctive biconcave disc shape that undergoes dramatic changes as the cell squeezes through capillaries during blood circulation and yet returns to normal when not compressed. This property of the red cell appears to be the result of the molecular interaction between the cytoskeleton and integral membrane proteins which has been biochemically characterized in detail (reviewed by Branton *et al.*, 1981).

Red cell plasma membranes are easily isolated in the form of ghosts by hypotonic lysis of whole cells to remove the hemoglobin (Dodge *et al.*, 1963). Further extraction in the presence of Triton X-100 produces a detergent-insoluble matrix or cytoskeleton which still retains the approximate size and shape of the intact red cell (Yu *et al.*, 1973). The red cell cytoskeleton consists predominantly of actin and the actin-binding protein spectrin, the latter accounting for 75% of the protein mass (Lux *et al.*, 1976). Spectrin is composed of two nonidentical polypeptides, termed α-spectrin (M_r 240,000) and β-spectrin (M_r 220,000), that self-associate to form an $(\alpha\beta)_2$ tetramer (for references, see Branton *et al.*, 1981). The spectrin tetramer binds to and cross-links F-actin (Brenner and Korn, 1979) and, through the β-subunit, is attached to another protein, termed ankyrin, which is itself directly bound to the anion transporter which is embedded in the lipid bilayer (Tyler *et al.*, 1979; Bennett and Stenbuck, 1979a,b; 1980). These specific protein interactions are thought to confer two important properties on the plasma membrane. First, the interaction between actin, spectrin, and other cytoskeletal proteins forms a highly cross-linked protein matrix which provides high tensile strength to the plasma membrane. Second, the interaction between this matrix and the anion transporter provides a topologically invariant anchorage site for the cytoskeleton on the plasma membrane and results in the immobilization of the anion transporter in the plane of the membrane. The molecular continuity thus established confers structural support to the plasma membrane that allows reversible deformation of the cell during blood circulation.

3. THE MOLECULAR ORGANIZATION OF THE RED CELL CYTOSKELETON: UNIQUE OR A PARADIGM FOR NONERYTHROID CELLS?

The possibility that proteins, other than actin, structurally and functionally analogous to the well-characterized proteins of the red cell cytoskeleton are present in nonerythroid cells has always been theoretically attractive (see discussions by Branton et al., 1981; Bennett, 1982). This is due to the obvious parallels between the membrane-associated functions of the cytoskeleton in the red cell and those in nonerythroid cells. However, a molecular basis for this idea has only recently been established experimentally, as proteins antigenically related to spectrin have been shown to be expressed in nonerythroid cells.

4. SPECTRIN IN NONERYTHROID CELLS: STRUCTURAL AND BIOCHEMICAL PROPERTIES SIMILAR TO ERYTHROCYTE SPECTRIN

Two experimental approaches have been taken to identify proteins analogous to erythrocyte spectrin in non-erythroid cells. First, antibodies have been raised against purified α- and β-subunits of avian erythrocyte spectrin and used as probes to identify antigenically related proteins in avian tissues. Second, high-molecular-weight proteins have been purified and characterized from several tissues and subsequently shown to be structurally related and functionally analogous to erythrocyte spectrin (see Lazarides and Nelson, 1982). These studies have established that spectrin is expressed in almost all tissues and cells with the possible exception of smooth muscle, the microvilli of the intestinal brush border, and tracheal cilia (Table I) (Repasky et al., 1982).

Nonerythroid spectrin has many structural and functional properties similar to those of erythrocyte spectrin. It forms an extended rod consisting of two parallel strands with an overall length of about 200 nm in most cells (Glenney et al., 1982a; Bennett et al., 1982) but 264 nm in the intestinal terminal web (Glenney et al., 1982a). This structure is consistent with biochemical data in-

TABLE I. Distribution and Subunit Composition of Spectrin in Cells and Tissues[a]

Spectrin subunit	α-Spectrin (240,000)	TW260 (260,000)	γ-Spectrin (235,000)	β'β-Spectrin (225/220,000)
Erythrocytes	+	−	−	+
Central nervous system	+	−	+	+[b]
Peripheral nervous system	+	−	+	ne
Intestinal epithelium	+	+	+	−
Tracheal epithelium	+	ne	ne	−
Proventriculus epithelium	+	ne	ne	−
Liver epithelium	+	ne	+	−
Kidney epithelium	+	ne	ne	−
Lens epithelium	+	−	+	−
Endothelial cells	+	ne	+	−
Neuroblastoma	+	ne	+	ne
Macrophages	+	ne	+	ne
Lymphocytes	+	ne	+	−
Skeletal muscle	+	ne	−	+
Cardiac muscle	+	ne	−	+
Fibroblasts	+	ne	+	−
Myoblasts	+	−	+	−
Gizzard	−	ne	ne	−

[a] Data compiled from Repasky et al. (1982); Nelson and Lazarides (1983a,b); Glenney et al. (1982a); Burridge et al. (1982); Lehto and Virtanen (1983); and Bennett et al. (1982). Molecular weight in parentheses. ne, Not examined.
[b] In the cell bodies of granule and Purkinje cells and cells of the cerebellar nuclei; also retinal ganglion cells and cones.

dicating that spectrin exists *in situ* as a tetramer formed from the head-on association of two identical dimers (Bennett *et al.*, 1982). Nonerythroid spectrin tetramers also bind and cross-link F-actin (Glenney *et al.*, 1982a,b; Bennett *et al.*, 1982; Burridge *et al.*, 1982; Burns *et al.*, 1983) and are closely associated with the cytoplasmic face of the plasma membrane (Repasky *et al.*, 1982; Bennett *et al.*, 1982; Burridge *et al.*, 1982; Levine and Willard, 1981; Lehto and Virtanen, 1983; Nelson and Lazarides, 1983a). Thus, erythrocyte spectrin and the spectrin analogs characterized in nonerythroid cells appear to have similar biophysical properties indicating that they may have similar biological functions, namely to cross-link actin filaments in the subcortical cytoplasm and to mediate linkage of actin filaments to the plasma membrane. In addition, spectrin binds calmodulin (Glenney *et al.*, 1982c; Kakiuchi *et al.*, 1982) and, therefore, may play a role in regulating the concentration of Ca^{2+} in the subcortical cytoplasm, as well as its own structural dynamics in a Ca^{2+}-sensitive manner.

5. SUBUNIT COMPOSITION OF SPECTRIN: POLYMORPHISM CORRELATED WITH CELL-TYPE SPECIFICITY

All avian nonerythroid cells, in which spectrin analogs have been identified, express a common spectrin subunit, which has an identical molecular weight (Repasky *et al.*, 1982; Palfrey *et al.*, 1982) and peptide map (Glenney *et al.*, 1982c; Nelson *et al.*, 1983a) to avian erythrocyte α spectrin; this subunit is referred to as α-spectrin. Mammalian nonerythroid cells also express a common α-subunit of spectrin, but which has a completely different peptide map than mammalian erythrocyte α-spectrin (Bennett *et al.*, 1982). This indicates that mammalian erythrocyte spectrin has diverged considerably during evolution of the red cell from its counterpart in nonerythroid cells and from avian erythrocytes, which are nucleated and contain an intermediate filament network (Granger *et al.*, 1983).

The spectrin subunit expressed in association with the α-subunit to form the spectrin tetramer is polymorphic and exhibits cell-type specificity in its cellular expression and distribution. Three polymorphic subunits have been identified: a M_r-260,000 subunit, termed TW260, which is unique to the terminal web of the intestinal brush border (Glenney *et al.*, 1982a); a M_r-220/225,000 subunit, termed β/β'-spectrin, whose expression and distribution are limited to erythrocytes, adult cardiac and skeletal muscle, and certain areas of the central nervous system (Nelson and Lazarides, 1983a; Lazarides and Nelson, 1983a); and a M_r-235,000 subunit, termed γ-spectrin (Nelson *et al.*, 1983b), which is found in all other cells where α-spectrin is expressed (Levine and Willard, 1981; Lehto and Virtanen, 1983; Burridge *et al.*, 1982; Nelson *et al.*, 1983a,b; Lazarides and Nelson, 1983a). These polymorphic subunits are antigenically distinct and have different peptide maps (Glenney *et al.*, 1982c; Nelson and Lazarides, 1983a; Nelson *et al.*, 1983a), indicating they are the products of different genes.

Although the subunit composition of spectrin is clearly cell-type specific, each combination of subunits, as noted earlier, forms a tetramer with similar biophysical properties. Thus the question arises as to the functional significance of the cell-type specificity of the polymorphic subunit. A possible answer comes from our understanding of the molecular organization of the erythrocyte cytoskeleton, in which the spectrin–actin cytoskeleton is linked to the membrane-bound ankyrin–anion transporter complex through the β-subunit of the spectrin tetramer; the α-subunit does not appear to play a direct role in the binding reaction (see Section 2). Therefore, the functional significance of the polymorphic subunit may be to regulate the assembly of the spectrin–actin cytoskeleton onto subunit-specific membrane-bound receptor proteins, which may themselves be associated with a cell-type-specific function(s) analogous to the anion transporter in erythrocytes.

6. ASSEMBLY OF SPECTRIN DURING TERMINAL DIFFERENTIATION: A MODEL SYSTEM TO UNDERSTAND THE ASSEMBLY OF DISTINCT MEMBRANE–CYTOSKELETON DOMAINS

Spectrin has two features that make it an ideal model system to analyze during cellular differentiation as a means of understanding how membrane–cytoskeleton domains are assembled and established. First, spectrin is linked to the plasma membrane and binds actin and is, therefore, a marker for the expression and distribution of the membrane-associated cytoskeleton. Second, the expression of spectrin subunits is cell-type specific, which can be used to correlate the assembly and establishment of spectrin-based membrane–cytoskeleton domains with the terminal differentiation of particular cell types.

The assembly of spectrin has been analyzed in detail in three differentiation systems: erythropoiesis (Blikstad et al., 1983; Moon and Lazarides, 1983), myogenesis (Nelson and Lazarides, 1983b) and neuronal terminal differentiation (Lazarides and Nelson, 1983a,b). Each of these systems provides a different perspective of how the spectrin subunit composition is regulated during terminal differentiation and the mechanism by which spectrin is assembled onto the plasma membrane; together they suggest a general mechanism for the assembly of membrane–cytoskeleton domains.

6.1. Erythropoiesis: Assembly of a Topologically Uniform Membrane–Cytoskeleton Domain

For the purposes of this review, the membrane–cytoskeleton of the erythrocyte is considered to comprise a single homogenous domain, whose molecular composition and topological distribution are uniform. Therefore, the analysis of spectrin assembly during erythropoiesis provides the simplest example of the assembly of a membrane–cytoskeleton domain.

The stable assembly of a spectrin tetramer onto the erythrocyte plasma membrane during erythropoiesis is regulated by two assembly-limiting steps (Blikstad et al., 1983; Moon and Lazarides, 1983). First, the extent of assembly of the α- and β-subunits onto the membrane is limited by the amount of β-spectrin synthesized; α-spectrin is not limiting since it is synthesized in an approximately threefold excess of β-spectrin throughout erythropoiesis. Second, the assembly of β-sectrin onto the plasma membrane is limited by the availability of binding sites (receptors) on the membrane, which in the erythrocyte comprise the ankyrin–anion transporter complex. Preliminary studies indicate that the assembly-limiting component at this step is most likely the anion transporter. The assembly of newly synthesized subunits onto the plasma membrane is very rapid and results in the stabilization of the spectrin (cytoskeleton) complex. All components not assembled onto the plasma membrane, including any excess α-spectrin, are rapidly degraded in the cytosol. These observations allow the conclusion that the plasma membrane, and its constituent cytoskeleton–receptor proteins, regulate the stable assembly of the two spectrin subunits during erythropoiesis (Blikstad et al. 1983; Moon and Lazarides, 1983; for details see Chapter 5 by Moon et al., in this volume).

As noted in Section 2, the primary function of the erythrocyte cytoskeleton is to structurally support the plasma membrane. In order to sustain this function during forced changes in cell shape, the spatial distribution of the cytoskeleton must be uniform over the entire cytoplasmic face of the plasma membrane. This may be established during erythropoiesis by virtue of the initial distribution in the plane of the membrane of the assembly-limiting membrane receptor protein, the anion transporter. After synthesis and transport to the membrane, the distribution of the anion transporter in the plane of the membrane may be limited initially only by the fluidity of the surrounding lipid matrix in regions of the plasma membrane that contain little or no spectrin, as has been shown for concanavalin A receptors (Tokuyasu et al., 1979); this establishes a random distribution of the anion transporter in the membrane. However, the gradual assembly of spectrin onto the ankyrin–anion transporter complex and the subsequent cross-linking of the cytoskeletal matrix during erythropoiesis causes the immobilization of the anion transporter in the membrane. This may also have a secondary effect on other membrane proteins not directly linked to the cytoskeleton, whose mobility in the plane of the membrane has been shown to decrease during erythropoiesis (Shekman and Singer, 1976).

Thus, during the establishment of the interconnection of the membrane and the cytoskeleton, not only is the extent of assembly but also the distribution of the cytoskeleton limited by the membrane receptor. There is, however, a subsequent reciprocal effect of the cytoskeleton on its receptor, which is to limit the mobility of the receptor in the plane of the membrane.

6.2. Distribution and Assembly of Spectrin in Muscle Cells

6.2.1. Distribution of Spectrin in Muscle: Control of Membrane Deformability

Morphological observations of contracted and relaxed muscle over the past 30 years have revealed two important but unexplained features of skeletal muscle cell structure. First, the axial register of the Z lines and elements of the contractile apparatus across the myofiber is maintained at all times during contraction and relaxation. Second, the sarcolemma exhibits deformability such that in contracted muscle it forms outfoldings or scallops of membrane away from the myofiber in the region in between Z lines, whereas regions of the sarcolemma directly adjacent to the Z lines remain connected by filaments (Tiegs, 1954; Pierbon–Bormioli, 1981; Street, 1983). From these observations, it has been suggested that there exists a framework within the myofiber responsible for maintaining the structural and functional integrity of the cell during cycles of contraction and relaxation and for directly attaching the myofiber to the sarcolemma.

The existence of such a myofiber framework has been established in molecular terms only recently through the biochemical and immunological characterization of intermediate filament proteins and spectrin in muscle cells, which together have been shown to comprise the myogenic cytoskeleton (Granger and Lazarides, 1978, 1979; Lazarides, 1980; Repasky *et al.*, 1982; Nelson and Lazarides, 1983a,b). These studies indicate that the myogenic cytoskeleton is composed of two distinct domains (Fig. 1). The first is a transcytoplasmic domain composed predominantly of intermediate filaments, actin, and small amounts of spectrin which form a scaffold around the periphery of individual Z discs (Fig. 1) (Granger and Lazarides, 1978, 1979; Lazarides and Granger, 1983). It has been proposed that this Z-disc cytoskeletal scaffold mechanically integrates and unifies the contractile actions of the myofiber by physically linking together myofibrils laterally at their Z discs (Granger and Lazarides, 1978, 1979; Lazarides, 1980). The second cytoskeleton domain is composed of the erythroid form of spectrin ($\alpha\beta$-spectrin) and is located predominantly on the cytoplasmic surface of the sarcolemma. In relaxed skeletal muscle, the distribution of spectrin has been shown to be as an intricate gridlike network of prominent rings surrounding the myofiber, which are localized as foci on the sarcolemma directly adjacent to the Z lines and, to a lesser extent, the M lines. These rings are interconnected by regularly spaced lines of spectrin that extend along the sarcolemma in the long axis of the myofiber (Fig. 1) (Repasky *et al.*, 1982; Nelson and Lazarides, 1983a). In contracted muscle, the network of spectrin is still apparent, although

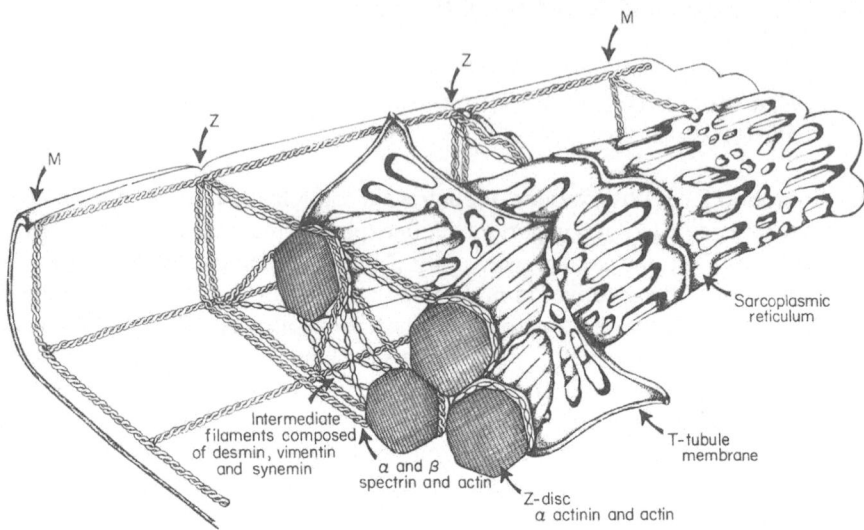

FIGURE 1. Schematic representation of a skeletal muscle myofiber with the interior exposed to show the distribution of the intermediate-filament proteins and spectrin in relationship to the Z discs and the sarcolemma.

the interval between the rings adjacent to the Z lines appears to decrease (Nelson and Lazarides, 1983a).

The distribution of spectrin on the sarcolemma, together with the fact that in erythrocytes it binds actin filaments and can become extensively cross-linked (see Section 2), strongly indicates that in muscle cells spectrin plays an important role in the maintenance of the shape and structural integrity of the sarcolemma during contraction and relaxation. We hypothesize that filaments emanating from the Z lines are bound to spectrin, which, through the attachment of spectrin to proteins in discrete areas of the sarcolemma adjacent to the Z lines, results in the establishment of a topologically invariant and inextensible anchorage site for the transcytoplasmic Z-disc cytoskeletal scaffold on the sarcolemma. As the myofiber shortens during muscle contraction and the interval between the Z line decreases, areas of the sarcolemma that contain little spectrin pucker to form outfoldings, whereas spectrin-containing foci directly adjacent to the Z lines remain attached to the myofiber (Fig. 2). Thus spectrin may be involved in the control of membrane deformability and in establishing a molecular continuity

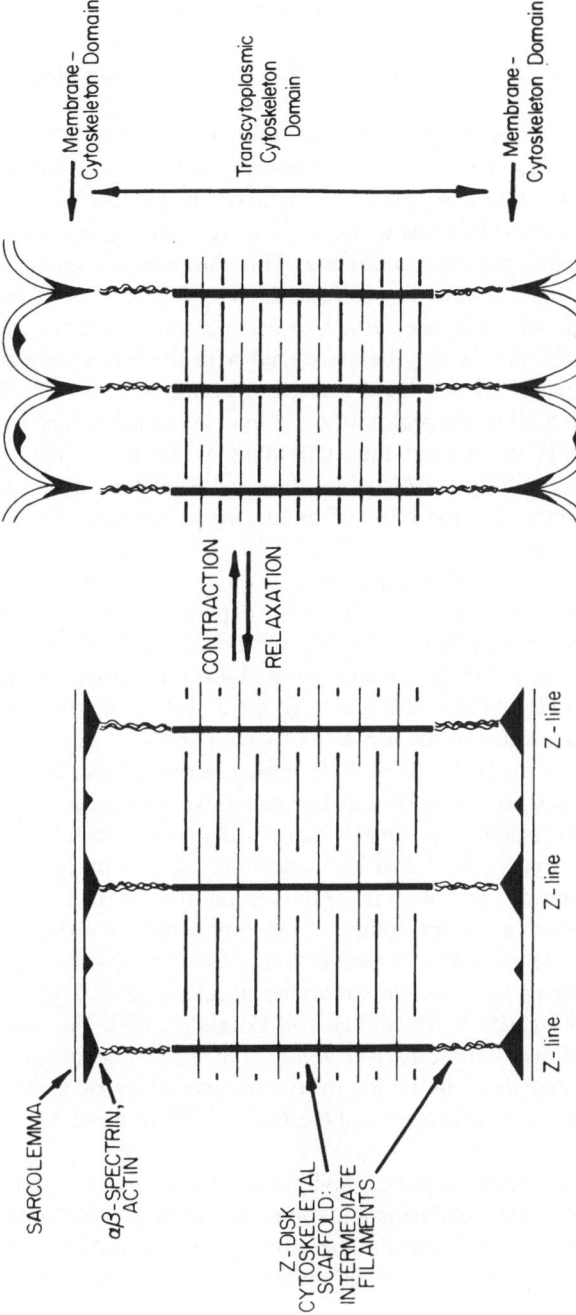

FIGURE 2. Molecular basis for membrane deformability and mechanical integration of the muscle fiber during contraction and relaxation (for details, see text).

between the myofiber at the level of the Z-line and the surrounding sarcolemma (Fig. 1).

If, as we suggest, such a molecular continuum of structural proteins exists around each sarcomeric unit, then it is possible that the resulting framework may additionally serve as a support upon which tension can be developed by the contractile apparatus. For this to work, however, the framework itself must develop tension during muscle contraction. This idea seems reasonable in light of the finding that muscle cells do, in fact, develop lateral tension during contraction in the areas where Z lines attach to the membrane (Street, 1983). This type of tension development may be associated with the sarcomeric framework and may be modulated by the local effect of calmodulin on the intracellular levels of Ca^{2+} and ATP in the region of the Z line. As noted earlier (see Section 3), calmodulin binds to spectrin thus providing a simple mechanism for the colocalization of calmodulin and elements of the cytoskeleton on the sarcolemma adjacent to the Z lines. On the basis of its known properties, calmodulin may modulate the biophysical properties of the sarcomeric framework in several ways. First, the interaction between spectrin and actin may be perturbed as a result of the phosphorylation of spectrin by a calmodulin-dependent spectrin kinase similar to that identified previously in erythrocytes (Huestis et al., 1981) and neuronal synapses (Sobue et al., 1982). Second, calmodulin may affect the local ionic environment and ATP levels in the region of the Z line by regulating the Ca^{2+} ATPase pump. Activiation of the ion pump by calmodulin is known to result in the depletion of intracellular levels of ATP, which, together with the Ca^{2+} influx prior to muscle contraction, may lead to the cross-linking of cytoskeletal proteins into high-molecular weight complexes that results in the stabilization of the sarcolemma Z-disc cytoskeletal scaffold connection as tension is developed by the contractile apparatus. This may make the region of the Z line less extensible than others and result in the festooning of the sarcolemma between the Z lines as the myofiber shortens. In this respect it should be noted that the cross-linking of cytoskeletal proteins has been demonstrated in ATP-depleted, Ca^{2+}-enriched erythrocytes (Palek et al., 1978; Sieting and Lorand, 1978). Furthermore, calmodulin has been shown indirectly to be involved in regulating erythrocyte shape; it has been postulated that this is due to its multiple levels of interaction with spectrin, a spectrin-specific kinase, and the Ca^{2+} ATPase pump (G. A. Nelson et al., 1983).

These studies, therefore, have provided for the first time a basis for understanding the molecular mechanism involved in the mechanical integration of the contractile actions of the muscle fiber and a new approach to elucidate the mechanism of tension development and membrane deformability in contracting muscle.

6.2.2. Myogenesis: Assembly of Two Interconnected Cytoskeleton Domains during Myogenesis

The spectrin phenotype expressed in fully differentiated muscle cells has been shown to be exclusively $\alpha\beta'\beta$-spectrin, which is similar to that expressed in the erythrocyte, although there are slight differences in the relative electrophoretic mobilities and ratio of erythrocyte β' : β-spectrin and the antigenically related proteins in muscle (Nelson and Lazarides, 1983a). The analysis of the spectrin phenotype of mitotic myoblasts reveals that these cells, on the other hand, express $\alpha\gamma$-spectrin and little or no β-spectrin (Nelson and Lazarides, 1983b), which appears to be similar to the phenotype expressed in other mitotic cells and also in brain tissue (Levine and Willard, 1981; Bennett et al., 1982; Burridge et al., 1982; Glenney et al., 1982c; Nelson et al., 1983b). We have found that the expression of the fully differentiated spectrin phenotype is developmentally regulated and requires the gradual switch in the subunit composition of spectrin from $\alpha\gamma \rightarrow \alpha\beta$-spectrin during myogenesis (Nelson and Lazarides, 1983b).

Pulse-labeling experiments followed by immunoprecipitation of newly synthesized spectrin subunits have revealed that as the myogenic cell becomes postmitotic and enters the phase of terminal differentiation the synthesis of $\beta'\beta$-spectrin is initiated. During terminal differentiation, the amount of β-spectrin in the cell gradually increases whereas that of γ-spectrin decreases, even though the synthesis of γ-spectrin is continuous throughout myogenesis. This results eventually in the predominance of the $\alpha\beta$-spectrin phenotype in the adult muscle cell. Pulse-chase experiments have shown that this differential accumulation of β-spectrin over γ-spectrin appears to result from the gradual stabilization of $\alpha\beta$-spectrin and the relatively rapid degradation of newly synthesized γ-spectrin. α-Spectrin does not appear to play a significant role in the switch of the polymorphic spectrin subunit, since its rate of synthesis and stability are similar initially to those of γ-spectrin in the myoblast, and later to those of β-spectrin in the mature myotube (Nelson and Lazarides, 1983b).

From these data, and by analogy to the spectrin assembly-limiting steps in erythrocytes, we have postulated that the assembly and differential stabilization of the two spectrin phenotypes is regulated at the posttranslational level by the availability on the plasma membrane of receptors specific for γ- and β-spectrin, respectively (Nelson and Lazarides, 1983b). Thus in the myoblast, the $\alpha\gamma$-spectrin phenotype is expressed as a result of the coordinate synthesis of γ-spectrin and its specific membrane receptor which allows the assembly and stabilization of newly synthesized $\alpha\gamma$-spectrin on the membrane. Subsequently, the synthesis of γ-spectrin and its specific membrane receptor at the onset of, and during, terminal differentiation results in the gradual assembly and stabilization of αβ-spectrin. During the same period, the amount of the γ-spectrin-specific membrane receptor declines thus causing newly synthesized, but unassembled, γ-spectrin to be degraded relatively rapidly. An essential feature of this model is that the assembly and stabilization of αβ- or αγ-spectrin is independent of the relative rates of synthesis and the size of the pools of unassembled γ- and β-spectrin and is driven solely by the affinity of the membrane receptor for β- or γ-spectrin, respectively; α-spectrin is not limiting since it is present in sufficient amounts to saturate all available β- and γ-subunits on the membrane. Hence the cell can assemble αβ-spectrin and begin to switch its spectrin phenotype, even though αγ-spectrin is present initially in vast excess over αβ-spectrin (Nelson and Lazarides, 1983b). Experimental support for some of these ideas comes from *in vitro* reconstitution experiments with erythroid spectrin, which indicate that the assembly of α-spectrin onto the erythrocyte membrane is limited by the availability of β-spectrin. The assembly of both subunits is concentration independent (i.e., is not driven by mass action) but is most likely dependent on the high affinity of β-spectrin for its membrane receptors (ankyrin–anion transporter) (Moon and Lazarides, 1983; see also Section 6.1). However, in the case of the assembly of muscle spectrin, the presence of subunit-specific, assembly-limiting membrane receptors for spectrin remains to be proven experimentally.

The expression and accumulation of the αβ-spectrin phenotype during myogenesis coincides with a dramatic reorganization of the myogenic cytoskeleton. Initially, this takes the form of a reorganization of spectrin as elements of the spectrin matrix begin to coalesce to form a gridlike network on the plasma membrane (Fig. 3) (Nelson and Lazarides, 1983b). By analogy to the erythrocyte, this may be the result of increased cross-linking of adjacent spectrin tetramers by actin as more αβ-spectrin accumulates on the sarcolemma, which may also cause a gradual decrease in the observed mobility and concomitant localized clustering of other proteins in the plane of the membrane (e.g., acetylcholine receptors; Axelrod *et al.*, 1976). Eventually, the spectrin network forms ring structures around each Z line, which are interconnected by lines of spectrin in the long axis of the myofiber.

Concomitant with the assembly of the spectin network on the sarcolemma, the intermediate filament network also undergoes a reorganization in its subunit composition and spatial distribution (Fig. 3). Initially, in the mitotic myoblast, intermediate filaments are composed predominantly of vimentin and are randomly distributed throughout the cytoplasm. However, upon terminal differentiation, the synthesis of desmin is augmented such that during myogenesis there is a gradual but incomplete switch in the predominant intermediate-filament subunit

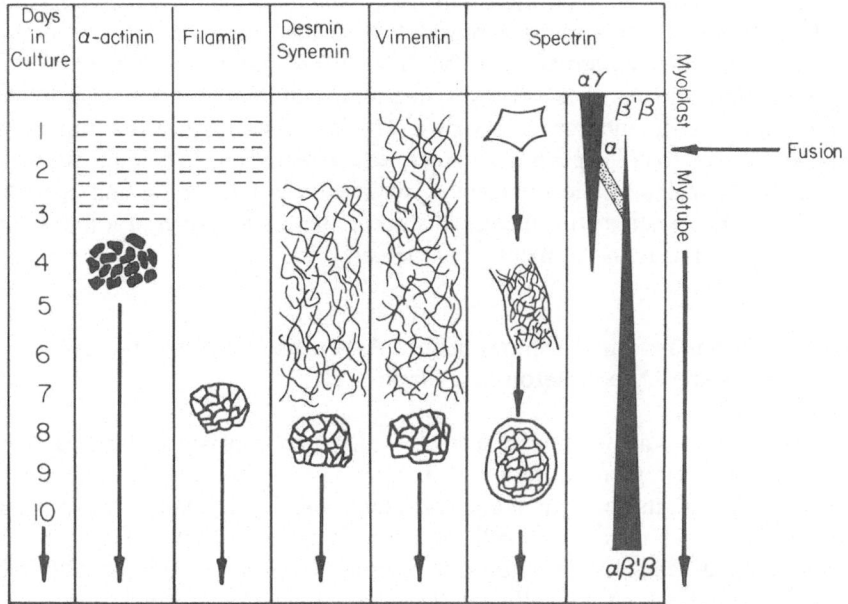

FIGURE 3. Diagram indicating the presence and morphology of various components of the myogenic cytoskeleton during myogenesis. Dashed lines indicate the speckled appearance of antigens
on actin filament bundles; blanks indicate the absence of antigen; and wavy lines indicate a random
distribution of antigen on the membrane (spectrin) or in a filamentous form in the cytoplasm (vimentin,
desmin, and synemin). Solid patterns indicate that the antigen is found in the interior of adult Z
discs; open patterns indicate the localization of the antigen at the periphery of adult Z discs; solid
lines for spectrin indicate localization of the antigen on the plasma membrane. The solid triangles
represent the relative amounts and accumulation of $\alpha\gamma$- and $\alpha\beta'\beta$-spectrin during myogenesis; note
that α-spectrin is common to both phenotypes (speckled band). The diagram is based on data from
Gard and Lazarides (1980); Granger and Lazarides (1980); Gomer and Lazarides (1981); and Nelson
and Lazarides (1983b).

from vimentin to desmin (Gard and Lazarides, 1980). Later in myogenesis, and
after the formation of the α-actinin-containing Z-line striations, vimentin and
desmin gradually disappear from the cytoplasm and become associated with the
peripheries of the Z discs (Fig. 3) (Gard and Lazarides, 1980).

Thus, the temporal and spatial assemblies of the spectrin and intermediate
filament cytoskeleton domains have several similar characteristics: (1) there is
a gradual switch of the predominant subunit upon terminal differentiation from
the phenotype expressed in the mitotic cell to that of the postmitotic, fully

234 W. JAMES NELSON and ELIAS LAZARIDES

differentiated cell; and (2) during terminal differentiation there is a reorganization of the networks from a random distribution in the mitotic cell into a highly ordered interconnected structure in the fully differentiated cell. The molecular trigger(s) that initiates these changes in subunit composition and distribution upon terminal differentation is unknown. However, we suggest that this event may be common to both spectrin and intermediate filaments, so that the assembly of the two domains of the myogenic cytoskeleton is temporally and spatially coordinated to coincide with the establishment and maintenance of a functional contractile apparatus in the mature muscle cell.

6.3. Neuronal Terminal Differentiation: Assembly of an Anisotropic Membrane Cytoskeleton in a Single Cell

The neuron is an example of a cell in which the cytoplasm is clearly divided into two distinct regions in terms of morphology and function. One region is the cell body with its dendritic processes, which contains the nucleus and protein-synthesizing machinery of the cell and is the main region onto which synaptic connections are established with other neurons. Emanating from the cell body is the second region of the cell comprising the axon, which extends away from the cell body until it establishes a terminal synaptic connection(s) with another neuronal cell body or its dendritic processes. One of the main functions of the axon is the transport of proteins and organelles synthesized and assembled in the cell body to the terminal synapse(s).

The molecular basis for establishing and maintaining these two morphologically and functionally distinct regions of the neuron is poorly understood. Recently, we have taken a new approach to this problem by analyzing the distribution and composition of spectrin as a well-defined marker to probe for heterogeneity in the molecular composition of the neuronal membrane-associated cytoskeleton.

The distribution of spectrin in neurons is most clearly demonstrated in the cerebellum in which distinct layers of axons and cell bodies can be distinguished. These layers comprise, from the outside of the cerebellar cortex inward, the molecular, Purkinje cell, and granule cell layers (Fig. 4) (Jacobson, 1978). We have shown that cerebellar neurons contain two forms of spectrin at steady state, $\alpha\beta'\beta$-spectrin (erythroid spectrin) and $\alpha\gamma$-spectrin (also termed fodrin or brain spectrin), which are topologically segregated on the plasma membrane (Lazarides and Nelson, 1983a). Whereas $\alpha\gamma$-spectrin is found in all discernible cell bodies and axons that populate all three layers of the cerebellar cortex, $\alpha\beta'\beta$-spectrin is confined to the plasma membrane of cell bodies of Purkinje and granule cells and the neurons of the cerebellar nuclei (Fig. 4). In addition, $\alpha\beta'\beta$-spectrin is present in the initial portion of the dendritic trunks that emanate from the Purkinje cell bodies and extend a short distance into the molecular layer. $\alpha\beta'\beta$-Spectrin is not found in the rest of the axons and dendritic processes that populate the molecular layer, or in the axonal processes of mossy and climbing fibers, or in the axons of Purkinje cells present in the white matter (Fig. 4) (Lazarides and Nelson, 1983a).

These results provide, for the first time, evidence of an anisotropy in the molecular composition of the neuronal membrane cytoskeleton. To determine whether the establishment of this anisotropy is correlated with the onset of synaptogenesis and neuronal terminal differentiation, we have analyzed the temporal expression of the two forms of spectrin during development of the cerebellum (Lazarides and Nelson, 1983b). Neuronal differentiation in the cerebellum occurs in three phases. An initial proliferative phase is followed by a postmitotic phase during which time the cell body of each cell type migrates to a specific area of the cortex. Once the cell body reaches its designated position, the cell undergoes terminal differentiation with the establishment of functional synaptic connections with other neurons (Jacobson, 1978). We have observed that the expression of $\alpha\gamma$-spectrin is constitutive during all phases of neuronal differentiation in both mitotic and postmitotic cells. On the other hand, $\alpha\beta'\beta$-spectrin is absent from all mitotic cells and is expressed only when these cells become postmitotic and, specifically, when they have entered their phase of terminal differentiation; at this time $\alpha\beta'\beta$-spectrin accumulates on the plasma membrane of the cell bodies but is absent from the axon (Lazarides and Nelson, 1983b). In fact, the temporal and spatial appearance of $\alpha\beta'\beta$-spectrin coincides remarkably with the establishment of synaptic contacts on the cell bodies and initial portion of the dendritic trunks of Purkinje and granule cells and the neurons of the cerebellar nuclei (Table II). These contacts constitute the two major input and one major output relay synapses of the cerebellar cortex, respectively (Fig. 4) (Jacobson, 1978; Mugnaini, 1969; Shimono *et al.*, 1976). Thus, the molecular event(s) that triggers the establishment of the anisotropy in the molecular composition of spectrin appears to be associated with the formation of a functional synaptic contact on the cell body. We have postulated that this event is marked by the appearance of a membrane receptor specific for the β-subunit of spectrin at the site of a synaptic contact, which drives the stable assembly of newly synthesized $\alpha\beta'\beta$-spectrin onto the plasma membrane of the cell body (Lazarides and Nelson, 1983b) (Fig. 5). On the other hand, the membrane receptor for γ-spectrin is not confined to the cell body, and thus, $\alpha\gamma$-spectrin is free to enter the axon during axonal transport (Fig. 5), as was shown initially by Levine and Willard (1981).

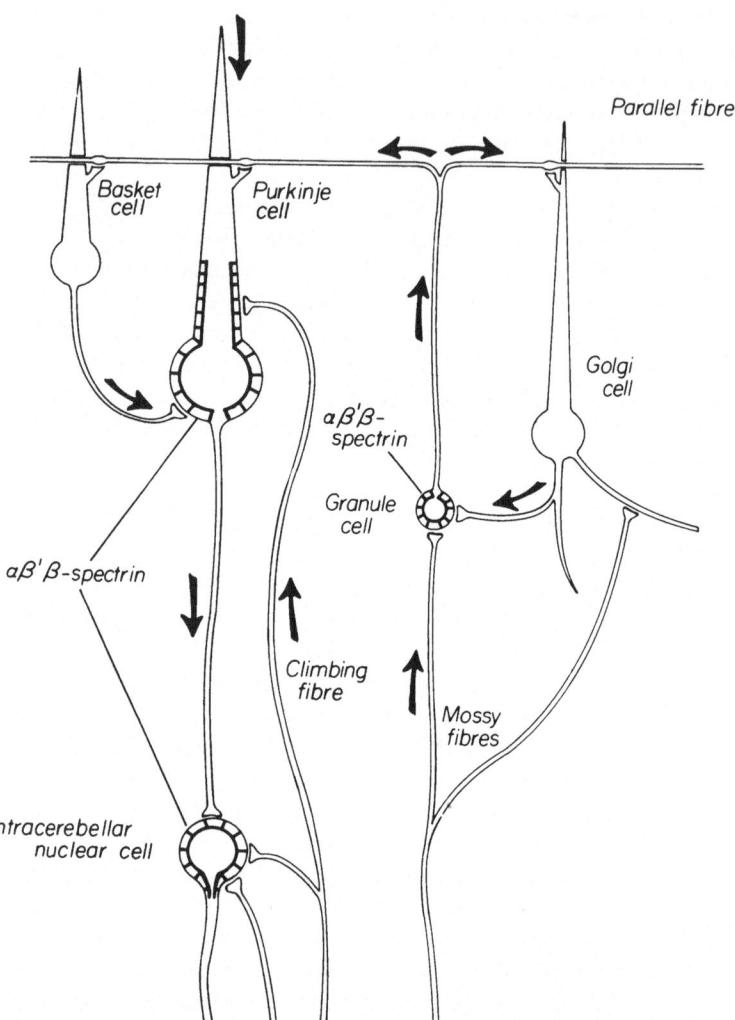

FIGURE 4. Schematic representation of the distribution of $\alpha\beta'\beta$-spectrin in the cortex of a fully developed adult cerebellum. The position of $\alpha\beta'\beta$-spectrin is indicated by a hatched lining on the inside of the Purkinje and granule cell bodies and the cell bodies of the intracerebellar nuclei. The distribution of $\alpha\gamma$-spectrin is thoughout all structures. The diagram also indicates the synaptic contacts between climbing and mossy fibers and the main neuronal elements of the cerebellar cortex and the neurons of the intracerebellar nuclei (adapted from Warwick Williams, 1973; based on data from Lazarides and Nelson, 1983a,b).

TABLE II. Temporal Expression of αγ- and αβ'β-Spectrin during Development of the Cerebellum in Chick Embryos[a]

Embryonic day	Cell type	Developmental position	α-Spectrin (γ-Spectrin)	β'β-Spectrin
15	Granule cell	External granule layer, dividing, no migration	+	−
	Purkinje cell	Migrated to Purkinje layer, no contact with incoming climbing fibers	+	−
16–17	Granule cell	A few cells postmitotic and migrating past Purkinje cell layer to form internal granule layer	+	−
	Purkinje cell	Contacts established with incoming climbing fibers	+	+, as foci close to cell body plasma membrane
17–19	Granule cell	Many cells in internal granule layer; contacts established with incoming mossy fibers	+	+, as foci close to cell body plasma membrane
	Purkinje cell	As day 16–17	+	+, as day 16–17
	Neurons of the cerebellar nuclei	Contacts established with axons of Purkinje cells and climbing fibers	+	+, as foci close to cell body plasma membrane

[a] From Lazarides and Nelson (1983b).

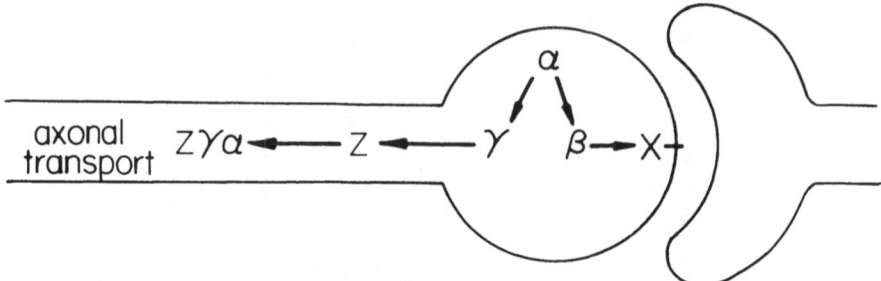

FIGURE 5. Receptor-driven spectrin subunit segregation in neurons. Upon synaptic contact, the appearance of a membrane protein (X) on the plasma membrane of the neuronal cell body drives the stable assembly of β-spectrin, and an equimolar amount of α-spectrin, onto the membrane, which results in the retention of αβ-spectrin in the cell body. On the other hand, receptor protein (Z), which may or may not be attached to the membrane, drives the stable assembly of γ-spectrin and an equimolar amount of α-spectrin. The resulting Z–αγ-spectrin complex is then transported down the axon. This results in an anisotropic distribution of αγ- and αβ-spectrin observed in neurons of the cerebellum (for details, see text).

6.4. Regulation of Spectrin Assembly during Terminal Differentiation: The Function of Membrane Receptors

The analysis of the subunit composition of spectrin during terminal differentiation of erythrocytes, muscle cells, and neurons clearly shows that the expression of different spectrin phenotypes is developmentally regulated (Fig. 6). The predominant, if not exclusive, spectrin phenotype of mitotic precursor cells is αγ-spectrin. On the other hand, postmitotic cells, which have entered the phase of terminal differentiation, express αβ-spectrin; in these cells, αγ-spectrin is either absent (erythrocytes), replaced by αβ-spectrin (muscle), or coexpressed but spatially segregated from αβ-spectrin (neurons). In this respect, it is interesting to note that epithelial cells of the intestinal brush border express the TW260-spectrin subunit together with α-spectrin in the terminal web (Glenney *et al.*, 1982a). On the basis of the changes in the spectrin phenotype of other terminally differentiating cells, we would expect that prior to and during terminal differentiation of the intestinal epithelium, these cells will express αγ-spectrin. However, upon commitment to differentiate, the synthesis of TW260 would be expected to commence and result in the accommodation of αTW260 in addition to αγ-spectrin.

Given the fact that αγ- and αβ-spectrin appear to have similar biophysical properties (see Section 4), what is the functional significance in the expression

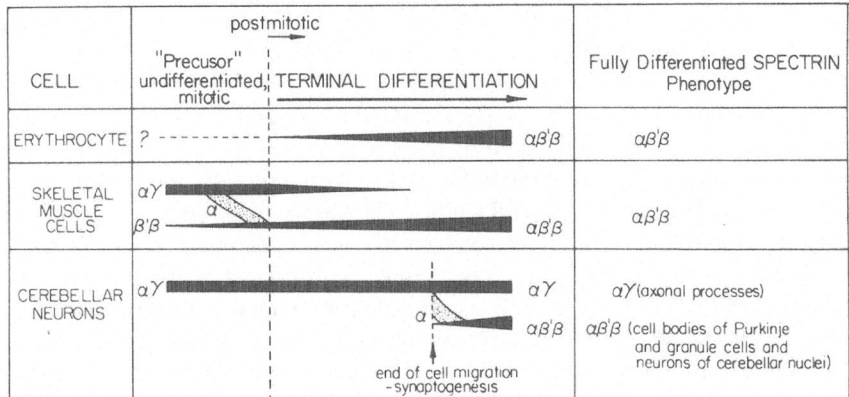

FIGURE 6. Diagram summarizing the expression and temporal accumulation of different spectrin phenotypes during erythropoiesis (Blikstad *et al.*, 1983), myogenesis (Nelson and Lazarides, 1983b), and neuronal terminal differentiation (Lazarides and Nelson, 1983b). The solid bars represent the relative accumulation and decline of spectrin phenotypes. It should be noted that where more than one phenotype is expressed, the α-subunit is common to each spectrin complex (speckled bars).

of αβ-spectrin in terminally differentiated cells? An answer to this question may be in the identity and function of the hypothesized membrane receptor(s) for β-spectrin. We have postulated previously that the switch in the subunit composition of muscle spectrin (Nelson and Lazarides, 1983b) and the segregation of αγ- and αβ-spectrin in neurons (Lazarides and Nelson, 1983a) are based on the expression and distribution of different membrane receptors specific for each of the polymorphic spectrin subunits. Although the identification of these receptors has not yet been established, we believe that this postulate is based on a logical interpretation of all existing data on the expression and assembly of spectrin during terminal differentiation (see Sections 6.1, 6.2, and 6.3). We suggest that the membrane receptor(s) for muscle and neuronal β-spectrin, by analogy to the anion transporter in erythrocytes, has a catalytic function in or on the plasma membrane, whose expression is required in a localized region of the terminally differentiated cell. We speculate that this function(s) may involve the regulation of intracellular ion fluxes and, perhaps, involves the release of Ca^{2+} ions in-response to extracellular stimuli. Thus the function of αβ-spectrin in muscle and neurons may be twofold, as it is in the erythrocyte (see Section 2): (1) to mediate the attachment of actin, and perhaps intermediate filaments, to the membrane; and (2) to limit the mobility of its membrane receptor in the plane of the membrane, which may result in the localization of the receptor in specific areas of the cell where its catalytic function is required, such as the postsynaptic area of neurons and at the area of the Z line in muscle cells.

The distribution and properties of αγ-spectrin during differentiation indicate that a membrane receptor for γ-spectrin must be relatively mobile in the plane of the membrane and may function, therefore, to provide a mobile anchorage site for the spectrin cytoskeleton on the plasma membrane. Experimental support for this notion comes from several studies: (1) during ligand-induced cell surface receptor capping in lymphocytes, αγ-spectrin has been shown to patch and cap coordinately with the cell surface receptors (Nelson et al., 1983b; Levine and Willard, 1983); (2) αγ-spectrin is a component of axonal transport (Levine and Willard, 1981); and (3) αγ-spectrin is found almost exclusively in cells or regions thereof where proteins are relatively mobile in the plane of the membrane (lymphocytes, myoblasts, fibroblasts, and neuronal axons) but is absent from cells or regions thereof where proteins are fixed in the plane of the membrane (mature erythrocytes, adult muscle cells). With these facts in mind, we suggest that the role of αγ-spectrin, in addition to cross-linking the cytoskeleton in the subcortical cytoplasm, may be to provide a localized site for the modulation of Ca^{2+} and ATP levels by calmodulin. This may affect the activity of the actomyosin ATPase (Shimo–Oka and Watanabe, 1981) and the phosphorylation of myosin light chain, which are known to be involved in initiating the interaction between actin and myosin to produce tension in the subcortical cytoplasm. This may be an important mechanism, for instance, in regulating the movement of cell surface receptors and the transport of proteins and organelles down the neuronal axon.

Clearly the identification of the membrane receptors for spectrin will help to elucidate the molecular mechanism for the assembly and establishment of spectrin in specific regions of the plasma membrane during terminal differentiation. Furthermore, together with the analysis of spectrin they may provide the first evidence for a mechanism involved in maintaining clusters of enzymes and receptor proteins in specific regions of the cell.

7. IMPLICATION OF SPECTRIN ASSEMBLY ON THE ASSEMBLY OF A MEMBRANE–CYTOSKELETON DOMAIN IN GENERAL: THE RECEPTOR CASCADE MODEL

The biochemical and morphological analysis of spectrin in the three differentiation systems outlined previously has provided, for the first time, a model system to understand the assembly and segregation of one molecular aspect of the membrane–cytoskeleton during cellular differentiation. Some of the main principles that have emerged from the study of spectrin assembly during differentiation may be applicable to the assembly and establishment of membrane–cytoskeleton domains in general. These include the following:

1. Assembly is regulated primarily at the posttranslational level.
2. The principal, if not exclusive, component that limits *stable* assembly is the plasma membrane (membrane receptors).
3. Assembly onto the membrane is a rapid, ordered process involving specific protein–protein interactions.
4. Assembled components on the membrane are stable.
5. Unassembled components are rapidly degraded in the cytosol (fluid phase catabolism).
6. The cellular distribution of the membrane-assembled proteins is determined initially by the membrane receptor, whose final distribution and/or appearance is determined as the cell becomes postmitotic and enters the phase of terminal differentiation.
7. The fully assembled cytoskeletal components have a subsequent reciprocal effect on the membrane receptor by limiting its mobility and distribution in the plane of the membrane.

With these principles in mind, we suggest a model for the assembly of a membrane–cytoskeleton domain, which we have termed the receptor cascade model (Fig. 7).

At a specific time during terminal differentiation, an integral membrane protein is synthesized and transported to the plasma membrane where it may perform a catalytic function and act also as a specific receptor for a cytoplasmic protein complex, in which there is a strict hierarchical order of components. The appearance of the membrane receptor initiates an assembly cascade of these proteins onto the membrane from small soluble pools in the cytoplasm. Owing to the hierarchical order of interactions, each newly assembled protein acts as a receptor for, and thus limits the assembly of, the next protein in the cascade. All excess, unassembled components of the receptor cascade are removed by fluid phase catabolism. The protein composition of a receptor cascade depends on the specificity of the membrane receptors for the first protein in the assembly cascade and, subsequently, the specificity of each newly assembled protein as a receptor for the next protein. In this way, the receptor cascade may include a discrete number of specific proteins, as in the case of the erythrocyte anion transporter–ankyrin–spectrin–actin cascade, or have the potential to include a variety of proteins depending on the relative sizes of the soluble cytoplasmic pools of newly synthesized proteins that can bind to each newly assembled component of the cascade. Each completed, membrane-stabilized complex forms a membrane–cytoskeleton domain which may be topologically segregated from

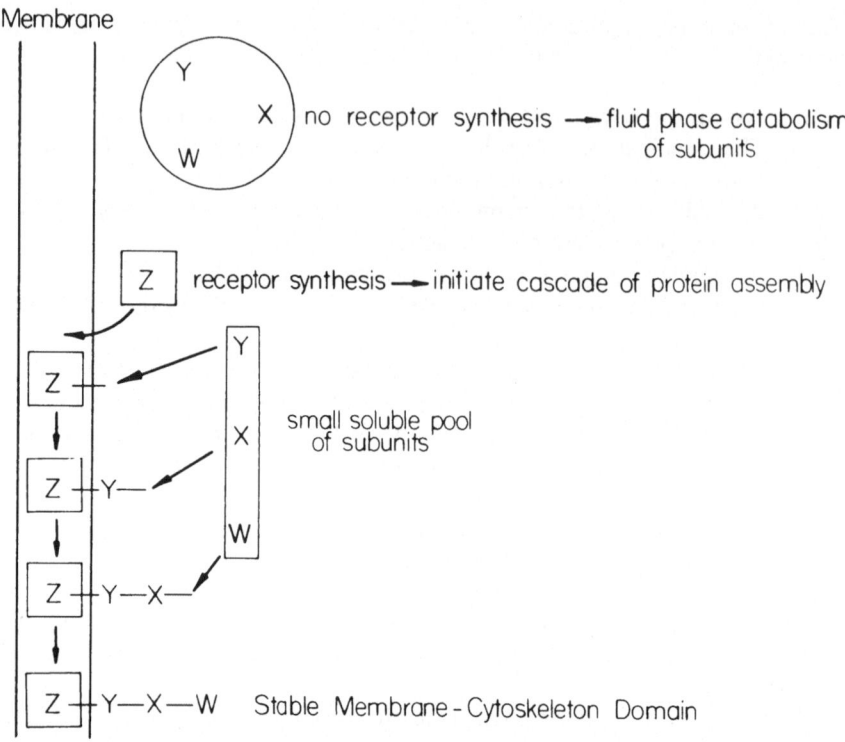

FIGURE 7. Assembly of a membrane–cytoskeleton domain: the receptor cascade model. The cytoplasmic components of the cascade are represented by Y, X, and W. In the absence of the membrane receptor Z, the cytoplasmic subunits are removed by fluid phase catabolism. The synthesis and insertion into the membrane of the receptor Z initiate a cascade of assembly in which each cytoplasmic component is assembled from a small soluble pool in the correct order: $Y \rightarrow X \rightarrow W$. The assembled components are then stable and constitute a membrane–cytoskeleton domain (for further details, see text).

or integrated with other domains in the same cell. The segregation of domains is brought about initially by differences in the lateral diffusion of the integral membrane receptor proteins in the lipid bilayer and subsequently by the cross-linking of elements of the cytoskeleton during the assembly of the domain. The latter process may play a role in maintaining a local concentration of integral membrane receptor proteins in particular areas of the cell where their catalytic function is required, as well as providing a topologically invariant anchorage site for the cytoskeleton on the plasma membrane.

We believe that the concept of receptor-driven protein assembly, as envisaged in the receptor cascade model, provides a useful approach to characterizing proteins in terms of their position and function(s) in a receptor cascade. A paradigm for this is the elucidation of the protein interactions involved in the receptor cascade that constitutes the red cell membrane–cytoskeleton, which has provided a molecular basis for understanding how the overall shape of the red cell is maintained. The characterization of receptor cascades in nonerythroid cells may, therefore, provide an insight into how structural and functional heterogeneity in the plasma membrane is established and maintained during terminal differentiation.

REFERENCES

Axelrod, D., Ravdin, P., Koppel, D. E., Schlesinger, J., Webb, W. W., Elson, E. L., and Podleski, T. R., 1976, Lateral motion of fluorescently labeled acetylcholine receptors in membranes of developing muscle fibers, *Proc. Natl. Acad. Sci. USA* **73**:4594–4598.

Bennett, V., 1982, The molecular basis for membrane–cytoskeleton association in human erythrocytes, *J. Cell. Biochem.* **18**:49–65.

Bennett, V., and Stenbuck, P., 1979a, Identification and partial purification of ankyrin, the high affinity membrane attachment site for human erythrocyte spectrin, *J. Biol. Chem.* **254**:2533–2541.

Bennett, V., and Stenbuck, P. J., 1979b, The membrane attachment protein for spectrin is associated with band 3 in human erythrocyte membranes, *Nature (London)* **280**:468–473.

Bennett, V., and Stenbuck, P. J., 1980, Association between ankyrin and the cytoplasmic domain of band 3 isolated from the human erythrocyte membrane, *J. Biol. Chem.* **255**:2540–2548.

Bennett, V., Davis, J., and Fowler, W. A., 1982, Brain spectrin, a membrane-associated protein related in structure and function to erythrocyte spectrin, *Nature (London)* **299**:126–131.

Ben–Ze'ev, A., Duerr, A., Solomon, F., and Penman, S., 1979, The outer boundary of the cytoskeleton: a lamina derived from plasma membrane proteins, *Cell* **17**:859–865.

Blikstad, I., Nelson, W. J., Moon, R. T., and Lazarides, E., 1983, Synthesis and assembly of spectrin during avian erythropoiesis: Stoichiometric assembly but unequal synthesis of α- and β-spectrin, *Cell* **32**:1081–1091.

Branton, D., Cohen, C. H., and Tyler, J., 1981, Interaction of cytoskeletal proteins on the human erythrocyte membrane, *Cell* **24**:24–32.

Brenner, S. L., and Korn, E. D., 1979, Spectrin-actin interaction. Phosphorylated and dephosphorylated spectrin tetramers crosslink F-actin, *J. Biol. Chem.* **254**:8620–8627.

Burns, N. R., Ohanian, V., and Gratzer, W. B., 1983, Properties of brain spectrin (fodrin), *FEBS Lett.* **153**:165–168.

Burridge, K., Kelly, T., and Mangeat, P., 1982, Nonerythrocyte spectrin: Actin–membrane attachment proteins occurring in many cell types, *J. Cell Biol.* **95**:478–486.

Dodge, J. T., Mitchell, C., and Hanahan, P. J., 1963, Preparation and chemical characteristics of hemoglobin-free ghosts of human erythrocytes, *Arch. Biochem. Biophys.* **100**:119–130.

Gard, D. L., and Lazarides, E., 1980, The synthesis and distribution of desmin and vimentin during myogenesis in vitro, *Cell* **19**:263–275.

Glenney, J. R., Glenney, P., Osborn, M., and Weber, K., 1982a, An F-actin and calmodulin-binding protein from isolated intestinal brush borders has a morphology related to spectrin, *Cell* **28:**843–854.

Glenney, J. R., Glenney, P., and Weber, K., 1982b, F-actin-binding and crosslinking properties of porcine brain fodrin, a spectrin-related molecule, *J. Biol. Chem.* **257:**9781–9787.

Glenney, J. R., Glenney, P., and Weber, K., 1982c, Erythroid spectrin, brain fodrin and intestinal brush border protein (TW260/240) are related molecules containing a common calmodulin-binding subunit bound to a variant cell type-specific subunit, *Proc. Natl. Acad. Sci. USA* **79:**4002–4005.

Gomer, R., and Lazarides, E., 1981, The synthesis and deployment of filamin in chicken skeletal muscle, *Cell* **23:**524–537.

Granger, B. L., and Lazarides, E., 1978, The existence of an insoluble Z disc scaffold in chicken skeletal muscle, *Cell* **15:**1253–1268.

Granger, B. L., and Lazarides, E., 1979, Desmin and vimentin coexist at the periphery of the myofibril Z disc, *Cell* **18:**1053–1063.

Granger, B. L. and Lazarides, E., 1980, Synemin: a new high molecular weight protein associated with desmin and vimentin filaments in muscle, *Cell* **22:**727–738.

Granger, B. L., Repasky, E. A., and Lazarides, E., 1983, Synemin and vimentin are components of intermediate filaments in avian erythrocytes, *J. Cell Biol.* **92:**299–312.

Huestis, W. H., Nelson, M. J., and Ferrell, J. E., Jr., 1981, Calmodulin-dependent spectrin kinase activity in human erythrocytes, in: *Erythrocyte Membranes: Recent Clinical and Experimental Advances* (W. C. Kruckeberg, J. W. Eaton, and G. J. Brewer, eds.), Liss, New York, pp. 137–152.

Jacobson, M., 1978, Comments on assembly of complex neuronal systems, in: *Developmental Neurobiology*, 2nd ed., Plenum Press, New York, pp. 75–92.

Kakiuchi, S., Sobue, K., Kanda, K., Morimoto, K., Tsukati, S., Tsukita, S., Ishikawa, H., and Kurokawa, M., 1982, Correlative biochemical and morphological studies of brain calspectin: A spectrin-like calmodulin-binding protein, *Biomed. Res.* **3:**400–410.

Koch, G. L. E., and Smith, M. J., 1978, An association between actin and the major histocompatibility antigen H-2, *Nature (London)* **273:**274–278.

Lazarides, E., 1980, Intermediate filaments as mechanical integrators of cellular space, *Nature* **283:**249–256.

Lazarides, E., and Granger, B. L., 1983, Transcytoplasmic integration in avian erythrocytes and striated muscle, *Modern Cell Biol.* **2:**143–162.

Lazarides, E., and Nelson, W. J., 1982, Expression of spectrin in nonerythroid cells, *Cell* **31:**505–508.

Lazarides, E., and Nelson, W. J., 1983a, Erythrocyte and brain forms of spectrin in cerebellum: distinct membrane-cytoskeletal domains in neurons, *Science* **220:**1295–1296.

Lazarides, E., and Nelson, W. J., 1983b, Erythrocyte form of spectrin in cerebellum: appearance at a specific stage in the terminal differentiation of neurons, *Science* **222:**931–933.

Lehto, V.-P., and Virtanen, I., 1983, Immunolocalization of a novel, cytoskeleton-associated polypeptide of M_r 230,000 daltons (p230), *J. Cell Biol.* **96:**703–716.

Levine, J., and Willard, M., 1981, Fodrin: axonally transported polypeptides associated with the internal periphery of many cells, *J. Cell Biol.* **90:**631–643.

Levine, J., and Willard, M., 1983, Redistribution of fodrin (a component of the cortical cytoplasm) accompanying capping of cell surface molecules, *Proc. Natl. Acad. Sci. USA* **80:**191–195.

Lux, S. E., John, K. M., and Karnovsky, M. J., 1976, Irreversible deformation of the spectrin-actin lattice in irreversibly sickled cells, *J. Clin. Invest.* **58:**955–963.

Mescher, M. F., Jose, M. J. L., and Balk, S. P., 1981, Actin-containing matrix associated with the plasma membrane of murine tumor and lymphoid cells, *Nature (London)* **289:**139–144.

Moon, R. T., and Lazarides, E., 1983, β-spectrin limits the assembly of α-spectrin onto membranes following synthesis in a chicken erythroid cell lysate. *Nature (London)* **305**:62–65.

Mugnaini, E., 1969, Ultrastructural studies on the cerebellar histogenesis. II. Maturation of nerve cell populations and establishment of synaptic connections in the cerebellar cortex of the chick, in: *Proceedings of the First International Symposium of the Institute for Biomedical Research: Neurobiology of Cerebellar Evolution and Development* (R. Llinas, ed.), Institute for Biomedical Research, Chicago, Illinois, 1969, p. 749–782.

Nelson, G. A., Andrews, M. L., and Karnovsky, M. J., 1983, Control of erythrocyte shape by calmodulin, *J. Cell Biol.* **96**:730–735.

Nelson, W. J., and Lazarides, E., 1983a, Expression of the β-subunit of spectrin in nonerythroid cells, *Proc. Natl. Acad. Sci. USA* **80**:363–367.

Nelson, W. J., and Lazarides, E., 1983b, Switching of the subunit composition of muscle spectrin during myogenesis in vitro, *Nature (London)* **304**:364–368.

Nelson, W. J., Granger, B. L., and Lazarides, E., 1983a, Avian lens spectrin: subunit composition compared with erythrocyte and brain spectrin, *J. Cell Biol.* **91**:1271–1276.

Nelson, W. J., Colaco, C. A. L. S., and Lazarides, E., 1983b, Involvement of spectrin in cell-surface receptor capping in lymphocytes, *Proc. Natl. Acad. Sci. USA* **80**:1626–1630.

Nicolson, G. Z., 1976, Transmembrane control of the receptors on normal and tumor cells. I. Cytoplasmic influence over cell surface components, *Biochim. Biophys. Acta* **457**:57–108.

Palek, J., Liu, S. C., and Liu, P. A., 1978, Crosslinking of the nearest membrane neighbors in ATP-depleted, calcium-enriched and irreversibly sickled red cells, in: *Erythrocyte Membranes: Recent Clinical and Experimental Advances* (W. C. Kruckeberg, J. W. Eaton, and G. J. Brewer, eds.), Liss, New York, pp. 75–88.

Palfrey, M. C., Schieber, W., and Greengard, P., 1982, A major calmodulin-binding protein common to various vertebrate tissues, *Proc. Natl. Acad. Sci. USA* **79**:3780–3784.

Pierbon–Bormioli, S., 1981, Transverse sarcomere filamentous systems: "Z- and M-lines," *J. Musc. Res. Cell Motil.* **2**:401–408.

Pollard, T. M., and Weihing, R. R., 1974, Actin and myosin and cell movement, *CRC Crit. Rev. Biochem.* **2**:1–65.

Repasky, E. A., Granger, B. L., and Lazarides, E., 1982, Widespread occurrence of avian spectrin in nonerythroid cells, *Cell* **29**:821–833.

Schliwa, M., 1981, Proteins associated with cytoplasmic actin, *Cell* **25**:587–590.

Shekman, R., and Singer, S. J., 1976, Clustering and endocytosis of membrane receptors can be induced in mature erythrocytes of neonatal but not adult humans, *Proc. Natl. Acad. Sci. USA* **73**:4075–4079.

Shimono, T., Nosaka, S., and Sasaki, K., 1976, Electrophysiological study on the postnatal development of neuronal mechanisms in the rat cerebellar cortex, *Brain Res.* **108**:279–294.

Shimo–Oka, T., and Watanabe, Y., 1981, Stimulatin of actomyosin Mg^{2+}-ATPase activity by a brain microtubule-associated protein fraction. High-molecular-weight actin-binding protein is the stimulating factor, *J. Biochem.* **90**:1297–1307.

Sieting, G. E., Jr., and Lorand, L., 1978, Ca^{++}-modulated crosslinking of membrane proteins in intact erythrocytes, in: *Erythrocyte Membranes: Recent Clinical and Experimental Advances* (W. C. Kruckeberg, J. W. Eaton, and G. J. Brewer, eds.), Liss, New York, pp. 25–32.

Singer, S. J., and Nicolson, G. L., 1972, The fluid mosaic model of the structure of cell membranes. *Science (Washington, D. C.)* **175**:720–731.

Street, S. F., 1983, Lateral transmission of tension in frog myofibers: A myofibrillar network and transverse cytoskeletal connections are possible transmitters, *J. Cell. Physiol.* **114**:346–364.

Sobue, K., Kauda, K., and Kakiuchi, S., 1982, Solubilization and partial purification of protein kinase systems from brain membranes that phosphorylate calspectin, *FEBS Lett.* **150**:185–190.

Tiegs, O. W., 1954, The flight muscles of insects—their anatomy and histology; with some obser-
    vations on the structure of striated muscle in general, *Roy. Soc. Lond. Phil. Trans. Ser. B*
    238:221–348.
Tokuyasu, K. T., Shekman, R., and Singer, S. J., 1979, Domains of receptor mobility and en-
    docytosis in the membranes of neonatal human erythrocytes and reticulocytes are deficient in
    spectrin, *J. Cell Biol.* **80:**481–486.
Tyler, J., Margreaves, W., and Branton, D., 1979, Purification of two spectrin binding proteins:
    biochemical and electron microscopic evidence for site-specific reassociation between spectrin
    and bands 2.1 and 4.1 to spectrin, *J. Biol. Chem.* **255:**7034–7039.
Warwick, R., and Williams, P. L. (eds.), 1973, *Gray's Anatomy.* Saunders, London, p. 871.
Weihing, R. R., 1979, The cytoskeleton and plasma membrane, in: *Methods and Achievements in
    Experimental Pathology; Cel Biology,* Volume 8 (G. Gabbiani, ed.), Karger, Basel, pp. 42–109.
Yu, J., Fischman, D. A., and Steck, T. L., 1973, Selective solubilization of proteins and phos-
    pholipids from red blood cell membrane by nonionic detergents, *J. Supramol. Struct.* **1:**233–248.

7

ROLE OF THE β-ADRENERGIC RECEPTOR IN THE REGULATION OF ADENYLATE CYCLASE

Murray D. Smigel, Elliott M. Ross, and Alfred G. Gilman

1. INTRODUCTION

Hormones circulating outside of cells influence the metabolic activities of enzymes within them. Although the existence of these interactions has been known, at least in rudimentary form, for close to a century, the identities of the proteins responsible for their mediation are just now being established. Among the many hormonally controlled systems that have been studied, the β-adrenergic receptor–adenylate cyclase complex holds an important place. Studies of it have, in many instances, shaped our ideas about the mechanisms of the signaling process. In part, this has been due to the great variety of chemical compounds that are known to either stimulate or block β-adrenergic receptors. The availability of these compounds has allowed the pharmacological definition of a variety of subclasses of adrenergic receptors. Prominent among them are the α and β receptors and the more recently established distinctions within each of these classes (Langer, 1977).

The development of radioactive ligands that interact with β-adrenergic receptors has allowed the identification and purification of macromolecules with affinities for β-adrenergic agonists and antagonists that constitute presumptive evidence for their assignment as at least a component of the β-adrenergic receptor. After a stormy beginning, the measurement of β-adrenergic receptors by radio-

Murray D. Smigel, Elliott M. Ross, and Alfred G. Gilman • Department of Pharmacology, University of Texas Health Science Center at Dallas, Dallas, Texas 75235.   Work supported by United States Public Health Service Grants NS18153 and GM 30355 and by American Cancer Society Grant BC240D.   Elliott M. Ross is an Established Investigator of the American Heart Association.

247

ligand binding techniques is now a routine part of the characterization of many tissues. The demonstration by Sutherland, Rall, and co-workers of β-adrenergic effects in broken cell preparations forms the foundation of many of the current biochemical investigations of the actions of adrenergic hormones and neuro-transmitters (Rall *et al.*, 1957; Murad *et al.*, 1962; Sutherland *et al.*, 1962). Their discovery of cyclic AMP as the mediator of many of the effects of β-adrenergic agonists has proven to be of fundamental importance. Although not all such effects may be so mediated (Maguire and Erdos, 1980), the emphasis of this review will be directed at understanding how β-adrenergic agonists control the synthesis of cyclic AMP.

1.1. Scope of This Review

The extent of the literature that relates to the β-adrenergic receptor far exceeds the range of our knowledge; no attempt will be made to summarize much of it. In particular, the reader is directed to Lands *et al.* (1967) and Furchgott (1972) for information on the pharmacology of receptor subtypes. The development of the methods used for measuring the binding of β-adrenergic ligands to membranes and cells is covered in depth by Maguire *et al.* (1977) and by Williams and Lefkowitz (1978). Only a superficial attempt will be made to deal with the earlier literature on adenylate cyclase; the reader is directed to the reviews by Perkins (1973) and by Ross and Gilman (1980) for further information. Regulation of the concentration and localization of receptors, although of unquestionable interest and importance, will not be addressed here (see Harden, 1983). The developments to be emphasized are those that shed light on the processes by which ligand binding influences the synthesis of cyclic AMP. We will therefore discuss in some detail the recent work that has defined mechanisms of activation of the stimulatory guanine nucleotide-binding regulatory protein (G_s) of adenylate cyclase. Understanding of these mechanisms is necessary to evaluate both the currently active models of the "coupling" process between ligand binding and enzyme activation and the recently developed methods used to assay the activity of β-adrenergic receptors in reconstituted membrane systems. Indeed, it is our firmly held belief that progress in understanding the β-adrenergic receptor is critically dependent on development of such methods for reconstitution. Study of the binding of ligands to membranes and cells is now a mature technology; the techniques needed to isolate the binding proteins of interest and the proteins with which they interact are also well developed. The current focus must be on functional assays that assess the competence of the isolated proteins as mediators of β-adrenergic responses. These considerations are the directing factors that have influenced our choice of topics in this review.

1.2. Overview of the Hormone-Sensitive Adenylate Cyclase System

In science, as in many other pursuits, hindsight can be a valuable aid to understanding. Although mindful of the possibility of biasing the reader toward our interpretations of the experiments that describe the organization and control of the β-adrenergic receptor–adenylate cyclase complex, we feel that a preliminary sketch of our conclusions will provide a framework for interpretation of the remainder of this discussion.

In an appropriate preparation of plasma membranes from a hormone-sensitive cell, adenylate cyclase activity can be stimulated by the addition of GTP plus one or more hormones, depending on what receptors are present. Activity can also be stimulated by fluoride or by nonhydrolyzable analogs of GTP, such as guanosine-5'-(β,γ,imino)triphosphate [Gpp(NH)p] or guanosine-5'-(3-0-thio)triphosphate (GTPγS). Other hormones promote the inhibition of adenylate cyclase, and the action of these agents also requires GTP.

The complex of receptors, regulatory proteins, and enzymes that mediates these phenomena resides within the plasma membrane of the hormone-sensitive cell and usually consists of the following components:

1. Receptors (e.g., β-adrenergic receptors) for one or more species of hormone, neurotransmitter, or autacoid that activate adenylate cyclase. These proteins bind stimulatory agonists and their antagonists.
2. A stimulatory, guanine nucleotide-binding regulatory protein, G_s, which mediates hormone receptor-induced stimulation of the catalyst of adenylate cyclase, C. G_s is a heterodimer of a 45,000-Da subunit (α), which binds guanine nucleotides, and a 35,000-Da subunit (β), which does not. The α subunit of G_s is the predominant substrate for the NAD-dependent ADP-ribosylation reaction that is catalyzed by cholera toxin. This covalent modification of the α subunit enhances the ability of G_s to activate the catalyst of adenylate cyclase. G_s is also the site of action of fluoride.* Exposure of G_s to nonhydrolyzable analogs of GTP or to fluoride endows it with the ability to stimulate the activity of the catalyst.
3. The catalytic protein of the adenylate cyclase system has not yet been purified. Although preparations of C that are devoid of G_s are almost totally unable to utilize MgATP as a substrate, some activity is observed in the presence of MnATP. Addition of the diterpene forskolin results in a dramatic stimulation of the catalytic activity of C.
4. Receptors for one or more hormones that inhibit adenylate cyclase may

---

* Sternweis and Gilman (1982) demonstrated that Al^{3+}, in μM concentrations, is an essential cofactor for activation of G_s by F^-. It is likely that the actual stimulatory ligand is AlF_4^-. The significance of this finding is totally obscure.

also be present. These proteins are the site of binding of inhibitory agonists and their antagonists.

5. A guanine nucleotide-binding regulatory protein, G_i, which is distinct from G_s, mediates guanine nucleotide-dependent inhibition of adenylate cyclase. This protein, which has been purified, is a heterodimer of a 41,000-Da (α) subunit and a 35,000-Da (β) subunit. The α subunit of G_i binds guanine nucleotides and is the substrate for ADP ribosylation catalyzed by a toxin from *Bordetella pertussis*. ADP ribosylation of G_i blocks the ability of inhibitory hormones to alter adenylate cyclase activity and often potentiates the action of stimulatory hormones. The β subunit of G_i appears to be identical to that of G_s.

The purification of G_s and G_i has allowed a detailed investigation of their biochemical behavior. This work has permitted the formulation of a model of the interactions of the components of the adenylate cyclase system. Although not all-inclusive, this model rationalizes much of the available information and permits predictions about the behavior of the system that should be valuable in the design of experiments that may cause its downfall. A fundamental tenet of the model is that both G_s and G_i normally exist in the membrane as intact heterodimers and that one result of the action of the appropriate hormone–receptor complex is the dissociation of their subunits. In our view, it is the free, guanine nucleotide-liganded α subunit of G_s that binds to the catalyst to form the active adenylate cyclase complex. This model predicts that the β subunit acts as an inhibitor of such activation. Since G_s and G_i appear to have identical β subunits, we propose that inhibitory hormones act by causing the dissociation of G_i and that the β subunit so released may associate with and thereby inhibit the activity of the α subunit of G_s.

The activation of adenylate cyclase by cholera toxin may be mediated, at least in part, by its promotion of the dissociation of the subunits of G_s. In contrast, the action of pertussis toxin in preventing receptor-mediated inhibition of adenylate cyclase is attributed to the increased affinity of the ADP-ribosylated α subunit of G_i for the β subunit. The stabilized heterodimer is unable to dissociate under the influence of the inhibitory agonist–receptor complex.

G_s catalyzes the hydrolysis of GTP. Although purified G_s is an extremely poor GTPase, its activity can be stimulated 10- to 20-fold by the addition of an agonist–receptor complex. Hormone-stimulated GTPase activities have been found in several preparations of cell membranes and presumably reflect the activities of the guanine nucleotide-binding regulatory proteins contained within them. Such activity is stimulated by hormones that activate adenylate cyclase, as well as by those that inhibit the enzyme; thus both G_s and G_i appear to be receptor-regulated GTPases. ADP ribosylation of either G_s or G_i by the appropriate toxin results in inhibition of hormone-stimulated GTPase activity.

2. MEASUREMENT OF LIGAND BINDING TO β-ADRENERGIC RECEPTORS

The β-adrenergic receptor was initially defined by its specificity for ligands, assayed first by the initiation of function by agonists and later by the blockade of function by antagonists. Many investigators have now probed the ligand-binding properties of the receptor more directly. Most commonly and to little end, radioligand binding assays have been used to "identify" β-adrenergic receptors in dozens of preparations of tissues previously known to display β-adrenergic regulation of function. Of more utility, binding assays have been used to assay receptors during their purification; to identify ligand-binding subunits by covalent labeling; to probe changes in affinity for agonists that reflect interaction with other molecules; and to monitor the synthesis, turnover, and sequestration of receptors in living cells.

Although the measurement of reversible binding of radioactive ligands to β-adrenergic receptors had a checkered start, Levitzki et al. (1974) were able to demonstrate appropriately specific binding of [³H]propranolol to turkey erythrocyte membranes a decade ago. Two more generally useful β-adrenergic ligands were introduced soon thereafter: [¹²⁵I](±)iodohydroxybenzylpindolol (IHYP) by Aurbach's group (Aurbach et al., 1974; Brown et al., 1976a,b) and [³H](−)dihydroalprenolol (DHA) by Lefkowitz and coworkers (Lefkowitz et al., 1974; Mukherjee et al., 1975). The evaluation of these compounds, their characteristics, and methods for their use have been reviewed elsewhere (Maguire et al., 1977; Wolfe et al., 1977; Williams and Lefkowitz, 1978). Both compounds are antagonists. Both are commercially available, although the iodination of HYP is easy. The advantages of [³H]DHA are the long half-life of the tritium label, its frequently low nonspecific binding, and its ability to bind reversibly to detergent-solubilized receptors (Mukherjee et al., 1975; Caron and Lefkowitz, 1976). Its low specific activity can be a major detraction when only small amounts of receptor are available, however, and it may interact with α-adrenergic sites (Guellaen et al., 1978). Although [¹²⁵I]IHYP offers high specific activity, it does not bind to soluble receptors (Haga et al., 1977a; Fleming and Ross, 1980; Witkin and Harden, 1981). The basis of the latter fact is unclear, but comparison of the binding of [¹²⁵I]IHYP and [³H]DHA has served as a useful criterion for the incorporation of receptors into membranes (Fleming and Ross, 1980; Pedersen and Ross, 1982). Since preparations of IHYP are racemic mixtures, care must be taken to ensure that the binding measured and the concentrations calculated are those of the active, presumably (−)-isomer (Bürgisser et al., 1981a,b).

Since the finding by Bearer et al. (1979) that [¹²⁵I]IHYP is iodinated on the indole ring (and not on the benzyl ring, as was originally presumed), two other pindolol-based ligands have been introduced: [¹²⁵I]iodopindolol (IPIN)

(Barovsky and Brooker, 1980; Ezrailson *et al.*, 1981) and [^{125}I]iodocyanopindolol (ICYP) (Engel *et al.*, 1981). Both are antagonists, both are commercially available or are readily prepared by the user, and both are of high specific activity. [^{125}I]IPIN is also available as the resolved (−)-isomer. [^{125}I]IPIN and [^{125}I]ICYP are probably most notable for their very low nonspecific binding when used to assay receptors in intact cells, a situation in which they are markedly more specific than either [^{3}H]DHA or [^{125}I]IHYP (Barovsky and Brooker, 1980). Witkin and Harden (1981) have also used [^{125}I]IPIN to assay soluble receptors.

Clearly, it would be of great value to utilize agonist, as well as antagonist, radioligands. Their development has been disappointing. Such ligands would greatly facilitate measurements of the kinetics of agonist binding, agonist-specific alterations in affinities, and related parameters. Williams and Lefkowitz (1977) demonstrated that the agonist [^{3}H]hydroxybenzylisoproterenol (HBI, Cc34) can be used as a specific label for β-adrenergic receptors if the proper precautions are taken. Nevertheless, the marginally useful affinity of HBI at physiological temperatures, its low radiochemical specific activity, and its preference for the β$_2$-adrenergic receptor subtype have limited its application. [^{3}H]Isoproterenol, used by Malchoff and Marinetti (1976), shares or surpasses the deficiencies of [^{3}H]HBI.

Covalent affinity labeling of β-adrenergic receptors has become increasingly useful. Atlas and co-workers synthesized the compound [^{3}H]N-[2-hydroxy-3-(1-naphthyloxy)-propyl]-N′-bromoacetyl-ethylenediamine (NHNP-NBE) and demonstrated its use as an affinity reagent for the receptor (Atlas *et al.*, 1976; Atlas and Levitzki, 1978a). The photoaffinity labels [^{3}H]acebutalolazide (Wrenn and Homcy, 1980; Homcy *et al.*, 1983), [^{125}I]iodoazidobenzylpindolol (Rashidbaigi and Ruoho, 1981), and two similar azidophenyl derivatives of [^{125}I]ICYP (Burgermeister *et al.*, 1982) have also been described. Each of these compounds is relatively selective for one or more peptides presumably related to the receptor, and acebutalolazide has been shown to label a polypeptide in a purified preparation of receptors from canine lung (Homcy *et al.*, 1983). Lefkowitz, Caron, and colleagues have utilized both labeled *p*-azidobenzylcarazolol and [^{125}I]*p*-aminobenzyliodocarazolol plus a photoactivated cross-linking reagent to label the receptor, with roughly similar results (Lavin *et al.*, 1981, 1982). Since none of the compounds described above has been widely used in multiple laboratories, it is difficult to compare them critically. All appear to label at most a few polypeptides in the 40,000- to 65,000-Da size range; labeling is blocked by β-adrenergic ligands. Since radioiodinated azidobenzylcarazolol is now available commercially, it is likely to receive the widest testing and utilization. However, the use of photoaffinity reagents generlly sacrifices high efficiency of labeling for ability to label specifically. These compounds are therefore poorly suited for quantitative studies.

A novel approach to equilibrium binding measurements was suggested by the synthesis of 9-aminoacridin-propranolol, a fluorescent β-adrenergic antagonist (Atlas and Levitzki, 1977). The binding of fluorescent ligands to proteins can frequently be detected as a change in emission polarization, quantum yield, or lifetime or by a blue shift in the emission spectrum. Such assays may provide continuous kinetic data on association and dissociation reactions, as well as convenient and nondestructive equilibrium data. Aminoacridin-propranolol is probably not the ligand of choice for such studies, however, since its affinity for the receptor is not great, its emission is not markedly enhanced when it is bound, and its quantum yield and extinction coefficient are modest. Furthermore, although it was used as a histochemical probe for receptors (Atlas and Levitzki, 1978b, and references therein), its selectivity in such applications has been questioned (Barnes et al., 1980). Propranolol, which is weakly fluorescent, has also been tested for such use (Cherksey et al., 1980), but the results obtained do not seem to be consistent with this application. Henis et al. (1982a) used NBD-alprenolol as a fluorescent probe to label β-adrenergic receptors on the surface of Chang liver cells. Binding was apparently fairly specific, the intensity of the NBD fluorophore is high, and its emission and absorption maxima are well separated from interfering cellular components. This compound has not been evaluated or used in equilibrium ligand binding assays. However, Henis et al. (1982a) used 6 μM NBD-alprenolol in their experiments, suggesting that the compound has a poor affinity for the receptor.

3. COMPONENTS OF HORMONE-SENSITIVE ADENYLATE CYCLASE

3.1. Resolution of the β-Adrenergic Receptor and Adenylate Cyclase

The hormone-sensitive adenylate cyclase system has been resolved into a number of protein components. In 1977 three groups reported the separation of the hormone binding and the catalytic components of the β-adrenergic–adenylate cyclase complex. Haga et al. (1977a) reported that mouse lymphoma S49 cell membranes could be labeled with [^{125}I]IHYP and that this antagonist remained bound to a protein component of the membrane after solubilization in the nonionic detergent Lubrol 12A9. Gel exclusion chromatography or sedimentation of the proteins in such detergent extracts through 5–20% sucrose gradients containing Lubrol resulted in resolution of peaks of protein-bound IHYP and adenylate cyclase activity. Similar resolution of activities was achieved by Limbird and Lefkowitz (1977). They solubilized frog erythrocyte plasma membranes in digitonin and used gel exclusion chromatography to resolve hormone binding and adenylate cyclase activities. This work is especially interesting in that they could show the binding of [³H]DHA to the solublized receptor (see also Caron and Lefkowitz, 1976). Resolution of activities by affinity chromatography of the receptor was also achieved in the turkey erythrocyte system by Vauquelin et al. (1977).

3.2. The Structure of the β-Adrenergic Receptor

For the most part, the β-adrenergic receptor has been defined, both conceptually and experimentally, by its functions: the binding site for ligands that elicit β-adrenergic responses and the mediator by which these ligands stimulate adenylate cyclase. Although it is clear that a reasonably complete description of the receptor's function is dependent on knowledge of its structure, few groups have approached this problem with serious attempts to purify the receptor. Lefkowitz's group has made the most consistent progress in this direction. The problem is quantity. The best membrane preparations contain about 1 pmole of β-adrenergic binding sites per milligram protein, but a pure receptor of molecular weight 50,000 has a theoretical capacity of 20 nmole/mg. Thus, a 20,000-fold purification is required to achieve homogeneity—a task that demands starting with 200 g of membrane protein to yield 1 mg of pure receptor in 10% yield.

Success at purification has depended on affinity chromatography of digitonin-solubilized receptors using antagonist-agaroses: alprenolol-agarose, developed by Caron et al. (1979) and Vauquelin et al. (1977; see also Durieu–Trautmann et al., 1980), and acebutolol-agarose, pioneered by Homcy et al. (1983). Both Homcy et al. (1983) and Shorr et al. (1981, 1982a,b) have preparations of receptor from canine lung or either turkey or frog erythrocytes that approach homogeneity. For each preparation the binding of radiolabeled ligands is close to theoretical, and published profiles of sodium dodecylsulfate–polyacrylamide gels or isoelectric focusing gels display a single prominent band that is also labeled by radioactive photoaffinity ligands. The polypeptide composition of each preparation is listed in Table I, as is a comparison of photoaffinity labeling data from several laboratories. The general trend that emerges is that the β-adrenergic receptor consists of a single polypeptide in the 40,000- to 60,000-Da size range. In general, the β_2 subtype displays a higher molecular weight than does the β_1 subtype (Lavin et al., 1982; Rashidbaigi and Ruoho, 1982). The good correlation between values of molecular weight obtained from purified preparations and those obtained by labeling of membranes suggests that differences between species and between β_1 and β_2 receptors are not merely artifactual. These conclusions remain arguable, however, because of the extreme susceptibility of β-adrenergic receptors to proteolysis and because the observed bands on sodium dodecylsulfate–polyacrylamide gels are often broad. Most investigators routinely use phenylmethylsulfonylfluoride to inhibit serine proteases, and Shorr et al. (1982a) have used a mixture of five different protease

TABLE I.   Molecular Weights of β-Adrenergic Receptor Ligand-Binding Subunits

Source	Subtype[a]	Method[b]	$M_r \times 10^{-4c}$	Reference
Lung, dog	2	P, A	5.2–5.3	Homcy et al. (1983)
Lung, rat	2 (80%)	A	3.6, 4.7, 6.2	Lavin et al. (1982)
Lung, rabbit	1 (80%)	A	3.8, 4.5, 6.5	Lavin et al. (192)
Erythrocyte, frog	2	P, A	5.8	Shorr et al. (1981, 1982a);
(R. pipiens)				Lavin et al. (1981)
Erythrocyte, frog	2	A	6.0–6.7	Rashidbaigi and Ruoho (1982)
Reticulocyte, rat	2	A	5.3–6.0	Lavin et al. (1982)
Myoblasts, rat (LP6)	—	A	3.7–4.1	Atlas and Levitzki (1978a)
Erythrocyte, turkey	1	P, A	3.9–4.0, 4.5	Shorr et al. (1982b)
Erythrocyte, turkey	1	A	4.0	Burgermeister et al. (1982)
Erythrocyte, turkey	1	A	4.35	Rashidbaigi and Ruoho (1982)
Erythrocyte, pigeon	—	A	4.5–4.6, 5.35	Rashidbaigi and Ruoho (1982)
Erythrocyte, duck	—	A	4.5–4.6, 5.35	Rashidbaigi and Ruoho (1981)

[a] As reported by the authors.
[b] P, Purified protein, visualized by stain or by labeling with ^{125}I; A, affinity labeling of membranes.
[c] Multiple bands are designated by values of M_r separated by commas; diffuse or variable values of M_r are shown as a range of values separated by dashes. Values of M_r are taken directly from references.

inhibitors (see also Benovic et al., 1983). We have also found the latter approach to be helpful in maintaining receptors during purification.

The observation of broad and diffuse bands on sodium dodecylsulfate–polyacrylamide gels probably also reflects the glycosylation of receptors. Stadel et al. (1981) reported that the receptor from frog erythrocytes is a glycoprotein, based on its binding to wheat germ lectin–Sepharose and the prevention of that binding of α-methylglucosamine. However, there are no reports to date on the ability of glycosidases to decrease the width of the protein band observed after electrophoresis. Rashidbaigi and Ruoho (1982) noted that band width is variable from tissue to tissue. The causes of band broadening, be it glycosylation or other modification, may therefore vary with the preparation.

Shorr and co-workers (1982b) have reported the existence of two distinct forms of β-adrenergic receptors isolated from turkey erythrocytes, presumably a homogeneous cell population. The peptides, with molecular weights of 40,000 and 45,000, both bind [^{125}I]ICYP and are labeled by p-azidobenzylcarazolol. The relationship of the two proteins is unknown. The authors were unable to convert the larger form into the smaller by exposure to proteases, but whether the two differ in carbohydrate content, by a polypeptide fragment, or by some other modification is unclear. Further modification of the receptor is suggested by the finding of Stadel et al. (1983) that receptors from frog erythrocytes are phosphorylated, perhaps at multiple sites. The phosphorylation appears to be associated with desensitization of the cells to β-adrenergic agonists (see Harden, 1983).

Few physicochemical data are available on the receptor. Some hydrody-namic studies are consistent with the existence of the detergent-solubilized re-ceptor as a monomer (Caron and Lefkowitz, 1976; Limbird and Lefkowitz, 1977; Haga *et al.*, 1977a; Shorr *et al.*, 1981), but difficulty in determination of the relevant effects of detergent binding mars interpretation. No data on amino acid or carbohydrate composition or on primary structure are yet available.

3.3. Resolution of the Catalytic and Regulatory Components of Adenylate Cyclase

Since the pioneering observations of Rodbell and co-workers, who dem-onstrated the requirement for both GTP and hormone for the activation of aden-ylate cyclase (Rodbell *et al.*, 1971a,b, 1975; Londos *et al.*, 1974; Rodbell, 1975; Rodbell *et al.*, 1975), there has been speculation that the system consists of both catalytic and GTP-binding regulatory domains. The interaction between the hor-mone receptor and the catalyst would then presumably occur through the in-tercession of this regulatory or transducing domain. This speculation has been confirmed in recent years, and the regulatory domain has taken form as a separate protein, which has been resolved and purified. Although conditions can be found under which adenylate cyclase activity can be solubilized and, to a limited degree, manipulated biochemically (Johnson and Sutherland, 1974; Neer, 1974; Haga *et al.*, 1977a; Neer, 1978; Stengel and Hanoune, 1980), it has become clear that at least two protein components are necessary for the expression of fluoride- and guanine nucleotide-sensitive adenylate cyclase activity. In 1977 Pfeuffer reported that he could partly resolve a guanine nucleotide-binding component of pigeon erythrocyte membranes from the catalytic component of adenylate cyclase (Pfeuf-fer, 1977). He bound the regulatory component to a GTP-affinity support and released it with Gpp(NH)p. This material stimulated the catalytic activity that flowed through the column.

Another line of evidence has come from investigations of the properties of genetic variants of the S49 mouse lymphoma cell line. These variants have been extremely valuable in the elucidation of the structure of the adenylate cyclase system (Bourne *et al.*, 1975a). S49 cells are killed when the intracellular con-centration of cyclic AMP is elevated. This cytocidal effect of cyclic AMP has allowed the isolation of numerous variant cell lines that are defective in various aspects of cyclic AMP metabolism and action (Coffino *et al.*, 1974; Bourne *et al.*, 1975b; Haga *et al.*, 1977b; Johnson *et al.*, 1979). One such variant line, cyc⁻, is phenotypically deficient in the synthesis of cyclic AMP upon stimulation by β-adrenergic agonists, prostaglandin E_1, or cholera toxin; all these agents markedly elevate cyclic AMP concentrations in wild-type cells. Furthermore, plasma membranes isolated from cyc⁻ cells are unable to synthesize cyclic AMP

in response to hormones, fluoride, or nonhydrolyzable analogs of guanine nucleotides, and the basal activity of adenylate cyclase is nearly undetectable. It
was thus assumed that cyc^- was deficient in adenylate cyclase. It was, however,
demonstrated that cyc^- S49 cells retain the complement of β-adrenergic receptors
that is characteristic of wild-type cells (Insel *et al.*, 1976).

Ross and Gilman then showed that a detergent extract of L-cell membranes,
which do not contain β-adrenergic receptors, could, when mixed with cyc^-
membranes, confer upon them the ability to synthesize cyclic AMP in response
to guanine nucleotides, fluoride, or β-adrenergic agonists (Ross and Gilman,
1977a; Ross *et al.*, 1978). Although the detergent extract did contain adenylate
cyclase activity, its inactivation by incubation at 37°C or by treatment with
sulfhydryl reagents did not eliminate the ability to reconstitute adenylate cyclase
activity in the cyc^- membranes. Further experiments quickly revealed that two
proteins were necessary for the expression of adenylate cyclase activity when
MgATP was the substrate: a labile, *N*-ethylmaleimide-sensitive component that
was retained in the cyc^- variant and another protein, stable to such treatment,
that could be extracted with detergent from the membranes of a variety of cells
(Ross and Gilman, 1977b; Ross *et al.*, 1978). This evidence, coupled with the
observation that cyclic AMP synthetic activity could be demonstrated in cyc^-
membranes with MnATP as substrate, prompted them to propose that the cyc^-
membrane retained the catalyst of adenylate cyclase but lacked a regulatory
protein that normally enables the catalyst to be stimulated by guanine nucleotides
or fluoride. We will refer to this regulatory protein as G_s. The core of a model
for regulation of adenylate cyclase activity thus became apparent:

$$G_s + C \rightleftharpoons G_sC \qquad (7.1)$$

3.4. Purification and Properties of G_s

Since it is now widely acknowledged that β-adrenergic regulation of adenylate cyclase activity is mediated by the interactions of G_s with the β-adrenergic
receptor and the catalyst, we will summarize here some of the information that
has been obtained about the regulatory protein. G_s has been purified from rabbit
liver plasma membranes (Northup *et al.*, 1980; Sternweis *et al.*, 1981) and from
turkey and human erythrocyte membranes (Hanski *et al.*, 1981; Hanski and
Gilman, 1982). The purification protocols for all three sources are similar, consisting of solubilization of membranes in sodium cholate, followed by ion exchange and gel filtration chromatography in the same detergent. The material is
then bound to a heptylamine–Sepharose column and is eluted with a gradient of
salt and cholate. This preparation is bound again to DEAE–Sephacel, the de-

tergent is switched to Lubrol 12A9, and G_s activity is eluted with salt. Further purification is sometimes achieved with an additional hydroxyapatite column. Overall, a several-thousand-fold purification is achieved with a 5–10% yield.

Gel electrophoresis in sodium dodecylsulfate shows that G_s from all three sources contains two prominent polypeptides with mobilities corresponding to those of proteins with molecular weights of 35,000 and 45,000. G_s from rabbit liver also contains variable amounts of a 52,000-Da polypeptide. The 45,000-Da and 52,000-Da subunits have been shown to have a binding site for guanine nucleotides; no guanine nucleotide binding activity has been demonstrated with the 35,000-Da subunit (Northup et al., 1982). Measurements of the slowly reversible binding of $[^{35}S]GTP\gamma S$ indicate that 1 mole of nucleotide is bound per 80,000 g of protein. The 45,000- and 52,000-Da polypeptides are, under the proper conditions, both specific substrates for ADP ribosylation catalyzed by cholera toxin (Northup et al., 1980; Schleifer et al., 1982). They appear to be, in sum, roughly stoichiometric with the 35,000-Da subunit, and measurements of the hydrodynamic properties of G_s in detergent solution indicate that it has a molecular weight of approximately 80,000; thus, we infer that G_s is a heterodimer of the 35,000-Da and either the 45,000-Da or 52,000-Da subunits. Much other work is consistent with this assignment (Smigel et al., 1982; Northup et al., 1983a,b). The 52,000-Da and the 45,000-Da polypeptides both appear to be missing in cyc⁻ membranes, and both are more acidic in another S49 cell variant that displays altered regulation of adenylate cyclase activity (UNC) (Schleifer et al., 1982). It thus seems possible that the two polypeptides are products of a single gene and that the larger of them is a precursor of the more prevalent 45,000-Da subunit of G_s.

3.5. Preparations and Properties of the Resolved Catalyst

Much less is known about the biochemistry of the catalytic moiety of adenylate cyclase, C, than about G_s. This fact is a reflection of the lability of the catalyst and the difficulty in its purification. The cyc⁻ cell line provides a convenient source of catalytic activity that is resolved genetically from the regulatory protein, and most of our information comes from work with cyc⁻ membranes. Studies of the catalyst solubilized from cyc⁻ membranes with various detergents indicate that the activity is extremely labile; typically more than 50% of the activity is lost during overnight storage at 0–4 °C. Several investigators have attempted to fractionate C in solutions containing Lubrol 12A9 (Lubrol PX), a gentle nonionic detergent. However, many anecdotal reports and several careful studies indicate that Lubrol (or Triton X-100) does not disrupt the C–G_s complex, and this detergent has not allowed effective resolution of C from G_s

(Pfeuffer, 1979; Howlett and Gilman, 1980; Neer and Salter, 1981). Although Storm and associates have reported a 50-fold purification of C in Lubrol-containing solutions, these procedures have evidently not yielded anywhere near a complete purification (Westcott et al., 1979).

These observations have prompted attempts to use ionic detergents to resolve G_s and C. Under the usual conditions of solubilization in cholate or deoxycholate, little catalytic activity is recovered. Recently, however, several preparations of chromatographically resolved catalyst have been described. Both Ross (1981) and Strittmatter and Neer (1980) succeeded in solubilizing catalytic activity from rabbit liver and bovine cerebral cortical plasma membranes using sodium cholate in the presence of high concentrations (0.5–0.6 M) of ammonium sulfate. The catalytic activity can then be resolved from G_s by gel exclusion chromatography in the presence of cholate and ammonium sulfate. A sulfobetaine derivative of cholic acid (CHAPS) (Hjelmeland, 1980) has also proven to be useful in solubilizing the catalyst in a relatively stable form (Bitonti et al., 1982). Unfortunately, none of these procedures has, to date, allowed a very effective purification of the catalytic protein.

A markedly more effective method for the purification of C has been reported recently by Pfeuffer and Metzger (1982). They have developed an affinity chromatographic procedure that utilizes forskolin coupled to a Sepharose support. (The properties of forskolin, a diterpene isolated from *Coleus forskolii*, are summarized below.) Catalytic activity in Lubrol extracts apparently binds selectively to forskolin-agarose and can then be eluted with detergent solutions containing forskolin. Although few details of this procedure were revealed, it will hopefully form the basis for the first large-scale purification of the catalyst.

Although it is not available as a pure preparation, some information has been obtained about the regulatory properties of the catalytic component of adenylate cyclase. It is likely that G_s is the principal physiological stimulator of the catalytic unit of adenylate cyclase, but other regulatory ligands are now known. Ross et al. (1978) noted that catalytic activity in cyc⁻ membranes was enhanced up to tenfold when assayed with MnATP, rather then with MgATP, as substrate. The chromatographically resolved catalyst is also stimulated by Mn^{2+} (Strittmatter and Neer, 1980; Ross, 1981). Some of the enhanced activity observed with Mn^{2+} may be due to increased stability of C in Mn^{2+}-containing solutions, but it is likely that Mn^{2+} acts as an allosteric activator of the catalytic unit (Johnson et al., 1976).

The finding that certain diterpenes from *Coleus forskolii* have effects that resemble those of β-adrenergic agonists prompted investigation of their actions on adenylate cyclase (see Seamon and Daly, 1981b). Forskolin activates the enzyme in both membranes and intact cells (Seamon et al., 1981). This activation equals or surpasses that achievable by reconstitution with G_s and, furthermore, does not require the regulatory protein (Seamon and Daly, 1981a). It is thus hypothesized that forskolin acts directly on C. Evidence in support of this notion comes from the work of Pfeuffer and Metzger (1982), since their forskolin affinity column does appear to remove catalytic activity from solution selectively. It certainly remains possible, however, that the binding site for the diterpene resides on a polypeptide associated with the catalyst. When present at low concentrations where its enhancement of catalytic activity is slight, forskolin enhances the stimulatory effects of G_s. Although some have thus presumed that forskolin also has effects on the regulatory protein, the logic of this argument escapes us, and there is no direct evidence to support it.

Adenosine is another compound that is thought to interact directly with the catalyst. Adenosine and certain related analogs can interact with cell surface receptors to either stimulate or inhibit adenylate cyclase activity by mechanisms that involve the relevant G proteins. In addition, however, adenosine and other analogs that retain an unmodified adenine moiety (e.g., 5'-deoxyadenosine and 2',5'-dideoxyadenosine) inhibit activity in the absence of GTP (Londos et al., 1979). This effect is noted in solubilized preparations, as well as in cyc⁻ membranes or with chromatographically resolved C (Johnson, 1982; Florio and Ross, 1983). It is likely that the locus of this purine group-specific inhibition, the "P site," is the catalytic unit—possibly the active site itself. However, understanding of the mechanism of P-site-mediated inhibition will probably await purification of the catalyst.

3.6. The Inhibitory Guanine-Nucleotide-Binding Regulatory Protein

Many hormones and neurotransmitters, including α_2-adrenergic and muscarinic agonists, opiates, and some prostaglandins, act at least in part by inhibiting adenylate cyclase. This hormone-mediated inhibition was shown to require guanine nucleotides, to be mediated by cell surface receptors, and to display many other phenomena that are characteristic of hormonal stimulation of the enzyme (see Jakobs et al., 1980, for a review). These and other observations led to the proposal that a guanine-nucleotide-binding regulatory protein mediates such inhibition of adenylate cyclase, and several lines of evidence argued against a role for G_s in the process. The most convincing demonstration of the existence of a separate regulatory protein came from the work of Ui and coworkers, who have investigated the mechanism of action of a toxin produced by Bordetella pertussis, the whooping cough pathogen. This toxin, which they named islet-activating protein (IAP), acts on adenylate cyclase systems in a wide variety of tissues to block the effects of hormones that inhibit adenylate cyclase and, in some cases, to potentiate the effects of hormones that activate the enzyme (Katada and Ui, 1979, 1980; Hazeki and Ui, 1981; Katada et al., 1982).

Katada and Ui (1982a,b) showed that IAP catalyzes the NAD-dependent ADP ribosylation of a 41,000-Da plasma membrane protein that is distinct from the substrate for cholera toxin. In membranes in which hormone-mediated inhibition of adenylate cyclase can be demonstrated, this inhibition is blocked under conditions where such ADP ribosylation occurs. The IAP substrate thus appeared to be a good candidate for the guanine-nucleotide-binding regulatory protein that mediates the inhibitory response.

Recently, the IAP substrate has been purified from rabbit liver plasma membranes. It is purified as a dimer of the 41,000-Da polypeptide in noncovalent association with a distinct 35,000-Da subunit (Bokoch *et al.*, 1983, 1984). The dimer copurifies with G_s through the initial DEAE and gel exclusion chromatographic steps. Amino acid analyses and peptide mapping have shown that the 35,000-Da (β) subunit of the IAP substrate is very similar or identical to that of G_s, whereas the α subunits of the two dimers are clearly homologous (Manning and Gilman, 1983). The IAP substrate contains one high-affinity binding site for guanine nucleotides, located on the α subunit. Studies of the functional properties of the IAP substrate (to be described below) indicate that it is, in fact, the inhibitory guanine-nucleotide-binding regulatory component of adenylate cyclase or G_i.

4. PERSISTENTLY ACTIVATED STATES: STABLE ACTIVATION OF G_s

When membranes that contain hormone-sensitive adenylate cyclase are assayed in the presence of GTP, the rate of synthesis of cyclic AMP rises quickly upon addition of hormone and declines very rapidly when hormone is removed or when antagonists of the hormone are added. Similarly, if membranes are first treated with hormone and GTP and the GTP is then removed by centrifugation, the membranes, when assayed, show the basal rate of cyclic AMP synthesis. When membranes are incubated with hormone and Gpp(NH)p or GTPγS (or with fluoride), the situation is markedly different. In these cases, even after the free ligands are effectively removed, a persistent elevation of adenylate cyclase activity is found. This pattern has been observed frequently in membranes from many different cells (Cuatrecasas *et al.*, 1975a,b; Schramm and Rodbell, 1975; Jacobs *et al.*, 1976; Sevilla *et al.*, 1976; Ross *et al.*, 1977). This persistently activated state is stable, sometimes for days, and is generally much more resistant to denaturation than is the nonactivated catalyst.

Measurement of the rate of formation of the persistently activated state have been of great importance in delineating the roles of guanine nucleotides and hormones in the activation of adenylate cyclase. Since formation of the irreversibly activated state allows a "trapping" of activated molecules and a concurrent depletion of nonactivated molecules, the effects of hormone on the rate of activation can be measured readily; in a rapidly reversible reaction no distinction can be made between the rate of activation of many molecules of adenylate cyclase and the degree of activation of a given molecule of the enzyme. (For a general discussion of this approach and its fruits, see Tolkovsky, 1983.)

With an increased understanding of the role of G_s in controlling the activity of the catalyst, questions arise as to the role of G_s in maintaining the persistently activated state. Among the possible mechanisms of persistent activation are a permanent activation of G_s with its reversible binding to the catalyst, a permanent association of G_s with the catalyst, and a transient association of G_s with the catalyst to cause its persistent activation. Although the conclusion is still not certain, the preponderance of data point to a stable G_s–C complex as the persistently activated species. Pfeuffer (1979) showed that the sedimentation properties of activated G_s were altered in the presence of C, indicating the GTPγS-dependent formation of a larger complex of the two. If there is an equilibrium binding interaction between activated G_s and C, the enhanced thermal stability of the persistently activated state suggests that the concentration of free (thermally labile) catalyst must be so low that the binding is essentially irreversible.

Resolved G_s mediates the activation of C by either guanine nucleotides or fluoride. This was first shown in experiments in which crude detergent extracts containing G_s were incubated with activating ligands in the absence of catalyst and were subsequently diluted out of the activating ligands and reconstituted with C. By these means it was shown that an "activated state" of G_s was produced that was only slowly reversible and that the association of activated G_s and C was even more stable (Howlett *et al.*, 1979). When purified G_s became available, Northup *et al.* (1982) demonstrated that a persistently activated form of the regulatory protein could be generated readily. This work has led to investigations that have clarified the molecular nature of this state of G_s (Northup *et al.*, 1983a,b).

4.1. Phenomenology of Binding of Guanine Nucleotides and Activation of G_s

A detailed investigation of the guanine-nucleotide-binding properties of G_s and the kinetics of its activation by such nucleotides has been performed. The purified regulatory protein was first incubated with stimulatory ligands for defined periods of time and then diluted into solutions containing GTP to terminate the activation process. GTP, in the absence of hormones, does not stimulate this reconstituted system, and GTP, when present during the initial incubation, blocks the activation of G_s by stable analogs of the nucleotide. The degree of activation of G_s achieved during the first incubation was then measured by its ability to activate the catalytic unit of adenylate cyclase present in cyc^- membranes (Sternweis *et al.*, 1981). When G_s is activated with GTPγS, it retains its ability to stimulate catalytic activity in cyc^- with little loss for as long as 1 day after the termination of the activation. The time course of activation of modestly concentrated G_s incubated in 10 mM Mg^{2+} is extremely slow, taking several hours to achieve a maximal extent. This process can be accelerated by increasing the concentration of Mg^{2+} or by dilution of G_s (Hanski *et al.*, 1981; Sternweis *et al.*, 1981; Northup *et al.*, 1982).

When G_s is incubated with [^{35}S]GTPγS, slowly reversible binding of the nucleotide to the protein can be demonstrated. Protein-bound [^{35}S]GTPγS can be quantitatively trapped on nitrocellulose filters, allowing determination of the time course of high-affinity binding under conditions identical to those used for the measurement of the activation process. This comparison showed that the kinetics of binding and activation were the same (Northup *et al.*, 1982). Similar measurements were made at increasing concentrations of GTPγS; the plateau levels of both binding and activation are saturable with similar $K_{1/2}$ values. These methods also allowed the determination of the relative potencies of various nonactivating nucleotides to prevent binding to and activation of G_s by GTPγS; again, there was good agreement between half-maximally effective concentrations for inhibition of both processes (Northup *et al.*, 1982).

Although the saturation behavior of binding and activation as a function of the concentration of GTPγS appears to be strictly hyperbolic, the lack of reversibility and the lack of dependence of the fractional initial rate of activation on the concentration of nucleotide argue strongly against a simple bimolecular model of association. These arguments are presented in detail by Smigel *et al.* (1982). These authors interpreted the activation process as a competition between irreversible steps leading to activation and denaturation, as follows:

$$G_s + N \overset{K_d}{\Longleftrightarrow} G_sN \overset{k^*}{\rightarrow} G_sN^*$$
$$k^+ \downarrow$$
$$G_s{}^+$$

where N is nucleotide and the superscripts indicate that G_s is in its activated ($*$) or denatured ($+$) state, respectively. Taking k^+ and k^* as the unidirectional first-order rate constants for denaturation and activation and K_d as the actual binding constant for nucleotide, they found that the concentration of nucleotide giving half-maximal saturation is given by:

$$K_{1/2} = K_d(k^+/k^*) \tag{7.2}$$

4.2. Mechanistic Aspects of the Activation Process

Although the formulation presented above accounts for the observed kinetics of the activation process, it does not shed any light on the nature of the activated state. Many types of experiments now indicate that the activation of G_s by stable guanine nucleotide analogs or by fluoride results from or reflects the dissociation of the 45,000-Da (α) and 35,000-Da (β) subunits of G_s (Northup *et al.*, 1983a). It has also been demonstrated that the appropriately liganded 45,000-Da subunit, free of the 35,000-Da subunit, is capable of activating the catalyst. We will summarize the evidence that led to these conclusions in this section.

Initial characterization of the hydrodynamic properties of G_s in crude extracts of membranes (Howlett and Gilman, 1980; Kaslow *et al.*, 1980) and later studies of purified G_s (Sternweis *et al.*, 1981; Hanski *et al.*, 1981) showed marked changes in the size of the protein upon activation. Activation of turkey erythrocyte G_s by either fluoride or GTPγS shifts its sedimentation coefficient from 4.6 S (basal) to 3.2 S (activated). These measurements, combined with estimates of Stokes radii from gel filtration, show that the basal species behaves like a particle with a molecular weight of 80,000; the activated species, on the other hand, seems to be about 30,000 Da smaller (Hanski *et al.*, 1981). Similar results are obtained with rabbit hepatic G_s.

The observation that the rate of activation of G_s is independent of the concentration of the activating guanine nucleotide and a large number of other experiments suggested that the rate-limiting step in the activation of G_s was the dissociation of its subunits. The data and arguments are presented in more detail elsewhere (Northup *et al.*, 1982; Smigel *et al.*, 1982). If the dissociation of subunits is thought to be the rate-limiting step for activation of G_s, it is predicted that G_s in dilute solution would be activated by GTPγS at a faster rate than would concentrated G_s. This effect is observed, as is the predicted decrease in the rate of activation, as the concentration of the β subunit is raised. Similarly, dilute G_s is much more labile than is concentrated G_s, and the activity can be stabilized by the addition of excess 35,000-Da subunit, suggesting that the dissociated, unliganded α subunit is the labile species. Although dilute G_s loses activity rapidly when held at 30°C in the presence of Mg^{2+}, it can be markedly stabilized by the addition of 10 nM β subunit. This stabilization by the β subunit parallels its ability to inhibit the rate of activation of α by guanine nucleotides. The activated state of G_s should be reversed by the addition of β subunit under conditions where association of β with α is favored. Since the dissociation of the subunits is thought to be enhanced strongly by Mg^{2+}, activated G_s should be most sensitive to the β subunit at low concentrations of the divalent cation. Such reversal is observed, and the stimulatory effect of β on this deactivation reaction forms the basis of a sensitive assay for this polypeptide (Northup et al., 1983b). It should be noted that chloride also influences the affinity of β for the liganded α subunit of G_s (P. C. Sternweis, personal communication).

The activation of G_s by either GTPγS or fluoride has allowed the separation of its subunits in active forms by high-performance gel filtration chromatography. The liganded, resolved 45,000-Da subunit can activate preparations of chromatographically resolved catalyst that are free of the 35,000-Da subunit. It can also activate the catalyst that is present in cyc^- membranes, although such membranes do contain the 35,000-Da subunit (apparently associated with the α subunit of G_i) (Northup et al., 1983b). The isolated 35,000-Da subunit is inactive in reconstituting either cyc^- membranes or chromatographically resolved catalyst, but it retains its ability to regulate the rate of activation of the α subunit both in detergent-containing solution and in membranes (Katada et al., 1984b). Thus, although the precise pathway of reactions that leads from basal G_s to the activated species has not been explored in detail, it is reasonably clear that basal G_s is an α · β heterodimer and that the active species is the appropriately liganded α subunit. The selection of the precise pathway of activation (i.e., dissociation preceding or following ligand binding) is probably a function of the concentrations of G_s, the activating ligand, Mg^{2+}, and Cl^-, as well as of the nature of the amphiphilic milieu in which the G_s is found. Pathway selection is never easy (Frost, 1916).

4.3. Comparison of G_s from Different Tissues

Although there are striking similarities among the preparations of G_s from widely differing cells and tissues, some of the differences bear comment. Most obvious is the presence of a 52,000-Da α subunit in rabbit liver G_s. The 52,000-Da protein is also visualized by cholera-toxin-catalyzed ADP ribosylation of membranes isolated from S49 mouse lymphoma cells, among others. It is found in addition to substantial amounts of the 45,000-Da protein that is also detected in these cells. The ratio of the two peptides is somewhat variable, and the 52,000-Da species can predominate, as in the RL-PR-C hepatoma cell (Reilly and Blecher, 1981). This situation is in contrast to that found in human and avian erythrocytes, where no detectable 52,000-Da subunit is found (Cassel and Pfeuffer, 1978; Kaslow et al., 1979; Hanski et al., 1981; Hanski and Gilman, 1982). Although little is known about the precise relationship between these polypeptides, genetic evidence, mentioned previously, and proteolytic mapping studies are consistent with, but do not prove, a precursor–product relationship (Manning and Gilman, 1983). Such a relationship is also consistent with the relative loss of the 52,000-Da protein during the maturation of rat reticulocytes (Larner and Ross, 1981).

Since hydrophobic chromatography allows a modest separation of 52,000/35,000 and 45,000/35,000 heterodimers, some biochemical characterization has been achieved. Studies of the rates of activation of the two species by GTPγS indicate that the 52,000/35,000 heterodimer requires less Mg^{2+} to accelerate its activation, presumably reflecting a more facile dissociation of its subunits (Sternweis et al., 1981). The activation behavior of the 45,000/35,000 heterodimer from liver is intermediate between that of the 52,000/45,000 heterodimer from liver and G_s isolated from turkey erythrocytes, which is a pure 45,000/35,000 dimer.

As will be discussed more fully below, the turkey erythrocyte plasma membrane shows very little adenylate cyclase activity when incubated with low concentrations of Mg^{2+} and with nonhydrolyzable analogs of GTP, but it displays high activity in the presence of such analogs plus either a β-adrenergic agonist or a high concentration of Mg^{2+} (Sevilla et al., 1976). The observation of extremely low activity in the presence of Gpp(NH)p alone seems to be limited to adenylate cyclases of avian erythrocyte membranes and, to a lesser extent, rodent erythrocyte membranes (Larner and Ross, 1981). Although adenylate cyclases in S49 cell membranes and in rabbit liver membranes show marked stimulation when incubated with nonhydrolyzable analogs of GTP, further stimulation of the rate of activiation by hormone can still be observed. When adenylate cyclase is measured in membranes derived from brain, basal activities are markedly elevated by nucleotides, and hormonal stimulation is frequently impossible to demonstrate. It is presumed that many of these differences reflect the properties of the G_s of the particular cell and/or species.

5. DEACTIVATION OF G_s: THE GTPase REACTION

Although it is clear that GTP and its nonhydrolyzable analogs can dissociate from G_s, this process is much slower than the observed rate of deactivation of adenylate cyclase when an antagonist is added to block hormonal stimulation. The observation that nonhydrolyzable analogs of GTP cause prolonged enhancement of catalytic activity led to the speculation that the primary mechanism of deactivation might involve hydrolysis of GTP to GDP, rather than the release of nucleotide (Schramm and Rodbell, 1975; Cuatrecasas *et al.*, 1975a). This is consistent with the inability of GDP to activate the enzyme in the presence or absence of hormone (Kimura and Shimada, 1983).

These ideas led Cassel and Selinger (1976) to probe for a GTPase activity in turkey erythrocyte plasma membranes that might be linked to regulation of adenylate cyclase activity. By using a combination of a low concentration of GTP to avoid abundant, high-K_m nucleoside triphosphatases, ATP and App(NH)p to block ATPases, and low pH, they were able to decrease a very substantial background activity and to observe a GTPase activity that was stimulated by hormone about twofold over the remaining background level. The activity of the catecholamine-stimulated GTPase was quite low, 5–10 pmole $\cdot$ min^{-1} $\cdot$ mg^{-1}, but it behaved as predicted with respect to its specificity for substrate, stimulation by β-adrenergic agonists, and sensitivity to sulfhydryl reagents (Cassel and Selinger, 1976).

A more striking aspect of the catecholamine-stimulated GTPase was its inhibition by prior incubation of the membrane with cholera toxin and NAD (Cassel and Selinger, 1977a). Cholera toxin had long been known to activate adenylate cyclase (see Gill and Meren, 1978), and many aspects of the regulation of adenylate cyclase in toxin-treated membranes could be explained by the hypothesis that the reaction of ADP-ribosylated G_s with GTP and Gpp(NH)p was equivalent; i.e., ADP-ribosylated G_s would have a diminished ability to hydrolyze the terminal phosphoryl group of GTP (Ross *et al.*, 1977; Johnson *et al.*, 1978; Ross and Gilman, 1980). The observation that treatment with cholera toxin promoted the GTP-dependent activation of adenylate cyclase in parallel with the inhibition of the GTPase strengthened Cassel and Selinger's argument that they were measuring the relevant GTPase activity. It also allowed them to formulate a fairly detailed hypothesis for the sequence whereby G_s controls the activity of adenylate cyclase. They suggested that hydrolysis of the GTP that is bound to adenylate cyclase (i.e., G_s) causes essentially immediate deactivation of the enzyme. Thus, the fractional activation of the enzyme simply reflects the steady-state fraction that is in the $G_s \cdot GTP$ form (as opposed to free G_s or $G_s \cdot GDP$). For hormone to increase both the steady-state concentration of this active species and the steady-state turnover of the GTPase itself, it must accelerate either the release of GDP or the binding of GTP.

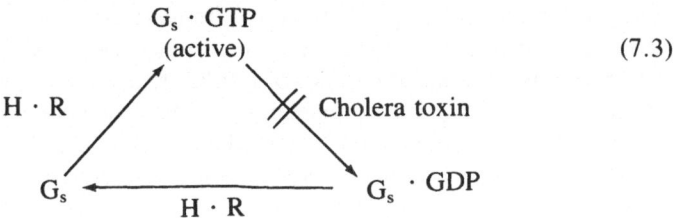

$$(7.3)$$

Cholera toxin, which activates adenylate cyclase but inhibits the GTPase, was proposed to inhibit the hydrolytic reaction itself. For several reasons, Cassel and Selinger (1977a,b, 1978) proposed that the hormone-receptor complex acts primarily to promote the release of GDP (see also Section 6 and Ross and Gilman, 1980).

Since 1977, several groups have measured the hormone-stimulated GTPase activity of G_s in membranes from different tissues (Lambert et al., 1979; Bitonti et al., 1980; Pike and Lefkowitz, 1980; Lester et al., 1982). The assayed activity is rarely higher than 20 pmole $\cdot$ min^{-1} $\cdot$ mg^{-1}, and the specific GTPase cannot be detected at all in several tissues because of a high background. Measured values, based on the increase caused by agonists, are probably only an approximation. Underestimation can occur because any "basal" GTPase activity of G_s is ignored and because of inhibition by the ATP and App(NH)p that are used to lower the background (Brandt et al., 1983). Overestimation can occur if the blank is inadequately suppressed. These technical problems have limited the analysis of the GTPase in native membranes, and most data speak only to the existence of this activity and to its correlation with G_s.

Despite the common assumption that G_s is the site of the regulatory GTPase, its catalytic activity has been demonstrated only recently. Northup et al. (1982) found that the GTPase activity of pure, rabbit hepatic G_s was unmeasurably low under the conditions utilized, below 0.06 moles of GTP hydrolyzed per mole of G_s per minute. These authors suggested either that the catalytic unit of adenylate cyclase or a hormone receptor might be needed for expression of hydrolytic activity. It had been assumed widely that C actually triggered the hydrolysis of GTP, thereby terminating activation of adenylate cyclase after a controlled "lifetime" of the active $G_s \cdot C$ species.

The GTPase activity of purified G_s was first measured by Ross' group (Brandt et al., 1983), who used a reconstituted system composed of rabbit hepatic G_s and turkey erythrocyte β-adrenergic receptors in unilamellar phospholipid vesicles. The GTPase activity of this preparation was stimulated up to 15-fold over the basal (no agonist) level. Although the molar turnover number of the reconstituted GTPase activity was still extremely low, originally less than 2 $\cdot$ min^{-1}, it was much greater than that observed by Northup et al. (1982) and correlated well with activities observed in native membranes (e.g., Cassel and Selinger, 1976; see Brandt et al., 1983, for comparison). These authors also noted that the hydrolysis of GTP might very well be limited by the rate at which the substrate binds to G_s; it certainly cannot be faster. In the receptor–G_s vesicles, the initial rate of agonist-stimulated binding of [^{35}S]GTPγS (to nucleotide-free G_s) is also in the range of 1-8 $\cdot$ min^{-1}, depending on the concentration of nucleotide and hormone. Thus, pending a more detailed analysis, it is reasonable to hypothesize that the binding of GTP to G_s is the hormone-stimulated, rate-limiting step in the GTPase cycle. The actual hydrolysis of GTP to GDP almost certainly is not. If receptor–G_s vesicles are incubated with $[\alpha^{32}P]GTP$, the only nucleotide found bound to the vesicles is GDP, not GTP (D. R. Brandt and E. M. Ross, unpublished data). Under what conditions, if any, the release of GDP may be rate limiting is not yet clear; crucial variables are presumably the concentrations of free nucleotide, hormone, and the 35,000-Da subunit of G_s.

It should be noted that, contrary to common expectation, the catalytic unit of adenylate cyclase does not appear to promote the steady-state hydrolysis of G_s-bound GTP (Brandt et al., 1983). The inclusion or elimination of the catalytic protein had no effect on the steady-state activity of the GTPase, although amounts of the catalyst that are stoichiometric with G_s have not yet been tested. It is interesting to speculate that C might actually inhibit the hydrolysis of GTP bound to G_s, since C may bind preferentially to $G_s \cdot$ GTP (rather than to $G_s \cdot$ GDP) and might be supposed, thermodynamically, to stabilize the active species.

It is also interesting to reconsider why detergent-solubilized G_s does not display appreciable GTPase activity. There are at least three reasons. First, Lubrol and cholate each inhibit the GTPase reaction (Brandt et al., 1983), both directly and by increasing the susceptibility of G_s to thermal denaturation. Second, when removed from a lipid bilayer, the reaction catalyzed by pure G_s slows after a few minutes, only in part owing to denaturation. Dissociation of the β subunit and the resultant slowing of GDP release might be involved. Third, stimulation of the reaction may be inadequte. Although the GTPase activity of pure G_s is stimulated by Mg^{2+} in the 10- to 50-mM range (as is GTPγS binding), the GTPase reaction in receptor–G_s vesicles is stimulated much less by Mg^{2+} than by hormone. This is consistent with the greater ability of hormone to increase the rate of binding of GTPγS (Ross et al., 1984). The control of the GTPase activity of G_s is still only poorly understood. A detailed study of the synthesis and degradation of each reaction intermediate—Gs, Gs $\cdot$ GTP, and Gs $\cdot$ GDP— will be required for the appreciation of its role in the regulation of adenylate cyclase activity.

6. MECHANISM OF RECEPTOR-MEDIATED INHIBITION OF ADENYLATE CYCLASE

The striking similarities between G_s and the substrate for IAP (G_i), mentioned previously, prompt consideration of their coordinated mechanisms of action. Questions that come to mind immediately include the following: (1) Is the IAP substrate in fact G_i; that is, does this protein serve as the coupling factor for receptor-mediated inhibition of adenylate cyclase? (2) Does the dissociation model that describes the activation of G_s in detergent solution apply to the IAP substrate as well? (3) If so, which of the subunits is responsible for the inhibition and what is the mechanism? We will consider these questions in order.

The development of the reconstitutive assay for G_s was aided greatly by the existence of a genetically resolved cell line, cyc⁻, which is devoid of G_s activity. Since no comparable variant that is deficient in G_i is available, Katada and co-workers explored the activities of the 41,000-/35,000-Da dimeric IAP substrate by its addition to membranes that had been rendered functionally deficient in inhibitory coupling factor activity by incubation with IAP and NAD (Katada *et al.*, 1984a,b,c). For example, such treatment reduces the ability of an α_2-adrenergic agonist (e.g., epinephrine) to inhibit platelet adenylate cyclase to a negligible level. After reconstitution with reasonable quantities of the 41,000-/35,000-Da dimer, full inhibition is restored (Katada *et al.*, 1984a). These experiments provide strong evidence that the IAP substrate functions as the inhibitory guanine-nucleotide-binding regulatory component of adenylate cyclase. Of considerable interest, if the reconstitution is performed after incubation of G_i with GTPγS, the protein becomes an inhibitor of adenylate cyclase in the absence of an inhibitory agonist. G_i can apparently be "activated" by binding of stable guanine nucleotide analogs, as can G_s.

Using techniques similar to those developed for G_s, Bokoch *et al.* (1983) showed that G_i behaves like a hydrodynamic particle with a molecular weight of approximately 80,000. When the protein is incubated with either GTPγS or F⁻, its apparent molecular weight is reduced by about 30,000. After such incubation the subunits of G_i can be resolved by either gel filtration or hydrophobic chromatography (Northup *et al.*, 1983b; Katada *et al.*, 1984a).

The subunits of G_i, resolved as described previously, have been tested in several systems. The majority of the inhibitory activity of GTPγS-treated G_i fractionates with the 35,000-Da subunit, whereas the GTPγS · 41,000-Da subunit complex displays inhibitory activity only at considerably higher concentrations. If the "activation" and resolution of subunits are done in the presence of F⁻, the inhibitory activity also fractionates with the 35,000-Da protein, whereas the 41,000-Da subunit actually stimulates adenylate cyclase activity.

The inhibitory effects of the 35,000-Da subunit of G_i are explained by its interaction with the α subunit of G_s. It is these data that indicate that the β subunits of the two dimers are functionally indistinguishable. Thus, the β subunit of G_i deactivates the α subunit of G_s in a manner that exactly duplicates the reaction of $G_{s\alpha}$ with $G_{s\beta}$. In addition, the α subunit of G_i markedly stabilizes the activated state of G_s. This effect is a reflection of the ability of $G_{i\alpha}$ to bind to $G_{s\beta}$. Indeed, the half-time of deactivation of G_s in the presence of excess $G_{i\alpha}$ is the same as that of resolved $G_{s\alpha}$.

M. Smigel (unpublished observations) has looked for direct effects of the subunits of G_i on the activity of the catalyst. The ability of $G_{i\beta}$ to inhibit catalytic activity is dependent on the presence of both G_s and GTP. The GTPγS-bound form of $G_{i\alpha}$ is a relatively impotent inhibitor of the catalyst; this effect may be the result of competition between $G_{s\alpha}$ and $G_{i\alpha}$. The ability of unliganded $G_{i\alpha}$ to stimulate the catalyst is dependent on the presence of G_s and is blocked by the addition of excess 35,000-Da protein.

In summary, the inhibitory activity of G_i appears to follow dissociation of its subunits, and it results from interactions at the level of the G proteins. The primary inhibitory reaction appears to be

$$\beta + \alpha_s \rightleftharpoons \alpha_s\beta \tag{7.4}$$

7. KINETIC MODELS FOR ACTIVATION OF ADENYLATE CYCLASE

The rich phenomenology presented by the adenylate cyclase system has spawned a lavish literature in which the kinetic properties of the system are described and analyzed. Although this review will not attempt to detail, or even politely to hint at, all the viewpoints held by the workers in the field, we will summarize some of the positions in order to provide a background for the questions to be raised in the following section. We refer the interested reader to reviews by Birnbaumer (1977), Boeynaems and Dumont (1977), Levitzki (1981), and Tolkovsky (1983) for guided tours of this literature.

Kinetic arguments that attempt to establish the multicomponent nature of adenylate cyclase are no longer needed; the biochemistry of the situation has been clarified considerably in the last few years. The questions that present themselves for analysis are ones concerning the state of association of the components in the membrane, how these associations change upon activation, and the nature of the rate-limiting step(s) of activation.

In the kinetic approach, one is allowed to vary the concentrations of the interacting components and the time of their interaction. In the early days of research on adenylate cyclase, when the nature and number of the protein components of the system were not defined, the experiments focused on variation of the concentrations of the soluble cofactors of the system: ATP, guanine nucleotides, hormones, and divalent cations. Although such measurements do not allow discrimination between many of the models of the activation process (see Boeynaems and Dumont, 1977; Levitzki, 1981), several important points emerged from their analysis: the requirement for guanine nucleotides for hormonal activation of adenylate cyclase, the role of hormone in increasing the rate of activation of the enzyme by stable guanine nucleotide analogs, and the role of hormone in lowering the K_{act} for divalent cations (Johnson and Garbers, 1977).

Let us look at these results more closely. Hormone plus GTP or Gpp(NH)p alone can activate adenylate cyclase. Careful measurements of the influence of hormone concentrations on the rate of activation of adenylate cyclase by Gpp(NH)p led to the proposal that the effect of hormone is to cause a more rapid activation of the enzyme (Cuatrecasas *et al.*, 1975a,b; Bennett and Cuatrecasas, 1976; Sevilla *et al.*, 1976; Levitzki, 1977; Tolkovsky and Levitzki, 1978a). Furthermore, in some systems the concentration of Gpp(NH)p has little effect on the rate of activation (Tolkovsky *et al.*, 1982). Therefore, the forward rate constant for binding of nucleotide is presumably not the rate-limiting step for activation. Several types of explanations have been proposed for these observations. They can be classified by the nature of the assumption that is made as to the nature of the rate-limiting step in the activation process.

7.1. Release of Tightly Bound Guanine Nucleotide

One such proposal is that although the rate of association of guanine nucleotide with an empty site on G_s is fast, the site is normally blocked by a bound, nonactivating species (presumably GDP) (Blume and Foster, 1976; Sevilla *et al.*, 1976; Cassel and Selinger, 1977a). In this model, it is the slow dissociation of nucleotide from its binding site that is the slow step. Hormone binding is thus postulated to speed activation of adenylate cyclase by facilitating the release of "tightly bound GDP" from G_s. This facilitation presumably results from a negative heterotropic binding interaction between hormone and nucleotide, an explanation that links a mechanism for activation of adenylate cyclase to the widely observed guanine-nucleotide-induced decrease in the affinity of receptors for agonists. Its experimental support and implications have been reviewed by Ross and Gilman (1980).

Although this proposal has had a pervasive influence on the thinking of workers in the field, its experimental support is by no means extensive. The core of this support is the observation that exposure to β-adrenergic agonists or to cholera toxin and NAD enhances release of guanine nucleotides that have been bound to avian erythrocyte membranes (Cassel and Selinger, 1977b; Cassel and Selinger, 1978; Pike and Lefkowitz, 1981; Burns *et al.*, 1982, 1983). The quantitative interpretation of these experiments is complicated by several considerations. Since the concentration of G_s in such membranes is very low, both in absolute terms (1 pmole/mg protein) and in comparison with the amount of guanine nucleotide that binds to such menbranes, it is difficult to establish that the nucleotide released is that which was bound to G_s. This problem has been obviated, to some extent, by studying only the nucleotide whose release is enhanced by incubation with agonists. Although this strategy has yielded consistent results and might seem to guarantee specificity, it virtually precludes absolute quantitation of the number of G_s molecules that are available to catalyze the GTPase reaction. Furthermore, the validity of this approach is based on the untested assumption that hormone receptors interact solely with the G_s coupling protein. The situation is further complicated by the discovery of G_i, the mediator of responses to inhibitory hormones. This protein is present in considerable excess over G_s, at least in several cell types, and it too is an avid binder of guanine nucleotides. Although the use of agonist-specific release of guanine nucleotide as the method for distinguishing nucleotide bound to G_s from that bound to G_i may be valid, the use of gel filtration of detergent solubilized proteins probably fails to separate the two.*

Other observations argue against a central role for slow release of guanine nucleotide in the kinetics of activation (Levitzki, 1980; Birnbaumer et al., 1980; Neer and Salter, 1981). The activation of purified G_s is slow (Sternweis et al., 1981; Hanski et al., 1981; Northup et al., 1982), and when the protein is reconstituted into membranes containing β-adrenergic receptors, it is accelerated by hormone (Citri and Schramm, 1980, 1982; Pedersen and Ross, 1982). G_s as isolated is free of guanine nucleotide. The purified coupling factor shows only a modest affinity for GDP (Northup et al., 1982); its binding is freely reversible. When the activation of turkey erythrocyte adenylate cyclase is measured at varying concentrations of Gpp(NH)p, the rate constant for activation is independent of the concentration of guanine nucleotide (Tolkovsky et al., 1982). This is exactly as found for the activation of purified G_s by GTPγS in detergent solution (Northup et al., 1982). These measurements argue strongly that neither association nor dissociation of guanine nucleotide is the rate-limiting step in activation.

7.2. Protein–Protein Association as the Rate-Limiting Step (Collision–Coupling)

By the mid-1970s it was recognized that the data being generated by the variation of the concentrations of substrates and cofactors of the adenylate cyclase complex were insufficiently discriminating to discern the mechanism of activation. Levitzki extended the kinetic methodology in a powerful way by utilizing techniques that allowed variation of the concentration of the protein components of the system. Rather than proceeding by purifying and reconstituting the proteins

---

* The situation is further complicated by the recent finding that β-adrenergic receptors can interact with G_i in preparations of reconstituted lipid vesicles containing the two proteins (T. Asano, T. Katada, A. G. Gilman, and E. M. Ross, unpublished observation).

involved, his laboratory developed an essentially pharmacological approach, in that they utilized inhibitors that were sufficiently selective to allow the directed inactivation of one or another of the components of an otherwise intact membrane system.

The use of affinity reagents that inactivate the β-adrenergic receptor (Atlas et al., 1976) allowed the independent variation of its concentration. Using Gpp(NH)p to trap whatever turkey erythrocyte adenylate cyclase became activated, Tolkovsky and Levitzki (1978a) showed that even after inactivation of 90% of the receptors, full activation of the enzyme could still be achieved. This result is in contrast to their results found with the stimulatory adenosine receptor in the same membrane (Tolkovsky and Levitzki, 1978b). They also showed that the apparent first-order rate constant for activation of the enzyme was proportional to the concentration of receptors in the membrane.

The sensitivity of the catalytic component of adenylate cyclase to inactivation by N-ethylmaleimide allows its selective depletion in intact membranes. Although this treatment no doubt inactivates many other proteins, conditions were found where most of the receptor and G_s remained active. Tolkovsky et al. (1982) showed that the kinetics of activation of such membranes is still first order, with an unaltered rate constant for activation. These data imply that, in this system, the rate of interaction of activated G_s with C is not rate limiting.

In a separate series of experiments, Levitzki and co-workers (Hanski et al., 1979; Rimon et al., 1978, 1980) showed that cis-unsaturated fatty acids could increase the efficiency with which β-adrenergic receptors increased the rate of the activation of the enzyme by Gpp(NH)p; this was an extension of the work of Orly and Schramm (1975). Increasing the temperature had an effect similar to that of adding fatty acid. Since both these perturbations decrease the "microviscosity" of the plasma membrane bilayer, as indicated by decreases in the fluorescence anisotropy of 1,6-diphenyl-1,3,5-hexatriene (DPH) (see Shinitzky and Barenholz, 1974), these authors proposed that the rate of activation of adenylate cyclase was limited by the diffusion-controlled collision of enzyme with receptor. The term "collision–coupling" describes well the phenomenology of activation of adenylate cyclase in the turkey erythrocyte system stimulated with β-adrenergic agonists. Whether diffusion is actually the rate-limiting process has been challenged, however. Henis et al. (1982b) showed that the rate of lateral diffusion of a fluorescently labeled phospholipid incorporated into turkey erythrocyte membranes does not correlate well with the measurements of microviscosity obtained with DPH; these results are in agreement with those of Kleinfeld et al. (1981). Furthermore, Citri and Schramm (1980) used a reconstituted system to show that the rate of activation of G_s did not have the dependence on concentrations of receptor and G_s that would be predicted by a collision–coupling model, even though first-order activation of G_s is observed in such systems over a wide range of protein concentrations (see also Pedersen and Ross, 1982; Brandt *et al.*, 1983). Thus, although Levitzki's clever approach and elegant experiments do depict the kinetic scheme of activation of G_s by hormone, some other step that is dependent on catalytic amounts of receptor and that is highly sensitive to the lipid environment may be rate limiting.

7.3. Release of an Active Species

In detailing the possible sequences of events leading from binding of agonist to receptor to activation of adenylate cyclase, Boeynaems and Dumont (1977) proposed a model in which the enzyme was inactive when bound to receptor but was released in an active form upon hormone binding:

$$L + RE <==> LRE <==> LR + E --> Response \qquad (7.5)$$

Although the rapid equilibrium formalism shown in this reaction is inconsistent with the kinetics of activation of turkey erythrocyte adenylate cyclase, the idea that activation involves the release of an active species from an inactive complex may well prove to be correct. In a revised version, it would be the 35,000-Da subunit that is released from G_s during the process of activation. Whether the kinetics of the interaction between receptor and G_s or that of release of the 35,000-Da subunit is the rate-limiting step may vary with the concentrations and mobilities of the various components in the membrane under study; it is important to remember that kinetic analysis is insensitive to actual mechanism; it only points to the rate-limiting step in the activation process.

8. ASSOCIATION OF COMPONENTS IN THE MEMBRANE: MODES OF COUPLING

The question of continuous association versus collision–coupling of the components of the hormone-sensitive adenylate cyclase system now must be thought of as two separate questions: is the receptor stably bound to G_s and/or is G_s stably bound to C?

The answers to these questions depend on the stoichiometry of the components in the membrane and on the dynamics of their association. For example, if G_s is in excess of C and if G_s and C form a 1 : 1 complex, then there must be G_s that is not bound to C in the membrane. In the few cases where estimates of stoichiometry have been made, it appears that G_s is considerably in excess of β-adrenergic receptors. Estimates of the concentration of C are extremely uncertain. However, when rabbit liver, S49 cell, or turkey erythrocyte plasma membranes are extracted with detergent, the G_s activity measured in that extract greatly exceeds the fluoride- or guanine-nucleotide-stimulated adenylate cyclase activity measured in the native membrane. It is not known whether this indicates that G_s is in great excess of C or whether this is simply a statement of their relative stabilities during the preparation of the membranes or extracts. In any case, the assumption of equimolar concentrations of receptor, G_s, and catalyst is unwarranted. As mentioned previously, attempts to estimate the concentration of G_s by measurement of binding of guanine nucleotide to intact membranes provide overestimates of the true amount, whereas attempts to utilize labeling with cholera toxin and [^{32}P]NAD will probably err on the low side.

Questions about the kinetics of the association of components also complicate the analysis. As another example, consider a membrane in which G_s is in excess of receptor and in which all of the receptor is "precoupled" to G_s. If there is an exchange process by which a receptor can associate with different molecules of G_s at different times, then kinetics consistent with a collision–coupling mechanism for activation of G_s may still be observed. When discussing the coupling of the components it is necessary to make clear the range of times about which one is talking. When we speak of "stable" association, we refer to the existence of complexes that have lifetimes significantly longer than that of the regulatory event in question.

8.1. Is the Receptor Stably Associated with G_s?

The answer to this question is almost certainly dependent on the presence or absence of regulatory ligands. Three lines of evidence speak toward the question. The measurements of Tolkovsky and Levitzki (1978a) on turkey erythrocyte membranes indicate that reduction of the concentration of receptor slows the rate, but not the extent, of activation of adenylate cyclase by Gpp(NH)p. Thus, either receptor is in great excess of G_s or one molecule of receptor can activate many molecules of G_s. In reconstituted vesicle systems, where the number of receptor and G_s molecules can be measured directly, Ross's group has shown that a single receptor molecule can, in fact, promote the activation of up to ten molecules of G_s within about 15 sec (Pedersen and Ross, 1982; Brandt et al., 1983; Ross et al., 1984; T. Asano and E. M. Ross, unpublished data). Taken together, these results argue strongly that G_s and receptor can and do associate and dissociate fairly rapidly in the presence of guanine nucleotide and agonist. These data are in contrast to the pattern observed by Braun and Levitzki (1979) for the stimulatory receptor for adenosine, where there is a marked decrease in the extent of activation of adenylate cyclase as the receptors are inactivated chemically. The latter data were interpreted to indicate the existence of a tight complex of adenosine receptor and G_s in the presence of adenosine and nucleotide; unfortunately, information on the stoichiometry of the relevant proteins is not available.

More direct data suggest that a receptor–G_s complex is stabilized in the membrane by the binding of agonist to the receptor. It had been known for some time that the affinity of receptor for agonists is frequently decreased in the presence of guanine nucleotides (Rodbell et al., 1971b; Maguire et al., 1976b). Ross et al. (1978) argued that the higher affinity is caused by the association of receptor with G_s (see also Sternweis and Gilman, 1979). These data suggested that the G_s–receptor–agonist complex is quite stable and that nucleotides dissociate the complex, presumably into the species nucleotide–G_s and receptor–agonist. This interpretation was supported by the work of Dufau et al. (1974) and of Limbird and Lefkowitz (1978), who showed that incubation of plasma membranes with agonist allowed the solubilization of receptors of a larger size than that obtained from untreated or antagonist-treated membranes. It was assumed that the large receptor is really the receptor–G_s complex. The increased size of the agonist–receptor complex was not observed in the presence of GTP. A review of the data on binding affinities and changes in the size of receptors is presented elsewhere (Ross and Gilman, 1980).

The association of receptor with G_s, as well as that of G_s with C, has also been studied by the technique of target size analysis by radiation-induced inactivation (Kempner and Schlegel, 1979), primarily by Rodbell and his coworkers (Schlegel et al., 1979) and by Martin (1983). These determinations have typically assigned very large sizes to putative complexes of the components of adenylate cyclase. Although these measurements have been taken to indicate the existence of preformed complexes of receptors, G_s, and catalyst, this interpretation is by no means unique. First, target size analysis is a procedure that is difficult to control internally for the effects of chemical artifacts. Second, the inactivation of a system wherein multiple components must interact to function may well mimic the high sensitivity to inactivation by irradiation shown by a monolithic system of much larger molecular weight (Swillens and Dumont, 1981; Kempner and Miller, 1983). We feel, therefore, that only the most cautious conclusions should be drawn from this work.

8.2. Is G_s Bound to C?

The data on the association of G_s and C in the membrane are scanty, and the answer probably varies with the nature of the ligand bound to G_s. When G_s is activated with a nonhydrolyzable guanine nucleotide, it appears to form a stable complex with C (Pfeuffer, 1979; Neer et al., 1980). In the presence of fluoride, a stably activated adenylate cyclase can also be resolved from free G_s, and this material can release free G_s upon denaturation of the catalyst in cholate (M. D. Smigel, unpublished observations). Thus, the active species presumably represents a G_s–C complex. The physiologically relevant state of G_s and C in the presence of GTP is considerably harder to investigate. Purified G_s can activate chromatographically resolved C in the presence of GTP, suggesting that the components have an appreciable affinity for each other. However, the impurity of the preparations of C that are currently available limits the interpretation of such experiments. Kinetic measurements from Levitzki's laboratory indicate that the rate of interaction between activated G_s and C is not the limiting step in the activation of turkey erythrocyte adenylate cyclase by Gpp(NH)p in the presence of a β-adrenergic agonist (Tolkovsky et al., 1982). This was interpreted to indicate the existence of a preformed C–G_s complex, but this suggestion is clearly not a unique one.

8.3. Consequences of the Association of Receptor and G_s for Agonist Binding Affinity

In the last few years it has been proposed that the binding of hormone to its receptor regulates the activity of adenylate cyclase by changing the affinities of other components of the system for their substrates and for each other (see Ross and Gilman, 1980, for review). In such a system of coupled, multiple ligand-binding equilibria, the requirement for the conservation of energy dictates that there must exist relationships among the equilibrium constants that govern the associations (Weber, 1975).

The first of these binding interactions to be found was the decrease in affinity of hormone binding that results from exposure of membranes to GTP or certain of its analogs (Rodbell et al., 1971b). Maguire et al. (1976a) found that this effect was specific for agonists (see also Maguire et al., 1977). The interpretation that this effect was a reflection of the interaction between G_s and receptor was strengthened by data that showed that it was abolished in membranes from cells that lacked functional G_s and that it could be restored by reconstitution of such membranes with the regulatory protein (Ross et al., 1978; Sternweis and Gilman, 1979).

Although it is frequently useful to view this effect as demonstrative of a negative heterotropic interaction between the binding of nucleotides to G_s and that of agonists to the receptor, recent information on the subunit composition of G_s and on its mechanism of activation make further discussion of these effects interesting. It should be remembered that the GTP-induced shifts in the affinity for agonists might arise through number of mechanisms. As an example, consider a thermodynamic cycle involving the subunits of G_s ($\alpha = 45{,}000$, $\beta = 35{,}000$), the receptor (R), hormone (H), and guanine nucleotide (N) in which the following reactions take place:

$$
\begin{array}{ccc}
\mathrm{H + R\beta\alpha + N} & \underset{K_1}{\rightleftharpoons} & \mathrm{HR\,\beta\alpha + N} \\[2pt]
\Big\updownarrow K_4 & & \Big\updownarrow K_2 \\[2pt]
\mathrm{R\beta\,\alpha N + H} & \underset{K_3}{\rightleftharpoons} & \mathrm{HR\beta\alpha N} \\[2pt]
\Big\updownarrow K_7 & & \Big\updownarrow K_5 \\[2pt]
\mathrm{R\beta + \alpha N + H} & \underset{K_6}{\rightleftharpoons} & \mathrm{HR\,\beta + \alpha N}
\end{array}
\tag{7.6}
$$

If N is a guanine nucleotide with a prominent ability to cause dissociation of the subunits of G_s, its addition will cause the binding of hormone to be controlled by K_6. In the absence of nucleotide, affinity for hormone is controlled by K_1. Thus, the effect of guanine nucleotide on hormone binding is given by

$$
\frac{K_1}{K_6} = \frac{K_4 \cdot K_5}{K_2 \cdot K_7}
\tag{7.7}
$$

The degree of shift in affinity for hormone that is induced by guanine nucleotide thus depends both on the extent to which hormone binding induces dissociation of the subunits of G_s (K_5/K_7) and on how much the binding of hormone weakens that of nucleotide (K_4/K_2). Although these free energy interactions are by no means all that can be proposed, they should inspire caution about facile attempts to interpret shifts in binding affinities in terms of mechanism.

8.4. Consequences of the Association of Receptor and G_s for the GTPase Activity of G_s

It is interesting, at this time, to reexamine the role of the GTPase reaction in the hormonal regulation of adenylate cyclase activity. Although data obtained with purified components are just now being gathered (see Section 5), it seems plausible that the actual GTPase activity of G_s is unregulated and that binding of GTP, release of GDP, or some other step determines the regulated steady-state rate of hydrolysis of GTP. Consequently, we present a brief development of the relationship between the GTPase activity and the dissociation equilibria of the subunits of G_s. (T = GTP, D = GDP, and the other abbreviations are as given previously.)

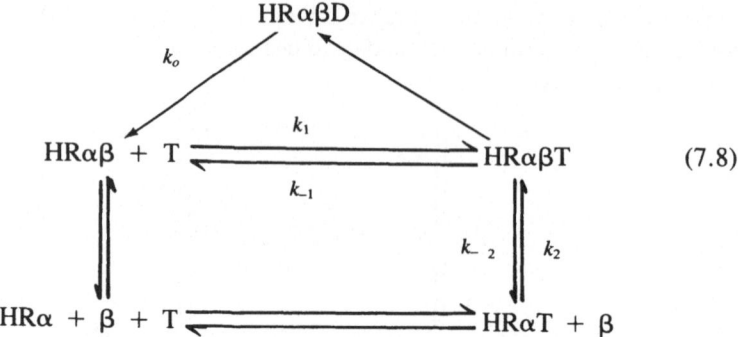

$$(7.8)$$

The catalytic constant for the GTPase reaction is k_g, and the dissociation rate constant of bound GDP is k_o. We propose here that the complex of liganded hormone receptor with undissociated G_s ($HR\alpha\beta T$) is an active GTPase (although $HR\alpha T$ may be assigned activity as well). We have analyzed the model shown under the assumption that the unidirectional rate constants for the binding of nucleotides are much faster than those that characterize release of nucleotide or the interactions of the protein subunits. If we make the further assumption that GTP is present at high concentrations, it can be shown that the steady-state level of GTP-liganded, dissociated 45,000-Da subunit (presumably the species that activates the catalyst of adenylate cyclase) is given by

$$HR\alpha T = \frac{G_{s,total}}{1 + \beta/\beta_{1/2}} \qquad (7.9)$$

where $G_{s,total}$ is the sum of all G_s-containing species and

$$\beta_{1/2} = \frac{k_{-1}}{k_2 (1 + k_g/k_o)} \qquad (7.10)$$

In this formalism, we see that β, the 35,000-Da subunit of G_s, decreases the activity of the system. Its potency as an inhibitor is determined by the ratio of the first-order rates of GTP hydrolysis and GDP release. This model again stresses the interactive behavior between hydrolysis of GTP, the dissociation of the subunits of G_s, and the regulation of adenylate cyclase activity. It is particularly interesting in that the guanine-nucleotide-binding protein that mediates inhibition of adenylate cyclase, G_i, shares a common 35,000-Da subunit with G_s and may regulate the system primarily by controlled release of this subunit (Smigel *et al.*, 1984).

9. RECONSTITUTION OF THE REGULATORY FUNCTIONS OF THE β-ADRENERGIC RECEPTOR

Hormonal regulation of adenylate cyclase depends on the presence of both receptors and G_s in or on a relatively unperturbed lipid bilayer (reviewed by Ross and Gilman, 1980). Study of the interaction of purified receptors and G_s thus requires their reconstitution either into a native membrane or, optimally, into a synthetic bilayer whose structure and composition are known. Reconstitutive strategies for the investigation of adenylate cyclase date to the cell fusion studies of Schramm and co-workers (Orly and Schramm, 1976; Schramm et al., 1977; Eimerl et al., 1980) and to the addition of solubilized G_s to genetically depleted membranes by Gilman and co-workers (Ross and Gilman, 1977a,b; Ross et al., 1978; Sternweis and Gilman, 1979; Howlett et al., 1979). These were analytical studies, designed to resolve the components of the system and to devise quantitative assays for them. The success of these experiments has provided the techniques necessary for developing more defined reconstituted systems that allow mechanistic studies.

9.1. Cell and Membrane Fusion

Orly and Schramm (1976) were the first to demonstrate that fusion of cells that are phenotypically complementary in their adenylate cyclase components allowed creation of a hybrid functional adenylate cyclase system in the membrane of the heterokaryon. They fused Friend erythroleukemia cells, which lack β-adrenergic receptors, with turkey erythrocytes that had been treated with N-ethylmaleimide to inactivate their adenylate cyclase. Membranes prepared from the erythrocyte–Friend cell heterokaryons displayed isoproterenol-stimulated adenylate cyclase activity, indicating that either adenylate cyclase from the Friend cell (C and/or G_s) or receptors from the erythrocytes could diffuse laterally in the plasma membrane of the heterokaryon and interact productively with the other component. Mixing of components was fairly rapid (≤ 2 min) (Schulster et al., 1978) and apparently highly efficient. Schramm (1979), Schwarzmeier and Gilman (1977), and Dufau et al. (1978) extended this technique to include membrane–membrane and membrane–cell fusions. These procedures have now been applied to a number of different analytical problems: the presence and activities of individual proteins in mutant cultured cells (Schwarzmeier and Gilman, 1977; Reilly and Blecher, 1981), proteolytic effects on G_s in membranes (Neufeld et al., 1980), comparison of activities of β_1 and β_2 receptors (Pike et

al., 1979), and assay of down-regulated β-adrenergic receptors after their sequestration in membrane vesicles (Strulovici et al., 1983). Fusion, usually mediated by polyethyleneglycol, appears to be a generally useful method for the
qualitative assay of components. It is surprisingly sensitive if an appropriate test
membrane is used, and this technique should receive wider use in physiological
and pharmacological studies of adenylate cyclase. It may be limited as a quantitative procedure, however, since determination of the efficiency of fusion is
difficult and because, to date, it has been impossible to relate observed reconstituted adenylate cyclase activities to chemical quantities of any single component of the system.

9.2. Reconstitution of Receptors and G_s in Artificial Membranes

The activities of membrane-bound proteins are generally influenced markedly by the composition and structure of their environment. Although C and G_s
retain activity in the presence of a variety of detergents, the ability of G_s to
stimulate C is facilitated by phospholipids, particularly phosphatidylcholine (Ross,
1982). The receptor–G_s interaction is even more sensitive to its surroundings,
since no coupling is observed in detergent solution unless the hormone–receptor–G_s complex has been formed prior to solubilization. Furthermore, many reagents that perturb membranes potently uncouple hormonal control
of the function of G_s (reviewed by Ross and Gilman, 1980; Houslay and Gordon,
1983). Therefore, a prerequisite to the study of hormonal regulation of G_s with
purified proteins is that these proteins be reconstituted into a suitable artificial
membrane, ideally unilamellar lipid vesicles. Initial progress toward this goal
was made by Hoffmann (1979a,b), who showed that dopamine-stimulated adenylate cyclase activity could be reconstituted in a crude cholate extract of membranes from the caudate nucleus by the addition of excess phospholipids and
removal of detergent. These studies also indicated that the individual components
of the system could be separated and recombined prior to reconstitution. A similar
approach was used by Keenan et al. (1982) to study a reconstituted β-adrenergic
adenylate cyclase system from turkey erythrocytes. These authors used Gpp(NH)p
to stabilize the activated C–G_s complex prior to solubilization and then demonstrated function of the receptor after reconstitution by measuring the hormone-
stimulated deactivation of adenylate cyclase in the presence of GTP. This strategy, precoupling of C and G_s, emphasizes one difficulty in reconstituting several
interacting proteins in vesicles: the preparation must be sufficiently concentrated
that at least one molecule of each protein is incorporated in the same vesicle.
As we have calculated (Pedersen and Ross, 1982), the random distribution of a
few molecules of G_s and receptor among many vesicles obviously tends to

decrease the apparent efficiency of coupling of the reconstituted preparation. In this light, it is remarkable that Hoffmann (1979a,b) had such success using the crude and dilute preparations that were available to him.

The problems of coordinate reconstitution of multiple proteins can be eased if one reconstitutes only two proteins at once, either C and G_s (Ross, 1982) or G_s and receptor. This strategy was initially used by Citri and Schramm (1980, 1982). They separately solubilized G_s and β-adrenergic receptors from turkey erythrocyte membranes, combined the two crude extracts with soy asolectin, and removed detergent by adsorption to styrene–divinylbenzene beads. This procedure restored the ability of the receptors to bind β-adrenergic ligands (cf. Fleming and Ross, 1980) and, more importantly, allowed agonist ligands to stimulate the activation of G_s by Gpp(NH)p, indicating that receptor–G_s coupling was also restored. Agonist was also shown to promote the deactivation of Gpp(NH)p-activated G_s in the vesicles (Citri and Schramm, 1982) in a manner similar to that observed by Keenan and co-workers (1982). Technically, these initial studies indicated the feasibility of reconstitution and suggested some general guidelines. Either gel filtration (Fleming and Ross, 1980; Pedersen and Ross, 1982; Brandt et al., 1983) or hydrophobic adsorption (Citri and Schramm, 1980, 1982) can be used to remove detergents, and the detergents that may be used include deoxycholate and cholate (Citri and Schramm, 1980), deoxycholate and Lubrol (Pedersen and Ross, 1982), cholate (Keenan et al., 1982), or a mixture of digitonin (for receptor), Lubrol (for G_s), and both bile salts (for lipids) (Brandt et al., 1983). The lipid requirements for restoration of the receptor's binding and regulatory functions are not yet well understood. Fleming and Ross (1980) and Pedersen and Ross (1982) supplemented receptors with a single lipid, dimyristoylphosphatidylcholine, but their receptor preparations were contaminated with endogenous lipids as well. They subsequently found that reconstitution of pure G_s and partly purified, lipid-depleted receptors required a more complex mixture of lipids. This requirement was met by dimyristoylphosphatidylcholine plus a crude polar lipid fraction from turkey erythrocyte plasma membranes (Brandt et al., 1983) or by a mixture of natural phosphatidylserine and phosphatidylethanolamine plus 5–10 mol % cholesterol (Asano et al., 1984). Cholesterol acts primarily to stabilize reconstituted receptors. Kirilovsky and Schramm (1983) found that when a crude receptor preparation was delipidated in deoxycholate, several natural phospholipids plus either cholesterol or cholesteryl-hemisuccinate could restore the binding properties of the receptor, but no one lipid or lipid mixture was better than soy asolectin. Coupling to G_s was not measured in this study.

Most of the studies cited were largely qualitative in their conclusions. The arguments of Citri and Schramm (1980) regarding mechanism of coupling are dependent on knowledge of both the molar amounts of receptor and G_s that were present in their vesicles and the extent of their interactions. These data are unavailable. The study of Brandt *et al.* (1983) on reconstituted GTPase activity was interesting primarily in that it showed that G_s plus receptor could carry out the hormone-stimulated GTPase reaction at near-physiological rates and that the catalytic unit of adenylate cyclase was not involved in regulation of the rate of hydrolysis of GTP.

However, the use of purified proteins and better assays have also allowed quantitative studies of reconstituted systems. Pedersen and Ross (1982) were able to calculate the distribution of G_s and receptor molecules among reconstituted vesicles and thereby estimate that a single hormone-liganded receptor could promote the activation of up to nine molecules of G_s on the surface of a vesicle. These calculations relied on ligand-binding assays to count receptors, a known reconstitutive specific activity to count G_s molecules (Northup *et al.*, 1982), and the average diameter of the vesicles obtained by electron microscopy. This estimate has been confirmed with more purified preparations, where the binding of [^{35}S]GTPγS was also used to measure both the kinetics of activation and, at long times and at high concentrations of GTPγS, the total concentration of G_s in the vesicles (Brandt *et al.*, 1983). This approach is now being extended to examine the individual steps in the G_s-catalyzed hormone-stimulated GTPase reaction, which is also restored after reconstitution of G_s and β-adrenergic receptors into phospholipid vesicles.

REFERENCES

Asano, T., Katada, T., Gilman, A. G., and Ross, E. M., 1984, Activation of the inhibitory GTP-binding protein of adenylate cyclase, G_i, by β-adrenergic receptors in phospholipid vesicles, *J. Biol. Chem.,* in press.

Atlas, D., and Levitzki, A., 1977, Probing of β-adrenergic receptors by novel fluorescent β-adrenergic blockers, *Proc. Natl. Acad. Sci. USA* **74**:5290–5294.

Atlas, D., and Levitzki, A., 1978a, Tentative identification of β-adrenoceptor subunits, *Nature* **272**:370–371.

Atlas, D., and Levitzki, A., 1978b, Fluorescent visualization of β-adrenergic receptors on cell surfaces, *FEBS Lett.* **85**:158–162.

Atlas, D., Steer, M. L., and Levitzki, A., 1976, Affinity label for β-adrenergic receptor in turkey erythrocytes, *Proc. Natl. Acad. Sci. USA* **73**:1921–1925.

Aurbach, G. D., Fedak, S. A., Woodard, C. J., Palmer, J. S., Hauser, D., and Troxler, F., 1974, β-Adrenergic receptor: Stereospecific interaction of iodinated β-blocking agent with high affinity site, *Science* **186**:1223–1224.

Barnes, P., Koppel, H., Lewis, P., Hutson, C., Blair, I., and Dollery, C., 1980, A fluorescent analogue of propranolol does not label β-adrenoceptor sites, *Brain Res.* **181**:209–213.

Barovsky, K., and Brooker, G., 1980, (−)-[^{125}I]Iodopindolol, a new highly selective radioiodinated β-adrenergic receptor antagonist: Measurement of β-receptors on intact rat astrocytoma cells, *J. Cyclic Nucleotide Res.* **6**:297–307.

Bearer, C. F., Knapp, R. D., Kaumann, A. J., Swartz, T. L., and Birnbaumer, L., 1979, Iodo-hydroxybenzylpindolol: Preparation, purification, localization of its iodine to the indole ring, and characterization as a partial agonist, *Mol. Pharmacol.* **17**:328–338.

Bennett, V., and Cuatrecasas, P., 1976, Irreversible activation of adenylate cyclase of toad erythrocyte plasma membrane by 5'-guanylylimidodiphosphate, *J. Membrane Biol.* **27**:207–232.

Benovic, J. L., Stiles, G. L., Lefkowitz, R. J., and Caron, M. G., 1983, Photoaffinity labelling of mammalian β-adrenergic receptors: Metal-dependent proteolysis explains apparent heterogeneity, *Biochem. Biophys. Res. Commun.* **110**:504–511.

Birnbaumer, L., 1977, The actions of hormones and nucleotides on membrane-bound adenylyl cyclases: an overview, in: *Receptors and Hormone Action*, Volume I (B. O'Malley and L. Birnbaumer, eds.), Academic Press, New York, pp. 485–548.

Birnbaumer, L., Swartz, T. L., Abramowitz, J., Mintz, P. W., and Iyengar, R., 1980, Transient and steady-state kinetics of the interaction of guanyl nucleotides with the adenylyl cyclase from rat liver plasma membranes, *J. Biol. Chem.* **255**:3542–3551.

Bitonti, A. J., Moss, J., Tandon, N. N., and Vaughan, M., 1980, Prostaglandins increase GTP hydrolysis by membranes from human mononuclear cells, *J. Biol. Chem.* **255**:2026–2029.

Bitonti, A. J., Moss, J., Hjelmeland, L., and Vaughan, M., 1982, Resolution and activity of adenylate cyclase components in a zwitterionic cholate derivative [3-[(3-cholamidopropyl)dimethylammonio]-1-propanesulfonate], *Biochemistry* **21**:3650–3653.

Blume, A. J., and Foster, C. J., 1976, Neuroblastoma adenylate cyclase: role of 2-chloroadenosine, prostaglandin E, and guanine nucleotides in the regulation of activity, *J. Biol. Chem.* **251**:3399–3404.

Boeynaems, J. M., and Dumont, J. E., 1977, Models of dissociable receptors applicable to cyclic AMP-dependent protein kinases and membrane receptors, *Mol. Cell Endocrinol.* **7**:275–295.

Bokoch, G. M., Katada, T., Northup, J. K., Hewlett, E. L., and Gilman, A. G., 1983, Identification of the predominant substrate for ADP-ribosylation by islet activating protein, *J. Biol. Chem.* **258**:2072–2075.

Bokoch, G. M., Katada, T., Northup, J. K., Ui, M., and Gilman, A. G., 1984, Purification and properties of the inhibitory guanine nucleotide-binding regulatory component of adenylate cyclase, *J. Biol. Chem.* **259**:3560–3567.

Bourne, H. R., Coffino, P., Melmon, K. L., Tomkins, G. M., and Weinstein, Y., 1975a, Genetic analysis of cyclic AMP in a mammalian cell, *Adv. Cyclic Nucleotide Res.* **5**:771–786.

Bourne, H. R., Coffino, P., and Tomkins, G. M., 1975b, Selection of a variant lymphoma cell deficient in adenylate cyclase, *Science* **187**:750–752.

Brandt, D. R., Asano, T., Pedersen, S. E., and Ross, E. M., 1983, Reconstitution of catecholamine-stimulated GTPase activity, *Biochemistry* **22**:4357–4362.

Braun, S., and Levitzki, A., 1979, Adenosine receptors permanently coupled to turkey erythrocyte adenylate cyclase, *Biochemistry* **18**:2134–2138.

Brown, E. M., Aurbach, G. D., Hauser, D., and Troxler, F., 1976a, β-Adrenergic receptor interactions: Characterization of iodohydroxybenzylpindolol as a specific ligand, *J. Biol. Chem.* **251**:1232–1238.

Brown, E. M., Fedak, S. A., Woodard, C. J., Aurbach, G. D., and Rodbard, D., 1976b, β-Adrenergic receptor interactions: Direct comparison of receptor interaction and biological activity, *J. Biol. Chem.* **251**:1239–1246.

Burgermeister, W., Hekman, M., and Helmreich, E. J. M., 1982, Photoaffinity labeling of the β-adrenergic receptor with azide derivatives of iodocyanopindolol, *J. Biol. Chem.* **257**:5306–5311.

Bürgisser, E., Hancock, A. A., Lefkowitz, R. J., and DeLean, A., 1981a, Anomalous equilibrium binding properties of high-affinity racemic radioligands, *Mol. Pharmacol.* **19**:205–216.

Bürgisser, E., Lefkowitz, R. J., and DeLean, A., 1981b, Alternative explanation for the apparent "two-step" binding kinetics of high-affinity racemic antagonist radioligands, *Mol. Pharmacol.* **19:**509–512.

Burns, D. L., Moss, J., and Vaughan, M., 1982, Choleragen-stimulated release of guanyl nucleotides from turkey erythrocyte membranes, *J. Biol. Chem.* **257:**32–34.

Burns, D. L., Moss, J., and Vaughan, M., 1983, Release of guanyl nucleotides from the regulatory subunit of adenylate cyclase, *J. Biol. Chem.* **258:**1116–1120.

Caron, M. G., and Lefkowitz, R. J., 1976, Solubilization and characterization of the β-adrenergic receptor binding sites of frog erythrocytes, *J. Biol. Chem.* **251:**2374–2384.

Caron, M. G., Srinivasan, Y., Pitha, J., Kociolek, K., and Lefkowitz, R. J., 1979, Affinity chromatography of the β-adrenergic receptor, *J. Biol. Chem.* **254:**2923–2927.

Cassel, D., and Pfeuffer, T., 1978, Mechanism of cholera toxin action: covalent modification of the guanyl nucleotide-binding protein of adenylate cyclase, *Proc. Natl. Acad. Sci. USA* **75:**2669–2673.

Cassel, D., and Selinger, Z., 1976, Catecholamine-stimulated GTPase activity in turkey erythrocytes, *Biochim. Biophys. Acta* **452:**538–551.

Cassel, D., and Selinger, Z., 1977a, Mechanism of adenylate cyclase activation by cholera toxin: an inhibition of GTP hydrolysis at the regulatory site, *Proc. Natl. Acad. Sci. USA* **74:**3307–3311.

Cassel, D., and Selinger, Z., 1977b, Catecholamine-induced release of [^{3}H]-Gpp(NH)p from turkey erythrocyte adenylate cyclase, *J. Cyclic Nucleotide Res.* **3:**11–22.

Cassel, D., and Selinger, Z., 1978, Mechanism of adenylate cyclase activation through the β-adrenergic receptor: Catecholamine-induced displacement of bound GDP by GTP, *Proc. Natl. Acad. Sci. USA* **75:**4155–4159.

Cherksey, B. D., Zadunaisky, J. A., and Murphy, R. B., 1980, Cytoskeletal constraint of the β-adrenergic receptor in frog erythrocyte membranes, *Proc. Natl. Acad. Sci. USA* **77:**6401–6405.

Citri, Y., and Schramm, M., 1980, Resolution, reconstitution, and kinetics of the primary action of a hormone receptor, *Nature (London)* **287:**297–300.

Citri, Y., and Schramm, M., 1982, Probing the coupling site of the β-adrenergic receptor. Competition between different forms of the guanyl nucleotide binding protein for interaction with the receptor, *J. Biol. Chem.* **257:**13257–13262.

Coffino, P., Bourne, H. R., and Tompkins, G. M., 1974, Somatic genetic analysis of cyclic AMP action: selection of unresponsive mutants, *J. Cell. Physiol.* **85:**603–610.

Cuatrecasas, P., Bennett, V., and Jacobs, S., 1975a, Irreversible stimulation of adenylate cyclase activity of fat cell membranes by phosphoramidate and phosphonate analogs of GTP, *J. Membrane Biol.* **23:**249–278.

Cuatrecasas, P., Jacobs, S., and Bennet, V., 1975b, Activation of adenylate cyclase by phosphoramidate and phosphonate analogs of GTP: Possible role of covalent enzyme-substrate intermediates in the mechanism of hormonal activation, *Proc. Natl. Acad. Sci. USA* **72:**1739–1743.

Dufau, M. L., Charreau, E. H., Ryan, D., and Catt, K. J., 1974, Soluble gonadotropin receptors of the rat ovary, *FEBS Lett.* **39:**149–153.

Dufau, M. L., Hayashi, K., Sala, G., Baukal, A., and Catt, K. J., 1978, Gonadal luteinizing hormone receptors and adenylate cyclase: Transfer of functional ovarian luteinizing hormone receptors to adrenal fasciculata cells, *Proc. Natl. Acad. Sci. USA* **75:**4769–4773.

Durieu–Trautmann, O., Delavier–Klutchko, C., Vauquelin, G., and Strosberg, A. D., 1980, Visualization of the turkey erythrocyte β-adrenergic receptor, *J. Supramol. Struct.* **13:**411–419.

Eimerl, S., Neufeld, G., Korner, M., and Schramm, M., 1980, Functional implantation of a solubilized β-adrenergic receptor in the membrane of a cell, *Proc. Natl. Acad. Sci. USA* **77:**760–764.

Engel, G., Hoyer, D., Berthold, R., and Wagner, H., 1981, (±)[^{125}Iodo]-cyanopindolol, a new ligand for β-adrenoceptors: Identification and quantitation of subclasses of β-adrenoceptors in guinea pig, *Naunyn Schmiedebergs Arch. Pharmacol.* **317:**277–285.

Ezrailson, E. G., Garber, A. J., Munson, P. J., Swartz, T. L., Birnbaumer, L., and Entman, M. L., 1981, [^{125}I]Iodopindolol: A new β-adrenergic receptor probe, *J. Cyclic Nucleotide Res.* 7:13–26.

Fleming, J. W., and Ross, E. M., 1980, Reconstitution of β-adrenergic receptors into phospholipid vesicles: Restoration of [^{125}I]iodohydroxybenzylpindolol binding to digitonin-solubilized receptors, *J. Cyclic Nucleotide Res.* 6:407–419.

Florio, V. A., and Ross, E. M., 1983, Regulation of the catalytic component of adenylate cyclase. Potentiative interaction of stimulating ligands and 2',5'-dideoxyadenosine. *Mol. Pharmacol.* 24:195–202.

Frost, Robert, 1916, The road not taken.

Furchgott, R. F., 1972, The classification of adrenoceptors (adrenergic receptors); an evaluation from the standpoint of receptor theory, in: *Handbook of Experimental Pharmacology,* Volume 33, *Catacholamines* (H. Blaschko, and E. Muscholl, eds.), Springer, Berlin, pp. 283–335.

Gill, D. M., and Meren, R., 1978, ADP-ribosylation of membrane proteins catalyzed by cholera toxin: Basis of the activation of adenylate cyclase, *Proc. Natl. Acad. Sci. USA* 75:3050–3054.

Guellaen, G., Yates–Aggerbeck, M., Vauquelin, G., Strosberg, A. D., and Hanoune, J., 1978, Characterization with [^{3}H]dihydroergocryptine of the α-adrenergic receptor of the hepatic plasma membrane: Comparison with the β-adrenergic receptor in normal and adrenalectomized rats, *J. Biol. Chem.* 253:1114–1120.

Haga, T., Haga, K., and Gilman, A. G., 1977a, Hydrodynamic properties of the β-adrenergic receptor and adenylate cyclase from wild type and variant S49 lymphoma cells, *J. Biol. Chem.* 252:5776–5782.

Haga, T., Ross, E. M., Anderson, H. J., and Gilman, A. G., 1977b, Adenylate cyclase permanently uncoupled from hormone receptors in a novel variant of S49 mouse lymphoma cells, *Proc. Natl. Acad. Sci. USA* 74:2016–2020.

Hanski, E., and Gilman, A. G., 1982, The guanine nucleotide-binding regulatory component of adenylate cyclase in human erythrocytes, *J. Cyclic Nucleotide Res.* 8:323–336.

Hanski, E., Rimon, G., and Levitzki, A., 1979, Adenylate cyclase activation by the β-adrenergic receptor as a diffusion-controlled process, *Biochemistry* 18:846–853.

Hanski, E., Sternweis, P. C., Northup, J. K., Dromerick, A. W., and Gilman, A. G., 1981, The regulatory component of adenylate cyclase: Purification and properties of the turkey erythrocyte protein, *J. Biol. Chem.* 256:12911–12919.

Harden, T. K., 1983, Agonist-induced desensitization of the β-adrenergic receptor-linked adenylate cyclase, *Pharmacol. Rev.* 35:5–32.

Hazeki, O., and Ui, M., 1981, Modification by islet-activating protein of receptor-mediated regulation of cyclic AMP accumulation in isolated rat heart cells, *J. Biol. Chem.* 256:2856–2862.

Henis, Y. I., Hekman, M., Elson, E. L., and Helmreich, E. J. M., 1982a, Lateral motion of β-receptors in membranes of cultured liver cells, *Proc. Natl. Acad. Sci. USA* 79:2907–2911.

Henis, Y. I., Rimon, G., and Felder, S., 1982b, Lateral mobility of phospholipids in turkey erythrocytes. Implications for adenylate cyclase activation, *J. Biol. Chem.* 257:1407–1411.

Hjelmeland, L. M., 1980, A nondenaturing zwitterionic detergent for membrane biochemistry: Design and synthesis, *Proc. Natl. Acad. Sci. USA* 77:6368–6370.

Hoffmann, F. M., 1979a, Solubilization and reconstitution of dopamine-sensitive adenylate cyclase from bovine caudate nucleus, *J. Biol. Chem.* 254:255–258.

Hoffmann, F. M., 1979b, A new method for removing nonionic detergent that allows reconstitution of dopamine-sensitive adenylate cyclase, *Biochem. Biophys. Res. Comm.* 86:988–994.

Homcy, C., Rockson, S. G., Countaway, J., and Egan, D. A., 1983, Purification and characterization of the mammalian β$_2$-adrenergic receptor, *Biochemistry* 22:660–668.

Houslay, M. D., and Gordon, L. M., 1983, The activity of adenylate cyclase is regulated by the nature of its lipid environment, *Curr. Top. Membrane Transport* 18:179–231.

Howlett, A. C., and Gilman, A. G., 1980, Hydrodynamic properties of the regulatory component of adenylate cyclase, *J. Biol. Chem.* **255**:2861–2866.

Howlett, A. C., Sternweis, P. C., Macik, B. A., Van Arsdale, P. M., and Gilman, A. G., 1979, Reconstitution of catecholamine-sensitive adenylate cyclase: association of a regulatory component of the enzyme with membranes containing the catalytic protein and β-adrenergic receptors, *J. Biol. Chem.* **254**:2287–2295.

Insel, P. A., Maguire, M. E., Gilman, A. G., Bourne, H. R., Coffino, P., and Melmon, K. L., 1976, β-Adrenergic receptors and adenylate cyclase: Products of separate genes? *Mol. Pharm.* **12**:1062–1069.

Jacobs, S., Bennett, V., and Cuatrecasas, P., 1976, Kinetics of irreversible activation of adenylate cyclase of fat cell membranes by phosphonium and phosphoramidate analogs of GTP, *J. Cyclic Nucleotide Res.* **2**:205–223.

Jakobs, K. H., Aktories, K., Lasch, P., Saur, W., and Schultz, G., 1980, Hormonal inhibition of adenylate cyclase, in: *Hormones and Cell Regulation*, European Symposium, Volume 4 (J. Dumont and J. Nunez, eds.), Elsevier Press, Amsterdam, pp. 89–106.

Johnson, G. L., Harden, T. K., and Perkins, J. P., 1978, Regulation of adenosine 3':5'-monophosphate content of Rous sarcoma virus-transformed human astrocytoma cells, *J. Biol. Chem.* **253**:1465–1471.

Johnson, G. L., Bourne, H. R., Gleason, M. K., Coffino, P., Insel, P. A., and Melmon, K. L., 1979, Isolation and characterization of S49 lymphoma cells deficient in β-adrenergic receptors: Relation of receptor number to activation of adenylate cyclase, *Mol. Pharmacol.* **15**:16–27.

Johnson, R. A., 1982, An approach to the identification of adenosine's inhibitory site on adenylate cyclase, *FEBS Lett.* **140**:80–84.

Johnson, R. A., and Garbers, D. L., 1977, An approach to the study of the kinetics of adenylyl cyclase, in: *Receptors and Hormone Action*, Volume I (B. O'Malley and L. Birnbaumer, eds.), Academic Press, New York, pp. 549–572.

Johnson, R. A., and Sutherland, E. W., 1974, Preparation of particulate and detergent-disperse adenylate cyclase from brain, in: *Methods in Enzymology*, Volume 38 (J. G. Hardman, and B. O'Malley, eds.), Academic Press, New York, pp. 135–143.

Johnson, R. A., Garbers, D. L., and Pilkis, S. J., 1976, Some kinetic and chromatographic properties of detergent-dispersed adenylate cyclase, *J. Supramol. Struc.* **4**:205–220.

Kaslow, H. R., Farfel, Z., Johnson, G. L., and Bourne, H. R., 1979, Adenylate cyclase assembled *in vitro*: Cholera toxin substrates determine different patterns of resolution by isoproterenol and guanosine 5'-triphosphate, *Mol. Pharmacol.* **15**:472–483.

Kaslow, H. R., Johnson, G. L., Brothers, V. M., and Bourne, H. R., 1980, A regulatory component of adenylate cyclase from human erythrocytes, *J. Biol. Chem.* **255**:3736–3741.

Katada, T., and Ui, M., 1979, Islet-activating protein. Enhanced insulin secretion and cyclic AMP accumulation in pancreatic islets due to activation of native calcium ionophores, *J. Biol. Chem.* **254**:469–479.

Katada, T., and Ui, M., 1980, Slow interaction of islet-activating protein with pancreatic islets during primary culture to cause reversal of α-adrenergic inhibition of insulin secretion, *J. Biol. Chem.* **255**:9580–9588.

Katada, T., and Ui, M., 1982a, ADP ribosylation of the specific membrane protein of C6 cells by islet-activating protein associated with modification of adenylate cyclase activity, *J. Biol. Chem.* **257**:7210–7216.

Katada, T., and Ui, M., 1982b, Direct modification of the membrane adenylate cyclase system by islet-activating protein due to ADP-ribosylation of a membrane protein, *Proc. Natl. Acad. Sci. USA* **79**:3129–3133.

Katada, T., Amano, T., and Ui, M., 1982, Modulation by islet-activating protein of adenylate cyclase activity in C6 glioma cells, *J. Biol. Chem.* **257**:3739–3746.

Katada, T., Bokoch, G. M., Smigel, M., Ui, M., and Gilman, A. G., 1984a, The inhibitory guanine nucleotide-binding regulatory component of adenylate cyclase. Subunit dissociation and the inhibition of adenylate cyclase in S49 lymphoma cyc⁻ and wild type membranes, *J. Biol. Chem.* **259**:3586–3595.

Katada, T., Bokoch, G. M., Northup, J. K., Ui, M., and Gilman, A. G., 1984b, The inhibitory guanine nucleotide-binding regulatory component of adenylate cyclase. Properties and function of the purified protein, *J. Biol. Chem.* **259**:3568–3577.

Katada, T., Northup, J. K., Bokoch, G. M., Ui, M., and Gilman, A. G., 1984c, The inhibitory guanine nucleotide-binding regulatory component of adenylate cyclase. Subunit dissociation and guanine nucleotide-dependent hormonal inhibition, *J. Biol. Chem.* **259**:3578–3585.

Keenan, A. K., Gal, A., and Levitzki, A., 1982, Reconstitution of the turkey erythrocyte adenylate cyclase sensitivity to 1-epinephrine upon re-insertion of the Lubrol-solubilized components into phospholipid vesicles, *Biochem. Biophys. Res. Commun.* **105**:615–623.

Kempner, E. S., and Miller, J. H., 1983, Radiation inactivation of glutamate dehydrogenase hexamer: Lack of energy transfer between subunits, *Science* **222**:586–589.

Kempner, E. S., and Schlegel, W., 1979, Size determination of enzymes by radiation inactivation, *Anal. Biochem.* **92**:2–10.

Kimura, N., and Shimada, N., 1983, GDP does not mediate but rather inhibits hormonal signal to adenylate cyclase, *J. Biol. Chem.* **258**:2278–2283.

Kirilovsky, J., and Schramm, M., 1983, Delipidation of a β-adrenergic receptor preparation and reconstitution by specific lipids, *J. Biol. Chem.* **258**:6841–6849.

Kleinfeld, A. M., Dragsten, P., Klausner, R. D., Pjura, W. J., and Matoyishi, E. D., 1981, The lack of relationship between fluorescence polarization and lateral diffusion in biological membranes, *Biochim. Biophys. Acta* **649**:471–480.

Lambert, M., Sroboda, M., and Christopher, J., 1979, Hormone-stimulated GTPase activity in rat pancreatic plasma membranes, *FEBS Lett.* **99**:303–307.

Lands, A. M., Arnold, A., McAuliff, J. P., Luduena, F. P., and Brown, T. G., Jr., 1967, Differentiation of receptor systems activated by sympathomimetic amines, *Nature* **214**:597–598.

Langer, S. Z., 1977, Presynaptic receptors and their role in the regulation of transmitter release. Sixth Gaddum Memorial Lecture, National Institute for Medical Research, Mill Hill (January 1977), *Br. J. Pharmacol.* **60**:481–497.

Larner, A. C., and Ross, E. M., 1981, Alteration in the protein components of catecholamine-sensitive adenylate cyclase during maturation of rat reticulocytes, *J. Biol. Chem.* **256**:9551–9557.

Lavin, T. N., Heald, S. L., Jeffs, P. W., Shorr, R. G. L., Lefkowitz, R. J., and Caron, M. G., 1981, Photoaffinity labeling of the β-adrenergic receptor, *J. Biol. Chem.* **256**:11944–11950.

Lavin, T. N., Nambi, P., Heald, S. L., Jeffs, P. W., Lefkowitz, R. J., and Caron, M. G., 1982, ¹²⁵I-labeled p-azidobenzylcarazolol, a photoaffinity label for the β-adrenergic receptor: Characterization of the ligand and photoaffinity labeling of β₁- and β₂-adrenergic receptors, *J. Biol. Chem.* **257**:12332–12340.

Lefkowitz, R. J., Mukherjee, C., Coverstone, M., and Caron, M. G., 1974, Stereospecific [³H](−)-alprenolol binding sites, β-adrenergic receptors and adenylate cyclase, *Biochem. Biophys. Res. Commun.* **60**:703–709.

Lester, N. A., Steer, M. L., and Levitzki, A., 1982, Prostaglandin-stimulated GTP hydrolysis associated with activation of adenylate cyclase in human platelet membranes, *Proc. Natl. Acad. Sci. USA* **79**:719–723.

Levitzki, A., 1977, The role of GTP in the activation of adenylate cyclase, *Biochem. Biophys. Res. Commun.* **74**:1154–1159.

Levitzki, A., 1980, Slow GDP dissociation from the guanyl nucleotide site of turkey erythrocyte membranes is not the rate limiting step in the activation of adenylate cyclase by β-adrenergic receptors, *FEBS Lett.* **115**:9–10.

Levitzki, A., 1981, The β-adrenergic receptor and its mode of coupling to adenylate cyclase, *CRC Crit. Rev. Biochem.*, **10**:81–112.

Levitzki, A., Atlas, D., and Steer, M. L., 1974, The binding characteristics and number of β-adrenergic receptors on the turkey erythrocyte, *Proc. Natl. Acad. Sci. USA* **71**:2773–2776.

Limbird, L. E., and Lefkowitz, R. J., 1977, Resolution of β-adrenergic receptor binding and adenylate cyclase activity by gel exclusion chromatography, *J. Biol. Chem.* **252**:799–802.

Limbird, L. E., and Lefkowitz, R. J., 1978, Agonist-induced increase in apparent β-adrenergic receptor size, *Proc. Natl. Acad. Sci. USA* **75**:228–232.

Londos, C., Salomon, Y., Lin, M. C., Harwood, J. P., Schramm, M., Wolff, J., and Rodbell, M., 1974, 5'-Guanylylimidodiphosphate, a potent activator of adenylate cyclase systems in eukaryotic cells, *Proc. Natl. Acad. Sci. USA* **71**:3087–3090.

Londos, C., Wolff, J., and Cooper, D. M. B., 1979, Action of adenosine on adenylate cyclase, in: *Physiological and Regulatory Functions of Adenosine and Adenine Nucleotides* (H. P. Baer and G. I. Drummond, eds.), Raven Press, New York, pp. 271–281.

Maguire, M. E., and Erdos, J. J., 1980, Inhibition of magnesium uptake by β-adrenergic agonists and prostaglandin E_1 is not mediated by cyclic AMP, *J. Biol. Chem.* **255**:1030–1035.

Maguire, M. E., Van Arsdale, P. M., and Gilman, A. G., 1976a, An agonist-specific effect of guanine nucleotides on binding to the β-adrenergic receptor, *Mol. Pharmacol.* **12**:335–339.

Maguire, M. E., Brunton, L. L., Wiklund, R. A., Anderson, H. J., Van Arsdale, P. M., and Gilman, A. G., 1976b, Hormone receptors and the control of cyclic AMP metabolism in parental and hybrid somatic cells, *Rec. Prog. Hormone Res.* **32**:633–667.

Maguire, M. E., Ross, E. M., and Gilman, A. G., 1977, β-Adrenergic receptor: Ligand binding properties and the interaction with adenylyl cyclase, *Adv. Cyclic Nucleotide Res.* **8**:1–83.

Malchoff, C. D., and Marinetti, G. V., 1976, Hormone action at the membrane level. V. Binding of (±)-[^{3}H]isoproterenol to intact chicken erythrocytes and erythrocyte ghosts, *Biochim. Biophys. Acta* **436**:45–52.

Manning, D. R., and Gilman, A. G., 1983, The regulatory components of adenylate cyclase and transducin. A family of structurally homologous guanine nucleotide-binding proteins, *J. Biol. Chem.* **258**:7059–7063.

Martin, B. R., 1983, The analysis of interactions between hormone receptors and adenylate cyclase by target size determinations using irradiation inactivation, *Curr. Top. Mem. Transport* **18**:233–254.

Mukherjee, C., Caron, M. G., Coverstone, M., and Lefkowitz, R. J., 1975, Identification of adenylate cyclase-coupled β-adrenergic receptors in frog erythrocytes with (−)-[^{3}H]alprenolol, *J. Biol. Chem.* **250**:4869–4876.

Murad, F., Chi, Y.—M., Rall, T. W., and Sutherland, E. W., 1962, Adenyl cyclase. III. The effect of catecholamines and choline esters on the formation of adenosine 3',5'-phosphate by preparations from cardiac muscle and liver, *J. Biol. Chem.* **237**:1233–1238.

Neer, E. J., 1974, The size of adenylate cyclase, *J. Biol. Chem.* **249**:6527–6531.

Neer, E. J., 1978, Size and detergent binding of adenylate cyclase from bovine cerebral cortex, *J. Biol. Chem.* **253**:1498–1502.

Neer, E. J., and Salter, R. S., 1981, Reconstituted adenylate cyclase from bovine brain. Functions of the subunits, *J. Biol. Chem.* **256**:12102–12107.

Neer, E. J., Echeverria, D., and Knox, S., 1980, Increase in the size of soluble brain adenylate cyclase with activation by guanosine 5'-(β,γ-imino)triphosphate, *J. Biol. Chem.* **255**:9782–9789.

Neufeld, G., Schramm, M., and Weinberg, N., 1980, Hybridization of adenylate cyclase components by membrane fusion and the effect of selective digestion by trypsin, *J. Biol. Chem.* **255**:9268–9274.

Northup, J. K., Sternweis, P. C., Smigel, M. D., Schleifer, L. S., Ross, E. M., and Gilman, A. G., 1980, Purification of the regulatory component of adenylate cyclase, *Proc. Natl. Acad. Sci. USA* **77**:6516–6520.

Northup, J. K., Smigel, M. D., and Gilman, A. G., 1982, The guanine nucleotide activating site of the regulatory component of adenylate cyclase: Identification by ligand binding, *J. Biol. Chem.* **257**:11416–11423.

Northup, J. K., Smigel, M. D., Sternweis, P. C., and Gilman, A. G., 1983a, The subunits of the stimulatory regulatory component of adenylate cyclase. Resolution of the activated 45,000-dalton α subunit, *J. Biol. Chem.* **258**:11369–11376.

Northup, J. K., Sternweis, P. C., and Gilman, A. G., 1983b, The subunits of the stimulatory regulatory component of adenylate cyclase. Resolution, activity, and properties of the 35,000-dalton β subunit, *J. Biol. Chem.* **258**:11361–11368.

Orly, J., and Schramm, M., 1975, Fatty acids as modulators of membrane functions: catecholamine-activated adenylate cyclase of turkey erythrocytes, *Proc. Natl. Acad. Sci. USA* **72**:3433–3437.

Orly, J., and Schramm, M., 1976, Coupling of catecholamine receptor from one cell with adenylate cyclase from another cell by cell fusion, *Proc. Natl. Acad. Sci. USA* **73**:4410–4414.

Pedersen, S. E., and Ross, E. M., 1982, Functional reconstitution of β-adrenergic receptors and the stimulatory GTP-binding protein of adenylate cyclase, *Proc. Natl. Acad. Sci. USA* **79**:7228–7232.

Perkins, J. P., 1973, Adenyl cyclase, *Adv. Cyclic Nucleotide Res.* **3**:1–64.

Pfeuffer, T., 1977, GTP-binding proteins in membranes and the control of adenylate cyclase activity, *J. Biol. Chem.* **252**:7224–7234.

Pfeuffer, T., 1979, Guanine nucleotide-controlled interactions between components of adenylate cyclase, *FEBS Lett.* **101**:85–89.

Pfeuffer, T., and Metzger, H., 1982, 7-O-Hemisuccinyl-deacetyl forskolin-sepharose: A novel affinity support for purification of adenylate cyclase, *FEBS Lett.* **146**:369–375.

Pike, L. J., and Lefkowitz, R. J., 1980, Activation and desensitization of β-adrenergic receptor-coupled GTPase and adenylate cyclase of frog and turkey erythrocyte membranes, *J. Biol. Chem.* **255**:6860–6867.

Pike, L. J., and Lefkowitz, R. J., 1981, Correlation of β-adrenergic receptor-stimulated [³H]GDP release and adenylate cyclase activation, *J. Biol. Chem.* **256**:2207–2212.

Pike, L. J., Limbird, L. E., and Lefkowitz, R. J., 1979, β-Adrenoceptors determine affinity but not intrinsic activity of adenylate cyclase stimulants, *Nature* **280**:502–504.

Rall, T. W., Sutherland, E. W., and Berthet, J., 1957, The relationship of epinephrine and glucagon to liver phosphorylase. IV. Effect of epinephrine and glucagon on the reactivation of phosphorylase in liver homogenates, *J. Biol. Chem.* **224**:463–475.

Rashidbaigi, A., and Ruoho, A. E., 1981, Iodoazidopindolol, a photoaffinity probe for the β-adrenergic receptor, *Proc. Natl. Acad. Sci. USA* **78**:1609–1613.

Rashidbaigi, A., and Ruoho, A. E., 1982, Photoaffinity labeling of β-adrenergic receptors: Identification of the β receptor binding site(s) from turkey, pigeon and frog erythrocyte, *Biochem. Biophys. Res. Commun.* **106**:139–148.

Reilly, T. M., and Blecher, M., 1981, Restoration of glucagon responsiveness in spontaneously transformed rat hepatocytes (RL-PR-C) by fusion with normal progenitor cells and rat liver plasma membranes, *Proc. Natl. Acad. Sci. USA* **78**:182–186.

Rimon, G., Hanski, E., Braun, S., and Levitzki, A., 1978, Mode of coupling between hormone receptors and adenylate cyclase elucidated by modulation of membrane fluidity, *Nature* **276**:394–396.

Rimon, G., Hanski, E., and Levitzki, A., 1980, Temperature dependence of β receptor, adenosine receptor, and sodium fluoride stimulated adenylate cyclase from turkey erythrocytes, *Biochemistry* **19**:4451–4460.

Rodbell, M., 1975, On the mechanism of activation of fat cell adenylate cyclase by guanine nucleotides: An explanation for the biphasic inhibitory and stimulatory effects of the nucleotides and the role of hormones, *J. Biol. Chem.* **250**:5826–5834.

Rodbell, M., Birnbaumer, L., Pohl, S. L., and Krans, H. M. J., 1971a, The glucagon-sensitive adenyl cyclase system in plasma membranes of rat liver. V. An obligatory role of guanyl nucleotides in glucagon action, *J. Biol. Chem.* **246**:1877–1882.

Rodbell, M., Krans, H. M. J., Pohl, S. L., and Birnbaumer, L., 1971b, The glucagon-sensitive adenyl cyclase system in plasma membranes of rat liver. IV. Effects of guanyl nucleotides on binding of ^{125}I-glucagon, *J. Biol. Chem.* **246**:1872–1876.

Rodbell, M., Lin, M. C., Salomon, Y., Londos, C., Harwood, J. P., Martin, B. R., Rendell, M., and Berman, M., 1975, Role of adenine and guanine nucleotides in the activity and response of adenylate cyclase systems to hormones: evidence for multisite transition states, *Adv. Cyclic Nucleotide Res.* **5**:3–29.

Ross, E. M., 1981, Physical separation of the catalytic and regulatory proteins of hepatic adenylate cyclase, *J. Biol. Chem.* **256**:1949–1953.

Ross, E. M., 1982, Phosphatidylcholine-promoted interaction of the catalytic and regulatory proteins of adenylate cyclase, *J. Biol. Chem.* **257**:10751–10758.

Ross, E. M., and Gilman, A. G., 1977a, Reconstitution of catecholamine-sensitive adenylate cyclase activity: Interaction of solubilized components with receptor-replete membranes, *Proc. Natl. Acad. Sci. USA* **74**:3715–3719.

Ross, E. M., and Gilman, A. G., 1977b, Resolution of some components of adenylate cyclase necessary for catalytic activity, *J. Biol. Chem.* **252**:6966–6969.

Ross, E. M., and Gilman, A. G., 1980, Biochemical properties of hormone-sensitive adenylate cyclase, *Ann. Rev. Biochem.* **49**:533–564.

Ross, E. M., Maguire, M. E., Sturgill, T. W., Biltonen, R. L., and Gilman, A. G., 1977, Relationship between the β-adrenergic receptor and adenylate cyclase. Studies of ligand binding and enzyme activity in purified membranes of S49 lymphoma cells, *J. Biol. Chem.* **252**:5761–5775.

Ross, E. M., Howlett, A. C., Ferguson, K. M., and Gilman, A. G., 1978, Reconstitution of hormone-sensitive adenylate cyclase activity with resolved components of the enzyme, *J. Biol. Chem.* **253**:6401–6412.

Ross, E. M., Asano, T., Pedersen, S. E., and Brandt, D. R., 1984, Reconstitution of the regulatory functions of β-adrenergic receptors, in: *Neurotransmitter Receptors: Mechanisms of Action and Regulation* (S. Kito, T. Segawa, K. Kuriyama, H. I. Yamamura, and R. W. Olsen, eds.), Plenum Press, New York, in press.

Schlegel, W., Kempner, E. S., and Rodbell, M., 1979, Activation of adenylate cyclase in hepatic membranes involves interactions of the catalytic unit with multimeric complexes of regulatory proteins, *J. Biol. Chem.* **254**:5168–5176.

Schleifer, L. S., Kahn, R. A., Hanski, E., Northup, J. K., Sternweis, P. C., and Gilman, A. G., 1982, Requirements for cholera toxin-dependent ADP-ribosylation of the purified regulatory component of adenylate cyclase, *J. Biol. Chem.* **257**:20–23.

Schramm, M., 1979, Transfer of glucagon receptor from liver membranes to a foreign adenylate cyclase by a membrane fusion procedure, *Proc. Natl. Acad. Sci. USA* **76**:1174–1178.

Schramm, M., and Rodbell, M., 1975, A persistent active state of the adenylate cyclase system produced by the combined actions of isoproterenol and guanylyl imidodiphosphate in frog erythrocyte membranes, *J. Biol. Chem.* **250**:2232–2237.

Schramm, M., Orly, J., Eimerl, S., and Korner, M., 1977, Coupling of hormone receptors to adenylate cyclase of different cells by cell fusion, *Nature* **268**:310–313.

Schulster, D., Orly, J., Seidel, G., and Schramm, M., 1978, Intracellular cyclic AMP production enhanced by a hormone receptor transferred from a different cell. β-Adrenergic responses in cultured cells conferred by fusion with turkey erythrocytes, *J. Biol. Chem.* **253**:1201–1206.

Schwarzmeier, J. D., and Gilman, A. G., 1977, Reconstitution of catecholamine-sensitive adenylate cyclase activity: Interaction of components following cell–cell and membrane–cell fusion, *J. Cyclic Nucleotide Res.* **3**:227–238.

Seamon, K., and Daly, J. W., 1981a, Activation of adenylate cyclase by the diterpene forskolin does not require the guanine nucleotide regulatory protein, *J. Biol. Chem.* **256**:9799–9801.

Seamon, K., and Daly, J. W., 1981b, Forskolin: A unique diterpene activator of cyclic AMP-generating systems, *J. Cyclic Nucleotide Res.* **7**:201–224.

Seamon, K., Padgett, W., and Daly, J. W., 1981, Forskolin: A unique diterpene activator of adenylate cyclase in membranes and in intact cells, *Proc. Natl. Acad. Sci. USA* **78**:3363–3367.

Sevilla, N., Steer, M. L., and Levitzki, A., 1976, Synergistic activation of adenylate cyclase by guanylyl imidophosphate and epinephrine, *Biochemistry* **15**:3493–3499.

Shinitzky, M., and Barenholz, Y., 1974, Dynamics of the hydrocarbon layer in liposomes of lecithin and sphingomyelin containing dicetylphosphate, *J. Biol. Chem.* **249**:2652–2657.

Shorr, R. G. L., Lefkowitz, R. J., and Caron, M. G., 1981, Purification of the β-adrenergic receptor. Identification of the hormone binding subunit, *J. Biol. Chem.* **256**:5820–5826.

Shorr, R. G. L., Heald, S. L., Jeffs, P. W., Lavin, T. N., Strohsacker, M. W., Lefkowitz, R. J., and Caron, M. G., 1982a, The β-adrenergic receptor: Rapid purification and covalent labeling by photoaffinity crosslinking, *Proc. Natl. Acad. Sci. USA* **79**:2778–2782.

Shorr, R. G. L., Strohsacker, M. W., Lavin, T. N., Lefkowitz, R. J., and Caron, M. G., 1982b, The β₁-adrenergic receptor of the turkey erythrocyte. Molecular heterogeneity revealed by purification and photoaffinity labeling, *J. Biol. Chem.* **257**:12341–12350.

Smigel, M. D., Northup, J. K., and Gilman, A. G., 1982, Characteristics of the guanine nucleotide-binding regulatory component of adenylate cyclase, *Recent Prog. Horm. Res.* **38**:601– 624.

Smigel, M. D., Katada, T., Northup, J. K., Bokoch, G. M., Ui, M., and Gilman, A. G., 1984, Mechanisms of guanine nucleotide-mediated regulation of adenylate cyclase activity, *Adv. Cyclic Nucleotide Res.* **17**:1–18.

Stadel, J. M., Shorr, R. G. L., Limbird, L. E., and Lefkowitz, R. J., 1981, Evidence that a β-adrenergic receptor-associated guanine nucleotide regulatory protein conveys guanosine 5'-O-(3-thiotriphosphate)-dependent adenylate cyclase activity, *J. Biol. Chem.* **256**:8718–8723.

Stadel, J. M., Nambi, P., Shorr, R. G. L., Sawyer, D. F., Caron, M. G., and Lefkowitz, R. J., 1983, Catecholamine-induced desensitization of turkey erythrocyte adenylate cyclase is associated with phosphorylation of the β-adrenergic receptor, *Proc. Natl. Acad. Sci. USA* **80**:3173–3177.

Stengel, D., and Hanoune, J., 1980, Solubilization and physical characterization of the adenylate cyclase from rat-liver plasma membranes, *Eur. J. Biochem.* **102**:21–34.

Sternweis, P. C., and Gilman, A. G., 1979, Reconstitution of catecholamine-sensitive adenylate cyclase. Reconstitution of the uncoupled variant of the S49 lymphoma cell, *J. Biol. Chem.* **254**:3333–3340.

Sternweis, P. C., and Gilman, A. G., 1982, Aluminum: A requirement for activation of the regulatory component of adenylate cyclase by fluoride, *Proc. Natl. Acad. Sci. USA* **79**:4888–4891.

Sternweis, P. C., Northup, J. K., Smigel, M. D., and Gilman, A. G., 1981, The regulatory component of adenylate cyclase: Purification and properties, *J. Biol. Chem.* **256**:11517–11526.

Strittmatter, S., and Neer, E. J., 1980, Properties of the separated catalytic and regulatory units of brain adenylate cyclase, *Proc. Natl. Acad. Sci. USA* **77**:6344–6348.

Strulovici, B., Stadel, J. M., and Lefkowitz, R. J., 1983, Functional integrity of desensitized β-adrenergic receptors. Internalized receptors reconstitute catecholamine-stimulated adenylate cyclase activity, *J. Biol. Chem.* **258**:6410–6414.

Sutherland, E. W., Rall, T. W., and Menon, T., 1962, Adenyl cyclase. I. Distribution, preparations, and properties, *J. Biol. Chem.* **237**:1220–1227.

Swillens, S., and Dumont, J. E., 1981, A pitfall in the interpretation of data on adenylate cyclase inactivation by irradiation, *FEBS Lett.* **134**:29–31.

Tolkovsky, A. M., 1983, The elucidation of some aspects of receptor function by the use of a kinetic approach, in: *Current Topics in Membranes and Transport*, Volume 18 (A. Kleinzeller and B. R. Martin, eds.), Academic Press, New York, pp. 11–41.

Tolkovsky, A. M., and Levitzki, A., 1978a, Mode of coupling between the β-adrenergic receptor and adenylate cyclase in turkey erythrocytes, *Biochemistry* **17**:3795–3810.

Tolkovsky, A. M., and Levitzki, A., 1978b, Coupling of a single adenylate cyclase to two receptors: Adenosine and catecholamine, *Biochemistry* **17**:3811–3817.

Tolkovsky, A. M., Braun, S., and Levitzki, A., 1982, Kinetics of interaction between β-receptors, GTP protein, and the catalytic unit of turkey erythrocyte adenylate cyclase, *Proc. Natl. Acad. Sci. USA* **79**:213–217.

Vauquelin, G., Geynet, P., Hanoune, J., and Strosberg, A. D., 1977, Isolation of adenylate cyclase-free β-adrenergic receptor from turkey erythrocyte membranes by affinity chromatography, *Proc. Natl. Acad. Sci. USA* **74**:3710–3714.

Weber, G., 1975, Energetics of ligand binding to proteins, *Adv. Prot. Chem.* **29**:1–83.

Westcott, K. R., LaPorte, D. C., and Storm, D. R., 1979, Resolution of adenylate cyclase sensitive and insensitive to Ca^{2+} and calcium-dependent regulatory protein (CDR) by CDR–Sepharose affinity chromatography, *Proc. Natl. Acad. Sci. USA* **76**:204–208.

Williams, L. T., and Lefkowitz, R. J., 1977, Slowly reversible binding of catecholamine to a nucleotide-sensitive state of the β-adrenergic receptor, *J. Biol. Chem.* **252**:7207–7213.

Williams, L. T., and Lefkowitz, R. J., 1978, *Receptor Binding Studies in Adrenergic Pharmacology*, Raven Press, New York.

Witkin, K. M., and Harden, T. K., 1981, A sensitive equilibrium binding assay for soluble β-adrenergic receptors, *J. Cyclic Nucleotide Res.* **7**:235–246.

Wolfe, B. B., Harden, K., and Molinoff, P. B., 1977, *In vitro* study of β-adrenergic receptors, *Ann. Rev. Pharmacol. Toxicol.* **17**:575–604.

Wrenn, S. M., Jr., and Homcy, C. J., 1980, Photoaffinity label for the β-adrenergic receptor: Synthesis and effects on isoproterenol-stimulated adenylate cyclase, *Proc. Natl. Acad. Sci. USA* **77**:4449–4453.

8

ACETYLCHOLINE RECEPTOR
Some Methods Developed to Study a Membrane-Bound Regulatory Protein

Susan E. Coombs and George P. Hess

1. INTRODUCTION

The acetylcholine receptor protein, found in the synaptic membranes of many nerve cells, at the neuromuscular junction, and in the electric organ of several fish, plays a key role in the transmission of nerve impulses (Katz, 1969; Nachmansohn, 1973; Nachmansohn and Neumann, 1975) and is being studied in many ways in many laboratories (Karlin, 1980; Changeux, 1981; Conti–Tronconi and Raftery, 1982). The electroplax of various fish (*E. electricus* and *Torpedo* spp.) has been used extensively in studies of the receptor protein. Here we will concentrate on methods developed in this laboratory to investigate the receptor and membrane-bound proteins in general. For studies of other membrane-bound proteins and the methods developed to study them, the reader is referred to Racker (1970), Kaback (1970), and Dewey and Hammes (1981).

The acetylcholine receptor from *Torpedo* species contains five major peptide chains, of molecular weight ranging from 40,000 for the α subunit to 65,000 for the γ subunit (Weill *et al.*, 1974; Reynolds and Karlin, 1978). The α subunit apparently carries the recognition site for chemical signals (Weill *et al.*, 1974; Reynolds and Karlin, 1978). The receptors from *Electrophorus electricus, Torpedo* spp., and mammalian muscle cells have antigenic properties in common (Claudio and Raftery, 1977; Patrick and Stallcup, 1979; Lindstrom *et al.*, 1980).

Susan E. Coombs and George P. Hess ● Section of Biochemistry, Molecular and Cell Biology, Division of Biological Sciences, Cornell University, Ithaca, New York 14853.

The properties of the receptor have been investigated in intact cells (Schoffeniels, 1957, 1959; Schoffeniels and Nachmansohn, 1957; Higman *et al.*, 1963; Karlin, 1967a; Sackmann and Neher, 1983), in membrane preparations (Kasai and Changeux, 1971a,b,c; Miledi *et al.*, 1971; Eldefrawi and Eldefrawi, 1973), and as the isolated protein (Karlsson *et al.*, 1972; Schmidt and Raftery, 1972; Biesecker, 1973; Eldefrawi and Eldefrawi, 1973; Klett *et al.*, 1973; Chang, 1974; Meunier *et al.*, 1974), and many reviews have appeared (Karlin, 1980; Changeux, 1981; Conti–Tronconi and Raftery, 1982; Hess *et al.*, 1983; Wennogle, 1984). An important primary event in the regulation of ion flux across many excitable membranes is believed to be the interaction between acetylcholine and its membrane-bound receptor (Katz, 1966, 1969). Several methods have been developed to study the relationship between the acetylcholine-binding process and the resulting translocation of inorganic ions through receptor-formed transmembrane channels and the associated inactivation of the receptor.

Membrane-bound rather than purified receptor was generally chosen for the studies to be described because the laboratory is mainly interested in the molecular processes involved in receptor-mediated changes in the permeability of cell membranes to inorganic ions (Hess *et al.*, 1983). The isolated acetylcholine receptor does not have a measurable function in solution, and changes in the ligand-binding properties are known to accompany the isolation process (Eldefrawi and Eldefrawi, 1973; Meunier *et al.*, 1974; Raftery *et al.*, 1975). A membrane preparation from the electric organ (electroplax) of *E. electricus* was chosen for most of the studies because it appeared to be uniquely suitable for finding a correlation between the interaction of chemical mediators with the receptor and the associated changes in the permeability of the membrane to specific inorganic ions. The chemical properties of the membrane (Nachmansohn, 1973; Nachmansohn and Neumann 1975) and the electrophysiological properties of single electroplax cells (Schoffeniels, 1957, 1959; Schoffeniels and Nachmansohn, 1957; Higman *et al.*, 1963; Karlin, 1967a,b; Lester *et al.*, 1975; Sheridan and Lester, 1975) had been investigated extensively, and Kasai and Changeux (1971a,b,c) had reported the preparation of electroplax membrane vesicles that exhibited receptor-mediated ion flux. For comparative purposes some studies were done with *Torpedo* spp. preparations.

Because the goal of the studies was an understanding of the relationship between the binding of acetylcholine to its receptor and the associated change in permeability of the membrane, the necessary initial step was to examine the equilibrium binding of activating and inhibitory ligands to the receptor protein, but first several methods, suitable for investigations of a membrane-bound protein, had to be developed. The equilibrium-binding studies indicated that, at least in *E. electricus* electroplax membrane preparations, there were partly interacting binding sites for activating and inhibiting ligands. In order to gain a better understanding of the interactions between the ligands and the membrane-bound receptor, the kinetics of the specific, irreversible reaction of an inhibitory ligand, α-bungarotoxin, with the receptor was then studied. It was possible to propose a simple model for the binding of ligands to the receptor that was consistent with all the equilibrium and kinetic data obtained. Because the reaction of toxin with the receptor was the best method available to compare results obtained in different laboratories, under different conditions, and with receptor protein obtained from different organisms, it was important to develop a simple, rapid, and reliable assay that could be used for both the solubilized and the membrane-bound receptor. The assay was then used to show that the reaction was multiphasic, regardless of whether the preparation was from *E. electricus* or from a species of *Torpedo*.

The binding of ligands to the receptor protein affects the function of the receptor. This was studied using the same *E. electricus* preparation in which Kasai and Changeux (1971a,b,c) had already demonstrated that receptor-mediated ion flux could be induced by activating cholinergic ligands. It became apparent that the preparation was functionally heterogeneous and it was necessary to fully characterize the vesicle populations. Once the preparation had been extensively characterized, it was possible to isolate the vesicles of interest on a preparative scale. During the characterization it was found that under conditions similar to physiological ones (high concentrations of K^+ internally and Na^+ externally) a very rapid ion movement occurred, induced by activating ligands, on a time scale that was comparable to the events observed physiologically, unlike the previous biochemical studies. It was particularly interesting to note that this rapid ion movement occurred only at acetylcholine concentrations much higher than was to be expected from the published equilibrium binding constants for acetylcholine (Karlin, 1980; Changeux, 1981). Several fast reaction techniques had to be adapted in order to measure the receptor function on the millisecond time scale. A minimum model was proposed, on the basis of the fast reaction measurements made with the *E. electricus* preparation, for the mechanism of action (both activation and inactivation) of the receptor. This was then extended to include the results obtained with a *T. californica* preparation, measured by using the inhibition of the receptor by its natural ligand, acetylcholine, to reduce flux rates that exceeded the time resolution of the fast reaction techniques. Recently single-channel currents have been recorded from the *E. electricus* electroplax cells from which the membrane vesicles are prepared. It was then possible to correlate the results obtained by chemical kinetics using vesicles with those obtained using electrophysiological techniques with cells.

The first methods to be described are those developed by Fu, Donner, and

Bulger to study the equilibrium binding of ligands (Fu *et al.*, 1974, 1977; Donner *et al.*, 1976; Bulger *et al.*, 1977).

2. RECEPTOR–LIGAND BINDING STUDIES

2.1. Equilibrium Studies of Ligand Binding in *E. electricus* Preparations

In order to study the binding isotherms of radioactive or chromophoric ligands to membrane-bound proteins, a number of problems must be solved. When binding isotherms are measured, it is necessary to use concentrations of the ligand being studied that approach the value of the dissociation constant of that ligand in order to ensure measurable differences between the concentration of the ligand bound to the protein on one side of the dialysis membrane and free ligand molecules on the other. This is not difficult to arrange when soluble proteins and ligands with high-affinity binding constants are used. However, a membrane-bound protein generally constitutes only a small part of the membrane and two problems arise. (1) Because the protein being studied forms only a small part of the membrane, solutions containing milligram quantities of the membrane must be used. This means that the volume occupied by the membrane becomes important because the ligand is excluded from this volume. The exclusion volume can introduce serious errors in an analysis of the binding isotherms of membrane-bound proteins (Donner *et al.*, 1976). However, corrections can be made. (2) The use of milligram quantities of membrane increases the problem of the ligand binding unspecifically to other components of the membrane that are necessarily also present. Often it is sufficient to determine the amount of labeled ligand bound in the presence of excess unlabeled ligand (Baxter and Tomkins, 1971). It is, however, possible to use the same experiments that give the corrections for the exclusion volume to correct for the unspecific (and low-affinity) binding for each ligand under investigation.

An approach that overcame these problems was developed. In the *E. electricus* preparation the receptor constitutes less than 2% of the membrane proteins (Kasai and Changeux, 1971a). Equilibrium dialysis or ultracentrifugation was used to study the equilibrium binding of [^{3}H]-decamethonium chloride to the receptor. This ligand initiates changes in electrical potential at nerve synapses and in single electroplax cells (Changeux *et al.*, 1969). Standard tritiated water was used to establish the efficiency of counting the radioactively labeled ligands. The efficiency of counting and the activity of known standard samples were determined under the same conditions as the experimental samples. The purity of the radioactive compounds was ascertained by thin-layer chromatography, and the specific activity of the compounds was used to convert the counts per minute to concentrations.

2.1.1. Equilibrium Dialysis Method

Microcells (Furlong *et al.*, 1972) consisting of two 100-μl compartments separated by a single thickness of pretreated dialysis tubing were used. Eighty microliters of the membrane suspension and 80 μl of a radioactively labeled ligand solution were allowed to equilibrate for 16 hr in slowly rotating (2 revolutions/minute) microcells which were hermetically sealed. The half-life for equilibration for each ligand was found to be less than 2 hr; thus complete equilibration was assured. More than 98% of the radioactive counts was recovered after equilibration, demonstrating that there was no significant loss due to binding to the dialysis cells or tubing. After equilibration, samples were taken from each compartment and the radioactivity counted (in duplicate) and the protein concentrations determined. The moles of ligand bound per milligram of membrane protein were calculated from the difference in the concentration of the radioactive ligand in the two chambers.

2.1.2. Ultracentrifugal Method

By the method of O'Brien and Gilmour (1969), membrane suspensions were allowed to equilibrate with an appropriate amount of radioactive ligand in polycarbonate tubes and shaken for 1 hr. In control experiments the equilibration time for decamethonium, the ligand to be used, was determined. The number of molecules bound was constant when incubated for 0.5–7 hr. Immediately before centrifugation triplicate samples were taken for counting, and one sample was used for protein determination. The tubes were spun for 1 hr at 100,000 g, and triplicate samples were withdrawn from the supernatant for counting. The difference in concentration before and after centrifugation equaled the amount of ligand bound by the known amount of membrane protein. Precautions were taken to avoid evaporation during centrifugation under vacuum.

2.1.3. Binding of a Radioactively Labeled Ligand

2.1.3a. Estimation of Unspecific Binding.　Membrane suspensions were treated with α-bungarotoxin for 1 hr and then Tetram for 0.5 hr. Tetram, a specific inhibitor of acetylcholinesterase, was added to prevent decamethonium binding to the enzyme. Decamethonium solutions in which the concentration of [³H]-decamethonium was varied were prepared. In the ultracentrifugal assay method, 250 μl of the pretreated membrane suspension were mixed with 50 μl of one of the various decamethonium solutions, which also contained Tetram, and allowed to stand for 16 hr. The specific radioactivity of decamethonium in the mixture of labeled and unlabeled decamethonium solutions was calculated as follows:

$$S_T(\text{mCi/mmol}) = [D^*] \times 418(\text{mCi/mmol})/([D^*] + [D])$$

$[D^*]$ is the concentration of labeled decamethonium with known specific radioactivity, and $[D]$ is the concentration of unlabeled decamethonium added, ranging from 0–200 μM. The specific radioactivity of the labeled decamethonium was 418 mCi/mmol. The total decamethonium concentration and the total amount of decamethonium bound to membrane vesicles per milligram of membrane protein was calculated using the specific radioactivity, S_T.

Typical results of equilibrium dialysis experiments are shown in Fig. 1a. The experiments were performed under identical conditions except that about half the specific receptor sites were irreversibly blocked by α-bungarotoxin in the experiment represented by curve 2. At high concentrations of decamethonium

---

→

FIGURE 1. (a) Binding of [³H]-decamethonium to membrane fragments. The data are plotted as r versus L: r is the number of moles of decamethonium bound per milligram of membrane-bound protein; L is the concentration of free decamethonium at equilibrium; r_O^S is the amount of specific binding sites for decamethonium, obtained by extrapolating the solid line to the ordinate [see Eq. (8.2)]; φ is a combined factor for unspecific binding of decamethonium and volume exclusion, obtained from the slope of the solid line. The points shown were obtained from membrane preparations from three eels. Curve 1, binding of decamethonium to membrane fragments treated with 0.1 mM Tetram: $r_O^S = 1.2 \times 10^{-11}$ mole/mg; φ = − 1.5 μl/mg membrane protein. Curve 2, binding of decamethonium to membrane fragments treated with 2 μM α-bungarotoxin and 1×10^{-4} M Tetram: $r_O^S = 0.6 \times 10^{-11}$ mole/mg; φ = 1.7 μl/mg membrane protein. (b) Estimation of volume occupied by membrane protein. [³H]-Inulin: 300 mCi/mmol. ●, Concentration of [³H]-inulin was kept constant at 0.4 μM; original protein content of membrane preparation was 10.6 mg/ml; ◇; concentration of [³H]-inulin was kept constant at 1.6 μM; □, △, ○, ▽, concentrations of [³H]-inulin were kept constant at 1.2, 2.4, 3.6, and 4.8 μM, respectively. Original protein content of membrane preparation was 18.2 mg/ml; specific activity of membrane-bound acetylcholinesterase was 0.9 mmol/mg/hr. Membrane preparations from three eels were used. The coordinates of the solid line were computed by a method of least squares. The slope of the line, V_E, is 3.1 ± 0.4 μl/mg membrane protein/ml solution. (c) Unspecific binding of [³H]-decamethonium to membrane fragments: △, original protein content of membrane preparation was 10.6–16.3 mg/ml, ultracentrifugal method; ○, original protein content of membrane preparation was 7.3–14.4 mg/ml, equilibrium dialysis method. Membrane preparations from four eels were used. Data obtained using the two experimental methods were analyzed by a linear least-squares computer program and gave $r_O^S = (1.2 \pm 0.1) \times 10^{-11}$ mole/ mg; φ = −(2.0 ± 0.1) μl/mg. Reprinted from Donner et al. (1976) with permission.

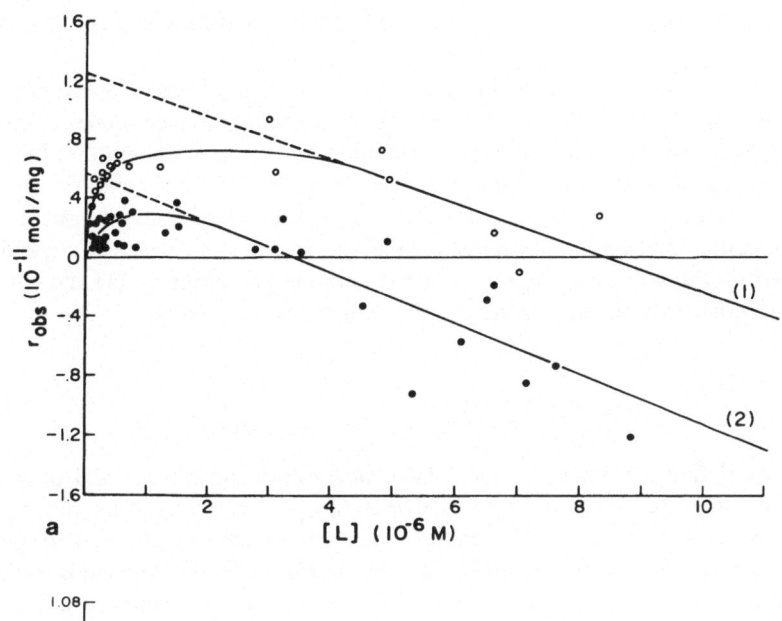

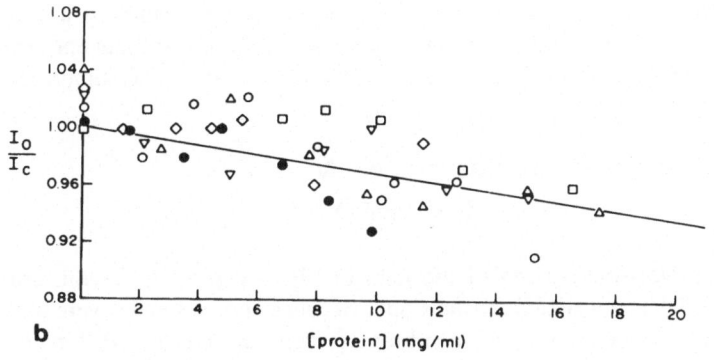

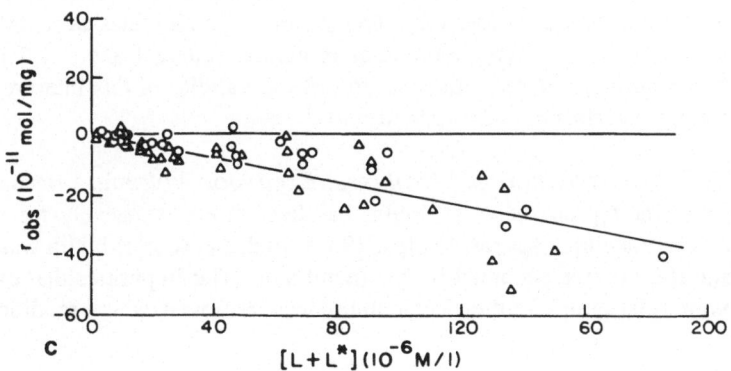

less ligand appears to be bound than at low concentrations. However, it was demonstrated that this reflects the volume occupied by the membrane fragments, a volume from which the ligand is excluded. It was also shown that the ordinate intercept of the solid line gives the concentration of specific decamethonium-binding sites. It was assumed that $r_{(obs)}$ in Fig. 1a depends on three factors: (1) specific binding, (2) unspecific binding, and (3) the moles of ligand excluded by the volume occupied by the membrane preparations. (1) and (2) contribute positively to $r_{(obs)}$, whereas (3) contributes negatively.

$$r_{(obs)} = r_o^S \frac{L}{L + K_S} + L\left(r_o^U \frac{1}{L + K_u} - V_E\right) \tag{8.1}$$

where r_o^S and r_o^U represent the total concentration of specific and unspecific binding sites, respectively, expressed as moles per milligram of membrane protein. K_S and K_U are the dissociation constants of the specific and unspecific binding sites, and V_E is the volume of solution from which the ligand is excluded per milligram of membrane protein. When experimental conditions are chosen so that $K_U > L > K_S$, and r_o^S becomes comparable to the subsequent terms of Eq. (8.1), the data shown in Fig. 1a are obtained and Eq. (8.1) simplifies to

$$r_{(obs)} = r_o^S + \phi L$$
$$\phi = (r_o^U/K_U) - V_E \tag{8.2}$$

The straight-line portion of the data in Fig. 1a gives r_o^S as the ordinate intercept, and ϕ is obtained from the slope. Because curves 1 and 2 were obtained with identical concentrations of membrane protein and the only difference was that about half the specific binding sites were blocked in the experiment represented by curve 2, the membrane exclusion volume and the amount of unspecific binding of the ligand were expected to be the same. Parallel lines were obtained with a ϕ value of -1.5 μl/mg membrane protein for curve 1 and -1.7 μl/mg membrane protein for curve 2. In order to test the validity of this interpretation the following experiments were carried out.

2.1.3b. Determination of Membrane Exclusion Volume. Because of its relatively inert properties [³H]-inulin has been used extensively for determination of extracellular spaces (Phelps, 1965). In this case [³H]-inulin was used to estimate the volume occupied by the membrane. The impurities that existed in the commercial sample of the labeled inulin were removed using a modification of Cohen's procedure (Cohen, 1969; Phelps, 1965; Donner et al., 1976). Eight suspensions containing different amounts of membrane proteins were prepared by dilutions with eel Ringer's solution (Keynes and Martins-Ferreira, 1953). One microliter of each solution was then mixed with 80 μl of the desired inulin solution in polycarbonate tubes. Samples were taken for radioactivity counting (in triplicate) and for protein determination. After centr..ugation at 100,000 g for 1 hr, triplicate samples were withdrawn from the supernatant for counting. The radioactive counts per minute and protein concentration of the samples before and after centrifugation were determined. The purified [³H]-inulin did not bind to the membrane suspension and remained in the supernatant. The protein assay (Lowry et al., 1951) indicated that all the membrane proteins were removed from the supernatant under these conditions. The ratio I_o/I_c can be determined as a function of both inulin and membrane protein concentration; I_o and I_c represent the radioactive counts from [³H]-inulin before and after centrifugation, respectively. If the membrane preparation does not bind or exclude inulin from the solution, I_o/I_c is 1. The relation between I_o/I_c and the exclusion volume is given by

$$I_o/I_c = 1 - V_E[P] \qquad (8.3)$$

where $[P]$ is the concentration of membrane protein per microliter of solution.

Experiments carried out in this way are illustrated in Fig. lb. V_E varies linearly with the protein concentration. The slope of the line gives V_E, and a value of 3.0 ± 0.3 μl/mg of membrane protein was thus obtained. The membrane preparation contained vesicles, and the value obtained using inulin is consistent with the apparent internal volumes measured using $^{22}Na^+$ and the same preparation (Kasai and Changeux, 1971c).

2.1.3c. Determination of Unspecific Binding of Decamethonium. An isotope dilution technique was used to investigate the specific and unspecific binding of decamethonium. The concentration of labeled decamethonium was held constant, while various amounts of unlabeled decamethonium were added. The amount of [³H]-decamethonium bound, $r^*_{(obs)}$ can be expressed as follows:

$$r^*_{(obs)} = \frac{r_o s}{1 + (K_s/L^*)[1 + (L/K_s)]}$$

$$+ \frac{r_o s}{1 + K_u/L^*(1 + L/K_U)} - V_E(L + L^*) \qquad (8.4)$$

$r^*_{(obs)}$ represents the moles of radioactive ligand, L^*, bound in the presence of unlabeled ligand, L, which acts as a competitive inhibitor of L^*.

The total amount of decamethonium bound, $r_{(obs)_T}$, can be calculated from the corrected specific activity and the relationship

$$\frac{r^*_{(obs)}}{r_{(obs)_T}} = \frac{L^*}{L + L^*} \tag{8.5}$$

Therefore, if $r_o{}^S$ and $r_o{}^U$ in Eq. (8.4) are multiplied by $(L^*)(L + L^*)^{-1}$, $r_{(obs)_T}$ is obtained. The conditions for the experiment shown in Fig. 1c were chosen so that $L > L^*$ and $K_U > K_S$, which leads to a simplification of Eq. (8.4), namely Eq. (8.2), except that ϕ is multiplied by $(L + L^*)$. When the data were analyzed by a least-squares method, it was found that the number of specific binding sites, obtained from the ordinate intercept of the graph, was $1.2 \pm 0.1 \times 10^{-11}$ mol/ mg of membrane protein. A value of $\phi = -2 \pm 0.1$ µl/mg membrane protein was obtained. Using the value of 3 µl/mg of membrane protein as the value for V_E, obtained from the experiments shown in Fig. 1b, the value for $r_o{}^U/K^U$ is 1×10^{-3}/mg of membrane protein [Eq. (8.2)]. This value of $r_o{}^U/K^U$ indicates that unspecific binding of decamethonium to the membrane suspension constitutes almost one third of the total binding when the concentration of free decamethonium is equal to the dissociation constant for the decamethonium : membrane site complex.

The experiments indicated that in equilibrium measurements corrections for unspecific binding and membrane exclusion volume can easily be obtained, in a single experiment for each membrane preparation and each ligand used, by determination of ϕ as illustrated in Fig. 1a. Two independent methods were used to demonstrate the validity of this approach. The value for ϕ of -2 ± 0.1 µl/ mg of membrane protein determined by an independent method was in good agreement with the ϕ values obtained with decamethonium directly (Fig. 1a). It must be emphasized that the curves shown in Fig. 1a are not normally seen, but the experimental conditions were chosen to obtain the straight-line portion of the graphs used for evaluating ϕ.

The binding of ligands, such as decamethonium, to electroplax membranes (Kasai and Changeux, 1971c; Fu et al., 1974; Meunier et al., 1972; Weber and Changeux, 1974) and other membrane preparations had been studied extensively. There had been little agreement as to the number of binding sites or whether the binding isotherms showed cooperativity or not (Kasai and Changeux, 1971c; Eldefrawi et al., 1971a,b; Weber and Changeux, 1974). Determination of the correction factor, ϕ, for each ligand used and correction of the binding isotherms made it possible to obtain interesting information that had been obscured in previous studies.

Because decamethonium and carbamoylcholine also bind to acetylcholinesterase, which is present in the membranes, some of the preparations used were extracted with 1 M NaCl before use. This procedure removes most of the esterase (Silman and Karlin, 1967). When binding of [^{3}H]-decamethonium to the membrane preparation was then measured, complex data were obtained. In Fig. 2a the number of moles of ligand bound per milligram of membrane protein is plotted as a function of the concentration of free ligand. Curve 1 was obtained when no inhibitors were present and curve 2 when the membrane-bound acetylcholinesterase was specifically inhibited by 3-hydroxyphenyltrimethylammonium iodide ((3-HOPTA) (Wilson and Quan, 1958) at a concentration that prevented the binding of decamethonium to the enzyme (Fu *et al.*, 1977). From this curve it can be seen that decamethonium binds to sites on the membrane other than the enzyme, presumably to the acetylcholine receptor.

It had already been shown (see above) that the appropriate correction terms for specific and unspecific binding to the membranes, and for the membrane exclusion volume, can be obtained from the linear portions of the graph (Fig. 2a, curve 2). The slope of line gives ϕ, which contains the correction terms. When curve 2 was corrected appropriately, curve 2a was obtained. If the data in curve 2 were not corrected and were presented in a Scatchard plot (Scatchard, 1949), more than one binding site or cooperative effects apparently existed (Fig. 2b).

When the corrected data are presented in a Scatchard plot (Fig. 2c), the ordinate intercept gives $r_o{}^S$, and the slope of the lines provides the value of the dissociation constant for the ligand : binding site complex. These data are consistent with the presence of one type of binding site. Similar results were obtained whether 6 μM 3-HOPTA or 0.1 mM Tetram was used as an enzyme inhibitor. The experimental results obtained with preparations from which almost 90% of the acetylcholinesterase had and had not been removed by salt extraction are summarized in Table I. The dissociation constant of the decamethonium : receptor site complex (K_D) is in the range 0.2–0.4 μM. The average concentration of decamethonium-binding sites (1×10^{-11} mol/mg of membrane protein) corresponds, within experimental error, to the number of moles of α-bungarotoxin-binding sites [$0.9 \pm 0.2 \times 10^{-11}$ mol/mg of membrane protein] (Bulger and Hess, 1973).

It can also be seen from Table I that decamethonium still binds to membrane vesicles that have been allowed to react stoichiometrically with α-bungarotoxin. In the presence of either 3-HOPTA or Tetram, toxin reduces the number of

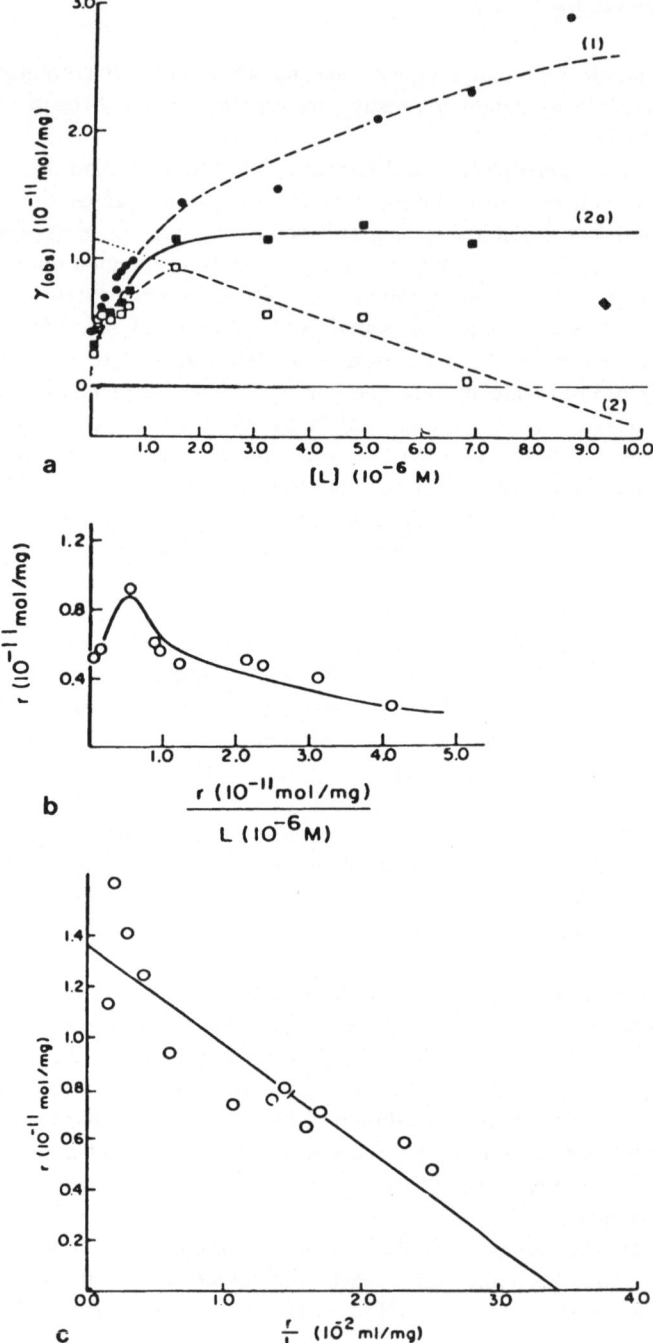

decamethonium-binding sites by only about 50% and changes the dissociation constants of the remaining sites by a factor less than 2 (Table I). It could also be shown that these results were not due to the binding of decamethonium to the enzyme [see Table I] (Fu *et al.*, 1977).

The following new information had been obtained by the methods described (Fu *et al.*, 1974, 1977; Donner *et al.*, 1976). The activating ligands decamethonium and carbamoylcholine occupy overlapping binding sites. The activating ligands and inhibitors (α-bungarotoxin and d-tubocurarine) compete for only one half of the sites available to them even though the stoichiometry of the sites is 1 : 1 as measured with decamethonium and α-bungarotoxin.

2.2. Kinetic Studies of Ligand Binding in *E. electricus* Preparations

Partly interacting binding sites for activating and inhibiting ligands, observed in the equilibrium binding measurements, had been observed in enzyme-catalyzed reactions. Different molecules, preexisting nonequivalent binding sites (MacQuarrie and Bernhard, 1971a,b), or an allosteric mechanism that involves ligand-induced conformational changes (Koshland *et al.*, 1966; Koshland, 1970; Conway and Koshland, 1968) are often invoked to account for such observations with well-characterized enzymes. Kinetic studies were undertaken by Bulger, Fu, Silberstein, and Donner, in the hope of further characterizing the interaction of ligands with the membrane-bound receptor. The kinetics of the specific and irreversible reaction of [^{125}I]-monoiodo-α-bungarotoxin (Chang and Lee, 1963; Lee and Chang, 1966; Lee *et al.*, 1967, 1972; Lee, 1972) with the membrane-bound receptor were studied.

$\leftarrow$
___

FIGURE 2.   (a) Binding of [^{3}H]-decamethonium to electroplax membrane preparations; pH 7.0, $\mu = 0.18$ M, eel Ringer's solution. $\bullet$, decamethonium binding (curve 1); $\square$, decamethonium binding in the presence of 6 μM 3-HOPTA (curve 2). $r_O^S = 1.2 \times 10^{-11}$ mole/mg; $\phi = 1.5$ μl/ mg. $\blacksquare$, data of curve 2 after corrections for unspecific binding and volume exclusion (curve 2a). (b) The data shown in curve 2 of Fig. 2a are replotted, according to Scatchard (1949), r versus r/L. Protein content was 14 mg of membrane protein per milliliter. The specific activity of the membrane-bound acetylcholinesterase was 2 mmol mg^{-1} h^{-1}, where mg represents milligrams of membrane protein. Equilibrium dialysis was used. (c) Binding data of [^{3}H]-decamethonium to electroplax membrane preparations, pH 7.0, 4°C, $\mu = 0.18$ M, eel Ringer's solution, are presented in the form of Scatchard plots. $\bigcirc$, decamethonium binding in presence of 0.1 mM Tetram. $r_O^S = (1.4 \pm 0.1) \, 10^{-11}$ mole/mg; $K_D = 0.4 \pm 0.06$ μM. Duplicate determinations were made at each decamethonium concentration used. For each membrane preparation ϕ values were obtained, and the measurements were corrected for unspecific binding and volume exclusion. The data points shown represent the average value obtained from experiments with membrane preparations from two eels. Protein content was 13–14 mg of membrane protein per milliliter in the various experiments. Equilibrium dialysis was used. Reprinted from Fu *et al.* (1977) with permission.

TABLE I. Specific Binding of Decamethonium to Membrane Fragments, Eel Ringer's Solution, pH 7.0, 4°C, $\mu = 0.18$ M[a]

Membrane preparation	Treatment	$r_O^S(10^{-11}$ mol/mg)	α-Bungarotoxin-binding sites (10^{-11} mol/mg)	$K_D^S(\mu M)$	No. of membrane preparations used[b]
Native	3-HOPTA (6 μM)	0.9 ± 0.03	0.9 ± 0.2[c]	0.3[d]	4
Native	3-HOPTA (6 μM), α-bungarotoxin (2 μM, 1 hr)	0.4 ± 0.02		0.5[d]	4
Native	Tetram (0.1 mM, 0.5 hr)	1.4 ± 0.1		0.4	2
Salt extracted	3-HOPTA (6 μM)	1.0 ± 0.1[d]		0.2[d]	2
Native		5.9[e]	10.8[e] 1–2[f]		

[a] Reprinted with permission from Fu et al. (1977).
[b] Each from a different eel.
[c] Bulger and Hess, 1973.
[d] Fu et al., 1974.
[e] Determined from the amount of bound [14C]-decamethonium displaced by α-bungarotoxin. The acetylcholinesterase was not inhibited. An ultracentrifugal method was used (Kasai and Changeux, 1971c).
[f] The number of α-bungarotoxin-binding sites was estimated from the titration curve of [14C]-decamethonium binding sites with α-bungarotoxin (Kasai and Changeux, 1971c).

The α fraction of the crude venom was isolated according to the method of Li (1968), and the α-bungarotoxin was iodinated (^{125}I) according to the method of Reif (1967) (Bulger et al., 1977). Several peaks containing radioactivity and protein were obtained. The major peaks corresponded to diiodotoxin, unlabeled unreacted toxin, and the monoiodotoxin which was used in the experiments. It is important in such experiments to use a homogeneous preparation of the iodinated ligand because of the necessity to determine the stoichiometry of the reaction and because of the possibility that the various toxin derivatives react differently.

The toxin concentration was determined by absorbance at 280 nm. The molar extinction coefficient of the toxin, based on the dry weight of the lyophilized powder and a molecular weight of 7904 (Clark et al., 1972), was 9500 M^{-1} cm^{-1} at 280 nm. A dilution technique was used to determine the specific activity in order to avoid the problems of labeled toxin binding unspecifically to the glassware. Five microliters of [^{125}I]-toxin were diluted with 25 μl of unlabeled toxin and 2 ml of bovine serum albumin (12 μg/ml). The syringe used for pipetting the [^{125}I]-toxin was equilibrated with the labeled toxin by withdrawing the samples at least 50 times. Ten-, twenty-, and thirty-microliter samples, in duplicate, of the diluted solution were counted. The counts/minute and the number of moles of [^{125}I]-toxin are linearly related, and the slope of the line gives the specific activity. A simulated ^{125}I standard was used to calculate the counting efficiency, which was taken as the ratio of observed counts/minute to expected disintegrations/minute. Using a $t_{1/2}$ for ^{125}I of 60 days, the specific activity could be determined for any time after the initial determination. Using the molar extinction coefficient and the specific activity of the toxin, it was determined that the labeled derivative contained 1 mole of ^{125}I per mole of toxin. The purity of the preparations used in the kinetic experiments was checked by electrophoresis on acrylamide gels according to the method of Davis (1964). Electrophysiological experiments were conducted by Drs. E. Bartels–Bernal and W. Niemi (Columbia University) on E. electricus electroplax and the response to the labeled and the unlabeled toxin was the same.

2.2.1. Measurement of Total Toxin Binding

The total amount of reversibly plus irreversibly bound toxin was determined by an ultracentrifugal method. Varying amounts of labeled toxin were added to a membrane suspension in eel Ringer's solution (Keynes and Martins–Ferreira, 1953) in a polycarbonate centrifuge tube in the presence or absence of various inhibitors of the binding reaction. The concentration of membrane protein ranged from 5.3–17 mg/ml. Toxin and membrane were incubated for the appropriate time, usually 60 min, during which duplicate samples were removed using pipettes with disposable plastic tips. The plastic tip and the sample were counted together in the same vial. At the end of the incubation period, the reaction mixture was centrifuged for 1 hr at 40,000 revolutions/minute. Triplicate samples of the supernatant were counted. More than 97% of the membrane protein was found in the pellet. From the difference in the concentration of toxin before and after centrifugation and the amount of protein sedimented, the concentration of bound toxin was determined and expressed as moles of toxin bound per milligram of protein. The concentration of free toxin was obtained from the reaction mixture *after* centrifugation.

2.2.2. Measurement of the Specific Irreversible Toxin Binding

About 18 mg of membrane protein were used to determine the concentration of labeled toxin that was bound irreversibly. An important aspect of the method is that in an experimental run the milligrams of membrane protein and the moles of labeled toxin were the same in each reaction vessel. Different volumes of eel Ringer's solution (against which the membrane preparation had been dialyzed) were added to each reaction tube. The volumes used were varied from 3–30 ml in order to obtain initial toxin concentrations in the 0.05- to 0.5-μM range. The concentration of labeled toxin was always ten times greater than the total number of toxin-binding sites in the solution, so that an approximately constant concentration of free toxin was maintained throughout the reaction. The reaction was initiated by addition of approximately 50 μl of labeled toxin to the membrane suspension. At appropriate time intervals (shorter at the beginning in order to obtain as many points during the initial fast phase as during the slower phase), samples containing approximately 0.6 mg of protein and 80 pmole of labeled toxin were transferred from the reaction mixture to a quench solution of 0.1 mM *d*-tubocurarine in eel Ringer's solution in polycarbonate centrifuge tubes. Either glass pipettes or pipettes with disposable tips were used. This dilution to a final volume of 8.5 ml resulted in a concentration of the free toxin in the quench solution of approximately 7 nM. The tubes were centrifuged for 1 hr at 105,000 g. The supernatant was counted, and the concentration of free toxin was determined. Experiments in which the free toxin concentration was found to be lower than 90% of the theoretical amount were discarded. The 10% loss was attributed to toxin that was bound specifically to the membrane-bound receptor sites and to experimental error. When control experiments were run with concentrations of 0.1 μM labeled toxin but in the absence of a membrane suspension, 60% of the toxin was apparently lost after 1.5 hr. In the absence of membrane or of bovine serum albumin, the toxin binds to glass and to the polycarbonate centrifuge tubes.

The amount of labeled toxin that was trapped in the pellet but which had not reacted with the membrane was determined. Samples of the membrane preparation, containing approximately 0.6 mg of membrane protein, were added to a quench solution with a final volume of 8.5 ml of eel Ringer's solution, 0.1 mM d-tubocurarine, and 7 nM [^{125}I]-toxin. The amount of labeled toxin per milligram of protein in the pellet was determined as described previously and this blank, usually about 5% of the total number of binding sites, was subtracted from each experimental determination of the amount of labeled toxin bound per milligram of protein. The total number of toxin-binding sites was determined for each membrane preparation used by incubating the preparation in 0.5–0.8 μM [^{125}I]-toxin for 5 hr. Control experiments lasting 24 hr indicated that 5 hr was adequate.

Bulger (Bulger and Hess, 1973) had shown that the irreversible reaction of 0.5 μM [^{125}I]-toxin (the highest initial toxin concentration used in the kinetic experiments) with membrane sites is prevented during 5 hr in the presence of 0.1 mM d-tubocurarine. The same result was obtained when d-tubocurarine was replaced by 0.16 mM decamethonium (Bulger and Hess, 1973).

To show irreversibility during the time of measurement, the following experiments were performed: (1) The membranes were incubated with 0.5 μM [^{125}I]-toxin for 3 hr and the amount of toxin bound was determined. Three other experiments were run in which after 3 hr of treatment with 0.5 μM [^{125}I]-toxin, the reaction mixture was diluted 100-fold with eel Ringer's solution containing (2) 2.2 μM toxin, (3) 0.16 mM decamethonium, and (4) 2.2 μM toxin and 0.16 mM decamethonium. After 15 hr, the amount of toxin bound irreversibly was essentially the same in the experiments.

The picomoles of toxin bound per milligram of membrane protein were (1) 4.8, (2) 4.7, (3) 4.5, and (4) 4.7 indicating the irreversibility of the reaction. The concentration of toxin-binding sites was found to be $0.9 \pm 0.2 \times 10^{-11}$ mol/mg of membrane protein, which corresponds, within experimental error, to the number of moles of binding sites for a reversible ligand, decamethonium, (Fu et $al.$, 1974) (see Section 2.1.3c and Table I).

The data in Fig. 3a show that the reaction blank depends on the concentration of membrane protein in the reaction. The experimental conditions were chosen so that the reaction blank corresponded to no more than 5% of the infinity point, or to less than 0.5% of the labeled toxin used in the experiments.

The time course of the reaction of [^{125}I]-toxin with membrane sites is shown in Fig. 3b. The fraction of unreacted membrane sites is plotted on a logarithmic scale as a function of time. Each curve corresponds to a different initial concentration of labeled toxin that was always ten times greater than that of the

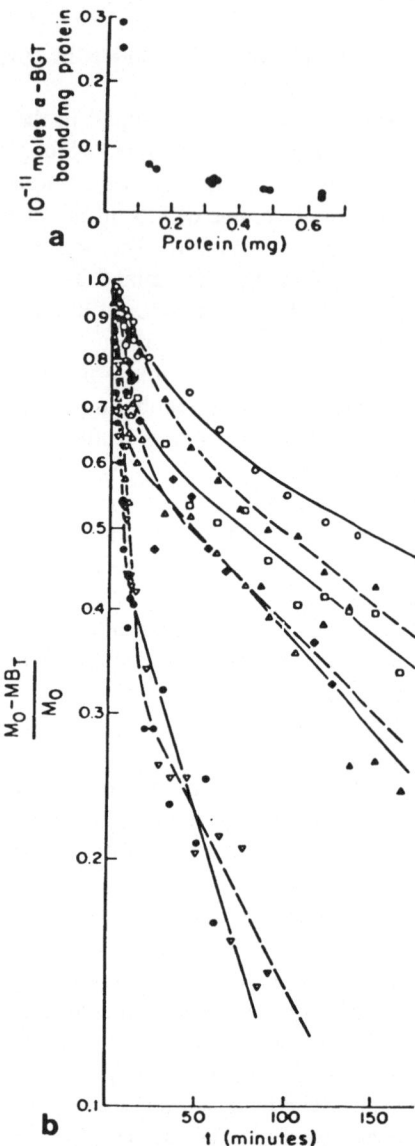

FIGURE 3. (a) Dependence of the pellet blank on the size of the pellet. The amount of $[^{125}I]$iodo-α-bungarotoxin trapped in the pellet but that had not reacted specifically with the membrane was determined by the addition of an aliquot of membrane preparation and sufficient $[^{125}I]$iodo-α-bungarotoxin to give a final concentration of 7 nM in the quench solution (see text). After centrifugation at 105,000 g for 1 hr, the amounts of toxin and protein in the pellet were determined. (b) The irreversible reaction of $[^{125}I]$iodo-α-bungarotoxin with *E. electricus* membrane preparations pH 7.0, 4°C, eel Ringer's solution. The amount of irreversibly bound $[^{125}I]$iodo-α-bungarotoxin was determined as described. The data are plotted on a semilogarithmic scale as the fraction of unreacted

$$B_0 + \; (M) \; \underset{}{\overset{K_1}{\rightleftharpoons}} \; (MB_1) \; \underset{k_{32}}{\overset{k_{23}}{\rightleftharpoons}} \; (\overline{MB_1}) \; \underset{}{\overset{K_2}{\rightleftharpoons}} \; (\overline{MB_2})$$

$$\Big\downarrow K_1 \qquad\qquad \xrightarrow{\; k_I \;} \qquad\qquad \Big\downarrow k_{II}$$

$$(MB_2) \qquad\qquad\qquad\qquad\qquad MB_T$$

$$MB_T = M_0\left(1 - (1 - f_s)e^{-k_I^\bullet t} - f_s\, e^{-k_{II}^\bullet t}\right) \qquad (I)$$

$$k_I^\bullet = \frac{2B_0}{K_1}(k_{23} + k_I') = k_I^* B_0 \qquad (II)$$

$$k_{II}^\bullet = k_{II}\frac{B_0}{B_0 + 2K_2} \qquad (III)$$

$$f_s = \frac{k_{23}\dfrac{2K_1}{B_0}}{k_I + \dfrac{2K_1}{B_0}(k_{23} + k_I)}$$

C

membrane sites versus time. M_o represents the initial concentration of receptor sites and MB_t represents the irreversibly formed toxin–receptor complex at time t. Various initial toxin concentrations were used: $\bigcirc$, 0.05; $\blacktriangle$, 0.07; $\square$, 0.08; 0.1; $\blacklozenge$, 0.18; $\triangledown$, 0.32; $\bullet$, 0.50 μM. The observed time course of the reaction was fitted to the sum of two exponentials using a nonlinear least-squares computer program, which yielded the coordinates of the solid lines. (c) Schematic representation of the allosteric model for the interaction of α-bungarotoxin with the membrane-bound acetylcholine receptor. The minimum model proposed (Hess *et al.*, 1975) required that the receptor molecule has at least two subunits. The squares and circles designate different subunit conformations. B_o represents the initial bungarotoxin concentration and M the concentration of free receptor sites. K_1 is the dissociation constant of the low-affinity site and K_2 the dissociation constant of the high-affinity site. k_I and k_{II} represent the rate constants for the formation of the irreversible receptor:toxin complexes, and k_{23} and k_{32} are the rate constants for the protein isomerization. Although considered in the development of the rate equation, the irreversible formation of MB_1 and $\overline{MB}_1$ complexes is not shown for aesthetic reasons. (d) The dependence of k_I^0, and k_{II}^0, and f_s on initial toxin concentration. Membrane preparations from four eels were used. Curves, such as those shown in Fig. 3b, were constructed for each initial concentration of [^{125}I]iodo-α-bungarotoxin used, and the exponentials and the 0 time intercept of the slow phase of the reaction (f_s) were calculated using a nonlinear least-squares computer program. Shown in d(A), d(B), and d(C) are the parameters plotted as a function of initial toxin concentration. The coordinates of the solid lines were obtained by using a linear least-squares computer program. The four eels are designated by different symbols. (A) The dependence of k_I^0 on initial toxin concentrations. The slope of the line has a value of 7×10^5 M^{-1} min^{-1}. (B) The evaluation of K_2 and k_{II} from the dependence of k_{II}^0 on initial toxin concentration. The intercept, reflecting k_{II}, has a value of 0.012 min^{-1}; the slope, reflecting K_2, has a value of 0.1 μM. The points in parentheses were not included in calculations of parameters of the solid line. (C) The dependence of $1/f_s$ on initial toxin concentration. The slope of the line has a value of 1.5 μM^{-1}. The value of the intercept is 1.3. Reprinted from Bulger *et al.* (1977) with permission.

toxin-binding sites. The reaction falls into two steps, an initial fast phase and a subsequent slower one. The ordinate intercepts, obtained by extrapolating the progress curve of the slow phase of the reaction to zero time, give the fraction of sites that react slowly (f_S) and the fraction reacting in the fast phase $(1 - f_S)$. The kinetic data were analyzed by fitting the time course of the reaction to the sum of two exponentials [Fig. 3c, Eq. (I)]. In this equation, MB_t represents the membrane site : toxin complexes formed irreversibly at time t and M_o the initial concentration of receptor-binding sites. $k_I{}^o$ and $k_{II}{}^o$ are the observed rate constants for the fast and slow phase of the reaction, respectively. The coordinates of the solid lines in Fig. 3 were obtained by using a nonlinear least-squares computer program that gave the values of the two exponentials and f_S. The data show that both phases of the reaction, and the fraction of the reaction that goes by the slow phase, depend on the initial toxin concentration, B_o. The concentration dependence of $k_I{}^o$, $k_{II}{}^o$, and f_S is illustrated in Fig. 3d. For each initial concentration of $[^{125}I]$-toxin used in the reaction, the extent of the irreversible reaction was determined as a function of time, as shown in Fig. 3b, and $k_I{}^o$, $k_{II}{}^o$, and f_S were evaluated. Figure 3d(a) shows the dependence of $k_I{}^o$ on the initial toxin concentration. The data are consistent with either a bimolecular reaction or a rapid formation of a reversible, low-affinity toxin : receptor complex preceding the irreversible reaction. These alternatives could not be resolved because one

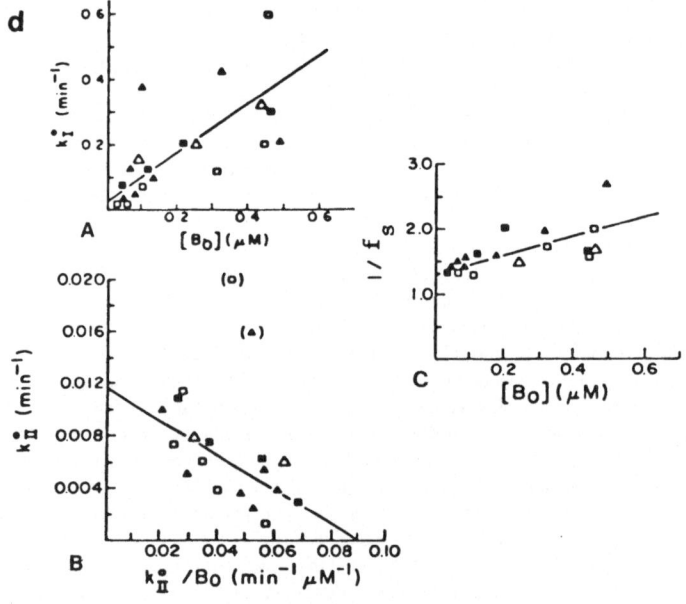

FIGURE 3. (*Continued*)

is limited in the initial toxin concentrations that can be used experimentally. In both cases, the observed rate constant has the same dimensions and can be expressed as $k_I^o = k_I''B_o$ [see Fig. 3c, Eq. (II)]. The slope of the line in Fig. 3d(a) gives a value of 7×10^5 M^{-1} min^{-1} for k_I''. k_{II}^o does not increase linearly with increasing toxin concentration but reaches a limiting value (Fig. 3d(b). A minimum mechanism describing the data is shown in Eq. (8.6).

$$M_o + B_o \xrightarrow{K_2} \overline{MB} \xrightarrow{k_{II}} \overline{MB}_t \qquad (8.6)$$

$\overline{MB}$ represents a reversibly formed toxin : receptor complex with dissociation constant K_2. The formation of the complex is considered to be rapid compared to formation of the irreversible complex $\overline{MB}_t$, a process characterized by the rate constant k_{II}.

For the mechanism shown in Eq. (8.6), the exponential k_{II}^o has the form given in Fig. 3c, Eq. (III). In Fig. 3d(b), the data are replotted according to a linear form of Eq. (III) in Fig. 3c. As predicted by the mechanism, a straight line is obtained with the slope and intercept corresponding to $2K_2 = 0.1$ μM and $k_{II} = 12 \times 10^{-3}$ min^{-1}, respectively.

Fig. 3d(c) shows a plot of f_S^{-1} as a function of the initial toxin concentration. This plot demonstrates an unusual feature of the data: the fraction of the reaction that proceeds by the slow phase, f_S, and that is characterized by a high affinity for the toxin, decreases with increasing initial toxin concentration; the fraction of the reaction $(1 - f_S)$ that proceeds rapidly and that is characterized by a low affinity for the toxin increases correspondingly.

The corrections that need to be made in equilibrium measurements are avoided by the kinetic approach illustrated here. The approach is similar to one used previously in investigations of chymotrypsin- and lysozyme-catalyzed reactions in which specific chromophoric inhibitors with high-affinity binding constants were used to measure interactions with the substrate-binding sites for colorless substrates that bind only poorly to the enzyme (Hess et al., 1971; Holler et al., 1975a,b).

In the reaction of labeled toxin with the membrane-bound receptor, two factors needed to be considered: (1) reversibly and unspecifically bound toxin (Bulger and Hess, 1973) and (2) the tendency of toxin to stick to surfaces. The reversibly bound toxin, which constitutes a major portion of the total toxin bound (Bulger and Hess, 1973), had not been considered in several reports in the literature (Kasai and Changeux, 1971c; Barnard et al., 1971; Miledi and Potter, 1971; Miledi et al., 1971) determining the total number of toxin-binding sites. This could explain why the number of toxin molecules bound was apparently larger than the number of decamethonium molecules displaced (Kasai and Changeux, 1971c), whereas in the experiments described here the ratio of toxin sites to decamethonium sites was 1 : 1 (Fu *et al.*, 1974). The method developed results in low and reproducible blanks and allows one to determine the material balance of both protein and radioactivity.

The model proposed on the basis of the kinetic studies (Fig. 3c) is consistent with all the kinetic (Bulger *et al.*, 1977) and equilibrium data (Fu *et al.*, 1977). The requirement for a ligand-induced conformational change is attractive because of the phenomena to be explained: the inactivation of the receptor produced by the interaction of acetylcholine with the receptor. The model is similar to one proposed earlier by Katz and Thesleff (1957) on the basis of electrophysiological measurements with cells and to the mechanism that adequately accounts for the mode of action of well-characterized regulatory enzymes (Monod *et al.*, 1965; Koshland *et al.*, 1966; Eigen, 1967; Hammes and Wu, 1974).

2.3. A Simple Quantitative Assay of Toxin Binding to the Solubilized and the Membrane-Bound Receptor

Because assays for the receptor often depend on the formation of an essentially irreversible complex between the protein and the toxin, a variety of methods have been used to separate the complex from the unreacted species. The detergent-solubilized receptor and/or the solubilized complex has been assayed by adsorption onto DEAE–cellulose filter paper discs (Schmidt and Raftery, 1973b; Fulpius *et al.*, 1972). The method is convenient and useful when high concentrations of receptor sites exist, as in *Torpedo* spp. membranes. It is, however, not readily applicable to the low concentrations that exist in *E. electricus* membranes. At high toxin concentrations, when high concentrations of membrane protein must be used to obtain a measurable incorporation of labeled toxin, the filter technique gives a reaction blank several times higher than the blank obtained in the centrifugation assay (Bulger and Hess, 1973). The filter disc assay is also not suitable when the protein concentration in the toxin-labeled membrane preparation must be determined. Because the determination of the number of toxin-binding sites was the best method yet available to compare preparations of the acetylcholine receptor in different forms, from different sources, and between different laboratories, a simple, quantitative method to assay the relatively low concentrations of the toxin-binding sites on solubilized and membrane-bound acetylcholine receptor was developed by Kohanski, with Andrews, Wins, and Professors A. T. and M. E. Eldefrawi (Kohanski *et al.*, 1977). The method had to satisfy the following requirements: (1) complete recovery of reacted and unreacted toxin in relatively small volumes; (2) efficient isolation of the specific and irreversible receptor : labeled toxin complex when the assay demands a large excess of unlabeled toxin for quenching (a 20-fold excess of unlabeled over labeled toxin); (3) isolation of the toxin : receptor complex in order to measure the protein concentrations—this is a necessity in experiments covering a wide range of receptor concentrations; and (4) a low blank over a wide range of binding-site concentrations.

Solubilized receptor purified from *T. ocellata* electroplax (Eldefrawi and Eldefrawi, 1973) was used in 5 mM Na^+ phosphate buffer, pH 7.4. The specific binding activity showed a maximum concentration of 12 nmol of actylcholine-binding sites per milligram of protein. Protein concentrations varied between 0.06 and 0.12 mg/ml. The membrane-bound receptor from *E. electricus* electroplax was reacted with toxin in buffer II (0.9 M sucrose, 190 mM CsCl, 1 mM Na_2HPO_4, pH 7.0). The concentration of protein was typically 6.0 mg/ml at $0.7 \pm 0.1 \times 10^{-11}$ mol of toxin-binding sites per milligram of protein. The salt was removed by exhaustive dialysis against buffer II but without CsCL. Duplicate aliquots (300–600 μl) were applied to carboxymethylcellulose columns. Before use carboxymethylcellulose was equilibrated with buffer I (100 mM Na_2HPO_4, 0.01% (v/v) Triton X-100, 0.03% NaN_3 (w/v), pH 702) until the pH of the supernatant was constant and equal to that of the buffer. Minicolumns were prepared from disposable Pasteur pipettes; the thin tips were removed and fitted with a brittle glass wool plug. Each column contained 1.25 ml of carboxymethyl cellulose.

2.3.1. Measurement of the Toxin : Receptor Complex and Determination of the Blank

The receptor and labeled toxin were incubated under a variety of conditions (to be described later) and added to the columns. In this method, the toxin : receptor complex appeared in the eluate, and the unreacted toxin remained associated with the carboxymethylcellulose. The eluate and the resin were counted separately, as well as samples of the incubation mixture. To overcome fluctuations in counting efficiency, samples of the original stock solution of labeled toxin were counted with each experiment to determine the current specific activity.

2.3.2. Membrane-Bound Receptor

Assuming a concentration of toxin-binding sites of approximately 10 pmole/mg of protein (Hess *et al.*, 1975), the membrane-bound receptor (less than 0.3 mg of protein per milliliter in buffer II) was incubated with a tenfold excess of labeled toxin for 2 hr. The salt was removed by dialysis for 4 hr against a 1000-fold excess of buffer II without CsCl. Between 0 and 4% of the labeled toxin was lost during dialysis. Duplicate aliquots were applied to separate minicolumns and eluted with buffer II. The same assay was done on samples in which the buffer was exchanged for Ringer's solution (Keynes and Martins–Ferreira, 1953) using a Sephadex G-25 column and on vesicles prepared in 0.9 M sucrose, 1 mM Na_2HPO_4, pH 7.4, with and without dialysis. The former condition was used to determine the effects of sucrose on toxin binding and the latter to exclude a change in the amount of toxin bound after dialysis. Complete titrations were done using less than 0.6 mg membrane protein per incubation mixture. The blanks for each assay were determined by preincubation of the receptor with unlabeled toxin at a concentration fivefold greater than that of the assumed sites before the addition of labeled toxin. (See Fig. 4a for details of the conditions.)

2.3.3. Solubilized Receptor

(See Fig. 4b for details of the conditions.) From the quenched reaction, for each time point, aliquots were applied to the minicolumns and eluted with 1.1 ml of buffer II. In order to calculate the number of toxin-binding sites occupied at each point, the eluate, minicolumn, and unassayed quenched reaction (with the pipette tip used for transfer) were counted separately. This was done because the protein in the eluate attributable to the receptor was too small a fraction of the total protein to be assayed accurately by the methods normally used.

The blank was determined by preincubating the receptor in a fourfold molar excess of unlabeled toxin for 3 hr and then adding labeled toxin to give the final concentration of the corresponding experimental incubation. When a second receptor–ligand was used, it was added to the blank incubation just before the addition of labeled toxin. Aliquots of the blank were treated in exactly the same way as those from the experimental incubation.

Use of the cation-exchange resin with the membrane-bound receptor requires reduction of the high salt concentrations sometimes used in the purification of vesicles from *E. electricus* electroplax. High salt concentrations, the presence of sucrose, or removal of salt by dialysis did not significantly alter the total amount of toxin bound to the column, and salt or sucrose did not affect the efficacy of the assay method. The blank was found to be less than 2% of the total amount of labeled toxin applied. Reversible, nonspecific binding of labeled toxin to the membrane does not significantly add to the blank owing to the column efficiency.

The efficiency of the carboxymethylcellulose minicolumns was investigated (Fig. 4c). The greatest retention (lowest blank) was obtained when less than 10 pmol of labeled toxin were applied to the minicolumn. However, quantities of toxin greater than 40 pmol/minicolumn may be used if the receptor : toxin ratio is greater than 0.7 or if the settled-bed volume of the minicolumns is increased from 1.25 ml.

Kinetic experiments were performed under pseudo-first-order conditions, a

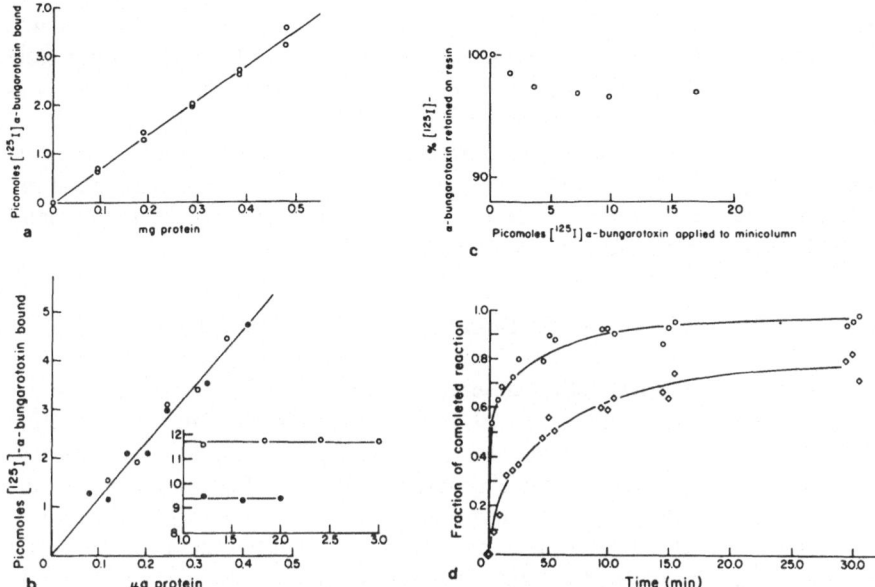

FIGURE 4. (a) Titration of membrane-bound receptor with $[^{125}I]$-α-bungarotoxin. Aliquots, 200 μl, were taken from incubations (250 μl each) in buffer I (no CsCl) with a constant concentration of labeled toxin ($5.1 \times 10^{-8}M$) and increasing quantities of membrane-bound receptor protein. Eluate volumes were 1.3 ml. Each assay point was corrected for the blank. The slope of this line gives a value of 7 pmol of $[^{125}I]$-labeled α-bungarotoxin bound per milligram of protein. (b) Titration of solubilized receptor with ^{125}I-labeled α-bungarotoxin. Aliquots, 100 μl, were taken from a series of incubation mixtures (125 μl each) with a constant concentration of labeled toxin ($\bigcirc$, 12 μM; $\bullet$, 0.98 μM) and increasing quantities of solubilized receptor protein. Eluate volume was 1.2 ml. The quantity of toxin bound was determined from the radioactivity of the eluate and was corrected for a blank. The slope gives the concentration of toxin-binding sites per microgram of protein at 11.6 pmol. The inset shows titration points where the concentration of binding sites exceeded the concentration of $[^{125}I]$-α-bungarotoxin in the incubation. The blank for these points is due to apparent retention of radioactivity, amounting to 7% of the total applied toxin, and is independent of the receptor concentration. (c) Percentage of applied toxin (without receptor) retained by the carboxymethylcellulose (1.25 ml settle-bed volume) as a function of the quantity applied. Points not shown in this figure include total applied toxin between 75 and 305 pmol with greater than 90% retention. Aliquots of ^{125}I-labeled α-bungarotoxin solution were applied using Centaur pipettes with disposable tips. Final eluate volumes were 1.5 ml. The average flow rate was approximately 1 ml/14 min. (d) The time course of the reaction of $[^{125}I]$-α-bungarotoxin with the purified receptor in buffer II at 0°C. In the absence of any cholinergic effector ($\bigcirc$), 0.14 nM receptor (by toxin-binding sites) was reacted with 14 μM $[^{125}I]$-α-bungarotoxin in a polyethylene vial. The reaction was complete within 1 hr and was within 9% of the endpoint calculated from the initial concentrations. This reaction was repeated in the presence of 0.1 μM d-tubocurarine ($\diamondsuit$) using a different preparation of purified receptor. The endpoints were taken after 14 hr and were within 7% of the calculated endpoint. For each time point in these experiments, 50-μl aliquots were quenched in 650 μl of 0.21 μM α-bungarotoxin; 500 μl of these quenched reactions was assayed by the minicolumn method. Reprinted from Kohanski *et al.* (1977) with permission.

tenfold excess of labeled toxin over receptor sites (Fig. 4d). The binding was quenched in a 20-fold molar excess of unlabeled toxin. Aliquot volumes were adjusted so that the concentration of labeled toxin in the quenched reaction was 0.4 ± 0.15 pmol/500 μl. In Fig. 4d, the upper and lower curves represent the reaction in the absence and presence, respectively, of d-tubocurarine. These data show clearly the biphasic nature of the complex formation and the inhibitory effect of d-tubocurarine on the rates, as was observed previously with the membrane-bound receptor (Bulger et al., 1977).

2.4. Kinetic Studies of Toxin Binding to *T. californica* Receptor

As a result of the studies described in Section 2.2, this group reported (Bulger and Hess, 1973; Fu et al., 1974; Hess et al., 1975b; Bulger et al., 1977) that in the case of receptor-containing vesicles prepared from the electroplax of *E. electricus*, the specific and irreversible reaction of α-bungarotoxin with the receptor proceeds in two phases: an initial fast step followed by the slower one. The dependence of the reaction on the initial toxin concentration indicated that the reversible formation of a toxin : receptor complex precedes the irreversible step and that two toxin molecules bind to the receptor before the irreversible reaction occurs. The reaction of α-bungarotoxin with the membrane-bound acetylcholine receptor isolated from the electroplax of *T. californica* (Quast et al., 1978; Blanchard et al., 1979) appeared to be different.

Because the toxin reaction forms the basis of many biochemical studies of the receptor, Leprince and Noble investigated the reaction of the toxin with the receptor in *T. californica* membrane preparations to determine whether the reported differences were due to variation in the receptor or the toxin preparation, in the methods used for measuring the receptor–toxin reaction, in the experimental design, or in the interpretation of the results.

Two methods that are frequently used to separate free and bound toxin (see Section 2.3), the CM-52 cellulose minicolumn assay and a DE-81 filter disc assay, were compared. Both techniques gave essentially the same results (Leprince et al., 1981).

In order to exclude differences that could originate from the iodinated toxin or the *Torpedo* membrane preparations, Leprince used the procedure described by Blanchard et al. (1979) and obtained similar results. Only a single slow phase of the reaction was observed (Fig. 5a). These results show that neither the assay technique nor the membrane or toxin preparations *per se* explained the different results obtained with *E. electricus* and *Torpedo* preparations. When, however, a reaction blank was subtracted from the toxin-binding data, both a fast and a slow phase of the reaction could be observed (Fig. 5b, lower curve). About 33% of the reaction had gone to completion before the first measurement could be made by the methods used. This initial fast phase was followed by a slower

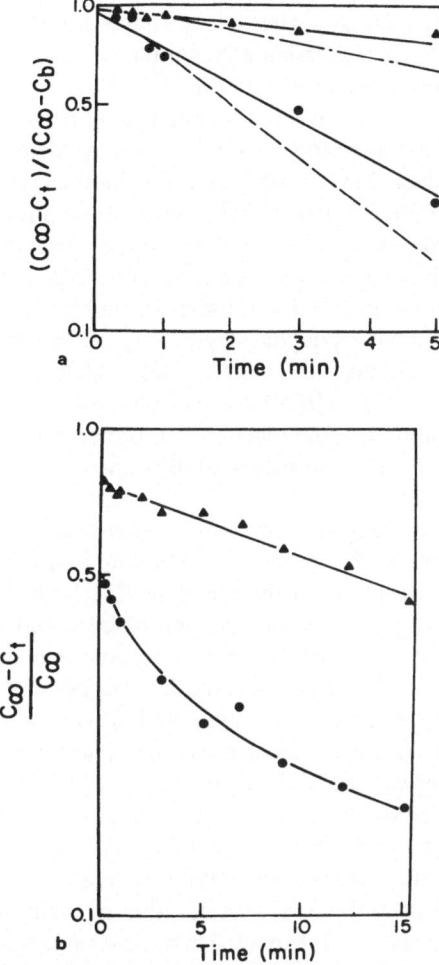

FIGURE 5. Binding of [^{125}-I]-α-bungarotoxin to acetylcholine receptor in *T. californica* vesicles in *Torpedo* Ringer's solution, pH 7.5, 0°C. *T. californica* vesicles (50 nM toxin-binding sites) were incubated, for the time shown, in *Torpedo* Ringer's solution with 100 (▲) or 500 nM (●) [^{125}I]-α-bungarotoxin. Free and bound toxin were separated on a DE-81 filter. (a) Plot of the time course of the toxin according to Blanchard *et al.* (1979). The background value, C_b, was obtained by extrapolating the time-dependent toxin reaction to zero time. C_t and C_∞ represent the toxin binding at a time t and after 6 hr of reaction, respectively. The coordinates of the solid line are determined by our measurements. For purposes of comparison, the data of Blanchard *et al.* (1979) are also shown. The dashed line was obtained with 500 nM toxin and the dashed and dotted line with 125 nM toxin. (b) Replot of the time course of the reaction shown above. Unspecific binding, measured after incubation with a tenfold excess of unlabeled toxin, has been subtracted from these curves. Reprinted from Leprince *et al.* (1981) with permission.

phase, which occurred within the time resolution of the methods used. A simple bimolecular reaction scheme suggested by Blanchard *et al.* (1979) is ruled out by these results, which are essentially similar to those obtained with the *E. electricus* preparation (Hess *et al.*, 1975b; Bulger *et al.*, 1977).

The difference between the results obtained by Blanchard *et al.* (1979) and those obtained in this laboratory arises because the data here is corrected for unspecific toxic binding. In the experiments of Blanchard *et al.* (1979), a reaction blank was not determined independently but the slow phase of the reaction was extrapolated to 0 time, and this interpolated value was subtracted from all measurements. The initial phase of the reaction, which is too fast to be followed by the DEAE filter assay, is thereby subtracted, and the incorrect conclusion is reached that the reaction is monophasic with *Torpedo* membranes. The difference between the experimentally determined blank and the blank obtained by the method of Blanchard *et al.* (1979) corresponds to the initial fast phase of the toxin reaction. In addition, if the reaction is followed for a longer period of time than was the case in the experiments of Blanchard *et al.* (1979), a third and slower phase is also observed.

The reaction of α-bungarotoxin with *T. marmorata* membrane preparations was treated as a bimolecular process by Franklin and Potter (1972), and they observed an initial fast phase in the reaction. The limited concentration range of reactants they used precluded observation of the complexity of the reaction. A multiphasic kinetic behavior for the interaction of α-bungarotoxin with rat diaphragm muscle has also been observed by Brockes and Hall (1975a,b).

At low concentrations of toxin the reaction with the receptor, in both *E. electricus* and *Torpedo* membranes, proceeds almost completely by the slow phase (Fig. 5b, upper curve). In the case of *E. electricus,* this result is predicted by the equations based on the minimum reaction proposed (Fig. 3c, Eq. I) (Hess *et al.*, 1975b). In the case of *T. californica,* this result is in agreement with results obtained in other laboratories (Weiland *et al.*, 1976, 1977; Delegeane and McNamee, 1980); at the low concentration of toxin used in these studies, only one phase of the reaction is expected to be observed. Also, the concentrations of toxin-binding sites and receptor sites were nearly equal in those experiments so that the concentration of free toxin changed while the measurements were made, which complicated the analysis of the data. Interpretations of the kinetic measurements of the reaction of toxin with *E. electricus* membranes were facilitated by (1) using a concentration of toxin that was in excess of that of the receptor sites so that the concentration of free toxin during the measurements remained essentially constant (Bulger and Hess, 1973; Bulger *et al.*, 1977) and (2) investigating the reaction over a wide range of initial toxin concentrations (Bulger *et al.*, 1977) and taking a proper reaction blank.

The initial time course of the reaction of toxin with *Torpedo* membranes is too fast to be measured by presently available techniques, and a detailed analysis of models for this reaction is, therefore, not possible. The important point is that a comparison of the methods and toxin used by different laboratories indicates that the toxin reaction with *T. californica* membranes and with *E. electricus* membranes is similar.

3. RECEPTOR-CONTROLLED ION FLUX IN VESICLES

Having examined the mechanism by which ligands bind to the membrane-bound acetylcholine receptor, the next step was to examine the receptor-controlled flow of inorganic ions across the membrane that is induced by the binding of activating ligands, and this was done by Andrews and Struve (1975a).

Kasai and Changeux (1971a,b,c) had demonstrated that electroplax vesicles exhibited receptor-mediated ion flux. The vesicles were the same preparation as was used in the studies of ligand binding to the membrane-bound receptor. The complexity of the efflux, the apparent inefficiency of the ligand-induced efflux, and the variability from preparation to preparation suggested that the parameters they measured reflected only partly the underlying process (Hess *et al.*, 1975a). Therefore, the flux of sodium ions from the vesicles was examined and it was found that the receptor-controlled flux, which is only a small part of the total observed flux, is obscured by efflux from the vesicles without receptor channels. The solid lines in Fig. 6a were obtained in an analysis of the flux of $^{22}Na^+$ from vesicles. A nonlinear least-square computer program was used to fit the data to the sum of two exponentials. The upper curve in Fig. 6a appears identical to the lower curve (carbamoylcholine present) except for a very rapid initial release of $^{22}Na^+$ in the presence of carbamoylcholine. This suggested that only a small fraction of the vesicles are active. It was subsequently shown that the active vesicles were much less permeable to sodium ions in the absence of carbamoylcholine than were the inactive vesicles (Hess, *et al.*, 1977). This observation was used in the design of the experiment illustrated in Fig. 6b. The results showed that after efflux is initiated by diluting $^{22}Na^+$-loaded vesicles 70-fold with buffer in the absence of carbamoylcholine (Fig. 6b, upper curve), most (about 85%) of the tracer ion inside the vesicles has equilibrated with nonradioactive sodium ions in the dilution buffer. After 120 min, addition of carbamoylcholine to this solution induces a rapid efflux of $^{22}Na^+$ (Fig. 6b) that follows a single exponential rate law:

$$[^{22}Na^+]_t = [^{22}Na^+]_{t=0}e^{-k_{obs}t} \qquad (8.7)$$

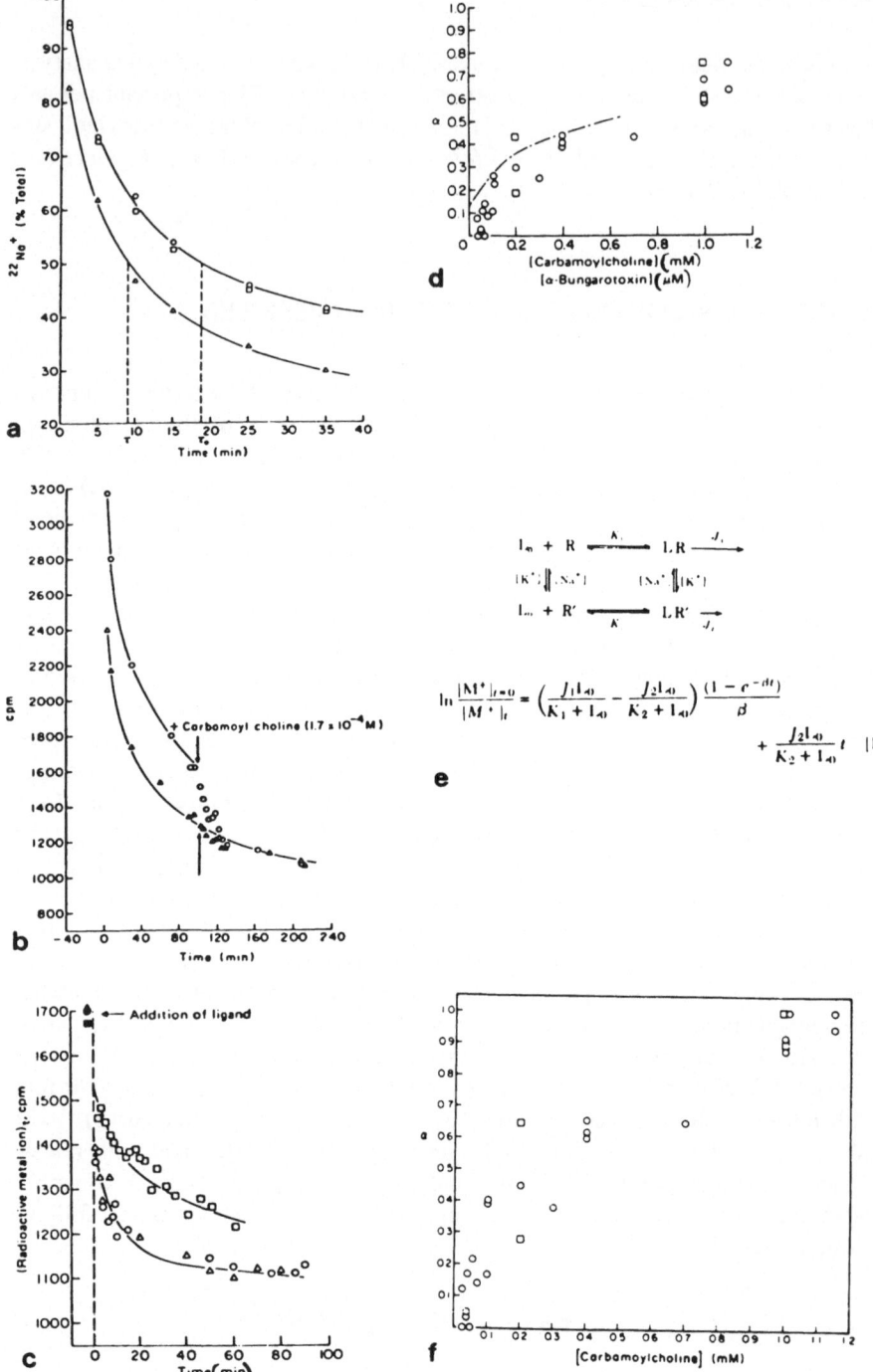

In this equation $[^{22}Na^+]$ represents the concentration of radioactive ions inside the vesicles. k_{obs} is an overall flux rate constant that reflects the movement of ions across the vesicle membrane. The diffusion of $^{22}Na^+$ into the vesicles and its concentration inside the vesicles at $t = \infty$ can be neglected, however, because the volume of the external solution is large compared with the internal volume

←——

FIGURE 6. (a) $^{22}Na^+$ efflux from *E. electricus* electroplax membrane vesicles, diluted 70-fold according to the procedure of Kasai and Changeux (1971b), pH 7.0, 22°C. The time for half-equilibration of the vesicles with the dilution buffer is given for the control curve (τ_0) and carbamoylcholine-mediated flux curve (τ). ○, Efflux in the absence of effector (control); △, efflux in the presence of 0.1 mM carbamoylcholine; □, superposition by a parallel shift of the carbamoyl-choline-mediated flux curve upon the control curve. (b) Permeability of vesicles incubated in the presence or absence of carbamoylcholine for 40 min and then diluted 100-fold, pH 7.0, 4°C. ○, Efflux of $^{22}Na^+$ from vesicles incubated in absence of carbamoylcholine, with further addition of carbamoylcholine as indicated; △, flux of $^{22}Na^+$ from vesicles incubated in the absence of carbamoyl-choline, with further addition of carbamoylcholine as indicated. (c) $^{22}Na^+$ and $^{86}Rb^+$ efflux as a function of time under physiological conditions. For the $^{86}Rb^+$ efflux, sucrose-free solutions of vesicles were prepared by the procedure of Fu *et al.* (1977). The vesicles, resuspended in a potassium Ringer's solution (150 mM KCl, 9.4 mM NaCl, 3 mM CaCl₂, 1.5 mM MgCl₂, 1 mM phosphate, pH 7.0), were incubated overnight with $^{86}RbCl$ (35 μCi/ml), 4°C. Efflux was initiated by dilution into a sodium eel Ringer's solution (169 mM NaCl, 5 mM KCl, 3 mM CaCl₂, 1 mM phosphate, pH 7.0), 4°C. After 120 min, ligand was added. ○, $^{86}Rb^+$ efflux, 1 mM carbamoylcholine; △, $^{86}Rb^+$ efflux, 10 μM acetylcholine (vesicles were preincubated with 56 μM Tetram); □, $^{22}Na^+$ efflux, 1 mM carbamoylcholine. Solid symbols represent the respective average of several experimental points taken before the addition of ligand. The lines were computed according to Eq. (8.2). (d) Fraction of ligand-induced efflux due to the fast process plotted versus carbamoylcholine concentration. Vesicles incubated overnight with 100 mM KCl, 0.4 M sucrose, 35 μCi of $^{86}RbCl$ per milliliter, were diluted into a medium containing 100 mM NaCl, 0.4 M sucrose, 1 mM phosphate, pH 7.0, 4°C, to initiate efflux. Conditions for $^{22}Na^+$ efflux were the same except that NaCl was substituted for KCl, and $^{22}NaCl$ was substituted for $^{86}RbCl$. After 120 min carbamoylcholine was added. ○, $^{86}Rb^+$ efflux; ○, $^{22}Na^+$ efflux. For comparison, the dashed line is included and indicates the fraction of the fast phase of the reaction of $[^{125}I]\alpha$-bungarotoxin with the receptor of the vesicle. (e) In this model, L_O represents the initial concentration of ligand, R and R' the concentrations of the two conformations of the receptor, K_1 and K_2 the receptor–ligand dissociation constants, and the vertical arrows the rate constants for the isomerization of receptor conformations. The initial fast efflux of inorganic ions, characterized by the rate constant J_1, is associated with the receptor conformation R. The slow phase of the efflux, characterized by the rate constant J_2, is associated with the receptor conformation R'. A feature of this model is that the interconversion between receptor conformations is in part a function of external Na^+ and K^+ concentration. This is a working model that allows one to estimate the efficiency of the receptor-mediated translocation of inorganic ions. It served as a basis for further experiments. The equation for the efflux of inorganic ions, $[M^+]$, from vesicles, based on the model, is given by Eq. (1) where β contains the ligand concentration-dependent rate constants for the isomerization of receptor forms. (f) Corrected plot of the fraction of the ligand-induced $^{86}Rb^+$ efflux due to the fast process (α) versus carbamoylcholine concentration. The raw data were taken from Fig. 6. Reprinted with permission from Hess *et al.* (1975) (a and b), Hess *et al.* (1978) (c–e), Kim and Hess (1981) (f).

of the vesicles. The amount of ^{22}Na$^+$ released in response to carbamoylcholine varied from preparation to preparation, presumably reflecting the size of the fraction of the total preparation that contained the vesicles with active receptors. However, the k_{obs} values obtained were identical and did not vary between preparations for any given concentration of carbamoylcholine. No evidence for cooperativity was obtained, in contrast to the dose-response curves obtained in electrophysiological measurements and in flux measurements made by Kasai and Changeux (1971b).

Rübsamen, in collaboration with Professors A. T. and M. E. Eldefrawi, used purified *T. ocellata* receptor to show that the activating ligands acetylcholine, carbamoylcholine, and decamethonium displaced a fluorescent lanthanide (terbium) from the calcium-binding sites of the receptor whereas *d*-tubocurarine did not (Rübsamen *et al.*, 1978). In addition, decamethonium was shown to displace about half the displaceable Tb^{3+} from the receptor–toxin complex. This was in agreement with the model proposed by Fu *et al.* (1977) and Bulger *et al.* (1977). Andrews and Struve (1975b) were then able to show that *d*-tubocurarine was a noncompetitive inhibitor of decamethonium- and carbamoylcholine-induced efflux of ^{22}Na$^+$ from the *E. electricus* vesicles (Hess *et al.*, 1976), and Lipkowitz found (Lipkowitz, Coombs, and Hess, unpublished observations) that Tb^{3+} inhibited the efflux.

When flux experiments were performed at a constant ionic strength, but in varying mole fractions of NaCl and KCl, it was found that potassium ions had a small but definite inhibitory effect on the receptor-controlled flux of sodium ions.

Lipkowitz and Struve investigated the effect of sodium and potassium ions on the receptor-controlled flux of ions (Hess *et al.*, 1978). The efflux assay was done the same way as previously except that the composition of the solutions with which vesicles were incubated and diluted was varied. The solutions used for incubation and dilution contained only NaCl and KCl. The ionic strength of NaCl and KCl together was held constant both inside and outside the vesicles. All solutions also contained 0.4 M sucrose so that the vesicles could be used directly after purification by sucrose density-gradient centrifugation (Fu *et al.*, 1977). Physiological conditions (a high concentration of potassium internally and of sodium externally) led to a very rapid initial efflux followed by a slower, single exponential efflux (Fig. 6c). The rapid initial phase was over within 20 sec of the addition of an activating ligand, before the first measurement could be made. It was necessary to show that the rapid initial phase was not due to breakage of vesicles nor to a change in the amount of vesicles retained on the filter. When the vesicles were incubated with ^{86}RbCl or [^{14}C]-sucrose, ^{86}Rb$^+$ was released in the rapid phase whereas the labeled sucrose was not. In addition both *d*-tubocurarine and α-bungarotoxin inhibited the rapid phase.

The data in Fig. 6d show that the extent of the initial fast process depends on the carbamoylcholine concentration and is about 65% of the total efflux when the ligand concentration is 1.2 mM. The results suggest that the binding of ligands to the receptor results in the interconversion of receptor conformations, one being associated with the rapid phase and one with the slower phase. This inactivation (desensitization) phenomenon had not previously been observed in experiments with vesicles. The kinetic measurements indicated a relationship between flux rates and receptor conformations and that the interconversions between receptor conformations are partially a function of the Na^+ and K^+ concentrations. The model, and the equation based on it, are shown in Fig. 6e (Hess *et al.*, 1978).

3.1. Isolation and Characterization of Functional Vesicles

3.1.1. Inherent Difficulties in Efflux Measurements

Kim subsequently identified several intrinsic properties of receptor-rich membrane vesicles (Kim and Hess, 1981) that led to new interpretations of the results shown in Fig. 6d. It was found that some ions were exchanged only slowly and that this process was not related to the receptor-controlled ion flux. The time dependence of the "slowly exchanging ions" process could be fitted to a single exponential with a $t_{1/2}$ value of about 3 min. This slow exchange of ions accounts for 30–35% of the efflux observed and is seen in the presence of both carbamoylcholine and gramicidin (Kim and Hess, 1981). When the data in Fig. 6d were corrected for this slow efflux phase, it could be seen that at high concentrations of carbamoylcholine all tracer ions are exchanged rapidly (Fig. 6f).

The problem was avoided in all future experiments: the vesicles were not equilibrated for long periods of time with tracer ions. Instead influx measurements were made in a time region that avoided the slow exchange of tracer ions. In addition, the scheme in Fig. 6e could be simplified in that it was no longer necessary to suggest that the inactivated form of the receptor (*LR'* in Fig. 6e) gives rise to an open channel.

3.1.2. Isolation and Characterization of Functional Vesicles

Andrews approached the possibility of isolating the functional vesicles from vesicles that do not respond to the addition of receptor–ligands by efflux of ions and that constitute 85% of the total population (Hess and Andrews, 1977). The method is based on the kinetic analysis of the efflux data, which indicated that when the vesicle population is equilibrated with NaCl, the vesicles with non-functional receptors (inactive vesicles) can be filled with CsCl whereas the vesicles with functional receptors (active vesicles) will remain filled with NaCl. The vesicle populations can then be separated from each other on the basis of their density difference. The data shown in Fig. 7a were obtained when a heterogeneous mixture of vesicles was equilibrated with ^{22}NaCl. The vesicles were diluted with nonradioactive buffer containing CsCl, and the radioactivity associated with the vesicles was measured at various time intervals. The first part of the efflux curve represents isotopic exchange in absence of carbamoylcholine. It was known that the rate coefficients associated with this exchange, and the fraction of the total ^{22}Na$^+$ that is exchanged in this manner, are independent of the addition of acetylcholine analogs (Hess et al., 1975a). This receptor-independent efflux ceases after about 120 min. Addition of carbamoylcholine to the mixture after this 2-hr period (Fig. 7a) induced a rapid efflux of ^{22}Na$^+$. An analysis of the exchange between internal ^{22}NaCl and CsCl in the external medium shown in Fig. 7 is given in Table II, which lists the half-times of the equilibration processes. Acetylcholine does not affect the $t_{1/2}$ of the nonfunctional vesicles but does reduce it in the functional vesicles by a factor of more than 100.

$\longrightarrow$

FIGURE 7. (a) ^{22}Na$^+$ flux from vesicles, pH 7.0, 4°C. The vesicles were equilibrated with 1.1 M sucrose, 190 M NaCl, and 0.27 μM ^{22}NaCl (stock solution 6.24 Ci/mg of ^{22}Na$^+$) and diluted 50-fold with 1.1 M sucrose, 190 M CsCl, and 1 mM Na phosphate buffer, pH 7.0. At the times indicated the solutions were made 1 mM in carbamoylcholine. ●--●, Mixture of heterogeneous vesicles; ▲--▲, active vesicles isolated from the mixture. Inset is a first-order plot for ^{22}Na$^+$ efflux in the presence of carbamoylcholine. ▲, 1 mM carbamoylcholine-induced efflux determined with the mixture of vesicles using a published kinetic technique for measuring only the specific efflux (Hess et al., 1975). ●, 1 mM carbamoylcholine-induced flux from vesicles with functional receptors. The k_{obs} value was determined from the slope of the linear least-squares fit of the experimental points obtained in both experiments; $k_{obs} = 0.12 \pm 0.01$ min^{-1}. (b) Purification of functional acetylcholine receptor-rich vesicles by sucrose–190 mM cesium chloride density-gradient centrifugation. Four milliliters of vesicles (0.75 mg of membrane protein per milliliter, 1.1 M sucrose, 190 mM CsCl, 1 mM sodium phosphate buffer, pH 7.0, 4°C) were placed on the bottom of a continuous gradient and centrifuged in a Beckman SW-27 swinging bucket rotor for 2.5 hr at 4°C. Two-milliliter fractions were collected and analyzed. The abscissa gives the molarity of sucrose in the fractions. The results of two experiments obtained with vesicle preparations from different eels are shown. The solid lines and closed symbols represent experiments in which the vesicles with functional receptors were mainly filled with 190 mM NaCl and the vesicles with nonfunctional receptors with 190 mM CsCl. The broken line and open symbols represent a control experiment in which all vesicles were equilibrated with 190 mM CsCl. ●, % of membrane protein applied to column; ▲, pmol of α-bungarotoxin-binding sites per milligram of membrane protein; ■-■, □---□, specific efflux (% total). Reprinted from Hess and Andrews (1977) with permission.

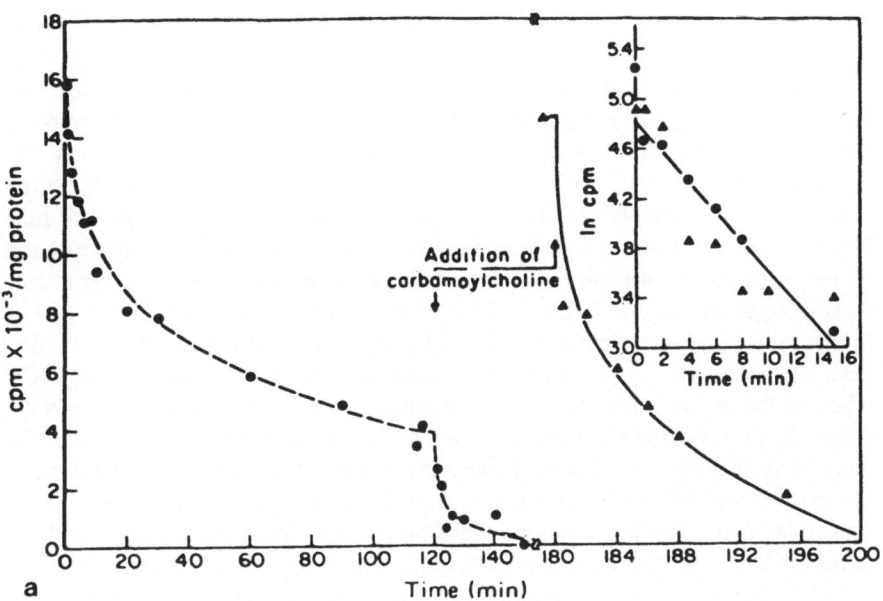

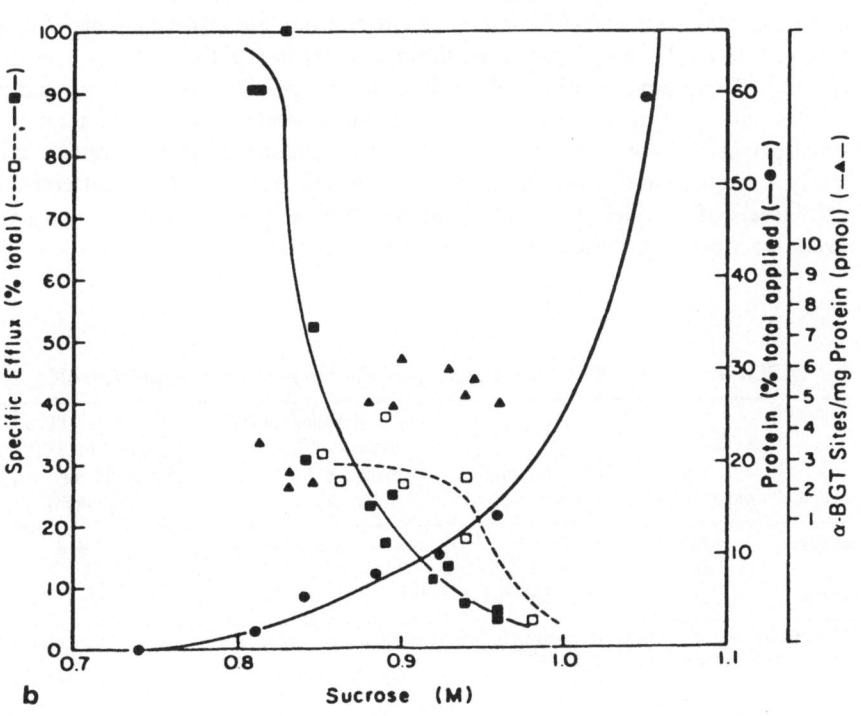

The following protocol for separating active from inactive vesicles was based on the data in Table II: (1) A discontinuous sucrose density-gradient centrifugation to collect vesicles rich in receptor sites and of similar density. (2) Equilibration of vesicles with 190 mM NaCl for 18 hr, long enough for all the vesicles to have reached equilibrium (see Table II). (3) Exchange of 190 mM NaCl for 190 mM CsCl using a Sephadex G-25 column and allowing equilibration of the vesicles for 40 min before centrifugation in a sucrose–CsCl gradient. According to the data in Table II, 35% of the inactive vesicle fraction ($t_{1/2}$ = 3 min) equilibrated completely with CsCl in this time period, 65% of the inactive vesicle fraction ($t_{1/2}$ = 40 min) exchanged half their NaCl with denser CsCl, whereas the active vesicles ($t_{1/2}$ = 330 min) still retained most of the NaCl originally present. (4) Separation of the NaCl-filled functional vesicles from the inactive vesicle fraction filled with denser CsCl by centrifugation in a continuous sucrose–190-mM-CsCl density gradient. After centrifugation at 25,000 revolutions/minute for 2.5 hr, fractions were taken. The fractions were assayed for toxin-binding sites, protein concentration, and acetylcholinesterase and Na^+/K^+ ATPase activities by the methods of Kohanski et al., (1977), Lowry et al., (1951), Ellman et al., (1961), and Bonting et al., (1961), respectively.

It can be seen (Fig. 7b) that the vesicles in the fractions that are 0.8 M in sucrose and 190 mM in CsCl comprise approximately 5% of the total membrane protein. The total $^{22}Na^+$ content of these vesicles is rapidly exchanged in the presence of carbamoylcholine. By this criterion this fraction contains active vesicles only. An interesting result is that the receptor site concentration (α-bungarotoxin-binding sites per milligram of membrane protein) is rather uniformly distributed in the membrane preparation. There are significant differences in the levels of acetylcholinesterase and Na^+/K^+ ATPase between the inactive vesicle fractions and the active vesicles (see Table II).

TABLE II. $^{22}Na^+$ Efflux from Electroplax Membrane Vesicles, pH 7.0, 4°C[a]

Vesicles	% Total radioactivity	Half-equilibration time (min)	Acetylcholinesterase (mmol of ATCH hydrolyzed, $hr^{-1}.mg^{-1}$ of protein)	Na^+/K^+ ATPase (μmol of ATP hydrolyzed, $hr^{-1}.mg^{-1}$ of protein)
Inactive	26	3	1.5	9
	49	40		
Active	25	330	0.04	<1

[a] The values given pertain to the data shown in Fig. 7a (dashed line). ATCH, acetylthiocholine. Reprinted with permission from Hess and Andrews (1977).

In Fig. 7a the flux of $^{22}Na^+$ from the unfractionated vesicle preparation is compared with that from purified active vesicles obtained from the same preparation. The amplitude of the carbamoylcholine-stimulated efflux is about four times greater in the vesicles isolated from the sucrose–CsCl density gradient than in the original vesicles. It can be seen from the inset to Fig. 7a that k_{obs} for Na^+ efflux is the same for active vesicles in the mixture before and after separation.

A correlation between receptor-mediated ion flux through vesicle membranes and through the plasma membranes of living cells requires a determination of the number of inorganic ions that pass through receptor-mediated channels per unit time (Katz and Miledi, 1972). This requires a knowledge of both the rate of the ion translocation process and the number of receptor sites per unit volume of solution inside the membrane vesicles, R_O. R_O is defined as moles of receptor per liter of solution inside the vesicles. The separation of membrane vesicles that contain functional receptors from vesicles that do not, as accomplished by the procedure here, is an important step in determining the value of R_O. The isolation procedure also makes it possible to characterize the active vesicles and to compare them to the inactive vesicles.

3.1.3. Large-Scale Isolation of Specific Vesicles

Noble, Sachs, and Lenchitz used the results obtained by Andrews to develop a method for the isolation, on a large scale, of vesicles that contain active vesicles on the basis of their different density (Sachs et al., 1982). The E. electricus vesicles were prepared by a modification (Fu et al., 1977) of the procedure of Kasai and Changeux, (1971a), and the following modifications of Sachs et al. (1982): (1) a discontinuous sucrose gradient consisting of three 30-ml layers, of 1.5 M sucrose, 0.9 M sucrose, and 0.4 M sucrose was used; and (2) instead of an ultracentrifuge, a low-speed centrifuge with 30 times greater capacity was used. After centrifugation for 10 hr at 11,000 revolutions/min, the receptor-rich vesicles were in a region of the gradient that was between 0.7 and 1.3 M in sucrose (see Fig. 8a). The average concentration of α-bungarotoxin sites in the fraction used for further purification (Fig. 8a, fractions 6–14) was 10.6 pM/mg membrane protein. This is in good agreement with the value of $9 \pm .2$ pM/mg membrane protein obtained with the earlier technique (Fu et al., 1977).

If the protein responsible for the movement of ions across a membrane is uniformly distributed on the membrane surface, then the exchange rate of the inorganic ions and the density of the vesicle are both proportional to the inverse of the vesicle diameter. By allowing the exchange of inorganic ions to occur for different periods of time, vesicles that all contain receptors but that have different receptor : vesicle volume ratios can be selectively enriched with inorganic ions of different densities (for instance Na^+ and Cs^+) and separated by centrifugation in a continuous sucrose–cesium chloride density gradient.

The technique is illustrated in Fig. 8b. The vesicles rich in receptor sites, isolated by the procedure illustrated in Fig. 8a (Sachs et al., 1982), were first equilibrated with 190 mM NaCl ($\sim$24 hr) and subsequently for 90 min with a 190 mM CsCl solution (Hess and Andrews, 1977) that also contained 15% Percoll (Sachs et al., 1982). In this time all the vesicles that do not contain functional receptors ($t_{1/2}$ for exchange $\sim$35 min) (Hess et al., 1975b; Hess and Andrews, 1977) equilibrate with CsCl, and therefore the vesicles that contain functional receptors ($t_{1/2}$ for exchange $\sim$5.5 hr) are separated from the other vesicles in 30 min by centrifugation at 15,000 revolutions/minute in the CsCl–Percoll solution, which forms a gradient during this time. The isolation step leads close to the theoretically possible fourfold purification of the vesicles containing functional receptors (Fig. 8b, fraction 37), as judged by determinations of the internal volume. The yield is about 100-fold higher than the yield from the sucrose–CsCl density gradient previously used (Hess et al., 1977). In the experiment illustrated in Fig. 8b the most purified fraction is 37. The active vesicle fraction, together with the vesicle protein peak, shifts to the left or right of fraction 37 depending on the density of the Percoll® and the ratio of proteins to lipids in any particular preparation (Sachs et al., 1982).

The total (vesicle and soluble) protein concentration of each fraction was determined using the Fluram method (Böhlen et al., 1973); the Folin–Lowry method cannot be used when Percoll is present (Peterson, 1979). The Fluram and the Folin–Lowry methods were correlated by obtaining standard curves using both methods with bovine serum albumen and with vesicles in the absence of Percoll. The influx of inorganic ions, the internal volume of the vesicles, and the concentration of toxin-binding sites were measured. A considerable fraction of the toxin-binding sites was associated with material that is not retained by the Millipore filter.

3.1.4. Internal Volume of the Vesicle Preparation

Measurements of the extent of the influx of $^{86}Rb^+$ over a wide range of $^{86}Rb^+$ concentrations at saturating concentrations of acetylcholine indicated that the amount of ions bound to the vesicles was insignificant (Kim and Hess, 1981). When the inorganic ion composition is the same both inside and outside the vesicles equilibrated with $^{86}Rb^+$, knowledge of the $^{86}Rb^+$ content of the vesicles

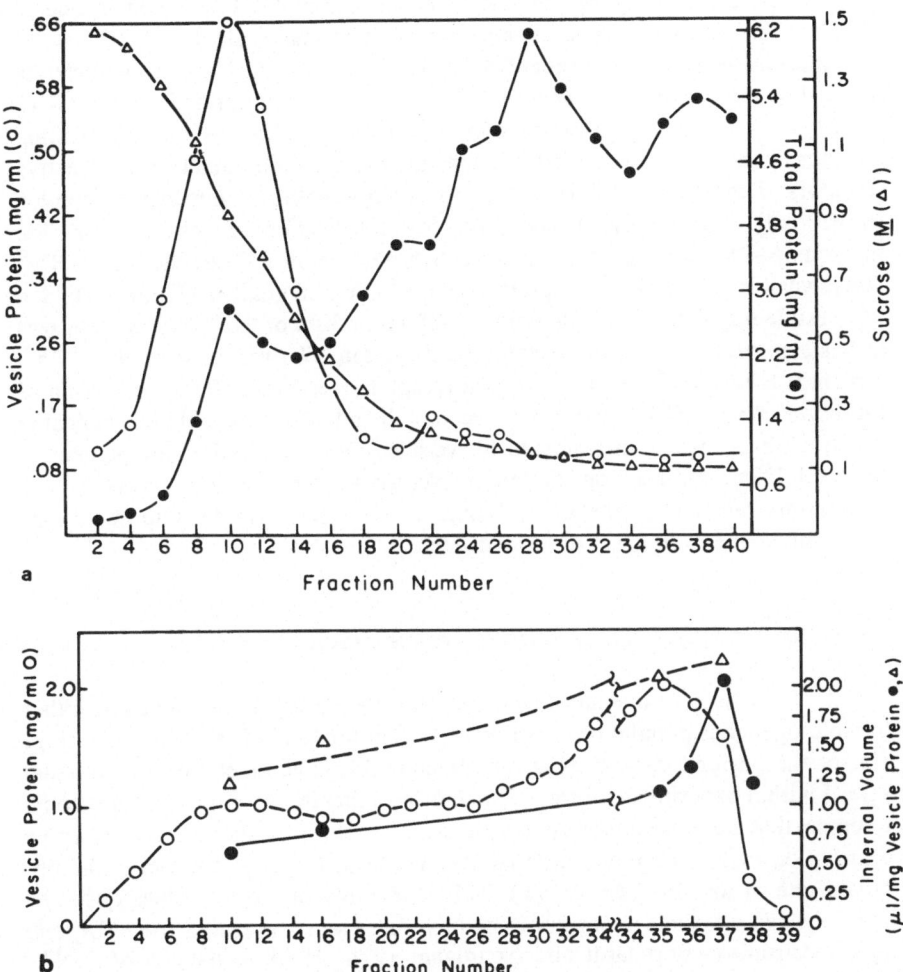

FIGURE 8. (a) Separation of the *E. electricus* vesicles preparations on a discontinuous sucrose gradient. △, density of sucrose; ○, mg vesicle protein per milliliter; and ●, mg total protein per milliliter. (b) Separation of vesicles containing functional receptors from a heterogeneous preparation using a 15% Percoll–190 mM CsCl density-gradient centrifugation. Vesicles (30–35 mg) were incubated for 90 min in 15% Percoll–190 mM CsCl and then centrifuged for 30 min at 15,000 revolutions/minute in a Sorvall RC2B centrifuge with a SS-34 rotor. Internal volumes were measured as described. ○, mg vesicle protein per milliliter; △, microliters of total internal volume; ●, microliters of internal volume associated with vesicles containing functional receptors. Reprinted from Sachs *et al.* (1982) with permission.

and the specific activity of the ^{86}Rb$^+$ in the external solution allows one to calculate the internal volume of the vesicles (Kim and Hess, 1981).

Internal volume measurements were used (Sachs et al., 1982) to characterize the vesicle population. The vesicles were allowed to equilibrate (a) with ^{86}Rb$^+$ for 24 hr or (b) in the presence of saturating concentrations of an activating receptor ligand for a few seconds to load only those vesicles that contained active receptors. The percent of ^{86}Rb$^+$ in the population with active receptors is given by $[(b)/(a)] \times 100$. For instance, the first purification of vesicles (Fig. 8a, fractions 6–14) yielded a fraction in which only about 25% of the total ^{86}Rb$^+$ was in vesicles with active receptors. The CsCl–Percoll gradient (Fig. 8b) yielded two vesicle populations. In fractions 10–35 about 50% of the ^{86}Rb$^+$ content was in vesicles with functional receptors; in fraction 37 over 80% of the ^{86}Rb$^+$ content was in vesicles with functional receptors (see Table III). Comparison of equilibration with ^{86}Rb$^+$ induced by receptor ligands or by gramicidin D indicated that fraction 37 was contaminated by vesicles that contained a few percent of the total ^{86}Rb$^+$ of fraction 37 but no detectable functional receptors. These vesicles are rather impermeable to inorganic ions and do not take up tracer ions in the influx measurements described later.

3.1.5. Kinetic Properties of Various Vesicle Fractions

A comparison of quench flow measurements (described below) using either the heterogeneous population, fractions 10–30 of the CsCl–Percoll gradient (Fig. 8b), or the purified vesicles (Fig. 8b, fraction 37) gave kinetic parameters that agreed within experimental error. This indicates that the vesicles with functional receptors that are present in the unfractionated vesicle population contain a narrow distribution of the receptor : internal volume ratio, R_O, that determines the observed rate of ion flux (see below). This uniformity of the R_O values observed in E. electricus vesicles does not exist in vesicles so far prepared from Torpedo spp. electroplax (Bernhardt and Neumann, 1978, 1980; Walker et al., 1981; Hess et al., 1982).

3.1.6. Determination of the Value of R$_O$, the Moles of Receptor per Liter Vesicle Internal Volume

The concentration of α-bungarotoxin-binding sites (Hess et al., 1981; Sachs et al., 1982) in the most active vesicle fractions was 4.8 ± 1.7 pM/mg membrane protein. The stoichiometry of toxin- and receptor-ligand-binding sites was shown to be 1 : 1 in E. electricus (Bulger et al., 1977; Fu et al., 1977; Karlin, 1980);

it was also shown that the binding of two molecules of an activating ligand is required to initiate ion translocation (Cash and Hess, 1980). Accordingly, the concentration of receptor sites is 2.4 pM/mg membrane vesicle protein. The internal volume of the vesicles (Fig. 8b) is 2.0 ± 0.3 μl/mg, and therefore the receptor concentration per liter vesicle internal volume, R_O, is 1.2 μM (Sachs et al., 1982).

3.1.7. Determination of the Size of the Vesicles

Electron micrographs revealed that the heterogeneous, unfractionated population contained vesicles of numerous sizes and shapes and that the fractionated vesicles became increasingly homogeneous toward the upper part of the gradient. The average radius of vesicles in the fraction containing vesicles with functional receptors (Fig. 8b, fraction 37) was 1600 ± 300 Å (based on 480 determinations). From this and the above results, an average of 12 receptor sites/vesicle was calculated (Sachs et al., 1982).

3.1.8. The Relation between α-Bungarotoxin-Binding Sites and Functional Receptors

The binding of ligands to the membrane-bound receptor protein has been measured (Eldefrawi and Eldefrawi, 1977; Heidman and Changeux, 1978; Karlin, 1980) in preparations which, although generally high in toxin sites, have not been assayed for receptor activity. It is, therefore, of interest to note that most of the toxin sites in the preparation used in this laboratory are associated either with vesicles that do not exhibit flux activity or with membrane fragments that are not retained by the Millipore filter used in the flux assay (Table III). The vesicle separation technique employed showed that the internal volume of the population that contains essentially only vesicles with functional receptors is about the same as that of the other fractions of the gradient (Fig. 8b, Table III). The ratio of α-bungarotoxin-binding sites to internal volume is similar in the unfractionated and the purified vesicle populations (Table III). Measurements of the receptor-controlled influx of tracer ions indicated that the observed rates were the same in the unfractionated as in purified preparations, but that the tracer ion content due to the receptor-controlled influx was about four times larger in the purified vesicles (Table III) (Sachs et al., 1982). Similar results were obtained when sucrose instead of Percoll was used (Hess and Andrews, 1977). These results imply that in heterogeneous populations the rates of receptor-controlled ion flux are not related to the concentration of toxin sites. Heterogeneous vesicle

TABLE III. Means and SD of Internal Volumes and α-Bungarotoxin-Binding-Site Concentrations Obtained in 12 Different Vesicle Preparations Separated on a CsCl–Percoll Density Gradient[a]

	Total internal volume (μl/mg vesicle protein)	Internal volume associated with vesicle-containing functional receptors (μl/mg vesicle protein)	α-Bungarotoxin-binding-site concentration (pM/mg vesicle protein)
Heterogeneous population of membrane vesicles obtained in the discontinuous sucrose gradient	1.9 ± 0.3	0.5 ± 0.2	5.5 ± 1.6
Membrane vesicles containing functional acetylcholine receptors isolated by the CsCl–Percoll density gradient centrifugation	2.4 ± 0.3	2.0 ± 0.3	4.8 ± 1.7

[a] Internal volumes and α-bungarotoxin-binding site concentrations were measured as described. Reprinted with permission from Sachs et al. (1982).

populations prepared from *Torpedo* spp. also show a lack of correspondence between the rates of receptor-controlled ion translocation and the toxin-site concentration (Walker *et al.*, 1981; Hess *et al.*, 1982).

Do the toxin sites that apparently do not contribute to receptor-controlled ion flux bind receptor–ligands such as acetylcholine, carbamoylcholine, and *d*-tubocurarine? The stoichiometry between binding sites for toxin and for activating ligands has been demonstrated with unfractionated membrane preparations from both *E. electricus* (Fu *et al.*, 1977) and *Torpedo* spp. (Eldefrawi *et al.*, 1975; Neubig and Cohen, 1979). With *E. electricus* vesicles obtained from a CsCl–Percoll density gradient it was specifically shown with fractions both high and low in ion flux activity that all the toxin reaction, and not just a part, is inhibited by both carbamoylcholine and acetylcholine (Leprince *et al.*, 1981).

There are also differences in the passive permeability to inorganic ions and in the acetylcholinesterase and Na^+/K^+ ATPase content between vesicles that contain receptor sites that are active in the ion flux assay and those that are not (see Table II) (Hess and Andrews, 1977). Although inactive receptors appear to be mainly in membranes that appear to be different from those associated with functional receptors, it could not be determined whether they also reside in the purified vesicle preparations. In ligand-binding experiments with *Torpedo* spp. and *E. electricus* vesicles it is important to be aware that the measurements can reflect the properties of receptors that are apparently incapable of forming ion-conducting channels.

3.2. Measurements of Ion Flux in the Millisecond Time Region

Various techniques for measuring ion translocation across vesicle membranes in subsecond time regions will be described in this section. Fast reaction techniques are necessary because inactivation of the receptor on exposure to ligand occurs rapidly and because, with the number of receptors and the size of the *E. electricus* vesicles, the flux process is complete within a few hundred milliseconds with saturating concentrations of carbamoylcholine (Cash and Hess, 1981) or acetylcholine (Cash *et al.*, 1981). Similar considerations are probably applicable to ion translocations mediated by other receptors.

3.2.1. The Use of Tracer Ions to Detect Flux in a Quench Flow Apparatus

In this section is described the application (Cash and Hess, 1981) to receptor-containing vesicles of a quench flow technique covering reaction times from 5 milliseconds to minutes (Fersht and Jakes, 1975) using small sample volumes and low pressures to obtain mixing. Techniques for rapid mixing of solutions, including quench flow techniques, have been used extensively with enzymes and proteins to follow reactions occurring in the millisecond–second range (Chance, 1974; Chance et al., 1964; Gibson, 1966; Hammes, 1982a; Hammes and Wu, 1974; Fersht and Jakes, 1975; Lymn and Taylor, 1970; Martinosi et al., 1974; Froehlich and Taylor, 1976; Roughton and Chance, 1963). Transmembrane ion transport has been demonstrated using quench flow techniques in studies of ATP synthesis by submitochondrial particles (Thayer and Hinkle, 1975) and of calcium uptake by sarcoplasmic reticulum vesicles (Kurzmack et al., 1977; Sumida et al., 1978; Thayer and Hinkle, 1975; Verjovski–Almeida and Inesi, 1979). With continuous mode quench flow systems, the reaction time depends on the flow rate and the distance between an initiation and a quenching event. Thus as the reaction time increases with increasing length of the reaction tube, the recovery of reactant decreases owing to its increased retention in the tube. This was overcome by the pulsed quench flow system of Fersht (Fersht and Jakes, 1975) in which the reaction mixture is stationary between initiation and quenching and the quantity of sample recovered is independent of the reaction time (Fig. 9). The initiation and quenching events are controlled independently of each other, and the reaction time is preset with an electronic timer. The lower limit of the reaction time is equal to the syringe stroke time.

A combination of these two forms of quench flow allows a wide range of reaction times to be used. The shortest possible reaction time is limited by the mixing of the solutions and is about a millisecond. In practice the time resolution of the instrument is given by the precision of determination of the dead time (Cash and Hess, 1981). When receptor-controlled ion flux is studied in vesicles, a measurement can be obtained 5 msec after the reaction is initiated. Cash developed a quench flow apparatus in which small volumes of vesicle suspensions and a wide range of incubation times could be used (Cash and Hess, 1981). Diagrams of the continuous and pulsed quench flow apparatus are shown in Fig. 9a and b respectively. Both have been described in detail together with the measurements that were made to study the characteristics of the instrument (Cash and Hess, 1981).

The instrument is of the multiple-mixer type incorporating the pulsed quenched-

$\longrightarrow$

FIGURE 9.   Diagram of the quench flow apparatus. (a) A partial schematic representation showing three syringes and the plumbing for continuous mode quench flow. The unit shown is mounted in one column. A machine capable of two pulsed incubations has three such columns mounted on the base. (b) Plumbing arrangement for two pulsed incubations with three columns. Reagents from syringes 1 and 2 are incubated in tube 1 before being displaced by solution from syringes 4 and 5, mixed with reagent from syringe 6, and incubated in tube 2 before being displaced by solution from 7 and 8 and mixed with reagent from syringe 9. Reprinted from Cash and Hess (1981) with permission.

(a)

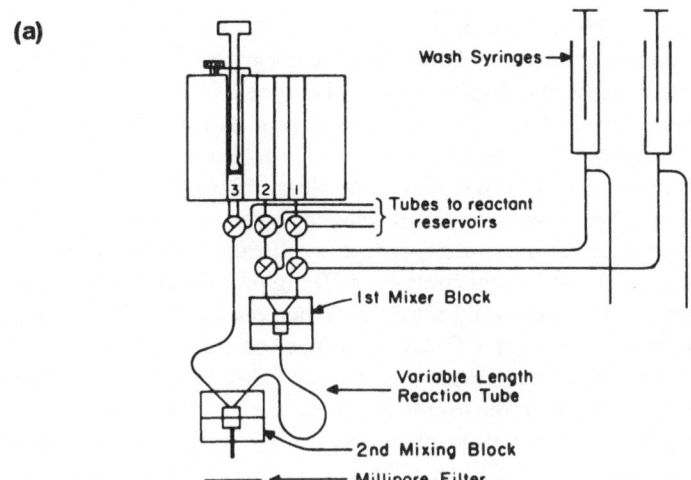

Wash Syringes →

Tubes to reactant reservoirs

1st Mixer Block

Variable Length Reaction Tube

2nd Mixing Block

Millipore Filter

(b)

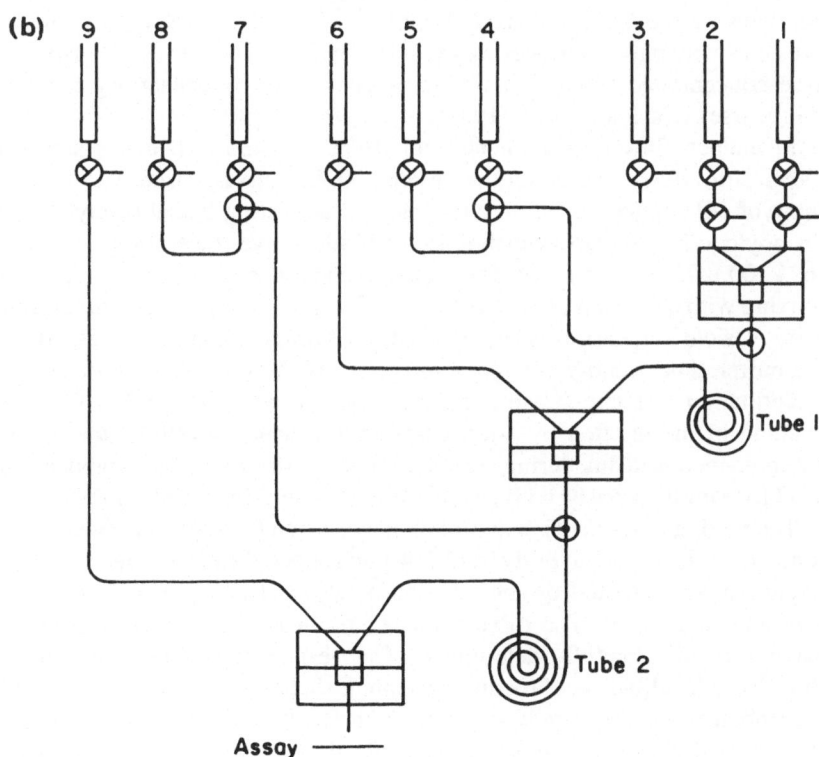

Tube 1

Tube 2

Assay

flow modification of Fersht (Fersht and Jakes, 1975). The pistons are driven pneumatically and are activated electronically. A linear stroke is obtained up to a stroke time of 6 sec/2 in. displacement. The speed is adjusted with flow restrictors and a reducing valve. Linear displacement as a function of time is recorded on a storage oscilloscope. Aoshima and Cash (Hess *et al.*, 1979, 1980, 1981; Aoshima *et al.*, 1980, 1981; Cash and Hess, 1980; Cash *et al.*, 1980, 1981; reviewed in Hess *et al.*, 1983) developed the use of this apparatus to measure, in the millisecond time range, receptor-controlled ion flux in vesicles.

Vesicle suspensions (typically containing 10 mg protein per milliliter) were used immediately or stored in liquid nitrogen. When needed the stored samples were allowed to thaw slowly at 4°C and were kept below 4°C throughout the experiments. The vesicles were passed down a Sephadex G-25 (coarse) column in eel Ringer's solution, to remove any nonvesicular material and remaining sucrose; this improved the precision of the assay. The vesicle suspension was diluted to 800 μg protein per milliliter. The vesicle concentration, expressed in terms of protein, was measured by determining the optical density at 520 nm and calibrating a linear plot of optical density versus protein concentration as determined by the Lowry method (Lowry *et al.*, 1951). Prolonged contact of the vesicles with plastic surfaces was avoided. If necessary the vesicle suspension can be concentrated without loss of receptor activity by adding dry Sephadex G-25 (coarse) (Cash and Hess, 1981).

In quench flow measurements, eel Ringer's solution (Keynes and Martins–Ferreira, 1953) containing the vesicles (0.225 ml) is mixed with an equal volume of eel Ringer's solution containing an activating ligand (acetylcholine, carbamoylcholine, or suberyldicholine) and radioactive tracer ion (for instance $^{86}Rb^+$, 100 μCi/ml). After a predetermined reaction time, the reaction is quenched by mixing with a solution of *d*-tubocurarine chloride (10 mM final concentration) and the vesicle suspension is expelled onto a Millipore filter and washed with 20 ml buffer. The quantity of tracer ions in the vesicles on the filter is determined by scintillation counting (Cash and Hess, 1981; Hess *et al.*, 1979). Complete quenching of the reaction by *d*-tubocurarine was demonstrated by the absence of receptor-mediated flux during very short reaction times or when *d*-tubocurarine was added simultaneously with an activating ligand (Hess *et al.*, 1979).

The mixing of reactants in the continuous quench flow apparatus was studied by observing the rate of hydrolysis of 2,4-dinitrophenylacetate. The viscosity of the solution was adjusted by the addition of glycerol to equal the viscosity of the vesicle suspension. The mixing–quenching dead time for the vesicle suspension was also carefully determined. The vesicle membranes were labeled with [^{125}I]-iodosulfonic acid by diazo coupling (Sears *et al.*, 1971) to show that the membrane was completely retained on the Millipore filter. Control experiments showed that the vesicles were stable during the experiments. Vesicle disaggregation (Kim and Hess, 1981) was not important in the dilutions used. The breakage of vesicles in the apparatus was examined using the [^{125}I]iodosulfanilic-acid-labeled vesicles as an internal standard for ^{86}Rb$^+$ flux. Breakage was small with the pulsed mode and with continuous mode using short reaction tubes but was significant with longer reaction tubes in the continuous mode. The loss, when it occurred, was reproducible and could be corrected for.

When more than one mixer is used, vesicles can be preincubated using a pulsed or a continuous mode, or a mixture of the two, depending on the incubation times. Thus inactivation of the receptor during preincubation with a ligand for various times can be followed (Aoshima et al., 1980). By mixing unequal volumes, large dilutions of the ligand may be made, and concentration jump experiments performed. By initiating the ion influx assay at various times after such a dilution, the reactivation of the receptor can be studied after its inactivation with an activating ligand (Aoshima et al., 1980, 1981).

Both influx and efflux can be measured using this technique. Although efflux has the advantage of using less radioactive isotope, the corrections that must be made to account for leakage of the tracer ion from the active and inactive vesicles during the experiment and the corrections for the equilibration of tracer ions with a pool of slowly exchanging inorganic ions (Kim and Hess, 1981, and Section 3.1.1), make interpretation of the data more complicated. Experiments involving preincubation of the receptor with a ligand before the efflux is begun, in order to measure the rate of receptor inactivation, are not possible. They can be done when influx is measured.

3.2.2. Vidicon Flame Emission Spectroscopy to Measure Ion Flux with Nonradioactive Ions

In the fast reaction experiments described so far radioactive rubidium ions were used as a replacement for radioactive potassium ions because they are more stable and have been reported to be an effective substitute (Palfrey and Littauer, 1976). The wish to test the substitution of ^{86}Rb$^+$ for K$^+$ in the system used, the desirability of measuring the rates of movement of ions that were not available in a radioactive form, and the desirability of developing a technique that did not involve radioactive ions, all led to the design of another form of quench flow measurement.

Flame emission spectroscopy is a good method to measure trace amounts of alkali metal ions (Morrison, 1979). When a vidicon tube detector is placed at the exit slit of a flame spectrometer, it is possible to determine simultaneously multielement emission frequencies within a continuous wavelength region (Busch and Morrison, 1973; Busch and Malloy, 1979). Aoshima, with Ramseyer in

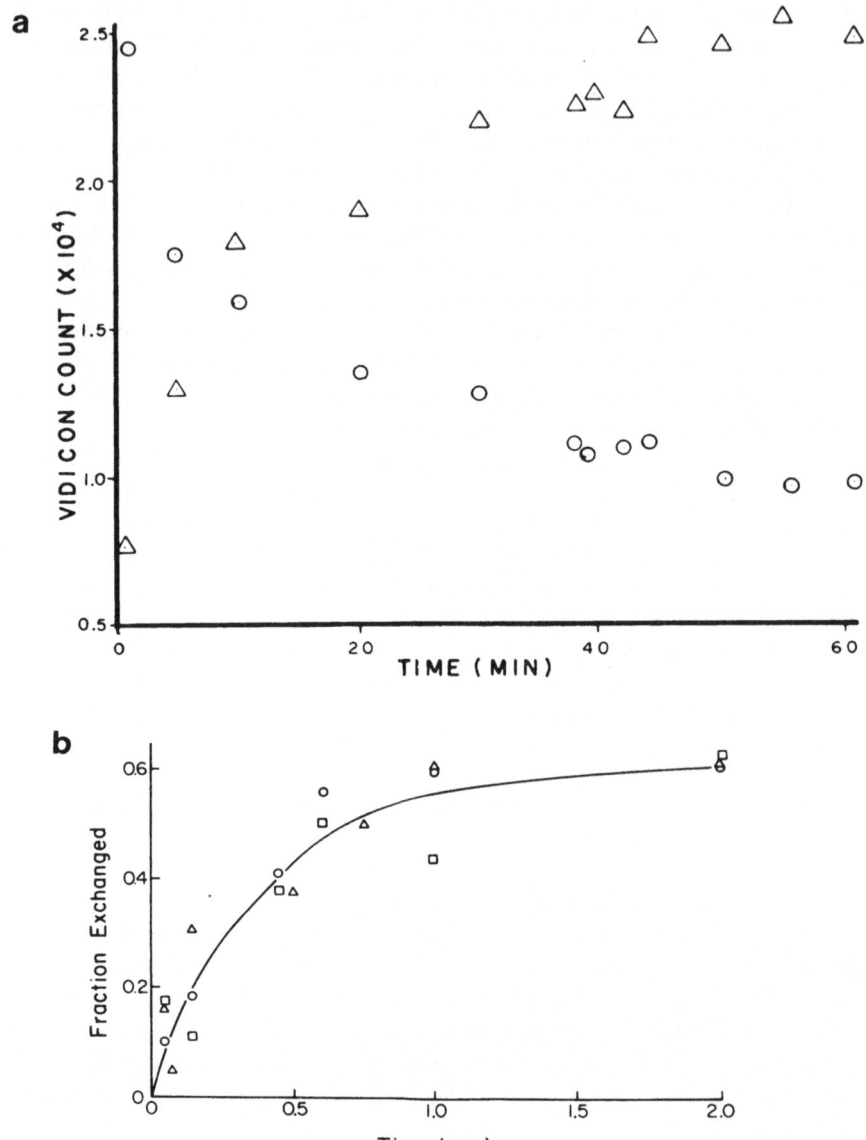

FIGURE 10. (a) Na$^+$ and K$^+$ fluxes through *E. electricus* membrane measured by vidicon flame emission spectroscopy. Vesicles (8.16 mg protein) incubated in high-Na$^+$ solution overnight were diluted 50-fold in high-K$^+$ solution. Influx of K$^+$ ($\triangle$) and efflux of Na$^+$ ($\bigcirc$) were observed by vidicon flame emission spectroscopy. One part per million Na$^+$ = 33,182 counts; 1 ppm K$^+$ = 31,452 counts. (b) Flux of Na$^+$ ($\bigcirc$), Li$^+$ ($\triangle$), and K$^+$ ($\square$) into vesicles in the presence of 1 mM

Professor Morrison's laboratory in the Department of Chemistry at Cornell, adapted this approach to the measurement of ion fluxes in vesicles. The vesicles were suspended in a high-K^+/high-Na^+ or an eel Ringer's solution. Twice-distilled water and Millipore filters (HATF) that were free of detergent were used for all experiments. The filters were placed in a plastic box containing 100 ml of carefully deionized water for several days before use. The water was changed at least three times. The filters were then placed for a few minutes in whatever solution (CsCl or NaCl) would be used for washing, and dried and stored in Beckman Poly A bottles. Ultrapure CsCl and HCl were used when needed.

The flux of Li^+, Na^+, and K^+ into receptor-rich vesicles was measured in the quench flow apparatus. Vesicles were incubated overnight with modified eel Ringer's solution in which KCl was replaced by NaCl. When Li^+ influx was measured, vesicles equilibrated with eel Ringer's solution were mixed with an eel Ringer's solution in which NaCl and KCl were replaced by LiCl. Control experiments using radioactive tracer ions were performed with eel Ringer's solution inside and outside the vesicles. For Na^+, K^+, and Li^+ flux measurements, vesicles emerging from the quench flow apparatus were collected on a Millipore filter, and the concentration of the alkali metal ions inside the vesicles was determined by flame emission spectroscopy.

In order to compare the tracer ion and the flame emission spectroscopic methods, the influx of $^{42}K^+$ (half-life, 12.4 hr) was measured in the quench flow apparatus in the usual way. Part of each sample was counted by liquid scintillation, and the remainder was stored for 10 days and then measured by flame emission.

K^+ efflux and Na^+ influx could be determined simultaneously (Ramseyer et al., 1981) in the same experiment (Fig. 10a). The experiments illustrate that (1) ion flux can be measured in the millisecond time region with cations that cannot be obtained in radioactive form, (2) efflux and influx measurements give the same rate coefficient (Fig. 10a), and (3) Na^+, K^+, Li^+, and $^{86}Rb^+$ are translocated at rates that appear to be equal, at least when measured at one concentration of carbamoylcholine (Fig. 10b). Electrophysiological measurements with cells indicate that the receptor-controlled translocation of these ions reflects the diffusion coefficient of the ions in an aqueous solution (Hille, 1976).

---

carbamoylcholine measured by the quench flow method and determined by vidicon flame emission spectroscopy. The solid curve was obtained from measurements of $^{86}Rb^+$. The curve was calculated as has been described (Cash and Hess, 1980; Aoshima et al., 1980) and represents measurements from five experiments using different membrane preparations. The rate coefficients were calculated from these measurements as previously described (Cash and Hess, 1980; Aoshima et al., 1980). $J_A = 1.6$ sec^{-1}, $J_I = 0.015$ sec^{-1}, $\alpha = 1.8$ sec^{-1}. Reprinted from Ramseyer et al. (1981) with permission.

The flame emission method is more susceptible to trace element contamination, particularly for Na^+ and K^+, and routine analysis is ideally carried out in a clean room.

3.2.3. Optical Detection of Cs^+ Flux Using Stopped-Flow Techniques

The use of fluorescence quenching to measure translocation of Tl^+ in vesicles by stopped-flow techniques was first introduced by Moore et al. (Moore and Raftery, 1980). Cs^+ has some advantage over Tl^+ in that (1) an equilibrium across vesicle membranes in the absence of an activating ligand occurs in hours with Cs^+ and in a few seconds with Tl^+ (Lymn and Taylor, 1970), (2) vesicles become permeable to other ions after short exposures to Tl^+; Cs^+ does not have this effect (Rabon and Sachs, 1981; Karpen et al., 1983); and (3) CsCl, in contrast to TlCl, is soluble in water allowing the use of the chloride ion.

Sachs prepared anthracene 1,5 disulfonic acid disodium salt (ADS) by the method of Kumar Ash and Rohatgi–Mukherjee (1979); the fluorescence of this dye is quenched by Cs^+. A method was developed for measuring the receptor-controlled flux of Cs^+ in vesicles in a modified Durrum–Gibson stopped-flow instrument (Hess et al., 1982; Karpen et al., 1983). Vesicles were loaded with ADS by modifying the normal method of preparation (Karpen et al., 1983). External dye was removed by passing the vesicles through a Sephadex G-25 (coarse) column equilibrated with the buffer used either in the experiments or for storing the dye-loaded vesicles in liquid nitrogen.

The flow system before the mixer in the stopped-flow spectrophotometer was adapted from the quench flow apparatus (see Section 3.2.1 and Fig. 9). Fluorescence measurements in the excitation maximum range that includes ADS (excitation maximum at 365 nm) are normally made with a xenon arc lamp as the light source. However, an inexpensive 150-W tungsten–halogen lamp is much more stable and provides light that is of sufficient intensity.

Control experiments showed that the rate of receptor-controlled ion flux is the same whether it is measured by stopped flow with ADS and Cs^+ or by quench flow with $^{86}Rb^+$ (Fig. 11). Figure 11b (inset) shows an oscilloscope trace obtained in a stopped-flow experiment in which E. electricus vesicles loaded with ADS were mixed with eel Ringer's solution in which NaCl and KCl were replaced by CsCl. The optical signal is due to Cs^+ quenching the fluorescence of the dye.

A similar, but as yet more qualitative, approach has been developed for use with cells in culture. The quenching of dye fluorescence by Cs^+ is observed in a fluorescence microscope (Sachs et al., 1983).

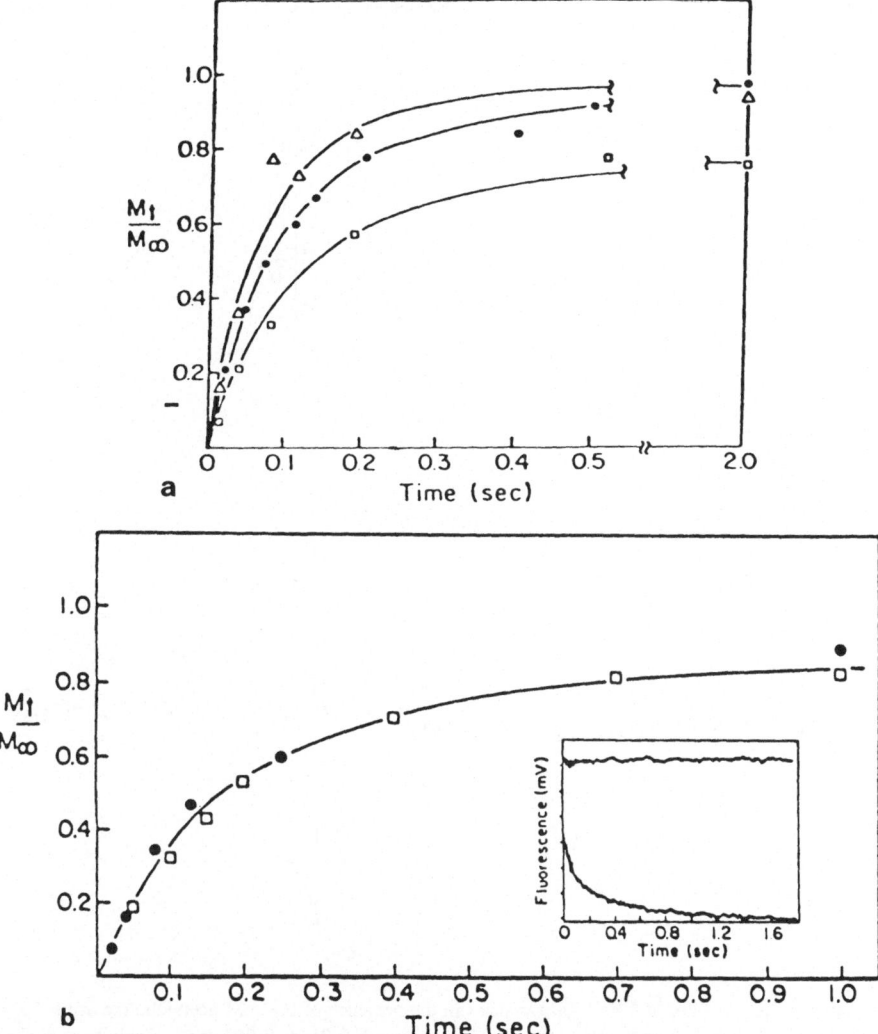

FIGURE 11. (a) Ion flux in vesicles, pH 7.0, 1°C, induced by 30 μM suberyldicholine (△, $J_A = 14.3$ sec^{-1}), mM acetylcholine (●, $J_A = 11.5$ sec^{-1}), or 10 mM carbamoylcholine (□, $J_A = 7$ sec^{-1}). These J_A values represent the maximum observed values. The data with acetylcholine were obtained using the stopped-flow technique and Cs$^+$ and the data with the other ligands using the quench flow technique and ^{86}Rb$^+$ as a tracer. (b) A comparison of influx measurements in the presence of 150 μM acetylcholine using the same preparation: ●, quench flow technique and ^{86}Rb$^+$; □, stopped-flow technique using Cs$^+$. Inset: Oscilloscope traces from stopped-flow experiments. Upper traces, Vesicles (400 μg/ml membrane protein) were mixed with an equal volume of Cs–eel Ringer's solution. Lower trace, As above, but the Cs–eel Ringer's solution was 300 μM (final concentration 150 μM) in acetylcholine. Ordinate: 2 mV/division. Reprinted from Hess *et al.* (1982) with permission.

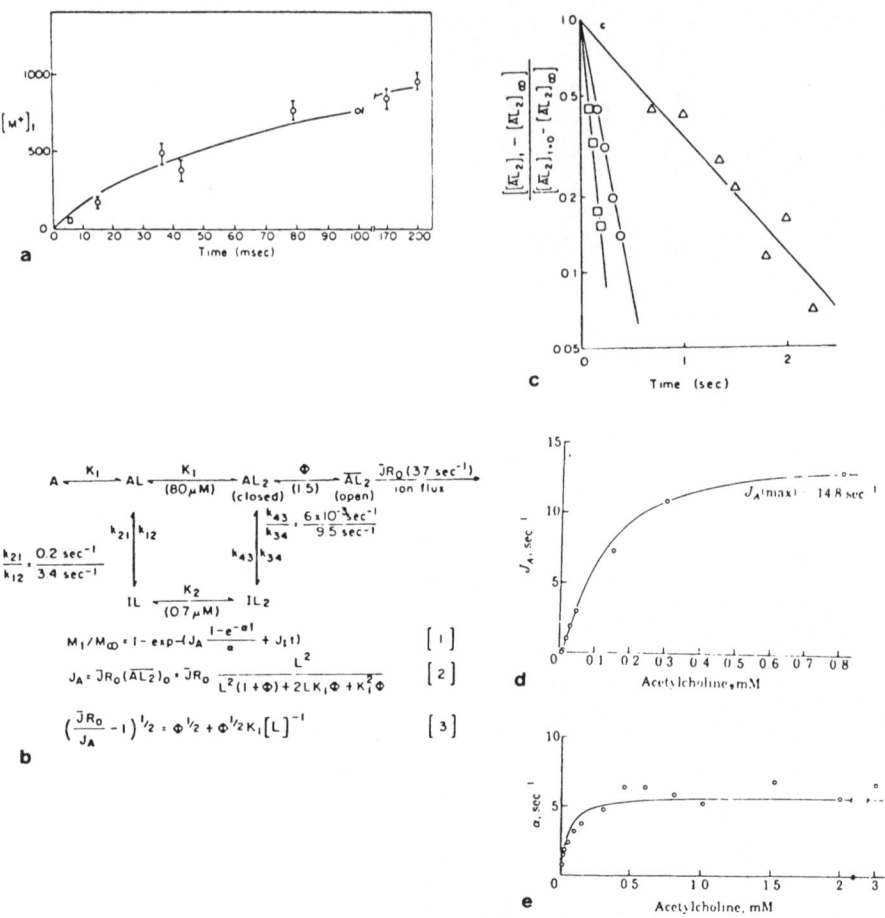

FIGURE 12. (a) Flux of $^{86}Rb^+$ into vesicles in the presence of 0.8 mM acetylcholine at pH 7.0, 1°C. $J_A = 14.5 \pm 2.1$ sec^{-1}; $\alpha = 5.6 \pm 0.7$ sec^{-1}; $J_I = 0.01 \pm 0.003$ sec^{-1}. Each point is the mean of three determinations. The coordinates of the solid lines were obtained using a nonlinear least-squares computer program. (b) Minimum mechanism (Cash and Hess, 1980; Aoshima *et al.*, 1980) to account for the rates of acetylcholine receptor-mediated cation translocation and of receptor inactivation and reactivation as a function of acetylcholine concentration measured using *E. electricus* vesicles. The active forms of the receptor (A) and the inactive forms (I) bind ligand (L) in rapidly achieved equilibria denoted by the microscopic equilibrium constants (K). Active receptor with two bound ligands (AL_2) converts rapidly to an open channel ($\overline{AL}_2$) with an equilibrium constant for channel opening ($1/\phi$). $\overline{AL}_2$ permits ion flux with a first-order-rate in a given vesicle system. The symbols containing A and I represent fractions of the receptor and have been defined previously in terms of ligand concentration and the constants in the minimum model (Cash and Hess, 1980). (c) Effect of carbamoylcholine concentration on the rates of receptor inactivation. Evaluation of the rate constant for the inactivaton reaction. The values of α obtained from the slope of the lines, at the

3.3. Quantitative Treatment of the Ion Translocation Process

3.3.1. Ion Translocation in E. electricus Vesicle Preparations

The kinetics of receptor-controlled flux of specific inorganic cations in the characterized $E.$ $electricus$ vesicle preparation were examined by Aoshima, Cash, Pasquale, and Takeyasu, and the approach has been reviewed in detail (Hess et $al.$, 1983). When the volume of the solution in which the vesicles are suspended is very large compared to the volume of solution inside the vesicles, the concentrations of inorganic ions in the external solution can be considered to be constant. The rate of ion translocation in the vesicles can then be given by

$$-\frac{d(M)}{dt} = \bar{J}R_O(\overline{AL_2})_t(M), \qquad (8.8)$$

where M represents the concentration of monovalent cations in the vesicles, R_O the concentration of receptor sites in moles per liter vesicle internal volume, $(\overline{AL_2})_t$ the fraction of the receptor in the open channel form at time t, and $\bar{J}$ the specific reaction rate of the receptor-controlled ion translocation. Measurement of the flux of $^{86}Rb^+$ into vesicles in eel Ringer's solution at pH 7.0 and 1°C (Fig. 12a) indicates the existence of three processes.

1. There is an initial rapid influx that occurs in the millisecond time region and comes to an end before the vesicles are completely filled (Fig. 12a). This flux is considered to be due to the active (A) form of the receptor and the rate coefficient pertaining to this process is designated J_A.
2. The observed conversion of a fast to a slow influx is considered to be due to the isomerization of the receptor from an active (A) state of the

---

different concentrations of carbamoylcholine, are as follows: 0.25 mM carbamoylcholine ($\triangle$), 0.9 $\pm$ 0.1 sec^{-1}; 2 mM carbamoylcholine ($\bigcirc$), 4.9 $\pm$ 0.1 sec^{-1}; 10 mM carbamoylcholine ($\square$), 8.8 $\pm$ 0.4 sec^{-1}. [$\overline{AL_2}$] represents the fraction of receptor in the open channel form; the subscripts refer to the time of measurements, where $t = 0$ and ∞ indicates the value of [$\overline{AL_2}$] before and after the inactivation (desensitization) reaction has gone to completion. (d) The relationship (Aoshima et $al.$, 1981) between [$\overline{AL_2}$] before onset of inactivation ($\bigcirc$), after equlibrium is reached (∞), and after exposure to acetylcholine for a period of time (t) is given by [$\overline{AL_2}$]$_t$ $-$ [$\overline{AL_2}$]$_\infty$ = ([$\overline{AL_2}$]$_0$ $-$ [$\overline{AL_2}$]$_\infty$)$e^{-\alpha t}$. The concentrations of the various species in this equation, determined by measuring acetylcholine-induced influx of $^{86}Rb^+$ for a constant period of time after variable periods of incubation with acetylcholine, have also been described (Aoshima et $al.$, 1981). First-order-rate constants for (d) influx of $^{86}Rb^+$ with receptor in the active state and (e) inactivation of the receptor. The lines were calculated by using the integrated rate equation pertaining to the scheme, and the values of the parameters listed in Table IV. The points were determined as described.

receptor to an inactive (I) state. The observed rate coefficient for this interconversion is designated α, which contains the rate constants for the interconversion.

3. The slow influx which occurs in the minute time region, is considered to be from an equilibrium mixture of A and I conformations. Only the A conformation can form ion-conducting channels, but the equilibrium mixture is in favor of the I conformation in the presence of an activating ligand. This equilibrium mixture is referred to as the inactive (desensitized) receptor and the observed rate coefficient associated with this phase as J_I.

A minimum mechanism that incorporates these observations is shown in Fig. 12b and the constants pertaining to the mechanism are shown in Table IV. The model relates the ion translocation rates to (1) the ligand-binding steps, (2) the equilibrium between the open and closed channel forms of the active receptor with two ligand molecules bound to it, $(\overline{AL_2})$ and (AL_2), respectively, and (3) the rates of interconversion between active and inactive forms of the receptor. The relationships between the various species of active and inactive forms of the receptor and the various constants used in the model are given in Fig. 12b and Table 4. Because the model can be treated quantitatively it can be tested over wide concentration ranges of various activating ligands (carbamoylcholine, acetylcholine, and suberyldicholine). The effects of the concentration of acetylcholine on J_A and α are shown in Fig. 12d and e. Inhibitors (cocaine, local anesthetics, and other pharmacological agents) were also used and Karpen, for instance, has shown that cocaine and phencyclidine both inhibit receptor-controlled flux, but by different mechanisms (Karpen *et al.*, 1982). Phencyclidine, unlike cocaine or procaine (a local anesthetic), increases the rate of receptor inactivation and changes the equilibrium between active and inactive receptor forms. Pasquale (Pasquale *et al*, 1984) studied the effect of suberyldicholine as an activating ligand. The influx rate coefficient, J_A, increased with increasing concentration of suberyldicholine, reached a plateau value, and then decreased when the suberyldicholine concentration was further increased. This inhibitory effect is not seen when one measures the effect of suberyldicholine concentration on the inactivation rate coefficient, α. The existence of regulatory binding sites had been proposed previously in order to explain the effect of local anesthetics (Adams, 1977; Heidmann and Changeux, 1978; Koblin and Lester, 1978; Neher and Steinbach, 1978), the inhibition caused by high concentrations of decamethonium (Adams and Sakmann, 1978) and of dansylcholine (Cohen and Changeux, 1973), and the inactivation (desensitization) of the receptor (Dunn and Raftery, 1982; Walker *et al.*, 1981).

It is also possible to vary the concentration and the type of inorganic ions

TABLE IV. Dynamic Properties of the Acetylcholine Receptor. Results Obtained with Acetylcholine, Carbamoylcholine, and Suberyldicholine

Parameter	Acetylcholine[a]	Carbamoylcholine[b]		Suberyldicholine[c]
Chemical kinetic measurements with membrane vesicles[d]				
$K_1 =$	80	1900		4.5
$K_2 =$	0.7	21		—
$K_{c1} =$	0.07	0.1		—
$K_{c2} =$	0.006	0.01		—
$\phi = (AL_2)/(\overline{AL_2})$	1.5	1.8		1.0
$J_A max = \overline{J}R_O/(1 + \phi),(sec^{-1})$	14.8	9.7		18.5
$k_{12}(sec^{-1})$	3.4	4.6		1.0
$k_{21}(sec^{-1})$	0.2	0.5		—
$k_{34}(sec^{-1})$	10	11		11.0
$k_{43}(sec^{-1})$	0.006	0.01		—
J, $(M^{-1}\ sec^{-1})$	3×10^7	3×10^7		3×10^7
Statistical methods[e] denervated frog muscle cells[f] and *E. electricus* electroplax cells[g]				
Average open time (msec)	11 ± 1.6[g]	3.9 ± 0.4[g]	1.5 ± 0.2[g]	19 ± 2.5[g]
Single channel conductance (pmho)	15 ± 1.8[g]	~15[g]	53[g]	~15[g]

[a] Cash and Hess (1981); Hess *et al.* (1980).
[b] Aoshima *et al.* (1981).
[c] Pasquale *et al.* (1983).
[d] *E. electricus* Ringer's solution, pH 7.0, 1°C. All influx measurements were normalized to an R_O value of 1.2 μM.
[e] Katz and Miledi (1972); Anderson and Stevens (1973).
[f] Data from Neher and Sakmann (1976b). Frog Ringer's solution, pH 7.1, 8°C, −80 mV. Properties of extrajunctional receptors obtained by noise analysis of voltage clamp current.
[g] *E. electricus* Ringer's solution, pH 7.0, 14°C, −85 mV. Data from Hess *et al.* (1984).

on either side of the membrane, the temperature, pH, and the transmembrane voltage. Takeyasu and Udgaonkar (Takeyasu *et al.*, 1983) have used the quench flow method to measure J_A and α under voltage clamp conditions over a wide range of acetylcholine and suberyldicholine concentrations. Na^+, K^+, and Cl^- in *E. electricus* vesicles exchange very slowly with ions in the external solution when no activating ligand is present (Hess and Andrews, 1977). When a monovalent cation is distributed asymmetrically across the membrane and the osmotic balance is maintained in the solutions by use of an impermeant cation, then making the vesicle membrane selectively permeable only to the asymmetrically distributed cation establishes a transmembrane voltage. Takeyasu and Udgaonkar found that at low ligand concentrations both α and J_A increased with increasing concentration. At high ligand concentrations a concentration-dependent decrease in the ion flux rate was observed without a concomitant change in the inactivation rate. This inhibitory effect had not been reported previously and was not observed with acetylcholine or carbamoylcholine in the absence of a transmembrane voltage. When the transmembrane potential is made more negative, the dissociation constant for binding of acetylcholine to the receptor before inactivation remains unchanged, whereas the dissociation constant for binding of acetylcholine to its regulatory site decreases dramatically. Shiono, Udgaonkar, Delcour, Takeyasu and Fujita have also shown, using the same approach, that procaine, a local anesthetic that is a non-competitive inhibitor of receptor function binds to an inhibitory site that is different from the regulatory site for acetylcholine. The regulatory site for acetylcholine can be fitted into the general model proposed to explain the activation, inactivation, and voltage-dependent modulation of the receptor function.

Aoshima (Aoshima *et al.*, 1980) paid special attention to the process of inactivation (desensitization) that occurs when the receptor is exposed to the ligand. Vesicles were first incubated with varying concentrations of a ligand for various periods of time before receptor activity was measured at saturating concentrations of the ligand. This allowed α, the rate coefficient for receptor inactivation, to be evaluated directly at all ligand concentrations (Fig. 12c). In addition, the reactivation of the inactivated receptor could also be measured. Although two ligand molecules must be bound to the receptor before the channel opens, inactivation occurs when one ligand molecule is bound. This accounts for the different effect of ligand concentration on the flux coefficient, J_A, and on the rate coefficient, α, associated with the inactivation and reactivation of the receptor [see Figs. 12d(a) and 12d(b)]. This has the effect, at least experimentally, that at low ligand concentrations the properties of only the inactivated receptor are observed, leading to the determination of dissociation constants that are much lower than those of the active form (Karlin, 1980).

The isolation of vesicles with active receptors (see Section 3) permitted the calculation of the moles of receptor sites per liter internal vesical volume, R_O. Chemical kinetic measurements, together with a knowledge of the value of R_O, made it possible to calculate the specific reaction rate, $\bar{J}$ ($\bar{J} = 3 \times 10^7$ M^{-1} sec^{-1}), of the receptor-controlled ion translocation (Hess *et al.*, 1981). The value of $\bar{J}$ can be used to calculate the value of the single-channel conductance, ($\Lambda = 33$ pS) (Hess *et al.*, 1984). Using the single-channel current recording technique (Neher and Sakmann, 1976a) with *E. electricus* electroplax cells, from which the membrane vesicles used in the chemical kinetic experiments are formed, a value for Λ of 53 pS was obtained (Hess *et al.*, 1984). Thus chemical kinetic measurements with vesicles and current measurements with cells give results that are in good agreement.

3.3.2. Ion Translocation from Torpedo spp. Vesicle Preparations

Receptor-conrolled ion flux has also been studied in vesicles prepared from the electroplax of *Torpedo* spp. (Hazelbauer and Changeux, 1974; Popot *et al.*,

1974, 1976; Sugiyama *et al.*, 1975; Miller *et al.*, 1978; Neubig and Cohen, 1980; Delegeane and McNamee, 1980; Moore and Raftery, 1980). Typical *Torpedo* spp. vesicles have about 1.5 nM of toxin sites per milligram of membrane protein (Karlin, 1980), which exceeds, by several orders of magnitude, the concentration in *E. electricus* vesicles. At high carbamoylcholine concentrations the observed influx rates exceed the time resolution of the quench flow and stopped-flow methods (Neubig and Cohen, 1980). *Torpedo* spp. vesicles of uniform size and containing functional receptors have not yet been separated from the heterogeneous preparation.

Walker, Pasquale, and McNamee developed a technique that avoids the use of inhibitors (Moore and Raftery, 1980; Neubig and Cohen, 1980) in measuring receptor-controlled influx rates that are too high to be measured with the usual quench flow method (Walker *et al.*, 1981; Hess *et al.*, 1982). Advantage is taken of the natural inactivation of the receptor by its own ligand and of the double-mixing technique to evaluate α, the inactivation rate constant (Eq 3 in Fig. 13c). When $(J_A)_T \gg \alpha$, $(J_A)_T$ can be evaluated from a single influx measurement (Hess *et al.*, 1982). With carbamoylcholine the $(J_A)_T$ values that must be known in order to evaluate $(J_A)_{T=0}$ and α were determined by first preincubating the vesicles for various periods of time (T) in the presence of various concentrations of carbamoylcholine and then measuring $^{86}Rb^+$ influx for 10 msec in the presence of 10 mM carbamoylcholine. A semilogarithmic plot of $(J_A)_T$ obtained after various periods of preincubation versus preincubation time is shown in Fig. 13a. The inactivation rate constant, α, is obtained from the slope of the line. All lines, corresponding to different concentrations of carbamoylcholine during the preincubation, have a common ordinate intercept because influx was always measured in the presence of 10 mM carbamoylcholine. The intercept gives the value of $(J_A)_{T=0}$ as 310 sec^{-1} in the absence of receptor inactivation. The initial rapid inactivation that occurs in *T. californica* spp. vesicles, which was detected in the experiments shown in Fig. 13a (Hess *et al.*, 1982), had escaped notice previously, but a slower process had been observed. After the first inactivation process has gone to completion, the flux rate still decreases with the length of the preincubation, but at only about 6% of the rate at which flux decreased during the first 2 sec of preincubation. A considerable amount of flux activity, characterized by the rate coefficient, J_I, is therefore still associated with the acetylcholine receptor after the first inactivation process has gone to completion. The value of the rate coefficient associated with this second slower inactivation process can be obtained from Eq. (4) (Fig. 13c) when the rate coefficient, $(J_I)_T$, is plotted on a logarithmic scale as a function of preincubation time (Fig. 13b). The slower inactivation process, characterized by the rate coefficient, β, also appears to follow a first-order-rate law. The value of β obtained from the slope of the line is about 1/15th the value of α at the same concentration of carba-

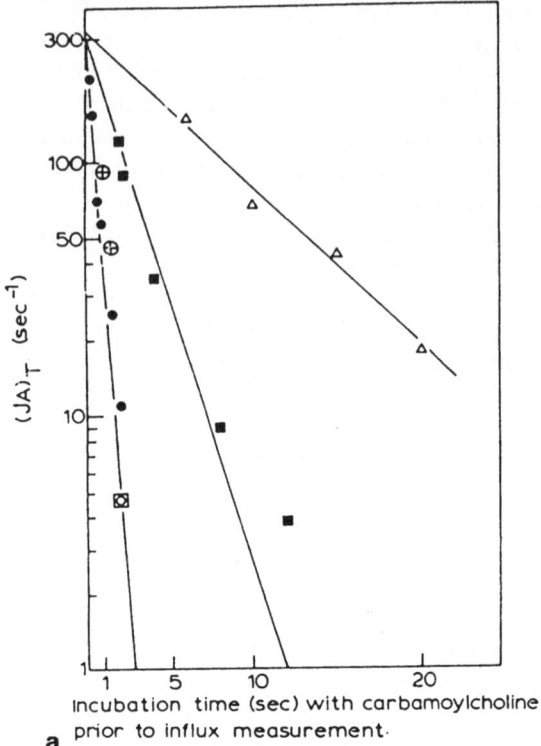

a

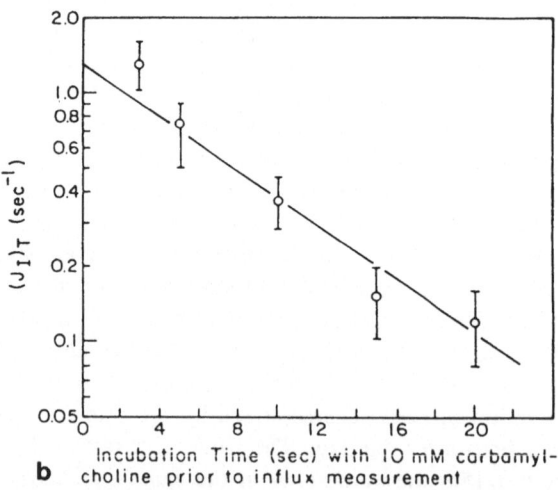

b

FIGURE 13. Rapid inactivation of the flux of $^{86}Rb^+$ into vesicles by different carbamoylcholine concentrations, pH 7.0, 1°C. (a) Evaluation of the inactivation rate coefficient, α, at various concentrations of carbamoylcholine and of the influx rate coefficient before inactivation, $(J_A)_{\tau=0}$, in 10 mM carbamoylcholine. α values at different concentrations of carbamoylcholine are 2 ± 0.2 sec

A

$$\begin{array}{c} \cdot \overset{K_I}{A \longleftarrow AL} \overset{K_I}{\longleftarrow AL_2} \overset{\phi}{\longleftarrow} \overset{\overline{JR_O}}{\overline{AL}_2} \rightleftharpoons \\ \text{(closed)} \quad \text{(open)} \\ \text{ion flux} \end{array}$$

$$K_{CI} \Updownarrow \qquad \Updownarrow K_{C2}$$

$$IL \underset{K_2}{\longleftarrow} IL_2$$

$$\overset{\beta}{\longleftarrow}$$

B

$$\left[(\overline{AL}_2)_{PE}\right]_T$$

$$\frac{M_t}{M_\infty} = 1 - \exp - \overline{J}R\left\{ \left[(\overline{AL}_2)_0 - (\overline{AL}_2)_\infty\right] \frac{1-e^{-\alpha t}}{\alpha} + (\overline{AL}_2)_\infty t \right\} \qquad [1]$$

$$\frac{M_t}{M_\infty} = 1 - \exp - \overline{J}R\left\{ \left[(\overline{AL}_2)_0 - (\overline{AL}_2)_{PE}\right] \frac{1-e^{-\alpha t}}{\alpha} + (\overline{AL}_2)_{PE} \frac{1-e^{-\beta t}}{\beta} \right\} \qquad [2]$$

C

$$(J_A)_T = (J_A)_{T=0} e^{-\alpha t} \qquad [3]$$

$$(J_I)_T = (J_I)_{T=0} e^{-\beta t} \qquad [4]$$

[10 mM ●, □]; 0.5 ± 0.04 sec^{-1} [3 mM ■]; 0.14 sec^{-1} ± 0.02 sec^{-1} [0.5 mM △]; an estimated value of 0.03 sec^{-1} is obtained at 0.25 mM carbamoylcholine. The mean value of $(J_A)_{T=0}$ obtained from the ordinate intercept of the solid lines is 310 ± 66 sec^{-1}. Solid lines were computed by using the same value of $(J_A)_{T=0}$ and values of α obtained with different concentrations of carbamoylcholine in the incubation solutions. Values of χ^2 were less than 1.0 in all determinations. The different symbols associated with experiments conducted at the same carbamoylcholine concentration denote different membrane preparations. (b) The second, slower, phase of receptor inactivation. Receptor-mediated flux of $^{86}Rb^+$ into vesicles in presence of 10 mM carbamoylcholine, pH 7.0, 1°C, after various relatively long periods of incubation with 10 mM carbamoylcholine. Evaluation of the inactivation rate coefficient, β, and the influx rate coefficient, $(J_I)_{T=0}$; the abscissa gives the incubation time with 10 mM carbamoylcholine prior to influx measurement. Vertical bars give the standard deviations. The ordinate intercept gives $(J_I)_{T=0} = 1.3 ± 0.2$ sec^{-1}, and the slope gives $\beta = 0.12 ± 0.03$ sec^{-1}; $\chi^2 = 0.8$. (c) The minimal mechanism to account for the rates of receptor-controlled cation translocation in *T. californica* membrane vesicles is based on measurements with *E. electricus* vesicles (left part of the scheme) (Cash and Hess, 1980; Hess *et al.*, 1983). On the right is an extension of the model to account for the measurements with *T. californica* vesicles. Three receptor states are involved in eel receptor: active (A), open channel ($\overline{A}$), and inactive (*I*), and two microscopic ligand dissociation constants: K_1 pertaining to the A state, and K_2 pertaining to the *I* state. Two conformational equilibrium constants are required, two pertaining to the equilibria between the A and *I* states, K_c, and one, ϕ, pertaining to the equilibrium between the closed (A) and open-channel ($\overline{A}$) active forms. The symbols containing A and *I* (representing fractions of the receptor) and α have been defined in terms of ligand (L) concentration and the constants of the minimal mechanism (Cash and Hess, 1980; Hess *et al.*, 1983), in which α is the rate coefficient for the isomerization between A and *I* forms. The major difference between the results obtained with the receptor from the two organisms is that a second inactivation step occurs in the case of *T. californica*. It is assumed that in this step the concentration of the open-channel form is reduced to a low final value, which is no longer detected in the flux assay. The rate coefficient for this inactivation step is designated β. The integrated rate equation [Eq. (1)] can be extended to accommodate a second inactivation reaction. When $\alpha \gg \beta$, $(\overline{AL}_2)$ [Eq. (1)] in the figure can be considered to be in preequilibrium with all the other receptor forms; we designate this species $[\overline{AL}_2]_{PE}$ and obtain Eq. (2), the integrated rate equation for *Torpedo* receptor-controlled ion flux. α and β can be evaluated using Eqs. (3) and (4), respectively. The subscript *T* refers to the preincubation time. The subscripts *t* and ∞ refer to time *t* and the end of the experiment, respectively. *M* and *R* represent concentrations of metal ion and receptor, respectively. Parts a, b, and c reprinted with permission from Hess *et al.* (1982).

moylcholine. The ordinate intercept gives a value of $(J_I)_{T=O}$, which is 300 times smaller than the value of $(J_A)_{T=O}$. McNamee at the University of California, Davis, with Walker and Takeyasu, has used the quench flow technique to study ion flux from reconstituted membranes. The membranes were reconstituted by cholate dialysis using soybean lipids and purified *T. californica* receptor. They have reported (Walker *et al.*, 1982) that a rapid receptor-controlled ion flux is observed and that two desensitization processes occur, one on a millisecond time scale and one of the order of seconds to minutes.

A minimum extension of the model derived from measurements with *E. electricus* vesicles will also accommodate the ion translocation process in *Torpedo* spp. vesicles. This model, and the corresponding integrated rate equation for ion flux, are shown in Fig. 13c. With caution, the model can be treated quantitatively. Because α and β may have different ligand dependencies, an interpretation of the influx measurements requires an independent knowledge of the constants. The integrated rate equation pertaining to the mechanism (Fig. 13c) is only valid under restricted conditions, for instance, $J_A \gg \alpha$ and $J_I \alpha \gg \beta$.

4. CONCLUSION

During the past 10 years the members of this laboratory have studied the mechanism of action of the acetylcholine receptor at the molecular level, using primarily the receptor-rich vesicles developed by Kasai and Changeux (1971a,b,c). First, equilibrium and kinetic methods were developed to characterize the binding of activating and inhibitory ligands to the receptor and particularly the stoichiometry of the activating ligand-binding and toxin-binding sites of the receptor protein. These studies were made easier by the earlier discovery of a toxin that binds specifically and irreversibly to the receptor (Chang and Lee, 1963). The toxin-binding properties of the receptor provide a measure for comparing the receptor in different forms and preparations and from different sources, that are used by different laboratories. It was, therefore, important to resolve apparent discrepancies and to develop assays that could be widely used.

Second, the mechanism of action of receptor-controlled ion flux, which can only be studied with the receptor bound in a membrane that separates two solution compartments, had to be studied and measured in an appropriate time region. The vesicles were isolated and characterized and rapid-mixing techniques were developed so that the function of the receptor in vesicles could be investigated in the same way as proteins in solution have been studied. A minimum mechanism, similar to those proposed for enzymes and regulatory proteins in solutions, has been proposed. The equilibrium and rate parameters have been measured over wide concentration ranges for three ligands that can activate the receptor.

The important regulatory mechanism of inactivation (desensitization) is included in the model, which can be related to the model proposed by Katz and Thesleff (1957). The model has been extended to include the regulation of receptor binding by voltage. The mechanism is now being further tested by studying the effects of compounds that are known to affect the cholinergic acetylcholine receptor. The results obtained by the chemical kinetic approach adopted in this laboratory can be compared to the results of the elegant electrophysiological studies of Neher and Sakmann (1976a,b) and Neher and Stevens (1977). Single-channel currents recorded from *E. electricus* electroplax cells (the same cells from which the vesicles were prepared) have been made, and the results from the two approaches are in agreement (Hess *et al.*, 1984). The chemical kinetic approach can provide information (the binding constants of the ligand to various receptor forms, the effect of ligand concentration on the activation and reactivation rates, the fraction of fully liganded receptor in the open channel form, and the effect of inhibitors on these constants) that cannot easily be obtained in electrophysiological experiments. Conversely, electrophysiology can provide information (the lifetime of the open channel and the effect of inhibitors on this lifetime) that cannot easily be obtained by chemical kinetics. The specific reaction rate, $\bar{J}$, and the single-channel conductance, λ, can be used to compare the results obtained with the two entirely different approaches.

Although the methods described were developed to study the acetylcholine receptor, they are applicable to the study of any regulatory protein that must be examined in a membrane-bound form for the function to be investigated, particularly if the function must be measured in a subsecond time region.

ACKNOWLEDGMENTS.   The methods described here were developed by John P. Andrews, Hitoshi Aoshima, James E. Bulger, Derek J. Cash, Mark J. Cooper, Anne H. Delcour, David B. Donner, Juain–juian L. Fu, Norihisa Fujita, Jeffrey W. Karpen, Peter S. Kim, Ronald A. Kohanski, Bernard Lenchitz, Pierre Leprince, Stanley Lipkowitz, Douglas E. Moore, Richard L. Noble, Elena B. Pasquale, Arturo Pece, Helga Rübsamen, Alan B. Sachs, Satoru Shiono, Gary E. Struve, Kunio Takeyasu, Jayant B. Udgaonkar, Jeffery W. Walker, and Charles Zacharchak.

Past and present members of the laboratory are grateful to many colleagues for helpful discussions and assistance: Professor D. Nachmansohn, Professor E. Bartels–Bernal, and Professor W. Niemi, Columbia University, for discussions and for performing some electrophysiological assays; Professor A. Fersht, Imperial College of Science and Technology, London University, for his advice in building the quench flow apparatus; Professor M. McNamee, University of California at Davis, and Professor A. Karlin, Columbia University, for instruction in making the original *E. electricus* and *T. californica* electroplax membrane preparations; Professor A. Lewis, Cornell University, for use of his computing facility, argon laser, and monochromator, and G. Perreault for instruction in their use; Professor R. D. O'Brien and Professor E. Racker, Cornell University, and Professor Unger, University of Edinburgh, for gifts of Tetram, tri-n-butyl tin chloride, and suberyldicholine, respectively; Professor H. L. Toor, Carnegie–Mellon University, and Professor R. S. Brodkey, Ohio State University, for helpful discussions on diffusion during mixing; Dr. C. Aikey and Dr. J. Telford, Cornell University, for preparing electron micrographs; Professor A. T. Eldefrawi and Professor M. E. Eldefrawi, University of Maryland Medical School, and Professor L. G. Abood, University of Rochester Medical School, for much assistance and advice; Professor M. Alfonzo, Cornell University, for determining the Na^+/K^+ ATPase activity of the vesicles. Mr. S. Carpenter, Cornell University, contributed greatly to the design and machining of the metal work for the quench flow and stopped-flow equipment. We are grateful to the National Science Foundation and the National Institutes of Health for financial support (G. P. Hess), to the U.S. Public Health Service (J. P. Andrews, J. E. Bulger, J.–j. L. Fu, G. E. Struve, J. Walker), the Muscular Dystrophy Association (R. L. Noble, E. B. Pasquale, A. Pece), and the Max Kade Foundation (H. Rübsamen) for postdoctoral fellowships, and to the Muscular Dystrophy Association for a Senior Investigator's Fellowship (H. Aoshima). D. J. Cash, D. E. Moore, and K. Takeyasu were supported by a Cancer Center Grant and P. Leprince and J. W. Karpen by a Training Grant awarded by the National Institutes of Health. H. Aoshima and D. E. Moore were on leave from the Universities of Yamaguchi and Sydney, respectively, and S. Shiono was on leave from, and supported by, the Mitsubishi Electrical Corporation, Japan. Some of this work formed the theses submitted by M. J. Cooper, S. Lipkowitz, P. S. Kim, R. Silberstein, A. B. Sachs, and C. Zacharchuk for award of A.B. (Honors) by Cornell University. Some of the work fulfilled requirements of the undergraduate research program of Cornell University (B. Lenchitz, S. E. Coombs). E. F. Hindy, M. Montgomery, L. Lapish, N. Leprince, and L. Tenney were technicians who made major contributions to the research. C. B. Scriber prepared illustrations for the papers referred to, and G. A. Celeste, E. Patterson, and C. Rooney contributed to the research in many invaluable ways.

REFERENCES

Adams, D. J., Gage, P. W., Hamill, O. P., 1977, Ethabol reduces excitatory postsynaptic current duration at a crustacean neuromuscular junction, *Nature* **266:**739–741.

Adams, P. R., and Feltz, A., 1977, Interaction of a fluorescent probe with acetylcholine-activated synaptic membrane, *Nature* **269:**609–611.

Anderson, C. R., and Stevens, C. F., 1973, Voltage clamp analysis of acetylcholine produced endplate current fluctuations at frog neuromuscular junction, *J. Physiol.* **235:**655–691.

Aoshima, H., Cash, D. J., and Hess, G. P., 1980, Acetylcholine receptor-controlled ion flux in electroplax membrane vesicles: A minimal mechanism based on rate measurements in the millisecond to minute time region, *Biochem. Biophys. Res. Commun.* **92**:896–904.

Aoshima, H., Cash, D. J., and Hess, G. P., 1981, Mechanisms of Inactivation (Desentization) of Acetylcholine Receptor. Investigations by Fast Reaction Techniques with Membrane Vesicles, *Biochemistry* **20**:3467–3474.

Barnard, E. A., Wieckowski, J., and Chiu, T. H., 1971, Cholinergic receptor molecules and cholinesterase molecules at skeletal muscle junctions, *Nature* **234**:207–209.

Baxter, J. D., and Tompkins, G. M., 1971, Specific cytoplasmic glucocorticoid hormone receptors in hepatoma tissue culture cells, *Proc. Natl. Acad. Sci. USA* **68**:932–937.

Bernhardt, J., and Neumann, E., 1978, Kinetic analysis of receptor-controlled tracer efflux from sealed membrane fragments, *Proc. Natl. Acad. Sci. USA* **75**:3756–3760.

Bernhardt, J., and Neumann, E., 1980, Physical factors determining gated flux from or into sealed membrane fragments, in: *Molecular Aspects of Bioelectricity* (E. Schoffeniels and E. Neumann eds.), Pergamon Press, Oxford, pp. 243–251.

Biesecker, G., 1973, Molecular properties of the cholinergic receptor purified from *Electrophorus electricus*, *Biochemistry* **12**:4403–4409.

Blanchard, S. G., Quast, U., Reed, K., Lee, T., Schimerlik, M. I., Vandlen, R., Claudio, T., Strader, C. D., Moore, H.–P. H., and Raftery, M. A., 1979, Interaction of [^{125}I]-α-bungarotoxin with acetylcholine receptor from *Torpedo californica*, *Biochemistry* **18**:1875–1883.

Böhlen, P., Stein, S., Dairman, W., and Udenfriend, S., 1973, Fluorometric assay of proteins in the nanogram range, *Arch. Biochem. Biophys.* **155**:213–220.

Bonting, S. L., Simon, K. A., and Hawkins, N. M., 1961, Studies on sodium-potassium-activated adenosine triphosphatase I. Quantitative distribution in several tissues of the cat, *Arch. Biochem. Biophys.* **95**:416–423.

Brockes, J. P., and Hall, Z. W., 1975a, Acetylcholine receptors in normal and denervated rat diaphragm muscle. I. Purification and interaction with [^{125}I]-α-bungarotoxin, *Biochemistry* **14**:2092–2099.

Brockes, J. P., and Hall, Z. W., 1975b, Acetylcholine receptors in normal and denervated rat diaphragm muscle. II. Comparison of junctional and extrajunctional receptors, *Biochemistry* **14**:2100–2106.

Bulger, J. E., and Hess, G. P., 1973, Evidence for separate initiation and inhibitory sites in the regulation of membrane potential of electroplax. I. Kinetic studies with α-bungarotoxin, *Biochem. Biophys. Res. Commun.* **54**:677–684.

Bulger, J. E., Fu, J.–j. L., Hindy, E. F., Silberstein, R. J., and Hess, G. P., 1977, Allosteric interactions between the membrane-bound acetylcholine receptor and chemical mediators. Kinetic studies, *Biochemistry* **16**:684–692.

Busch, K. W., and Malloy, B., 1979, in: *Multichannel Image Detectors* (Y. Talmi, ed.), American Chemical Society, Washington, D. C., pp. 27–58.

Busch, K. W., and Morrison, G. H., 1973, Simultaneous determination of electrolytes in serum using a vidicon flame spectrophotometer, *Anal. Chem.* **45**:712A–722A.

Cash, D. J., and Hess, G. P., 1980, Molecular mechanism of acetylcholine receptor-controlled ion translocation across cell membranes, *Proc. Natl. Acad. Sci. USA* **77**:842–846.

Cash, D. J., and Hess, G. P., 1981, Quenched flow technique with plasma membrane vesicles: acetylcholine receptor-mediated transmembrane ion flux, *Analyt. Biochem.* **112**:39–51.

Cash, D. J., Aoshima, H., and Hess, G. P., 1980, Acetylcholine-induced receptor-controlled ion flux investigated by flow quench techniques, *Biochem. Biophys. Res. Comm.* **95**:1010–1016.

Cash, D. J., Aoshima, H., and Hess, G. P., 1981, Acetylcholine-induced cation translocation across cell membranes and inactivation of the acetylcholine receptor: Chemical kinetic measurements in the millisecond time region, *Proc. Natl. Acad. Sci. USA* **78**:3318–3322.

Chance, B., 1974, in: *Techniques of Chemistry*, 3rd ed., Volume VI(II) (G. G. Hammes, ed.), p. 5.

Chance, B., Eisenhardt, R. H., Gibson, Q. H., and Lonberg-Holm, K. K. (eds.), 1964, *Rapid Mixing and Sampling Techniques in Biochemistry*, Academic Press, New York.

Chang, C. C., and Lee, C. Y., 1963, Isolation of neurotoxins from the venom of *Bungarus multicinctus* and their modes of neuromuscular blocking action, *Arch. Intern. Pharmacodyn. Ther.* **144**:241–257.

Chang, H. W., 1974, Purification and characterization of acetylcholine receptor from *Electrophorus electricus*, *Proc. Natl. Acad. Sci. USA* **71**:2113–2117.

Changeux, J.–P., 1981, AcChR, an allosteric protein, *Harvey Lect.* **75**:85–254.

Changeux, J.–P., Podleski, T. R., and Meunier, J.–C., 1969, On some structural analogies between acetylcholinesterase and the macromolecular receptor of acetylcholine, *J. Gen. Physiol.* **54**:225–244.

Clark, D. G., Macmurchie, D. D., Elliott, E., Wolcott, R. G., Landel, A. M., and Raftery, M. A., 1972, Elapid neurotoxins. Purification, characterization, and immunochemical studies of α-bungarotoxin, *Biochemistry* **11**:1663–1668.

Claudio, T., and Raftery, M. A., 1977, Immunological comparison of acetylcholine receptors and their subunits from species of electric ray, *Arch. Biochem. Biophys.* **181**:484–489.

Cohen, S. R., 1969, A rapid sensitive semimicro gel filtration procedure for detecting and removing low molecular weight fragments from [³H]- or [¹⁴C]-labeled inulin, *Analyt. Biochem.* **31**:539–544.

Cohen, J. B., and Changeux, J.–P., 1973, Interaction of a fluorescent ligand with membrane-bound cholinergic receptor from *Torpedo marmorata*, *Biochemistry* **12**:4855–4863.

Conti–Tronconi, B. M., and Raftery, M. A., 1982, The nicotinic cholinergic receptor: Correlation of molecular structure with functional properties, *Ann. Rev. Biochem.* **51**:491–530.

Conway, A., and Koshland, D. E., 1968, Negative cooperativity in enzyme action. The binding of diphosphopyridine nucleotide to glyceraldehyde 3-phosphate dehydrogenase, *Biochemistry* **7**:4011–4023.

Davis, B. V., 1964, *Ann. N.Y. Acad. Sci.* **121**:404.

Delegeane, A. M., and McNamee, M. G., 1980, Independent activation of the acetylcholine receptor from *Torpedo californica* at two sites, *Biochemistry* **19**:890–895.

Dewey, T. G., and Hammes, G. G., 1981, *Proc. Natl. Acad. Sci. U.S.A.* **78**:7422–7425.

Donner, D., Fu, J., and Hess, G. P., 1976, Equilibrium dialysis of the membrane-bound acetylcholine receptor: A simple method to avoid common errors, *Anal. Biochem.* **75**:454–463.

Dunn, S. M. J., and Raftery, M. A., 1982, Multiple binding sites for agonists on *Torpedo californica* acetylcholine receptor, *Biochemistry* **21**:6264–6272.

Eigen, M., 1967, Kinetics of reaction controls and information transfer in enzymes and nucleic acids, *Nobel Symp.* **5**:333–369.

Eldefrawi, M. E., and Eldefrawi, A. T., 1973, Purification and molecular properties of acetylcholine receptor from *Torpedo* electroplax, *Arch. Biochem. Biophys.* **159**:362–373.

Eldefrawi, M. E., and Eldefrawi, A. T., 1977, Acetylcholine receptors in: *Receptors and Recognition IV*, (A. P. Cuatrecasas and M. F. Greaves, eds.), Chapman and Hall, London, pp. 197–258.

Eldefrawi, M. E., Britten, A. G., and O'Brien, R. D., 1971a, Action of organophosphates on binding of cholinergic ligands, *Pest. Biochem. Physiol.* **1**:101–108.

Eldefrawi, M. E., Britten, A. G., and Eldefrawi, A. T., 1971b, Acetylcholine binding to *Torpedo* electroplax: Relationships to acetylcholine receptors, *Science* **173**:338–340.

Eldefrawi, M. E., Eldefrawi, A. T., and Shamoo, A. E., 1975, Molecular and functional properties of the acetylcholine-receptor, *Ann. N.Y. Acad. Sci.* **264**:183–202.

Ellman, G. L., Courtney, K. D., Andres, V. A., Jr., and Featherstone, R. M., 1961, A new and rapid colorimetric determination of acetylcholinesterase activity, *Biochem. Pharmacology* **7**:88–95.

Fersht, A. R., and Jakes, R., 1975, Demonstration of two reaction pathways for the aminoacylation of tRNA. Application of the pulsed quenched flow technique, *Biochemistry* **14**:3350–3356.

Franklin, G. I., and Potter, L. T., 1972, Studies of the binding of α-bungarotoxin to membrane-bound and detergent-dispersed acetylcholine receptors from *Torpedo* electric tissue, *FEBS Lett.* **28**:101–106.

Froehlich, J. P., and Tayler, E. W., 1976, Transient state kinetic effects of calcium ion on sarcoplasmic reticulum adenosine triphosphatase, *J. Biol. Chem.* **251**:2307–2315.

Fu, J.-j. L., Donner, D. B., and Hess, G. P., 1974, Half-of-the-sites reactivity of the membrane-bound *Electrophorus electricus* acetylcholine receptor, *Biochem. Biophys. Res. Commun.* **60**:1072–1080.

Fu, J.-j. L., Donner, D. B., Moore, D. E., and Hess, G. P., 1977, Allosteric interactions between the membrane-bound acetylcholine receptor and chemical mediators: Equilibrium measurements, *Biochemistry* **16**:678–684.

Fulpius, B., Cha, S., Klett, R., and Reich, E., 1972, Properties of the nicotinic acetylcholine receptor macromolecule of *Electrophorus electricus, FEBS Lett.* **24**:323–326.

Furlong, C. E., Morris, R. G., Kandrach, M., and Rosen, B. P., 1972, A multichamber equilibrium dialysis apparatus, *Analyt. Biochem.* **47**:514–526.

Gibson, Q., 1966, *Ann. Rev. Biochem.* **35**:435–456.

Hammes, G. G., 1982, *Enzyme Catalysis and Regulation*, Academic Press, New York.

Hammes, G. G., and Wu, C. W., 1974, Kinetics of allosteric enzymes, *Ann. Rev. Biophys. Bioeng.* **3**:1–33.

Hazelbauer, G. H., and Changeux, J.-P., 1974, Reconstitution of a chemically excitable membrane, *Proc. Natl. Acad. Sci. U.S.A.* **71**:1479–1483.

Heidmann, T., and Changeux, J.-P., 1978, Structural and functional properties of the acetylcholine receptor protein in its purified and membrane-bound states, *Ann. Rev. Biochem.* **47**: 315–357.

Hess, G. P., and Rupley, J. A., 1971, Structure and function of proteins, *Ann. Rev. Biochem.* **40**:1013–1044.

Hess, G. P., Bulger, J. E., Fu, J.-j. L., Hindy, E. F., and Silberstein, R. J., 1975b, Allosteric interactions of the membrane-bound acetylcholine receptor: Kinetic studies with α-bungarotoxin, *Biochem. Biophys. Res. Commun.* **64**:1018–1027.

Hess, G. P., Andrews, J. P., and Struve, G. P., 1976, Apparent cooperative effects in acetylcholine receptor-mediated ion flux in electroplax membrane preparations, *Biochem. Biophys. Res. Comm.* **69**:830–837.

Hess, G. P., and Andrews, J. P., 1977, Functional acetylcholine receptor-electroplax membrane microsacs (vesicles): Purification and characterization, *Proc. Natl. Acad. Sci. U.S.A.* **74**:482–486.

Hess, G. P., Lipkowitz, S., and Struve, G. E., 1978, Acetylcholine-receptor-mediated ion flux in electroplac membrane microsacs (vesicles): Change in mechanism produced by asymmetrical distribution of sodium and potassium ions, *Proc. Natl. Acad. Sci. USA* **75**:1703–1707.

Hess, G. P., Cash, D. J., and Aoshima, H., 1979, Acetylcholine receptor-controlled ion fluxes in membrane vesicles investigated by fast reaction techniques, *Nature* **282**: 329–331.

Hess, G. P., Cash, D. J., Aoshima, H., 1980, In *Molecular Aspects of Bioelectricity* (E. Schoffeniels and E. Neumann, eds.), pp. 233–242, Pergamon, Oxford.

Hess, G. P., Aoshima, H., Cash, D. J., and Lenchitz, B., 1981, Specific reaction rate of acetylcholine receptor-controlled ion translocation: A comparison of measurements with membrane vesicles and muscle cells, *Proc. Natl. Acad. Sci. U.S.A.* **78**:1361–1365.

Hess, G. P., Pasquale, E. B., Walker, J. W., and McNamee, M. G., 1982, Comparison of acetylcholine receptor-controlled cation flux in membrane vesicles from *Torpedo californica* and *Electrophorus electricus:* Chemical kinetic measurements in the millisecond region, *Proc. Natl. Acad. Sci. USA* **79**:963–967.

Hess, G. P., Cash, D. J., and Aoshima, H., 1983, Acetylcholine receptor-controlled ion translocation: Chemical kinetic investigations of the mechanism. *Ann. Rev. Biophys. Bioeng.* **12**: 443–473.

Hess, G. P., Kolb, H.-A., Läuger, P., Schoffeniels, E., and Schwartz, W. E., 1984, Acetylcholine receptor (from *E. electricus*): A comparison of single-channel current recordings and chemical kinetic measurements, *Proc. Natl. Acad. Sci. USA* **9**.

Higman, H. B., Podleski, T. R., and Bartels, E., 1963, Apparent dissociation constants between carbamylcholine, *d*-tubocurarine and the receptor, *Biochim. Biophys. Acta* **75**:187–193.

Hille, B., 1976, *Ann. Rev. Physiol.* **38**:139–152.

Holler, E., Rupley, J. A., and Hess, G. P., 1975a, Productive and unproductive lysozyme–chitosaccharide complexes. Equilibrium measurements, *Biochemistry* **14**:1088–1094.

Holler, E., Rupley, J. A., and Hess, G. P., 1975b, Productive and unproductive lysozyme–chitosaccharide complexes. Kinetic investigations, *Biochemistry* **14**:2377–2385.

Kaback, H. R., 1970, Transport, *Ann. Rev. Biochem.* **39**:561–598.

Karlin, A., 1967a, Chemical distinctions between acetylcholinesterase and the acetylcholine receptor, *Biochim. Biophys. Acta* **139**:358–362.

Karlin, A., 1967b, On application of "a plausible model" of allosteric proteins to the receptor for acetylcholine, *J. Theor. Biol.* **16**:306–320.

Karlin, A., 1974, The acetylcholine receptor: Progress report, *Life Sciences* **14**:1385–1415.

Karlin, A., 1980, Molecular properties of nicotinic acetylcholine receptors, in: *The Cell Surface and Neuronal Function* (C. W. Cotman, G. Poste, and G. L. Nicolson, eds.), Elsevier/North Holland, Amsterdam, pp. 191–260.

Karlin, A., Weill, C. L., McNamee, M. G., and Valderrama, R., 1975, Facets of the structures of acetylcholine receptors from *Electrophorus* and *Torpedo*, *Cold Spring Harbor Sympos.* **XL**:203–210.

Karlsson, E., Heilbronn, E., and Widlund, L., 1972, Isolation of the nicotinic acetylcholine receptor by biospecific chromatography on insolubilized *Naja naja* neurotoxin, *FEBS Lett.* **28**:107–111.

Karpen, J. W., Aoshima, H., Abood, L. G., and Hess, G. P., 1982, Cocaine and phencyclidine inhibition of the acetylcholine receptor: Analysis of the mechanisms of action based on measurements of ion flux in the millisecond-to-minute time region, *Proc. Natl. Acad. Sci. USA* **79**:2509–2513.

Karpen, J. W., Sachs, A. B., Cash, D. J., Pasquale, E. B., and Hess, G. P., 1983, Direct spectrophotometric detection of cation flux in membrane vesicles: Stopped-flow measurements of acetylcholine-receptor-mediated ion flux, *Analyt. Biochem.* **135**:83–94.

Kasai, M, and Changeux, J.-P., 1971a, *In vitro* excitation of purified membrane fragments by cholinergic agonists. I. Pharmacological properties of the excitable membrane fragments, *J. Membr. Biol.* **6**:1–23.

Kasai, M., and Changeux, J.-P., 1971b, *In vitro* excitation of purified membrane fragments by cholinergic agonists. II. The permeability change caused by cholinergic agonists, *J. Membr. Biol.* **6**:24–57.

Kasai, M., and Changeux, J.-P., 1971c, *In vitro* excitation of purified membrane fragments by cholinergic agonists. III. Comparison of the dose-response curves to decamethonium with the corresponding binding curves of decamethonium to the cholinergic receptor, *J. Membr. Biol.* **6**:58–80.

Katz, B., 1966, *Nerve, Muscle and Synapse*, McGraw–Hill, New York.

Katz, B. 1969, *The Release of Neural Transmiter Substances*, Liverpool University Press, Liverpool.

Katz, B., and Miledi, R., 1972, The statistical nature of the acetylcholine potential and its molecular components, *J. Physiol. (London)* **224**:665–699.

Katz, B., and Thesleff, S., 1957, A study of the "desensitization" produced by acetylcholine at the motor end-plate, *J. Physiol. (London)* **138**:63–80.

Keynes, R. D., and Martins–Ferreira, H., 1953, Membrane potentials in the electroplates of the electric eel, *J. Physiol. (London)* **119**:315–351.

Kim, P. S., and Hess, G. P., 1981, Acetylcholine receptor-controlled ion flux in electroplax membrane vesicles: Identification and characterization of membrane properties that affect ion flux measurements, *J. Membr. Biol.* **58**:203–211.

Klett, R. P., Fulpius, B. W., Cooper, D., Smith, M., Reich, E., and Possani, L. D., 1973, The acetylcholine receptor. I. Purification and characterization of a macromolecule isolated *Electrophorus electricus, J. Biol. Chem.* **248**:6841–6853.

Koblin, D. D., and Lester, H. A., 1979, Voltage-dependent and voltage-independent blockage of acetylcholine receptors by local anesthetics in *Electrophorus* electroplaques, *Mol. Pharmacol.* **15**:559–580.

Kohanski, R., Andrews, J., Wins, P., Eldefrawi, M., and Hess, G. P., 1977, A simple quantitative assay of ^{125}I-labeled α-bungarotoxin binding to soluble and membrane-bound acetylcholine receptor protein, *Anal. Biochem.* **80**:531–539.

Koshland, D. E., 1970, The molecular basis for enzyme regulation, *Enzymes* **1**:341–396.

Koshland, D. E., Jr., Nemethy, G., and Filmer, D., 1966, Comparison of experimental binding data and theoretical models in proteins containing subunits, *Biochemistry* **5**:365–385.

Kurzmack, M., Verjovski–Almeida, S., and Inesi, G., 1977, Detection of an initial burst of Ca^{2+} translocation in sarcoplasmic reticulum, *Biochem. Biophys. Res. Commun.* **78**:772–776.

Kumar Ash, S., and Rohatgi–Mukherjee, K. K., 1979, *Ind. J. Biochem. Biophys.* **16**:28–31.

Lee, C. Y., 1972, Chemistry and pharmacology of polypeptide toxins in snake venom, *Ann. Rev. Pharmacol.* **12**:265–286.

Lee, C. Y., and Chang, C. Y., 1966, *Memb. Inst. Butanton Symp. Int.* **33**:555.

Lee, C. Y., Tseng, L. F., and Chiu, T. H., 1967, Influence of denervation on localization of neurotoxins from clapin venoms in rat diaphragm, *Nature* **215**:1177–1178.

Lee, C. Y., Chang, S. L., Kau, S. T., and Luh, S.–H., 1972, Chromatographic separation of the venom of *Bungarus multicinctus* and characterization of its components, *J. Chromatogr.* **72**:71–82.

Leprince, P., Noble, R. L., and Hess, G. P., 1981, Comparison of the interactions of a specific neurotoxin (α-bungarotoxin) with the acetylcholine receptor in *Torpedo californica* and *Electrophorus electricus, Biochemistry* **20**:5565–5570.

Lester, H. A., Changeux, J.–P., and Sheridan, R. E., 1975, Conductance increases produced by bath application of cholinergic agonists to *Electrophorus electricus, J. Gen. Physiol.* **65**:797–816.

Li, N., 1968, *Arch. Pharm. Exp. Path.* **259**:360.

Lindstrom, J., and Patrick, J., 1974, Purification of the acetylcholine receptor by affinity chromatography, in: *Synaptic Transmission and Neuronal Interaction* (M. V. Bennett, ed.), Raven Press, New York, pp. 191–216.

Lindstrom, J., Cooper, J., and Tzartos, S., 1980, Acetylcholine receptors from *Torpedo* and *Electrophorus* have similar subunit structures, *Biochemistry* **19:1454–1458.**

Lowry, O. H., Rosebrough, N. J., Farr, A. L., and Randall, R. J., 1951, Protein measurements with the Folin phenol reagent, *J. Biol. Chem.* **193**:265–275.

Lymn, R. W., and Taylor, E. W., 1970, Transient state phosphate production in the hydrolysis of nucleoside triphosphate by myosin. *Biochemistry* **9**:2975–2983.

MacQuarrie, R. A., and Bernhard, S. A., 1971a, Mechanism of alkylation of rabbit muscle glyceraldehyde 3-phosphate dehydrogenase, *Biochemistry* **10**:2456–2466.

MacQuarrie, R. A., and Bernhard, S. A., 1971b, Subunit conformation and catalytic function in rabbit-muscle glyceraldehyde-3-phosphate dehydrogenase, *J. Mol. Biol.* **55**:181–192.

Martonosi, A., Lagwinska, E., and Oliver, M., 1974, Elementary processes in the hydrolysis of ATP by sarcoplasmic reticulum membranes, *Ann. N.Y. Acad. Sci.* **227**:549–567.

Meunier, J.–C., Olsen, R. W., Menez, A., Fromageot, P., Boquet, P., and Chaneux, J.–P., 1972, Some physical properties of the cholinergic receptor protein from *Electrophorus electricus* revealed by a tritiated α-toxin from *Naja nigricollis* venom, *Biochemistry* **11**:1200–1210.

Meunier, J.-C., Sealock, R., Olsen, R., and Changeux, J.-P., 1974, Purification and properties of the cholinergic receptor protein from *Electrophorus electricus* electric tissue, *Eur. J. Biochem.* **45**:371–394.

Miledi, R., and Potter, L. T., 1971, Acetylcholine receptors in muscle fibers, *Nature* **233**:599–603.

Miledi, R., Molinoff, P., and Potter, L. P., 1971, Isolation of the cholinergic receptor protein of *Torpedo* electric tissue, *Nature* **229**:554–557.

Miller, D. L., Moore, H.-P. H., Hartig, P., and Raftery, M. A., 1978, Fast cation flux from *Torpedo californica* membrane preparations: Implications for a functional role for acetylcholine receptor dimers, *Biochem. Biophys. Res. Commun.* **85**:632–640.

Monod, J., Wyman, J., and Changeux, J.-P., 1965, On the nature of allosteric transition: a plausible model, *J. Mol. Biol.* **12**:88–118.

Moore, H.-P. J., and Raftery, M. A., 1980, Direct spectroscopic studies of cation translocation by *Torpedo* acetylcholine receptor on a time-scale of physiological relevance, *Proc. Natl. Acad. Sci. USA* **77**:4509–4513.

Morrison, G. H., 1979, Elemental trace analysis of biological materials, *CRC Crit. Rev. Anal. Chem.* **8**:287–320.

Nachmansohn, D., 1973, in: *The Structure and Function of Muscle* 2nd ed., Volume 3 (G. H. Bourne, ed.), Academic Press, New York, pp. 32–116.

Nachmansohn, D., and Neumann, E., 1975, *Chemical and Molecular Basis of Nerve Activity*, Academic Press, New York.

Neher, E., and Sakmann, B., 1975a. Single-channel currents recorded from membrane of denervated frog muscle fibers, *Nature* **260**:779–802.

Neher, E., and Sakmann, B., 1976b, Noise analysis of drug-induced voltage clamp currents in denervated frog muscle fibres, *J. Physiol.* **258**:705–729.

Neher, E., and Steinbach, J. H., 1978, Local anesthetics transiently block currents through single acetylcholine-receptor channels, *J. Physiol.* **277**:153–176.

Neher, E., and Stevens, C. F., 1977, Conductance fluctuations and ionic pores in membranes, *Ann. Rev. Biophys. Bioeng.* **6**:345–381.

Neubig, R. R. and Cohen, J. B., 1979, Equilibrium binding of [^{3}H]-tubocurarine and [^{3}H]-acetyl-choline by *Torpedo* postsynaptic membranes: Stoichiometry and ligand interactions, *Biochemistry* **18**:5464–5475.

Neubig, R. R., and Cohen, J. B., 1980, Permeability control by cholinergic receptors in *Torpedo* postsynaptic membranes: Agonist dose-response relations measured at second and millisecond times, *Biochemistry* **19**:2770–2779.

O'Brien, R. D., and Gilmour, L. P., 1969, A muscarone-binding material in electroplax and its relation to the acetylcholine receptor, I. Centrifugal assay, *Proc. Natl. Acad. Sci. USA* **63**:496–503.

Palfrey, C., and Littauer, U. Z., 1976, Sodium-dependent efflux of K^+ and Rb^+ through the activated sodium channel neuroblastoma cells, *Biochem. Biophys. Res. Commun.* **72**:209–215.

Pasquale, E. B., Takeyasu, K., Udgaonkar, J. B., Cash, D. J., Severski, M. C., and Hess, G. P., (1984), Acetylcholine receptor: Evidence for a regulatory binding site in investigations of suberyldicholine-induced transmembrane ion flux in *Electrophorus electricus* membrane vesicles, *Biochemistry* **22**:5967–5978.

Patrick, J., and Stallcup, W. B, 1979, Immunological distinction between acetylcholine receptor and the α-bungarotoxin binding component on sympathetic neurons, *Proc. Natl. Acad. Sci. U.S.A.* **74**:4689–4692.

Peterson, G. L., 1979, Reviews of the Folin-Phenol protein quantitation method of Lowry, Rosebrough, Farr and Randall, *Anal. Biochem.* **100**:201–220.

Phelps, C. F., 1965, The physical properties of inulin solutions. *Biochem. J.* **95**:41–47.

Popot, J.-L., Sugiyama, H. J., and Changeux, J.-P., 1974, Demonstration de la densensibilisation pharmacologique du recepteur de l'acetylcholine *in vitro* avec des fragments de membranes excitable de Torpille, *C.R. Acad. Sci. Paris, Ser. D.* **279**:1721–1724.

Quast, U., Schimerlik, M., Lee, T., Witzmemann, V., Blanchard, S., and Raftery, M. A., 1978, Ligand-induced conformation changes in *Torpedo californica* membrane-bound acetylcholine receptor, *Biochemistry* **17**:2405–2414.

Rabon, E. C., and Sachs, G., 1981, *J. Membr. Biol.* **62**:19–27.

Racker, E. (ed.), 1970, *Membranes of Mitochondria and Chloroplasts,* Van Nostrand–Reinhold, New York.

Ramseyer, G. O., Morrison, G. H., Aoshima, H., and Hess, G. P., 1981, Vidicon flame emission spectroscopy of Li^+, Na^+, and K^+ fluxes mediated by acetylcholine receptor in *Electrophorus electricus* membrane vesicles, *Analyst. Biochem.* **115**:34–41.

Reif, A. E., 1967, A simple procedure for high efficiency radioiodination of proteins, *J. Nucl. Med.* **9**:148–155.

Reynolds, J., and Karlin, A., 1978, Molecular weight in detergent solution of acetylcholine receptor from *Torpedo californica, Biochemistry* **17**:2035.

Roughton, F. J. W., and Chance, B., 1963, in: *Technique of Organic Chemistry, 2nd ed.*, Volume 8 (S. L. Friess, E. S. Lewis, and A. Weissberger, eds.), Wiley, p. 2.

Rübsamen, H., Eldefrawi, A. T., Eldefrawi, M. E., and Hess, G. P., 1978, Characterization of the calcium-binding sites of the purified acetylcholine receptor and identification of the calcium-binding subunit, *Biochemistry* **17**:3818–3825.

Sachs, A., 1982, Honors Thesis, Cornell University. The acetylcholine receptor: Characterization of α-bungarotoxin sites in *Electrophorus electricus* membrane preparations, development of fluorescent quenching techniques to measure ion flux and the effects of phencyclidine on the receptor in living cells.

Sachs, A. B., Lenchitz, B., Noble, R. L., and Hess, G. P., 1982, A convenient large-scale method for the isolation of membrane vesicle permeable to a specific inorganic ion: Isolation and characterization of functional acetylcholine receptor-containing vesicles from the electric organ of *Electrophorus electricus, Analyt. Biochem.* **124**:185–190.

Sakmann, B., and Adams, P. R., 1979, *Advances in Pharmacology and Therapeutics 1*, Receptors. Biophysical aspects of agonist action at frog endplate.

Sakmann, B., and Neher, E. (eds.), (1983) *Single-Channel Recording*, Plenum Press, New York.

Scatchard, G., 1949, The attractions of proteins for small molecules and ions, *Ann. N.Y. Acad. Sci.* **50**:660–672.

Schmidt, J., and Raftery, M. A., 1972, Use of affinity chromotography for acetylcholine receptor purification, *Biochem. Biophys. Res. Commun.* **49**:572–578.

Schmidt, J., and Raftery, M. A., 1973, A simple assay for the study of solubilized acetylcholine receptors, *Anal. Biochem.* **52**:349–354.

Schoffeniels, E., 1957, An isolated single electroplax preparation. II Improved preparation for studying ion flux, *Biochim. Biophys, Acta* **26**:585–596.

Schoffeniels, E., 1959, Ion movements studied with single isolated electroplax, *Annals N.Y. Acad. Sci.* **81**:285–306.

Schoffeniels, E., and Nachmansohn, D., 1957, An isolated single electroplax preparation. I. New data on the effect of acetylcholine and related compounds, *Biochim. Biophys. Acta* **26**:1–15.

Sears, D. A., Reed, C. F., and Helmkamp, R. W., 1971, A radioactive label for the erythrocyte membrane. *Biochim. Biophys. Acta* **233**:716–719.

Sheridan, R. E., and Lester, H. E., 1975, Relaxation measurements on the acetylcholine receptor, *Proc. Natl. Acad. Sci. USA* **72**:3496–3500.

Silman, H. I., and Karlin, A., 1967, Effect of local pH changes caused by substrated hydrolysis on the activity of membrane-bound acetylcholinesterase, *Proc. Natl. Acad. Sci. USA* **58**:1664–1675.

Sugiyama, H., Popot, R. L., Cohen, J. B., Weber, M., and Changeux, J.-P., 1975, in: *Protein-Ligand interactions* (H. Sund and G. Blauer, eds.), De Guyter, Berlin, pp. 289–503.

Sugiyama, H. J., Popot, J.-L., and Changeux, J.-P., 1976, Studies on the electrogenic action of acetylcholine with *Torpedo marmorata* electric organ. III Pharmacological desensitization *in vitro* of the receptor-rich membrane fragments by cholinergic agonists, *J. Mol. Biol.* **106:**469–483.

Sumida, M., Wang, T., Mandel, F., Froehlich, J. P., and Schartz, A., 1978, Transient kinetics of Ca^{2+} transport of sarcoplasmic reticulum. A comparison of cardiac and skeletal muscle, *J. Biol. Chem.* **253:**8772–8777.

Takeyasu, K., Udgaonkar, J. B., and Hess, G. P., 1983, Acetylcholine receptor: Evidence for a voltage-dependent regulatory site for acetylcholine. Chemical kinetic measurements in membrane vesicles using a voltage-clamp, *Biochemistry* **22:**5973–5978.

Thayer, W. S., and Hinkle, P. C., 1975, Kinetics of adenosine triphosphate synthesis in bovine heart submitochondrial particles, *J. Biol. Chem.* **250:**5336–5342.

Verjovski–Almeida, S., and Inesi, G., 1979, Fast kinetic evidence for an activating effect of ATP on the Ca^{2+} transport of sarcoplasmic reticulum ATPase, *J. Biol. Chem.* **254:**18–21.

Walker, J. W., McNamee, M. G., Pasquale, E., Cash, D. J., and Hess, G. P., 1981, Acetylcholine receptor inactivation in *T. californica* electroplax membrane vesicles. Detection of two processes in the millisecond and second time processes, *Biochem. Biophys. Res. Commun.* **100:**86–90.

Weber, M., and Changeux, J.-P., 1974, Binding of *Naja nigricollis* [^{3}H]-α-toxin to membrane fragments from *Electrophorus* and *Torpedo* electric organs. II. Effect of cholinergic agonists and antagonists on the binding of the tritiated α-neurotoxins, *Mol. Pharmacol.* **10:**15–34.

Weiland, G., Georgia, B., Wee, V. T., Chignell, C. F., and Taylor, P., 1976, Ligand interactions with cholinergic receptor-enriched membranes from *Torpedo:* Influence of agonist exposure on receptor proteins, *Mol. Pharm.* **12:**1091–1105.

Weiland, G., Georgia, B., Lappi, S., Chignell, C. F., and Taylor, P., 1977, Kinetics of agonist-mediated transitions in state of the cholinergic receptor, *J. Biol. Chem.* **252:**7648–7656.

Weill, C. L., McNamee, M. G., and Karlin, A., 1974, Affinity-labeling of purified acetylcholine receptor from *Torpedo californica, Biochem. Biophys. Res. Commun.* **61:**997–1003.

Wennogle, L. P., 1984, The endplate acetylcholine receptor. Structure and function, in: *Handbook of Experimental Pharmacology* (D. A. Kharkevich, ed.), Springer, Berlin.

Wilson, I. B., Ginsburg, S., and Quan, C., 1958, Molecular complimentariness as basis for reactivation of akyl phosphate-inhibited enzyme, *Arch. Biochem. Biophys* **77:**286–296.

INDEX